Management of Electronic Waste

Management of Electronic Waste

Resource Recovery, Technology and Regulation

Edited by

Anshu Priya
City University of Hong Kong, Kowloon, Hong Kong

WILEY

Published by John Wiley & Sons, Inc., Hoboken, New Jersey.
Published simultaneously in Canada.

Library of Congress Cataloging-in-Publication Data

Names: Priya, Anshu, editor.
Title: Management of electronic waste : resource recovery, technology and regulation / edited by Anshu Priya.
Description: Hoboken, New Jersey : Wiley, [2024] | Includes index.
Identifiers: LCCN 2023046576 (print) | LCCN 2023046577 (ebook) | ISBN 9781119894339 (cloth) | ISBN 9781119894346 (adobe pdf) | ISBN 9781119894353 (epub)
Subjects: LCSH: Electronic waste–Management.
Classification: LCC TD799.85 .M36 2024 (print) | LCC TD799.85 (ebook) | DDC 621.3815028/6–dc23/eng/20231102
LC record available at https://lccn.loc.gov/2023046576
LC ebook record available at https://lccn.loc.gov/2023046577

Cover image(s): Wiley
Cover design: © liangpv/Getty Images

Set in 9.5/12.5pt STIXTwoText by Straive, Chennai, India

Dedicated to my dearest grandfather, Mr. Anand Swaroop Varma, for his countless blessings, immense love, and endless support. You are a great source of encouragement to me. Thank you for inspiring me always. I owe it all to you...

Contents

List of Contributors

Abhinav Ashesh
KPMG India
Gurgaon
Haryana
India

Mishra Bhawana
Department of Environmental Sciences
Central University of Himachal Pradesh (CUHP)
Shahpur
Dharamshala
Kangra
Himachal Pradesh
India

Amilton Barbosa Botelho Junior
Department of Chemical Engineering
Polytechnic School
University of Sao Paulo
Sao Paulo
Brazil

Marcelo P. Cenci
Materials Engineering Department
Federal University of Rio Grande do Sul
Porto Alegre/RS
Brazil

Mital Chakankar
Department of Biotechnology
Helmholtz Institute Freiberg for Resource Technology
Helmholtz-Zentrum Dresden-Rossendorf
Dresden
Germany

Venkata Ravi Sankar Cheela
Civil Engineering Department
MVGR College of Engineering (A)
Vizianagaram
Andhra Pradesh
India

Pranav Prashant Dagwar
Department of Environmental Science and Engineering
SRM University-AP
Amaravati
Andhra Pradesh
India

and

CSIR-National Environmental Engineering Research Institute (CSIR-NEERI)
Nagpur
Maharashtra
India

Shailesh R. Dave
Xavier Research Foundation
Loyola Centre for Research and Development
St. Xavier's College Campus
Ahmedabad
India

Pant Deepak
Department of Environmental Sciences
Central University of Himachal Pradesh (CUHP)
Shahpur, Dharamshala, Kangra
Himachal Pradesh
India

Deblina Dutta
Department of Environmental Science and Engineering
SRM University-AP
Amaravati
Andhra Pradesh
India

Tingyue Gu
Department of Chemical and Biomolecular Engineering
Institute for Sustainable Energy and the Environment
Ohio University
Athens
OH
USA

Subrata Hait
Department of Civil and Environmental Engineering
Indian Institute of Technology Patna
Patna
Bihar
India

Pg Rusydina Idris
Civil Engineering Programme Area
Universiti Teknologi Brunei
Gadong
Brunei Darussalam

Kaviul Islam
Department of Mechanical Engineering
Iowa State University
Ames
IA
USA

and

School of Science and Engineering
Canadian University of Bangladesh
Dhaka
Bangladesh

Rohan Jain
Department of Biotechnology
Helmholtz Institute Freiberg for Resource Technology
Helmholtz-Zentrum Dresden-Rossendorf
Dresden
Germany

Sharifa Khatun
Department of Mechanical Engineering
Rajshahi University of Engineering and Technology
Rajshahi
Bangladesh

Atul Kumar
Department of Veterinary Public Health and Epidemiology
CSK HP Agricultural University
Palampur
Himachal Pradesh
India

Sunil Kumar
CSIR-National Environmental Engineering Research Institute (CSIR-NEERI)
Nagpur
Maharashtra
India

Sabine Kutschke
Department of Biotechnology
Helmholtz Institute Freiberg for Resource Technology
Helmholtz-Zentrum Dresden-Rossendorf
Dresden
Germany

Franziska Lederer
Department of Biotechnology
Helmholtz Institute Freiberg for Resource Technology
Helmholtz-Zentrum Dresden-Rossendorf
Dresden
Germany

Mahadi Hasan Masud
Mechanical & Automotive Discipline
School of Engineering
RMIT University, Bundoora Campus
Melbourne
VIC
Australia

and

Department of Mechanical Engineering
Rajshahi University of Engineering and Technology
Rajshahi
Bangladesh

Nahid Imtiaz Masuk
Department of Mechanical Engineering
Rajshahi University of Engineering and Technology
Rajshahi
Bangladesh

Shivangi Mathur
Department of Biotechnology
President Science College
Gujarat University
Ahmedabad
Gujarat
India

Sabine Matys
Department of Biotechnology
Helmholtz Institute Freiberg for Resource Technology
Helmholtz-Zentrum Dresden-Rossendorf
Dresden
Germany

Pankaj Meena
CSIR-National Environmental Engineering Research Institute (CSIR-NEERI)
Nagpur
Maharashtra
India

Soumya V. Menon
Department of Chemistry and Biochemistry
School of Sciences
Jain (Deemed to be) University
Bengaluru
Karnataka
India

José C. Mengue Model
Materials Engineering Department
Federal University of Rio Grande do Sul
Porto Alegre/RS
Brazil

Monjur Mourshed
Mechanical & Automotive Discipline
School of Engineering
RMIT University, Bundoora Campus
Melbourne
VIC
Australia

and

Department of Mechanical Engineering
Rajshahi University of Engineering and Technology
Rajshahi
Bangladesh

Seyyed Mohammad Mousavi
Biotechnology Group
Chemical Engineering Department
Tarbiat Modares University
Tehran
Iran

and

Modares Environmental Research Institute
Tarbiat Modares University
Tehran
Iran

Daniel D. Munchen
Materials Engineering Department
Federal University of Rio Grande do Sul
Porto Alegre/RS
Brazil

Ashkan Namdar
Faculty of Materials Science and Engineering
Khajeh Nasir Toosi University of Technology
Tehran
Iran

Tannaz Naseri
Biotechnology Group
Chemical Engineering Department
Tarbiat Modares University
Tehran
Iran

Piotr Nowakowski
Silesian University of Technology
Faculty of Transport and Aviation Engineering
Katowice
Poland

Biswaranjan Paital
Redox Regulation Laboratory
Department of Zoology
Odisha University of Agriculture and Technology
Bhubaneswar
Odisha
India

Katrin Pollmann
Department of Biotechnology
Helmholtz Institute Freiberg for Resource Technology
Helmholtz-Zentrum Dresden-Rossendorf
Dresden
Germany

Anshu Priya
School of Energy and Environment
City University of Hong Kong
Kowloon
Hong Kong

Seyed Omid Rastegar
Department of Chemical Engineering
Faculty of Engineering
University of Kurdistan
Sanandaj
Iran

Rahul Rautela
CSIR-National Environmental Engineering Research Institute (CSIR-NEERI)
Nagpur
Maharashtra
India

and

Academy of Scientific and Innovative Research (AcSIR)
Ghaziabad
Uttar Pradesh
India

Lucas Reijnders
Faculty of Science
IBED
University of Amsterdam
Amsterdam
The Netherlands

Nirmaladevi Saravanan
Department of Biochemistry and Biotechnology
Avinashilingam University for Women
Coimbatore
Tamil Nadu
India

Shahriar Shams
Civil Engineering Programme Area
Universiti Teknologi Brunei
Gadong
Brunei Darussalam

Abhishek Sharma
Department of Veterinary Public Health and Epidemiology
CSK HP Agricultural University
Palampur
Himachal Pradesh
India

Ummul Khair Sultana
School of Chemical Engineering
Faculty of Engineering
Architecture and Information Technology
University of Queensland
Brisbane
QLD
Australia

Ningjie Tan
State Key Laboratory of Bioreactor Engineering
East China University of Science and Technology
Shanghai
China

Shital C. Thacker
Department of Microbiology and Biotechnology
School of Sciences
Gujarat University
Ahmedabad
India

and

Department of Microbiology
Silver Oak Institute of Sciences
Silver Oak University
Ahmedabad
India

Devayani R. Tipre
Department of Microbiology and Biotechnology
School of Sciences
Gujarat University
Ahmedabad
India

James Vaughan
School of Chemical Engineering
Faculty of Engineering
Architecture and Information Technology
University of Queensland
Brisbane
QLD
Australia

Hugo M. Veit
Materials Engineering Department
Federal University of Rio Grande do Sul
Porto Alegre/RS
Brazil

Anusha Vishwakarma
Department of Civil and Environmental Engineering
Indian Institute of Technology Patna
Patna
Bihar
India

Ismawi Yusof
Civil Engineering Programme Area
Universiti Teknologi Brunei
Gadong
Brunei Darussalam

Xu Zhang
State Key Laboratory of Bioreactor Engineering
East China University of Science and Technology
Shanghai
China

Preface

Sustainable living conditions are necessary for a developing society. In order to meet the sustainable development goals for the attainment of a circular economy, there is a need for judicial utilization of natural resources, efficient technologies, effective waste management, recycling tools and techniques to maintain stability and continuity of the supply chain, and conservation of resources. Resource conservation can be facilitated through resource circularization, recycling, and reuse, driving toward zero waste discharge. Waste electrical and electronic equipment (WEEE) or electronic waste (e-waste) is one of the fastest-growing waste streams in the urban environment worldwide. A large portion of e-waste remains unaccounted for and is deprived of treatment through appropriate recycling chains and methods. They are often found to be managed by the informal sector using rudimentary approaches that not only cause environmental pollution but also lead to serious health hazards for the workforce employed. Out of the total e-waste generated, only a small share is collected formally by the take-back system, while the rest is disposed of in waste bins, which finally reach incinerators, landfills, and open dump yards. E-waste is a significant source of base, precious, and toxic metals. The composition of e-waste is diverse, and it contains a complex array of metals in quantities even higher than their natural deposits. This high metallic content contributes perpetually and comprehensively to the economic vitality of the recycling of metals. In addition, the recovery of metals is also necessary for waste treatment. Therefore, management and recycling of e-waste is an important aspect, not only from the point of waste treatment to avert environmental pollution but also for resource recovery for economic development. The hierarchy of e-waste management recommends the reuse and/or remanufacture of the whole end-of-life electrical and electronic equipment, followed by the recovery of materials by recycling techniques. For the recovery of metals from e-waste, various traditional metallurgical techniques, viz., pyrometallurgy using heat treatments like roasting, smelting, and hydrometallurgy based on chemical leaching, are employed. These processing methods are often demonstrated to be energy intensive, costly, and associated with serious secondary pollution such as the release of toxic gases like dioxins, furans, metal dust, a high volume of hazardous lixiviants, and wastewater. In order to meet the environmental norms for hazardous waste disposal and conservation of natural resources around the world, there is a growing concern for the safe recycling of e-waste. The limitations of conventional techniques have led to a shift toward biotechnological approach. The integration of biotechnology into the hydrometallurgical process by the application of biocatalysts such as microorganisms and enzymes has evolved into a green, safe alternate metallurgical process known as biohydrometallurgy, biometallurgy, or bioleaching technique for e-waste recycling. In comparison to conventional technologies, bioleaching-based processes offer a number of advantages, such as low operation costs, environmental compatibility, relatively low energy consumption, and easy technology for metals extraction. Further, as the majority of the e-waste generated is being managed by the informal sector using rudimentary approaches, bioleaching technology provides an alternative to the metallurgical sector with no environmental and health hazards.

This book comprehensively presents various aspects of e-waste, such as generation, management techniques, role as secondary metal reservoir, different conventional as well as eco-friendly recycling approaches, biotechnological advances, and developments in resource recovery, along with the associated toxicity, risks, and scope for future research. The book aims to present hazards associated with conventional recycling methods and highlights environmentally compatible and economic approaches to resource recovery. A review of the progress in bioleaching technology focused on a sustainable future is also presented in the book. Further, the book fills in the gaps in understudied biotechnological recycling techniques and explores possibilities to mitigate environmental pollution caused by conventional and rudimentary e-waste management and recycling approaches. Some of the latest research and developments in the areas of e-waste management, biotechnological recycling processes, regulations, and policies are also highlighted in the book. A complete overview of all aspects related to the toxicity characterization of e-waste, resource recovery, recycling strategies, biotechnological advancements, and current perspectives on e-waste generation and management is covered in the book. Further, recycling and resource recovery aspects of e-waste under the current global scenarios are conceptualized in this book, adopting a qualitative approach that will facilitate future risk characterization and evaluation. This book is expected to serve as a guideline for producers, consumers, recycling industries, policy and law makers, academicians, and researchers. The book is also expected to contribute to society by creating awareness and knowledge among the general masses and the scientific community regarding pollution abatement and sustainable utilization of resources.

July 26, 2023

Dr. Anshu Priya
City University of Hong Kong
Hong Kong

Acknowledgment

I extend my deep sense of gratitude to John Wiley & Sons for giving me this wonderful opportunity to publish and spread awareness on one of the most serious issues of urban world – the e-wastes, their hazards, importance, resource recovery, reutilization, management, and regulations. My immense thanks goes to the researchers and scientists who helped me with their innovative suggestions, vast knowledge, scientific advice, and expertise that led to successful completion of this book. I would like to convey my heartfelt gratitude to all the authors and contributors of this book for their insightful articles, kind cooperation, and unstinted support. The love of my parents and family has always been outstanding. I owe my deepest and warmest gratitude to my grandparents, parents, brother, family members, and friends for their unparallel love, care, support, and encouragement. I convey my deep sense of appreciation to my husband, Dr. Ambresh Shivaji, for his motivation, consideration, and constant support. Words will not suffice to acknowledge their constant support throughout this work. A special thanks to all my well-wishers whose names I missed out here.

Above all, I express my profound devotion and deepest gratitude to The Almighty for bestowing His blessings upon me.

1

An Introduction to Electronic Waste

Anshu Priya

School of Energy and Environment, City University of Hong Kong, Tat Chee Avenue, Kowloon, Hong Kong

Abstract

Waste electrical and electronic equipment (WEEE) or electronic waste (e-waste) is one of the fastest-growing waste streams worldwide. The rich metallic content of e-waste makes them secondary reservoir of many base, precious, rare earth, and toxic metals. Additionally, several nonmetallic constituents, such as brominated flame retardants, in quantities higher than the permissible limit make them hazardous in nature. Thus, the management, treatment, and resource recovery of e-waste are essential not only for economic development but also to avert environmental pollution. The problem of e-waste generation is more prominent in developed and developing economies. Further, the problem gets compounded by the transboundary movement of e-waste despite the Basel Convention Prohibition (1992). The evaluation of their generation, flow in economy, and composition are essential elements for development and implementation of various eco-friendly recycling approaches, sustainable reprocessing technologies, and sound policies for a sustainable, green, and circular economy.

Keywords *e-waste; metal resources; recycling technologies; management approaches; sustainable and circular economy*

1.1 Introduction

The unprecedented digitalization and urbanization over the past few decades, coupled with a rapid product obsolescence rate, have led to enormous generation of waste electrical and electronic equipment (WEEE) or electronic waste (e-waste), ranking it among the fastest-rising waste streams worldwide. China was reported to generate the highest e-waste, with more than 10 million metric

Management of Electronic Waste: Resource Recovery, Technology and Regulation, First Edition.
Edited by Anshu Priya.

tons (Mt) in 2019. The United States followed China with around 7 Mt and was the second-largest e-waste generator. As per the report by the United Nations University (UNU 2015), 53.6 Mt of e-waste were estimated to be generated in the year 2019 (Baldé et al. 2022). The generation trend was further anticipated to increase to 74.7 Mt. by 2030 (Forti et al. 2020).

E-waste includes a wide range of components and constituents from polymers, glass, and ceramics to valuable resources like metals. Base metals, precious metals, toxic metals, as well rare earth metals (REMs) such as Cu, Zn, Cd, Pb, Hg, Ni, Co, Al, Fe, Sn, Au, Ag, Pd, Nd, Ce, and La are integral constituents of e-waste. It has been reported that the concentration of metals in e-waste is in sufficiently rich quantity, many times even higher than their natural repository (Konaté et al. 2022). The rich metallic content of e-waste thus makes waste a good secondary source of metals. Thus, the removal of metals is subject of utmost concern not only for treatment of waste but also for recovery of metallic fractions. Reprocessing of waste metals generated from electric and electronic equipment is vital, as it not only allows waste management, treatment, and resource recovery but also averts environmental pollution. However, management of e-waste is a challenge due to the fact that the current recycling technologies include either pyrometallurgical or hydrometallurgical processes, which are energy-concentrated, costly, and also associated with environmental problems linked to the release of toxic gases like dioxins, furans, metal dusts, and lixiviants (Dutta et al. 2023). Therefore, it is urgent to develop an appropriate recycling technology for reprocessing of WEEE in cost-effective and environmentally compatible manner.

Biotechnology has thus proven to be a dominating and versatile substitute for the existing conventional recycling methods for tackling various problems related to environmental pollution because of its low energy demand, less inputs, low waste generation, and emissions (Brindhadevi et al. 2023). To date, several biotechnologies have been subjugated in well-developed, mechanized systems that can be categorized under the heading "biohydrometallurgy." This technique exploits biological activities for efficient recovery of metal. Through these processes, metallic compounds are converted into their water-soluble states through a process called microbial leaching (Atlas and Bartha 1997). Over the past few years, studies related to biological leaching have shown to play a significant role in metal mobilization. Now there is a need for shifting the application of microbiological leaching process to extract metals from various wastes. There are many microorganisms that are reported to have significant role in metal recovery from industrial waste. Among the bacteria, sulfur-oxidizing bacteria, especially those of *Thiobacillus* spp. (also called *Acidithiobacillus* spp.), such as *Acidithiobacillus ferrooxidans*, *Acidithiobacillus thiooxidans*, *Leptospirillum ferrooxidans*, and *Sulfolobus* sp., are some of the well-known bioagents conducting bioleaching (Trivedi and Hait 2023; Qiu and Liu 2022; Trivedi et al. 2022). Metals present as leachate in the leached liquor are recovered using solvent extraction, adsorption,

ion exchange, precipitation, cementation, and electrowinning. Compared with conventional methods for removing toxic metals from leached liquor, the biosorption process has the advantages of biodegradability, low operational cost, reduced volume of chemicals and biological sludge to be disposed of, and high efficiency in detoxifying very dilute liquors (Rivas-Castillo et al. 2022).

The challenges in the management of e-waste using sustainable reprocessing technologies that are able to address the e-waste using economically and ecologically responsive methods have encouraged need of active attempts in the area. This book presents a comprehensive insight into the various metallurgical processes, policies, laws, hazards, life cycles, and techno-economic assessments involved in the recovery and management of e-waste.

1.2 Generation and Composition of E-Waste

The advent of new, sophisticated inventions, advanced technologies, and substantially shortened life span of many electronic commodities has kicked out many electronic items from the consumer's hands, which ultimately builds up e-waste. A drastic increase in e-waste generation is being observed with the rising economy and mounting demand for EEE. As per a study conducted by UNU, Asia produced the highest volume of e-waste with generation of 24.9 Mt, followed by the United States with 13.1 Mt and Europe with 12 Mt, while the contributions by Africa and Oceania were 2.9 and 0.7 Mt, respectively (Baldé et al. 2022). China ranked first in e-waste generation with 10.1 Mt, followed by the United States Of America with 6.9 Mt, and India with 3.2 Mt. Japan and Brazil share almost the same platform with 2.6 and 2.1 Mt, respectively. Table 1.1 presents e-waste generation data for the year 2019 for the major economies of the world (Baldé et al. 2022). The developed and emerging economies contribute to a significant portion of global e-waste. E-waste generation has been shown to have a direct correlation with both population and purchasing power of the masses. Income level and purchasing power are powerful drivers of the economy and govern e-waste production. With the increase in per capita purchasing power of developing economies, there is a significant transition toward industrialization and e-waste generation (Puntillo 2023).

The complexity and heterogeneity of e-waste owe to its various constituents. It has been estimated that one personal computer is usually composed of 40% steel, 21% plastic, 10% Al, and 10% other metals such as Cu, Cd, Au, Pt, and Ag (Brindhadevi et al. 2023). The lithium batteries, an important power source of electronic devices, contain 2–7% Li and 5–15% Co as basic constituents (Neumann et al. 2022). Co being an expensive metal is recycled from spent batteries. Printed circuit board (PCB) an important component of electrical and electronic goods contains the majority of metals; however, the actual composition varies with type of PCB. The complex compositions of e-waste make them a rich recovery source.

Table 1.1 E-waste generation for the major economies of the world for the year 2019.

Country/region	Total e-waste generated (Mt) in 2019
China	10.1
United States	6.9
India	3.2
Japan	2.6
Brazil	2.1
Russia	1.6
Indonesia	1.6
Germany	1.6
United Kingdom	1.6
France	1.4

Source: Adapted from Baldé et al. (2022).

The hazardous components present in e-waste are of serious concern and mainly include metals such as As, Cr(IV), Co, Hg, Pb, and Ni. These are considered hazardous as they are bioaccumulative, can concentrate in cells, and further carried on to the successive tropic level, where they produce deleterious impact (Brindhadevi et al. 2023). Recycling thus turns e-waste into a secondary reserve for the recovery and recycling of metals and nonmetals, thereby reducing their environmental and health hazards as well as finding a way out to solve metal crisis (Islam et al. 2020).

1.3 Present Status of E-Waste Management and Recycling

As per a study conducted by UNU, only 17% of the total global e-waste generated in 2019 was officially recognized to be collected and managed in an environmentally viable manner, thus ensuring the recycling of a gross value of US $9.4 billion of metals such as Fe, Au, and Cu. The fate of the remaining 44.3 Mt, making up about 83% of e-waste generated in 2019, was undocumented and not known (Baldé et al. 2022). Efficient management and handling of e-waste can help alleviate global warming and global greenhouse gas emissions. The high metallic content and heterogeneous and composite nature of e-waste make them hazardous and rich secondary sources for recovery of metals. Environmental pollution caused by the metal content of e-waste is very noticeable in dumping areas, landfills, and incinerators. The dumped discarded metals leach out, generate toxic emissions on burning, and contaminate air, water, and soil (UNEP 2009).

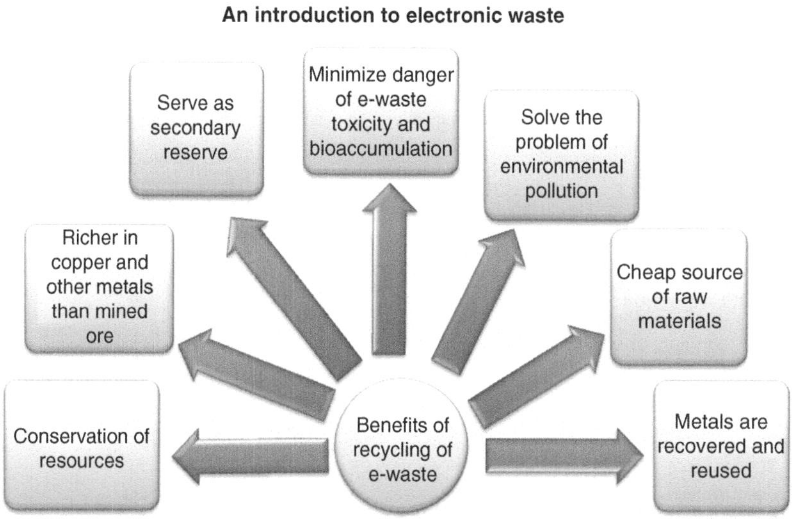

Figure 1.1 Benefits of e-waste recycling.

Metallic mercury, dimethyl mercury, Pb, and Cd are the most toxic leachates. Landfills and incinerators are also very much prone to uncontrolled fires and the release of harmful gases and toxic slag through which the metals reenter the environment (Murali et al. 2022). Taking all these into consideration, there is an urgent need to take up the "R" endeavors, i.e. Reduce, Reuse, and Recycle. The drift toward reducing the use of metals must be encouraged; however, the other two that deserve special attention are reuse and recycle. The benefits of metal recycling are depicted in Figure 1.1.

Recycling of e-waste from old, unused appliances or devices involves manual dismantling and segregation of the WEEE components like the metallic and non-metallic fractions. Following sorting, the metallic components are shredded and then subjected to recovery. The recovery of metals from e-waste involves a number of mechanical, physical, chemical, and biological processes (Islam et al. 2020).

1.3.1 Pyrometallurgical Process

Pyrometallurgical processes include thermal treatment of the shredded metallic fractions in controlled high-temperature environment to recover the metals of interest (Hartley et al. 2023). The heat treatment is given in specialized furnaces such as incinerators, plasma arc furnaces, or blast furnaces where reactions between the metallic oxides and the reducing agent such as charcoal or coke are carried out releasing carbon dioxide (CO_2). The nonmetallic fraction, i.e. gangue is heated with substance called flux to form slag. Slag is a molten mass that, being

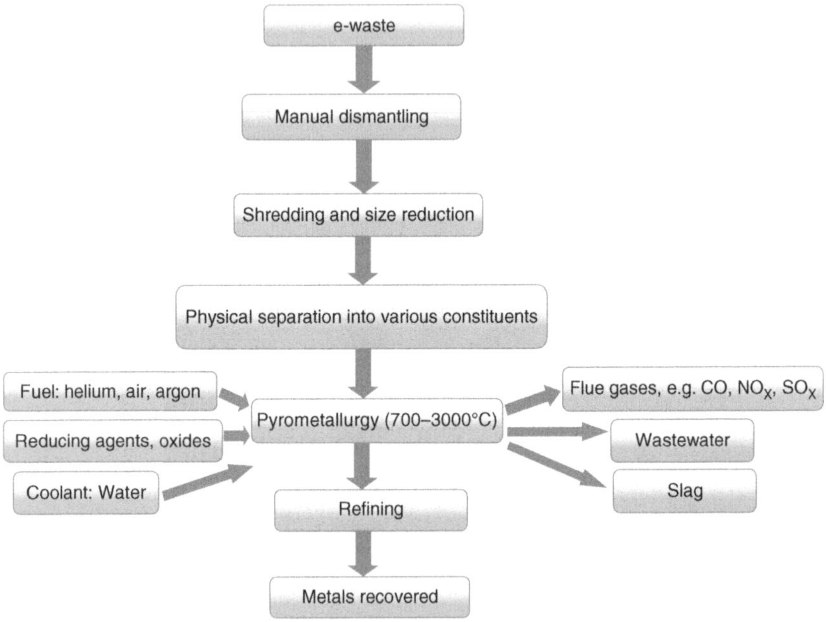

Figure 1.2 Generalized steps involved in pyrometallurgical recovery of metals from e-waste.

lighter than the metallic fraction, floats over it and can be easily decanted off (Ebin and Isik 2016).

A number of modifications have been proposed to the conventional pyrometallurgical processes to improve their yield as well as reduce their environmental hazardous impact. Wang et al. 2020 modified the gold refining process through reaction of gold scrap and chlorine. The resultant was washed with hydrochloric acid, ammonium hydroxide, and nitric acid to remove the impurities. Au with 99.9% purity was recovered by this process. Abdelbasir et al. (2018) applied vacuum pyrolysis and extracted Cu with 99% purity. Umicore's smelter trial recovered metals from electronic scrap by replacing the reducing agent coke and the energy source with plastic-rich WEEE materials (Castro and Bassin 2022). However, an important disadvantage linked to pyrometallurgical process is the emission of harmful gases and compounds, such as polybrominated dibenzodioxins (PBDD), polybrominateddibenzofurans (PBDF), phenol, naphthalene, biphenyl, anthracene or phenanthrene, dibromobenzene, dibenzofuran, dibenzo-*p*-dioxin, tribromobenzene, tetrabromobenzene, and many more, leading to serious environmental pollution (Hartley et al. 2023). Figure 1.2 shows generalized steps involved in pyrometallurgical metal recovery.

1.3.2 Hydrometallurgical Process

In conventional hydrometallurgical process, leaching of metals is carried out using suitable lixiviants. The crushed e-waste scrapes are subjected to various chemicals of acidic and basic nature such as aqua regia, nitric acid, sulfuric acid, thiourea, thiosulfate, cyanide, and halide to leach out the metals from them. Cyanide leaching is most widely used for effective and economic extraction of gold and can efficiently extract 1–3 g/t of Au, but the major drawback is that the chemical possesses environmental toxicity. Thiosulfate and thiourea also have fast leaching rate but low chemical stability. Acid leaching is conducted usually as first-stage leaching to recover base metals, especially Cu (Ashiq et al. 2019). Chemical leaching exploiting various chemical ligands such as ethylene diamine tetramaide (EDTA), diethylene triamine penta acetate (DTPA), and other chelators like oxalate and citric acid have also been exploited for metal extraction (Xavier et al. 2023). The metals are then purified through techniques such as electrorefining, solvent extraction, and adsorption. Figure 1.3 shows schematic representation of hydrometallurgical metals recovery. The wastewater generated in the whole process of recovery is extremely hazardous toxic and difficult to recycle (Islam et al. 2020).

Various improvements have been introduced in the existing pyrometallurgical process. Rath et al. 2012, coupled thermal plasma with acid leaching to improve metal recovery. This technique generates less effluent and offers fast separation of slag metal. Ultrasonic-assisted cleaner pyrometallurgical technique has also been shown to efficiently extract metals with less toxicant generation. The technique recovered Cu and Fe with an efficiency of 95.2–97.5% and 97.1–98.5%, respectively (Xie et al. 2009). In another approach, a combination of thermal and acid treatment was proposed for faster and more reliable gold dissolution and recovery (Mahapatra et al. 2019).

1.3.3 Biometallurgy

Biometallurgy, also known as biohydrometallurgy or bioleaching, is a novel technique aiming at efficient, green, and economic metal recovery (Figure 1.4). It is based on microbial solubilization of metals, through which they convert metals into soluble, extractable form. The leached metals in liquor are then adsorbed on suitable biological matrix and finally recovered. Microbes capable of secreting various organic and inorganic acids are exploited in this process to recover different metals from various low-grade sources (Arab et al. 2020; Dey et al. 2023). The frequently used microorganisms applied in bioleaching are acidophilic Fe- and sulfur-oxidizing bacteria such as *Thiobacillus ferrooxidans*, *T. thiooxidans*, *L. ferrooxidans,* and *Sulfolobus* spp., and fungi such as *Aspergillus niger* and *Penicillium* spp. to extract the target metals from waste (Dey et al. 2023). Apart from these, a number of heterotrophic bacteria, such as those of the genus *Bacillus* and

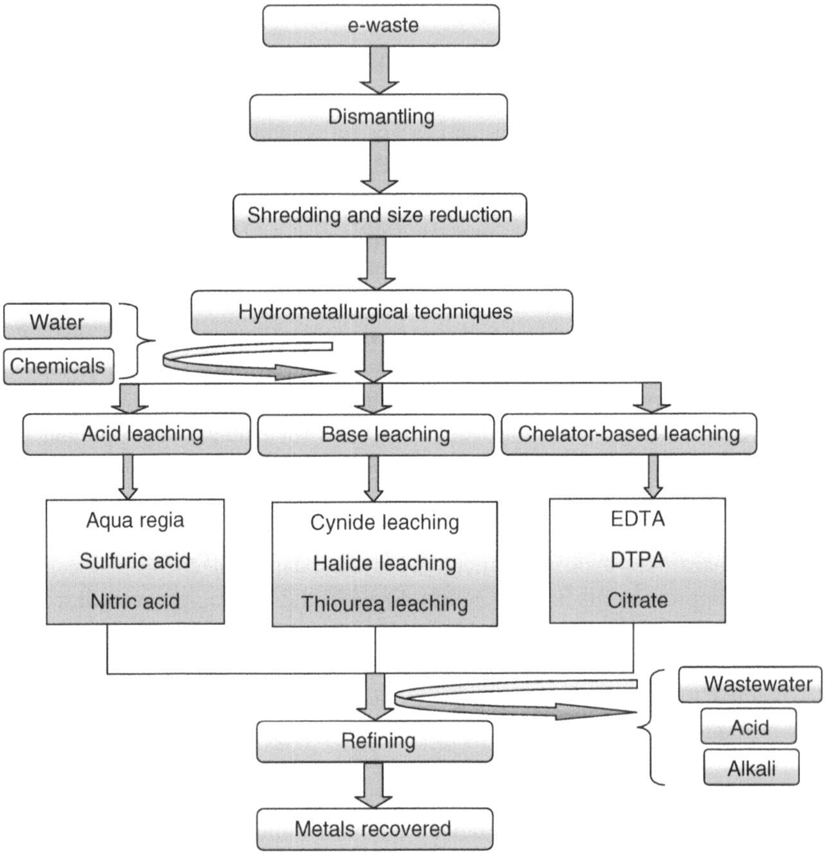

Figure 1.3 Schematic representation of hydrometallurgical recovery of metals from e-waste.

Pseudomonas, also have the potential to extract nonsulfidic minerals (Hubau and Bryan 2023).

Heterotrophic bioleaching microorganisms derive carbon and energy from organic carbon sources, and as a result of consumption, they produce a number of metabolic by-products such as citric acid, oxalic acid, and formic acid, which interact with the metallic waste to leach out metals (Dey et al. 2023; Hubau and Bryan 2023). However, there are several factors that regulate the metal bioleaching; these factors can be optimized to maximize the biological metal extraction. The parameters include biotic factors such as the characteristics and type of microorganisms used and size of inoculums, and abiotic factors such as growth environment, pH, temperature, aeration rate, particle size of extractable e-waste sources, incubation time, and nutritional composition of media (Valix 2017; Yang et al. 2009). Various researchers have extracted metals using bioleaching

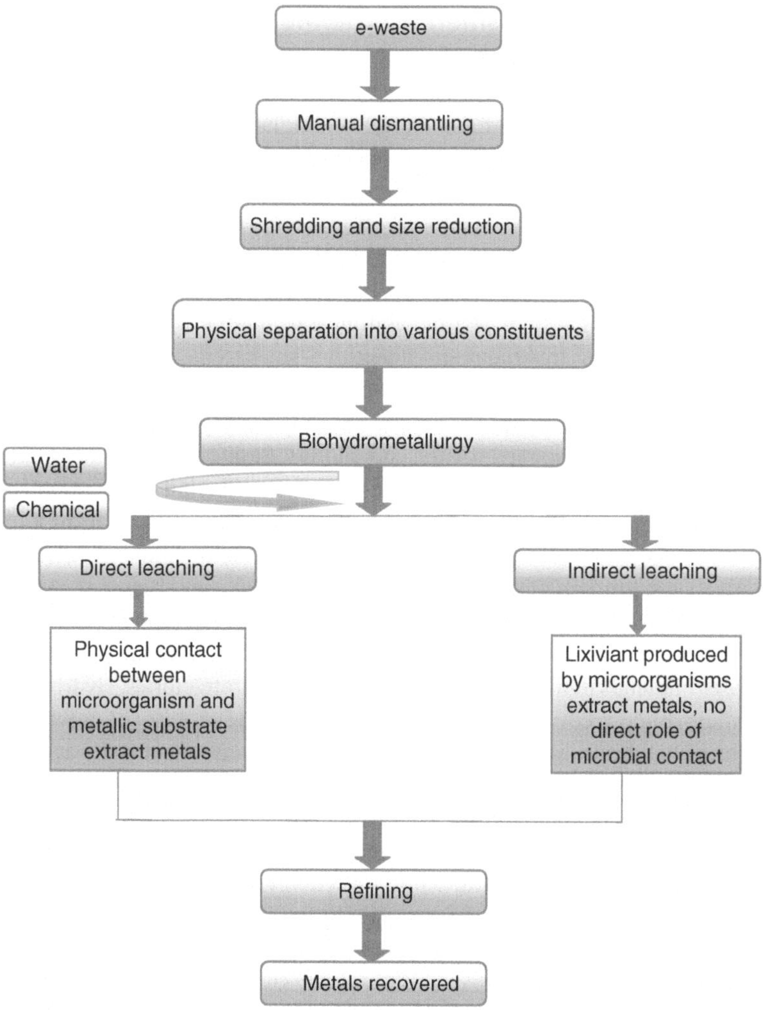

Figure 1.4 Schematic representation of biohydrometallurgical recovery of metals from e-waste.

reactions. Yang et al. (2009) observed the effects of concentrations of iron, pH, inoculum size, and concentration on Cu extraction by *A. ferrooxidans*. Ilyas et al. (2007) successfully used acidophilic heterotrophic bacteria to recover more than 80% of metals such as Al, Cu, Ni, and Zn from PCBs. Baldé et al. (2022) demonstrated the application of *T. thiooxidans* and *A. niger* in the recovery of metals from e-scrape. Faramarzi et al. (2004) applied cyanogenic *Chromobacterium violaceum* for bioextraction of Au from PCBs. Cu and Au are among the vital metals that are industrially extracted through biometallurgy (Ilyas et al. 2007).

1.4 Comparative Assessment of the Metallurgical Options for Metal Recovery

The three most widely employed metallurgical options, pyrometallurgy, hydrometallurgy, and biometallurgy are central apparatus for metal recovery. Choosing an appropriate metal recovery option is very much dependent on its extractability, economics, time, involvement of labor, and compatibility with the environment.

Extraction parameters such as environmental compatibility, economics, engagement of labor, expertise, waste type, and metal content in waste must be assessed before the selection of a metallurgical technique. Pyrometallurgical process involves thermal treatment and hence generates tremendous amount of hazardous gases and dust. Burning of plastics and other inorganic and organic waste constituents generates hazardous gases such as furans, dioxins, CO_2, CO, SO_2, NO_x, and various poisonous volatile compounds. Besides these, incineration produces ashes with high content of metals such as Cd, Pb, and As. Hydrometallurgical and biometallurgical treatments extract metals in soluble form and hence generate large amounts of wastewater; however, biometallurgy generates comparatively less amount of wastewater than hydrometallurgy (Dey et al. 2023). Types of wastewater include spent acids, cooling water, and washdowns.

Taking mass balance into consideration, pyrometallurgical method leads to higher losses of metals in the form of metal dust, slag, and refractories. Hydrometallurgy offers high recovery and less metal with relative ease of leaching. Biometallurgy exhibits a higher rate because of involvement of active bioagents, which can extract metals even from depleted, low-grade sources, and possibility of recirculation of leached liquor in the method (Trivedi and Hait 2023). A large amount of fuel and energy are also required in the pyrometallurgical process as compared to hydrometallurgical, while the energy requirement is the least in biometallurgy (Dey et al. 2023). Biometallurgical technique is the most time-consuming, labor-intensive, and requires expertise as compared to the other two metallurgical techniques (Hubau and Bryan 2023).

The expenses involved in current conventional metallurgical methods have necessitated need for exploration of efficient alternatives for waste management and metal recovery. Potential factors affecting metal recovery also include the cost of operation and maintenance, that is, the investment capital. The profitability of recycling methods drastically depends on the economics of the metallurgical technique.

1.5 Future Prospects

E-waste, especially PCBs, contains a wide range of metals in sufficient quantities, which get dumped when the equipment becomes obsolete. The conventional recovery techniques are costly, high energy consuming, and environmentally incompatible. Bioleaching, on the other hand, is economic and green approach for recovery of metals from e-waste and other low-grade metal repositories. However, the research in this area is very limited, on a laboratory scale, and has not yet developed at industrial scale. Process can be designed using agro- or food-industrial wastes as a growth medium for heterotrophic leaching microorganisms to carry out bioleaching. The limitation of bioleaching process is the inherently long operational time, which can be reduced through approaches such as genetic manipulation or the integration of other safe approaches such as physical and chemical approaches. Interdisciplinary research involving areas encompassing microbiology, chemical engineering, biotechnology, metallurgy, and biochemical engineering is required to take bioleaching to profitable commercial applications.

1.6 Conclusion

Environmental pollution by e-waste is very noticeable in dumping areas. The dumped metals leach out under the influence of physical, chemical, and biological factors of environment; their volume reduction by burning also generates toxic emissions and contaminates air, water, and soil. The high metallic content and heterogeneous and composite nature of e-wastes, however, make them rich secondary sources of metals. But the current recycling technologies, hydro, and pyrometallurgical techniques are environmentally incompatible, energy-concentrated, and uneconomic. Thus, there is an urgent need for recycling of metals, in eco-friendly, cost-effective way, to avert pollution with focus on economic metals recovery. Bioleaching of metals from e-waste emerged as a new approach for metal recovery in an environment-friendly manner. However, for maximizing the rate of recovery, certain biotic and abiotic factors must be taken into consideration and optimized so as to convert the process to a profitable one. Studies conducted on bioleaching show disparity in the process factors governing metal biodissolution. The same microorganism is shown to have different process requirements for metal recovery. Thus, there is a need to bridge the gaps present in the existing studies for efficient and profitable metal recovery by bioleaching.

References

Abdelbasir, S.M., El-Sheltawy, C.T., and Abdo, D.M. (2018). Green processes for electronic waste recycling: a review. *Journal of Sustainable Metallurgy* 4 (2): 295–311.

Arab, B., Hassanpour, F., Arshadi, M. et al. (2020). Optimized bioleaching of copper by indigenous cyanogenic bacteria isolated from the landfill of e-waste. *Journal of Environmental Management* 261: 110124.

Ashiq, A., Kulkarni, J., and Vithanage, M. (2019). Hydrometallurgical recovery of metals from E-waste. In: *Electronic Waste Management and Treatment Technology*, 225–246. Oxford: Butterworth-Heinemann.

Atlas, R.M. and Bartha, R. (1997). *Microbial Ecology: Fundamentals and Applications*. California: Benjamin Cummings.

Baldé, C.P., D'Angelo, E., Luda, V. et al. (2022). *Global Transboundary e-Waste Flows Monitor – 2022, United Nations Institute for Training and Research (UNITAR)*. Germany: Bonn.

Brindhadevi, K., Barceló, D., Chi, N.T.L., and Rene, E.R. (2023). E-waste management, treatment options and the impact of heavy metal extraction from e-waste on human health: scenario in Vietnam and other countries. *Environmental Research* 217: 114926.

Castro, F.D. and Bassin, J.P. (2022). Electronic waste: environmental risks and opportunities. In: *Hazardous Waste Management*, 421–458. Amsterdam: Elsevier.

Dey, S., Shekhawat, M.S., Pandey, D.K. et al. (2023). Microbial community and their role in bioremediation of polluted e-waste sites. In: *Metagenomics to Bioremediation*, 261–283. Cambridge: Academic Press.

Dutta, D., Rautela, R., Gujjala, L.K.S. et al. (2023). A review on recovery processes of metals from E-waste: a green perspective. *Science of the Total Environment* 859: 160391.

Ebin, B. and Isik, M.I. (2016). Pyrometallurgical processes for the recovery of metals from WEEE. In: *WEEE Recycling*, 107–137. Amsterdam: Elsevier.

Faramarzi, M.A., Stagars, M., Pensini, E. et al. (2004). Metal solubilization from metal-containing solid materials by cyanogenic Chromobacterium violaceum. *Journal of Biotechnology* 113 (1–3): 321–326.

Forti, V., Baldé, C.P., Kuehr, R., and Bel, G. (2020). The global e-waste monitor 2020: quantities, flows and the circular economy potential. In: *United Nations University (UNU)/United Nations Institute for Training and Research (UNITAR) – Co-Hosted SCYCLE Programme*. Bonn/Geneva/Rotterdam: International Telecommunication Union (ITU) & International Solid Waste Association (ISWA).

Hartley, J.M., Scott, S., Rivera, R.M. et al. (2023). Tailoring lixiviant properties to optimise selectivity in E-waste recycling. *RSC Sustainability* 1 (1): 107–116.

Hubau, A. and Bryan, C.G. (2023). Metal recovery from E-wastes. In: *Biomining Technologies*, 239–259. Cham: Springer.

Ilyas, S., Anwar, M.A., Niazi, S.B., and Ghauri, M.A. (2007). Bioleaching of metals from electronic scrap by moderately thermophilic acidophilic bacteria. *Hydrometallurgy* 88 (1–4): 180–188.

Islam, A., Ahmed, T., Awual, M.R. et al. (2020). Advances in sustainable approaches to recover metals from e-waste – a review. *Journal of Cleaner Production* 244: 118815.

Konaté, F.O., Ancia, P., Soma, F. et al. (2022). Waste electrical and electronic equipments as urban mines in Burkina Faso: characterization and release of metal particles. *Waste Management* 139: 17–24.

Mahapatra, R.P., Srikant, S.S., Rao, R.B., and Mohanty, B. (2019). Microwave heating and acid leaching processes for recovery of gold and other precious metals from e-waste. *Current Science* 116 (3): 463–468.

Murali, A., Plummer, M.J., Shine, A.E. et al. (2022). Optimized bioengineered copper recovery from electronic wastes to increase recycling and reduce environmental impact. *Journal of Hazardous Materials Advances* 5: 100031.

Neumann, J., Petranikova, M., Meeus, M. et al. (2022). Recycling of lithium-ion batteries—current state of the art, circular economy, and next generation recycling. *Advanced Energy Materials* 12 (17): 2102917.

Puntillo, P. (2023). Circular economy business models: towards achieving sustainable development goals in the waste management sector—empirical evidence and theoretical implications. *Corporate Social Responsibility and Environmental Management* 30 (2): 941–954.

Qiu, G.Z. and Liu, X.D. (2022). Biotech key to unlock mineral resources value. *Transactions of Nonferrous Metals Society of China* 32 (7): 2309–2317.

Rath, S.S., Nayak, P., Mukherjee, P.S. et al. (2012). Treatment of electronic waste to recover metal values using thermal plasma coupled with acid leaching–a response surface modeling approach. *Waste Management* 32 (3): 575–583.

Rivas-Castillo, A.M., Gómez-Ramírez, M., Lucas-Gómez, I.M. et al. (2022). A new technique to evaluate Acidithiobacillus thiooxidans growth during a bioleaching process based on DNA quantification. *Journal of Microbiological Methods* 198: 106494.

Trivedi, A. and Hait, S. (2023). Metal bioleaching from printed circuit boards by bio-Fenton process: optimization and prediction by response surface methodology and artificial intelligence models. *Journal of Environmental Management* 326: 116797.

Trivedi, A., Vishwakarma, A., Saawarn, B. et al. (2022). Fungal biotechnology for urban mining of metals from waste printed circuit boards: a review. *Journal of Environmental Management* 323: 116133.

UNEP (2009). *Recycling – from E-waste to Resources.* Paris: United Nations Environmental Programme (UNEP).

UNU (2015). *E-Waste World Map: Update to Quantitative Data and Legal Texts – STEP.* United Nations University (UNU).

Valix, M. (2017). Bioleaching of electronic waste: milestones and challenges. *Current Developments in Biotechnology and Bioengineering* 1: 407–442.

Wang, J., Zhang, M., and Lu, Y. (2020). Study of gold leaching behavior in the chlorination process from waste printed circuit boards. *ACS Sustainable Chemistry & Engineering* 9 (1): 284–290.

Xavier, L.H., Ottoni, M., and Abreu, L.P.P. (2023). A comprehensive review of urban mining and the value recovery from e-waste materials. *Resources, Conservation and Recycling* 190: 106840.

Xie, F., Cai, T., Ma, Y. et al. (2009). Recovery of Cu and Fe from printed circuit board waste sludge by ultrasound: evaluation of industrial application. *Journal of Cleaner Production* 17 (16): 1494–1498.

Yang, T., Xu, Z., Wen, J., and Yang, L. (2009). Factors influencing bioleaching copper from waste printed circuit boards by Acidithiobacillus ferrooxidans. *Hydrometallurgy* 97 (1–2): 29–32.

2

The Global Challenge of E-Waste Generation

Lucas Reijnders

Faculty of Science, IBED, University of Amsterdam, Science Park 904, Amsterdam 1090 GE, The Netherlands

Abstract

The worldwide generation of e-waste may be estimated at $>57 \times 109$ kg in 2022 and is increasing at a yearly rate of 3–5%. In dealing with e-waste, there are currently large losses of valuable resources, accompanied by substantial workplace hazards and pollution beyond the workplace. Presumably, about 10% of global e-waste is disposed of by landfilling and municipal waste incineration. Informal processing is estimated to apply to about 70% of global e-waste. Legal handling of e-waste (formal processing) may regard <18% of global e-waste. There is partial recycling of the major metals present in e-waste, but the conservation of quality is very limited. There is currently very limited high-quality material recycling of synthetic polymers and very little recycling of glass present in e-waste. Losses of geochemically scarce elements in processing e-waste are large. Minimizing the losses of valuable resources, the associated workplace hazards, and pollution beyond the workplace presents a major challenge.

Keywords *e-waste; steel; Al alloys; synthetic polymers; geochemically scarce elements; resource losses; pollution; collection; disposal; informal processing; formal processing*

2.1 Introduction

Currently, end-of-life electric and electronic equipment is generated in large amounts. In dealing with end-of-life electric and electronic equipment, the first choice from an environmental point of view is rather often (depending on product, electricity mix, and relative energy efficiencies) to focus on options such as refurbishment, remanufacture, and cascading use (Pérez-Belis et al. 2017; Lu et al. 2017; Baxter 2019; Graedel et al. 2019; Bovea et al. 2020). These options

Management of Electronic Waste: Resource Recovery, Technology and Regulation, First Edition.
Edited by Anshu Priya.

are currently applied to a limited amount of end-of-life electric and electronic equipment (Steuer et al. 2018; Tran et al. 2018; Cole et al. 2019; Vaccari et al. 2020; McMahon et al. 2021; Santana et al. 2021). Storage of end-of-life electronic equipment by users may also occur and can be sizable. This especially regards relatively small equipment, such as audio-visual and ICT equipment, and small household appliances (Navazo et al. 2014; Welfens et al. 2016; Owakowski 2019; Shaikh et al. 2020).

Commonly, end-of-life electric and electronic equipment is considered to be a waste: e-waste or waste electric and electronic equipment (abbreviated WEEE). The worldwide emergence of e-waste may be estimated at $>57 \times 10^9$ kg in 2022 (extrapolated from Forti et al. 2021 and Statista 2021) and is increasing at a yearly rate of 3–5% (Shittu et al. 2021).

E-waste contains a large variety of materials. Major metals present in e-waste are steel and Al alloys. On the basis of data provided by Ongondo et al. (2011), Buekens and Yang (2014), and Zhang and Xu (2016), the overall share of these metals in globally generated e-waste may be estimated at about 50 (weight)%, with steel predominating. Synthetic polymers are another major component of e-waste. Based on data provided by Ongondo et al. (2011), Buekens and Yang (2014), Alassali et al. (2019), Esposito et al. (2020), and Das et al. (2021), the overall share of synthetic polymers in current globally generated e-waste may be estimated at 20–30 (weight)%. Commonly applied polymers that may be found in e-waste are in listed Box 2.1.

Box 2.1 Synthetic polymers are commonly found in e-waste (Buekens and Yang 2014; Alassali et al. 2019; Esposito et al. 2020; Maisel et al. 2020).

Polyacrylonitrile-butadiene-styrene (ABS)
Polyamides (PA)
Polybutyleneterephtalate (PBT)
Polycarbonate (PC)
Polyethyleneterephthalate (PET)
Polymethylmethacrylate (PMMA)
Polyolefins (PE, PP)
Polyoxymethylene (POM)
Polyphenyleneoxide (PPO)
Polystyrene (PS)
Polystyrene-acrylonitrile copolymer (SAN)
Polyurethanes (PU)
polyvinylchloride (PVC)
Thermosetting resins (mainly epoxy resins)

Table 2.1 Relatively recently developed synthetic polymers are emerging in e-waste.

Application	Synthetic polymer	References
Liquid crystals for displays	Polymer of biphenyl- (e.g. phenyl-benzoate) monomers with CN, F, Cl, and Br in functional groups	Li et al. (2018), Li et al. (2021)
Transparent touchscreen and LED electrodes	Poly (3,4-ethylenedioxy-thiophene) doped with polystyrene sulfonic acid	Mochizuki et al. (2017), Luo et al. (2019)
Supercapacitors	Polyaniline, polypyrrole	Namsheer and Rout (2021)
OLED TVs and smartphones	Variety of conjugated polymers with derivatives of polyfluorene, polyphenylenevinylene, and polycarbazole	Inamuddin et al. (2020)

Relatively recently developed synthetic polymers that are emerging in e-waste are presented in Table 2.1.

A substantial component of e-waste is glass. Its contribution to global e-waste might be about 10 (weight)% (Buekens and Yang 2014; Wang and Xu 2014; Del Rio et al. 2022).

To the extent that geochemically scarce elements (with an upper crustal abundance of less than 0.025 (weight)%) are present in electric and electronic components of e-waste, their contribution to global e-waste may be 4–5 (weight)% (Bertram et al. 2002; Soulier et al. 2018). Geochemically scarce elements that can be present in the electric and electronic components of e-waste are in listed Box 2.2.

Conventional ways of dealing with e-waste are disposal of e-waste in dustbins or by dumping, informal processing, and formal processing (Parajuly et al. 2018; Steuer et al. 2018; Ádám et al. 2021; Mowla et al. 2021; Arain et al. 2022). Reliable statistics regarding global e-waste disposed of in dustbins or by dumping are not available, but in Europe it has been estimated that disposal in dustbins is about 5 (weight)% of e-waste (Forti et al. 2021). Case studies (Mihai et al. 2019) suggest that elsewhere, the percentage of e-waste disposed of in waste bins or by dumping can be much higher. For instance, in Mexico, the estimated percentage of e-waste ending up in legal and illegal landfills is about 50% and in the United States, the percentage ending up in incinerators and landfills is >50% (Mihai et al. 2019). It is likely that part of global e-waste disposed of in wastebins or by dumping is recovered, e.g. by scavengers or waste pickers (Andarani and Goto 2014; Bimir 2020). Taking this into account, I presume that the worldwide unrecovered disposal of e-waste by dumping, controlled landfilling, and municipal waste incineration is about 10 (weight)%. Such disposal is associated with the loss of valuable resources and pollution (Forti et al. 2021).

Box 2.2 Geochemically scarce elements that can be present in electric and electronic components of e-waste (Cesaro et al. 2018b; Andersson et al. 2019; Buechler et al. 2020; Reijnders 2022)

Ag	Li
As	Lu
Au	Mo
B	Nb
Be	Nd
Bi	Pb
Cd	Pd
Ce	Pt
Cr	Ru
Co	Sb
Cu	Sc
Dy	Sn
Eu	Ta
Ga	Tb
Gd	W
Ge	Y
In	Zn
Ir	
La	

Formal processing of e-waste meets the legal standards as to collection, treatment facilities, and handling residues of treatment (Forti et al. 2021). Formal processing has been estimated to regard <18 (weight)% of globally generated e-waste (Forti et al. 2021). In formal processing, there are often initial steps for the recovery of components such as batteries, lamps, refrigerants, oil, printed circuit boards, and phosphors (Sansotera et al. 2013; Ardente et al., 2015; Lecler et al. 2015; Salhofer et al. 2016; Yoshida et al. 2016). In further processing, shredding and sorting are often a centerpiece (Ardente et al. 2015; Salhofer et al. 2016; Andersson et al. 2019). After sorting, the metal outputs are commonly fed to smelters, but there are also smelters that can process intact small e-waste (Reijnders 2022). Shredder residues from formal processing are commonly landfilled or incinerated in a controlled way (Chagnes et al. 2016).

Relatively large amounts of e-waste are processed informally (Isimekhai et al. 2017; Ohajinwa et al. 2018; Dutta and Goel 2021; Mowla et al. 2021). Reliable statistics could not be found, but informal processing is very important in dealing with e-waste. For instance, in India, an estimated 95% of e-waste is reported to be informally processed (Dutta and Goel 2021). Estimates of the share of informal processing in dealing with global e-waste are in the order of 60–80% (Perkins et al. 2014; Ohajinwa et al. 2018; Aich et al. 2020; Ádám et al. 2021). It would seem that globally, the current share of informal processing may be estimated at about 70 (weight)%

One of the practices used in informal processing is poorly controlled incineration of e-waste to recover the major metals (steel and Al alloys) and Cu (Vats and Singh 2014). Often, informal processing of e-waste consists of dismantling with an eye on generating outputs that can be sold (e.g. components that can be used as spare parts, printed circuit boards, Al alloys, steel, and noble metals) (Pradhan and Kumar 2014; Isimekhai et al. 2017; Parajuly et al. 2018; Tran and Salhofer 2018; Vaccari et al. 2020; Mowla et al. 2021; Arain et al. 2022). After dismantling, further processing may occur. Parts of equipment such as printed circuit boards can be "cooked" or "baked" on indoor stoves or grills, or in special heating equipment, for de-soldering, which can facilitate the recovery of components and noble metals (Tran and Salhofer 2018; Guo et al. 2020). Cables containing Cu are commonly subjected to burning, and there may also be acid leaching to recover precious and base metals (Pradhan and Kumar 2014; Pascale et al. 2016; Parajuly et al. 2018; Vaccari et al. 2020; Lin et al. 2022). NaCN can be used in the extraction of Au (Terazono et al. 2017). There furthermore may be (poorly controlled) ovens for smeltig metals such as lead and copper (Salhofer et al. 2016).

Workplace hazards in the formal processing of e-waste can be substantial, but workplace hazards in the informal processing of e-waste tend to be larger (Okeme and Arrandale 2019; Ádám et al. 2021; Dutta and Goel 2021). Pollution beyond the workplace due to the informal processing of e-waste also tends to be larger than pollution beyond the workplace linked to formal processing (Ádám et al. 2021).

Workers in informal processing include women and children, who are relatively vulnerable to hazardous substances (Perkins et al. 2014; Terazono et al. 2017; Cesaro et al. 2019; Dutta and Goel 2021). Negative impacts on workers include respiratory disease, increased cancer risk, damage to the nervous system, and problems regarding pregnancy (Aich et al. 2020; Ádám et al. 2021). A study regarding teenage scavengers of e-waste in Nigeria (Alabi et al. 2020) found evidence of damage to DNA. The incineration of e-waste to recover metals is associated with a large workplace hazard (Vats and Singh 2014). In the case of informal processing by indoor dismantling and high-temperature treatment, there can be high indoor pollution by heavy metals, fine particulate matter, and hazardous organic compounds, corresponding with large worker hazards (Perkins et al. 2014; Ghosh et al. 2015; Awashti et al. 2016; Kaya 2016; Okeme and Arrandale 2019; Guo et al. 2020; Mowla et al. 2021; Lin et al. 2022). Indoor informal acid leaching can cause hazardous acid fumes (Singh et al. 2018). The use of NaCN in the extraction of Au is associated with toxic cyanide fumes (Terazono et al. 2017). Informal outdoor burning of cables to recover Cu is linked to negative impacts on worker health due to exposure to metal fumes, halogenated dioxins, and benzofurans (Pascale et al. 2016; Cesaro et al. 2019). As has also been implicated as a worker hazard in outdoor informal e-waste recycling (Okeme and Arrandale 2019; Yang et al. 2020).

Pollution beyond the workplace linked to the informal processing of e-waste can negatively affect birth outcomes and children's health (Song and Li 2014; Zeng et al. 2016). Substantial risks to public health are likely to be associated with informal dismantling and with poorly controlled incineration of e-waste plastics practiced in informal processing (Anh et al. 2017; Ohajinwa et al. 2018; Parajuly et al. 2018; Cardamone et al. 2021). Fine particulate matter and several halogenated compounds, including Cr, Hg, Pb, and Sb, can contribute to such risks (Anh et al. 2017; Ohajinwa et al. 2018; Guo et al. 2020; Singh et al. 2020; Ádám et al. 2021). Outdoor informal e-waste processing can cause severe soil and groundwater pollution (Pradhan and Kumar 2014; Terazono et al. 2017; Ohajinwa et al. 2018; Lin et al. 2022). Non-recovered materials remaining at outdoor sites for informal processing can also cause soil pollution (Isimekhai et al. 2017; Anselm et al. 2021). Residues from informal processing can be dumped outside the outdoor processing site (e.g. on riverbanks and in ponds) or incinerated in a poorly controlled way: activities that are associated with substantial pollution beyond the workplace and public health risks (Ohajinwa et al. 2018; Parajuly et al. 2018; Singh et al. 2018; Tran and Salhofer 2018; Vaccari et al. 2020; Arain et al. 2022).

Informal and formal processing are linked to large losses of valuable materials. In the next sections, the focus will be on the fate of the materials present in e-waste. Section 2 will deal with the fate of the major metals: steel and Al alloys. Section 3 will consider what happens to synthetic polymers. Section 4 regards the glass present in e-waste. In Section 5, the fate of geochemically scarce elements present in electric and electronic components of e-waste will be discussed. Section 6 considers what happens to Li-ion batteries, refrigerants, phosphors, and mercury in fluorescent lamps. Section 7 will present the conclusion of this chapter.

2.2 The Fate of Steel and Al Alloys

Bertram et al. (2017) estimated the global Al recycling rate for end-of-life Al in products at probably somewhat under 80%. The recycling rate for Al in e-waste may well be substantially lower than 80%. As pointed out above, about 10% of global e-waste is presumably disposed of in an unrecovered way. This negatively impacts the recycling rate. When collected e-waste is subjected to shredding, the recovery rate of Al in the output Al fraction may be estimated at values in the order of 44.6–68% (Bigum et al. 2012; Cesaro et al. 2017). There may be cases where the recovery of Al alloys by informal processing is better than 68% (Arain et al. 2022), but a substantially better Al recycling rate for worldwide informal processing would seem uncertain. The global recycling rate of end-of-life Fe has been estimated at about 80% (Helbig et al. 2020). The recovery rate for Fe in the Fe

Table 2.2 Geochemically scarce elements that may be functional, or dysfunctional in steel and Al alloy products and production (Reijnders 2016, 2018).

	Steel	**Al alloys**
Elements that may be functional	Cr. Co, Cu, Mo, Nb, Ni, Pb, Sn, V, W	B, Be, Bi, Cd, Cr, Cu, Ga, Gd, Hg, In, La, Li, Mo, Nd, Ni, Pb, Sb, Sc, Sn, V, Y, Yb, Zn, Zr
Elements that may be dysfunctional are dependent on their combination with other elements and concentration	Cu, Ni, Sn, Zn	B, Bi, Cd, Cr, Ga, In, Li, Mo, Ni, Pb. Sn, V, Zn, Zr

e-waste shredder output fraction has been estimated at 96% (Bigum et al. 2012). But the steel recycling rate in informal processing is uncertain.

Geochemically scarce elements present in steel and Al alloys can be functional (intentionally added as alloying elements), nonfunctional or dysfunctional. Geochemically scarce elements that may be functional or dysfunctional in steel and Al alloy products and production are in listed Table 2.2.

In current recycling by smelting the functionality of alloying elements in steel and Al alloys tends to be lost in a few life cycles (Reijnders 2016, 2018). Shredding followed by sorting, which is common in the formal processing of e-waste, can lead to substantial contamination of steel and Al outputs by hazardous elements such as Cd, Hg, and Pb (Goosey and Goosey 2019) and with elements that can be dysfunctional (Reijnders 2016, 2018). The latter can trigger the need for dilution with virgin metal to keep the concentration of tramp metals below the level where dysfunctionality becomes manifest (Reijnders 2016, 2018). In the case of wrought Al alloys, there is an incentive for recycling cast products, which tend to have higher tolerances for potentially dysfunctional elements than wrought products (Reijnders 2018). A comparison of Box 2.2 with Table 2.2 suggests, that in the case of steel derived from e-waste shredding and sorting operations, Cu, Sn, and Zn may emerge as elements that may be dysfunctional. In the case of Al alloys, the same holds for B, Bi, Cd, Cr, Ga, In, Li, Mo, Pb, Sn, and Zn.

2.3 The Fate of Synthetic Polymers

Commonly applied synthetic polymers derived from e-waste (cf. Box 2.1) obtained by disassembly or as shredding residue are often dumped, landfilled, or incinerated (Wang and Xu 2014; Nagajothi and Felixkala 2015; Amato et al. 2017; Chagnes et al. 2016; Cardamone et al. 2021; Das et al. 2021; Chaine et al. 2022). In the case of informal processing, dumping and poorly controlled incineration may regard >80% of e-waste polymers (Cardamone et al. 2021).

No commercial recycling options are available for the polymers presented in Table 2.1. Recycling options for the synthetic polymers derived from end-of-life e-waste presented in Box 2.1 are so-called mechanical recycling (involving size reduction and remelting) and feedstock or chemical recycling (Buekens and Yang 2014; Das et al. 2021). Feedstock recycling might also be considered for mixtures of polymers, such as those present in e-waste shredder residues (Esposito et al. 2020; Rieger et al. 2021). However, feedstock recycling is as yet not a commercial option (Maisel et al. 2020). From a by Liu et al. (2022), it may be estimated that in China, the mechanical recycling rate of e-waste-derived synthetic polymers is presumably <15%. From a study by Cardamone et al. (2021), it may be estimated that in the European Union, the mechanical recycling rate of e-waste plastics is <10%. Data from other studies also indicate that mechanical recycling of e-waste-derived synthetic polymers is practiced in a very limited way (Buekens and Yang 2014; Maisel et al. 2020; van Yken et al. 2021). Mechanical recycling of e-waste-derived synthetic polymers to high-quality products is limited by color, the presence of polybrominated fire retardants, Hg, Pb, and Sb, deterioration of polymers, and imperfect sorting (Buekens and Yang 2014; Sahajwalla and Gaikwad 2018; Wagner et al. 2019; Alassali et al. 2020; Das et al. 2021; Signoret et al. 2021). In practice, the safeguards against a significant presence of hazardous contaminants in recycled e-waste-derived polymers may well be insufficient. Following several geographically limited studies noting violations of legal requirements regarding brominated fire retardants in e-waste-derived black plastic and toys (Chaine et al. 2022), recycled e-waste-derived plastic has been found to be a probable source for high levels of decabromodiphenylether in new consumer products, including children's toys, sold in 18 countries in Africa, Asia, the Americas, and Europe (Kajiwara et al. 2022).

In end-of-life foams of synthetic polymers present in e-waste, halogenated blowing agents can be present. These are potent greenhouse gases and, when containing Cl, can negatively impact the ozone layer (Ardente et al. 2015). In informal processing, there is no recovery of halogenated blowing agents (Velásquez-Rodríguez et al. 2021). In the case of formal processing, they are partly recovered by adsorption on active coal, from which the blowing agents can be desorbed (Ardente et al. 2015; Keri 2019). No reliable data about their actual recovery rate could be found. In formal processing, blowing agents partly end up in shredder residue (Scheutz et al. 2010; Yazcici et al. 2014), which is commonly landfilled or incinerated in a controlled way. In the case of landfilling, blowing agents may be expected to be ultimately largely released (Yazcici et al. 2014). Recycling of recovered halogenated blowing agents may be considered controversial as these agents are being phased out (UNEP 2018), because less problematic blowing agents are available (Sovacool et al. 2021), and as there is the alternative of vacuum panels (Choi et al. 2021). Destruction, e.g. by controlled incineration or catalytic oxidation (Sovacool et al. 2021), of these compounds when adsorbed on, or desorbed from, active coal may be a fate preferable to recycling.

2.4 The Fate of Glass Present in E-Waste

Plate glass present in end-of-life refrigerators and glass from electric and microwave ovens cannot be co-recycled with glass bottles because such glass can negatively impact the quality of recycled glass (Deyer 2014; Heriyanto and Sahajwalla, 2018). Though there is scope for separate open-loop recycling of glass from refrigerators and ovens (Deyer 2014; Del Rio et al. 2022), such glass is often landfilled (Heriyanto and Sahajwalla, 2018). Limited recycling has been reported for glass used in photovoltaic systems (Del Rio et al. 2022). Glass fibers present in printed circuit boards are seldom recovered and are commonly lost from the economy (Del Rio et al. 2022).

Cathode ray tubes contain leaded glass (with 22–28 (weight)% Pb in funnel glass), from which hazardous Pb can leach (Meng et al. 2016; Cesaro et al. 2018a; Qi et al. 2019). Waste-leaded glass can be used for the production of new cathode ray tubes, but such recycling is phased out due to substitution by other types of displays (Qi et al. 2019; van Yken et al. 2021). Glass from cathode ray tubes is increasingly landfilled or dumped (Qi et al. 2019; van Yken et al. 2021).

Substantial amounts of Hg in fluorescent lamps end up in the waste glass matrix and have been reported to be partially recalcitrant to extraction (Rey-Raap and Gallardo 2012, 2013). Waste glass from fluorescent lamps from which Hg has been recovered has furthermore been reported to contain remnants of phosphors, metal, plastic, adhesives, and resins (Norias et al. 2016; Pitarch et al. 2021). Using such waste glass for the production of new glass is therefore problematical (Norias et al. 2016; Pitarch et al. 2021). Waste glass from fluorescent lamps is commonly landfilled (Pitarch et al. 2021).

The hazardous element As has been found in the glass of end-of-life monitors, displays of smartphones, and pre-2011 glass used in liquid crystal displays (Kim and Kim 2019; Hlavatska et al. 2021; Del Rio et al. 2022). End-of-life glass used in liquid crystal displays and smartphone displays can in addition contain Sb (Kim and Kim 2019; Del Rio et al. 2022). The presence of As and Sb in glass displays hinders recycling (Kim and Kim 2019; Del Rio et al. 2022). Furthermore, glass-based flat panel displays (including OLED displays) contain substantial amounts of In and Sn (Fontana et al. 2021). As there is as yet no recycling of such displays, these geochemically scarce elements are currently lost from the economy (Fontana et al. 2021). All in all, there is currently very little recycling of glass present in e-waste.

2.5 The Fate of Geochemically Scarce Elements in Electric and Electronic Components of E-Waste

Components of e-waste containing relatively high amounts of geochemically scarce elements may be 'mined' (selectively removed) from e-waste. Prominent among the geochemically scarce elements is Cu (Andersson et al. 2019). In, Europe the average Cu content of consumer e-waste may be in the order of 3–4.5% (Bertram et al. 2002; Soulier et al. 2018).

Waste printed circuit boards contain relatively large amounts of geochemically scarce elements (Priya and Hait 2018; Bookhagen et al. 2020) and are rather often used for the recovery of geochemically scarce elements. However, even in such printed circuit boards, the concentrations of specific geochemically scarce elements may be such that recovery is not commercially viable (Andersson et al. 2019; Fizaine 2020; Liu and Keoleian 2020). Formal processing of geochemically scarce elements may regard intact waste. For instance, the Umicore plant (Belgium) uses as an input for its smelting operations intact waste printed circuit boards and, after the removal of batteries, relatively small devices such as mobile telephones, digital cameras, and laptop computers (Hagelüken 2006b). It may be noted that this is associated with the non-recovery of synthetic polymers, Al, and steel. The Boliden Rönnskär smelter plant co-processes 'coarse' e-waste with Cu ore concentrate (Lennartsson et al. 2018). Shredded and sorted e-waste can also be co-processed with Cu ore concentrate (cf. Table 2.4).

Processing e-waste by shredding and sorting is currently associated with substantial losses of geochemically scarce elements. The more materials are interlinked and particles are interwoven in e-waste, the worse the results of shredding, and more sequential separations lead to larger losses (Hagelüken 2006a,b). Table 2.3 shows estimated losses of geochemically scarce functional elements in shredding and subsequent sorting.

Table 2.3 Estimated losses (as % of input) of geochemically scarce functional elements in shredding and subsequent sorting.

Element(s)	Loss (% of input)	Reference
Ag	88%	Bigum et al. (2012)
Au	75%	Bigum et al. (2012)
Cu	30–43%	Chancerel et al. (2009), Bigum et al. (2012), and Marra et al. (2018)
Pd	75%	Bigum et al. (2012)
Rare earths	>98%	Marra et al. (2018)

Table 2.4 Reported geochemically scarce elements produced by plants co-processing e-waste (Hagelüken and Corti 2010; Khaliq et al. 2014; Hagelüken 2015; Ariizumi et al. 2016; Salhofer et al. 2016; Shuva et al. 2016; Lennartsson et al. 2018; Andersson et al. 2019; Reijnders 2022).

Plant	Smelters involved	Type of e-waste	Reported output of relatively rare elements
Aurubis (Hamburg, Germany)	Cu and Pb smelter	Shredded e-waste and printed circuit boards	Ag, Au, Bi, Ir, Ni, Os, Pd, Rh, Ru, Pt, Sb, Se, Te, Zn
Boliden (Rönskär, Sweden)	Cu and Pb smelter	"course" e-waste, printed circuit boards, and shredded e-waste	Ag, Au, Cu, In, Ni, Pb, Pd, Pt, Sb, Se, Zn
Dowa (Kosaka, Japan)	Cu and Pb smelter	Shredded e-waste	Ag, Au, Bi, Cu, Ir, Ni, Os, Pb, Pd, Pt, Rh, Ru, Sb, Se, Sn, Te
Glencore (Horne, Canada)	Cu smelter	Shredded e-waste	Ag, Au, Ir, Ni, Os, Pd, Pt, Rh, Ru, Se, Te
Mitsubishi (Naoshima, Japan)	Cu smelter	Shredded e-waste	Ag, Au, Cu, Pb, Pt, Se
Umicore (Hoboken, Belgium)	Cu, Ni, and Pb smelter	Small intact e-waste (without batteries), printed circuit boards, and shredded e-waste	Standard run: Ag, Au, Bi, Cu, In, Ir, Ni, Pb, Pd, Pt, Rh, Ru, Sb, Se, Sn, Te Special runs: Co, Ga, rare earths

Current commercial recovery of geochemically scarce elements from e-waste uses pyrometallurgy, electrochemistry, and hydrometallurgy (Karhu et al. 2018; Zschiesche et al. 2018; Reijnders 2022). Both dedicated secondary smelters and primary smelters co-processing e-waste-derived materials may be used. Table 2.4 presents geochemically scarce elements produced by plants currently co-processing e-waste.

Relatively well studied is the overall recycling efficiency of Cu present in e-waste. The worldwide end-of-life Cu recycling rate for e-waste may be estimated at about 50%, with major losses originating from deficiencies regarding collection, sorting, and Cu recovery in informal processing (Klose and Pauliuk 2021). Table 2.5 presents the estimated Cu recovery percentage for the formal processing of e-waste in Organisation for Economic Cooperation and Development (OECD) countries. Table 2.5 also presents current recovery percentages for formal processing in OECD countries for a number of other geochemically scarce elements.

Table 2.5 Estimated current recoveries (in % of the amount present in e-waste) of geochemically scarce elements for formal e-waste processing in OECD countries (Bigum et al. 2012; Binnemans et al. 2013; Rollat et al. 2016; Nassar 2017; Ciacci et al. 2019; Althaf and Babbit 2021).

Geochemically scarce elements	Estimated current recovery as to formal processing in OECD countries (as % of the amount present in e-waste)
Ag	12%
Au	25%
Cu	57%
Ga	0%
In	0%
Pd	25%
Rare earths (Ce, Dy, Eu, Nd, Pr, Tb, Y)	<25%
Ta	Near 0%

2.6 What Happens to Other Significant Constituents of E-Waste?

Here the following constituents will be considered: Li-ion batteries, refrigerants, and phosphors and Hg used in fluorescent lamps.

2.6.1 Li-Ion Batteries

Rechargeable Li-ion batteries (containing substantial amounts of Co, Cu, Li, and Ni) have become important components in portable electronic equipment such as mobile phones, laptops, and tablets (Zubi et al. 2018). As noted in the Introduction, in formal processing such batteries are often extracted from equipment before shredding. The worldwide recycling rate for Li-ion batteries has been estimated at 5–7%, with a focus on the recovery of Co, Cu, and Ni (Pinegar and Smith 2019). In OECD countries, the collection and recycling rates of end-of-life Li-ion batteries can vary widely. For spent Li-ion batteries, the collection rate in Australia has been reported as about 2% (Boxall et al. 2018), whereas in an European there is a mandatory collection rate of 50% (Ciez and Whiteacre 2019). Collected Li-ion batteries are often disposed of in landfills, where they can release toxic substances and pose a fire and explosion-risk (Pinegar and Smith 2019). Mandatory recycling rates for end-of-life Li-ion batteries have only been reported for the European Union (25%) and several states in the USA (Ciez and Whiteacre 2019). The actual recycling rate in the United States as a whole has been estimated at <5% (Gaines 2019;

Pinegar and Smith 2019).). End-of-life batteries in equipment not subject to formal processing are likely to be dumped (Guevara-Garcia et al. 2012).

2.6.2 Refrigerants

Currently, common refrigerants (halocarbons, F-gases) are potent greenhouse gases and, when containing Cl, can negatively impact the ozone layer (UNEP 2018; Dauwe et al. 2018). In informal processing such refrigerants are likely to be disposed of by poorly controlled incineration or dumping (e.g. into rivers or illegal landfills) (Parajuly et al. 2018; Velásquez-Rodríguez et al. 2021). In formal processing, there is scope for very efficient recovery of refrigerants, but in practice, there is often only partial recovery from e-waste (Sansotera et al. 2013; Ohm et al. 2017; Dauwe et al. 2018; Keri 2019; Park et al. 2019; Velásquez-Rodríguez et al. 2021). In China, the refrigerant recovery rate is reportedly low; refrigerants are largely emitted from e-waste to the atmosphere (Duan et al. 2018; Liu et al. 2019; Wu et al. 2022).

In formal processing, there can be recycling of recovered refrigerants (Dauwe et al. 2018; Wu et al. 2022) or controlled incineration (Ohm et al. 2017). It may be noted that recycling of halogenated refrigerants is controversial, as recovered refrigerants are subject to a phase-out (UNEP 2018) and there are refrigerants with a much lower environmental burden (Roy and Halder 2020; Sovacool et al. 2021). Destruction of these refrigerants, e.g. by controlled incineration or catalytic oxidation (Sovacool et al. 2021), may be a preferable alternative to recycling.

2.6.3 Phosphors and Hg Used in Fluorescent Lamps

In the formal processing of fluorescent lamps, Hg is often recovered (Asari et al. 2008; Lecler et al. 2018; Chalkidis et al. 2020). Recovered Hg may be recycled, landfilled, or stored underground (Asari et al. 2008; Chalkidis et al. 2020). Landfilling may lead to emissions of volatile Hg (Chalkidis et al. 2020). Reported Hg recycling percentages are low. Asari et al. (2008) estimated the overall recycling rate of Hg from fluorescent lamps in Japan at about 7%, whereas for China Zhang et al. (2016) estimated a Hg recycling rate of <5% regarding domestically used fluorescent lamps. Phosphors are also recovered from e-waste in formal processing (van Broekhuizen et al. 2012; Lecler et al. 2018). Recovered phosphors may be disposed of (by landfilling or underground storage), which currently seems the predominant fate, or further processed, for reuse of purified phosphors or the recovery of geochemically scarce elements (Lee et al. 2015; Yoshida et al. 2016; Cesaro et al. 2018a; Rollat et al. 2016; Lecler et al. 2018). After informal processing, Hg and phosphors are reportedly disposed of by dumping (Yoshida et al. 2016; Chalkidis et al. 2020; De Farias et al. 2020; Anselm et al. 2021).

2.7 Conclusion: The Global Challenge of E-Waste

From the previous sections, it will be clear that the current handling of global e-waste leads to large losses of valuable resources, accompanied by substantial workplace hazards and pollution beyond the workplace. A similar conclusion was drawn more than a decade ago (Ongondo et al. 2011). Minimizing the losses of valuable resources, workplace hazards, and pollution beyond the workplace still presents a major challenge.

References

Ádám, B., Göen, T., Scheepers, T.J. et al. (2021). From inequitable to sustainable e-waste processing for reduction of impact on human health and the environment. *Environmental Research* 194: 110728. (7pp).

Aich, N., Kordas, K., Ahmed, S.I., and Sabo-Attwood, T. (2020). The hidden costs of e-waste: perspectives from environmental engineering, epidemiology, environmental health and human-computer interactions. In: *Transforming Global Health* (ed. K. Smith and P. Ram), 161–178. Cham: Springer.

Alabi, O.A., Adeolowa, V.M., and Bakare, A.A. (2020). Elevated serum Pb, Ni, Cd and Cr levels and DNA damage in exfoliated buccal cells of teenage scavengers at a major electronic waste dump site in Lagos, Nigeria. *Biological Trace Element Research* 194: 24–33.

Alassali, A., Abis, M., Fiore, S., and Kuchta, K. (2019). Classification of plastic waste originated from waste electric and electronic equipment based on the concentration of antimony. *Journal of Hazardous Materials* 380: 120874. (7 pp).

Alassali, A., Barout, A.D., Tirion, H. et al. (2020). Towards a high-quality recycling of plastics from waste electric and electronic equipmentthrough separation of contaminated fractions. *Journal of Hazardous Materials* 387: 121741. (8 pp).

Althaf, S. and Babbit, C.W. (2021). Disruption risk to material supply chains in the electronics sector. *Resources Conservation and Recycling* 167: 105348. (14pp).

Amato, A., Rochetti, L., and Beolchini, F. (2017). Environmental impact assessment of different end-of-life LCD management strategies. *Waste Management* 59: 432–441.

Andarani, P. and Goto, N. (2014). Potential e-waste generated from households in Indonesia using material flow analysis. *Journal of Material Cycles and Waste Management* 16: 306–320.

Andersson, M., Söderman, M.L., and Sandén, B.A. (2019). Challenges of recycling multiple scarce metals. The case of Swedish ELV and WEEE recycling. *Resources Policy* 63: 10143. (12pp).

Anh, M.Q., Nam, V.D., Tri, T.M. et al. (2017). Polybrominated diphenylethers in plastic products. Indoor dusts, sediment and fish from informal e-waste recycling sites in Vietnam: a comprehensive assessment of contamination, accumulation pattern, emissions, and human exposure. *Environmental Geochemistry and Health* 39: 935–954.

Anselm, O.H., Cavoura, O., Davidson, C.M. et al. (2021). Mobility, spatial variation and human health risk assessment of mercury in soil from an informal e-waste recycling site, Lagos, Nigeria. *Environmental Monitoring and Assessment* 193: 416. (13pp).

Arain, A.L., Neitzel, R.I., Nambunmee, K. et al. (2022). Material flows, economic and environmental life cycle performances of informal e-waste recycling in a Thai community. *Resources Conservation and Recycling* 180: 106129. (12pp).

Ardente, F., Pastor, M.C., Mathieux, F., and Talens-Peiró, L. (2015). Analysis of end-of-life treatments of commercial refrigerating appliances: bridging product and waste policies. *Resources, Conservation and Recycling* 101: 42–52.

Ariizumi, M, Takagi, M., Inoue, O. and Oguma, N. (2016). Integrated processing of e-scrap at Naoshima smelter and refinery. Proceedings Copper 2016. Kobe (Japan), November 13-16, 2016 6: RW 1-2

Asari, M., Fukui, K., and Sakai, S. (2008). Life cycle of mercury and recycle scenario of fluorescent lamps in Japan. *Science of the Total Environment* 393: 1–10.

Awashti, A.K., Zeng, X., and Li, J. (2016). Relation between e-waste recycling and human health risk in India: a critical review. *Environmental Science and Pollution Research* 23: 11509–11537.

Baxter, J. (2019). Systematic environmental assessment of end-of-life pathway for domestic refrigerators. *Journal of Cleaner Production* 208: 612–620.

Bertram, M., Graedel, T.E., Rechberger, M., and Spatari, S. (2002). The contemporary European copper cycle: waste management subsystem. *Ecological Economics* 42: 43–57.

Bertram, M., Ramkumar, S., Rechberger, H. et al. (2017). A regionally-linked, dynamic material flow modelling tool for rolled, extruded and cast aluminum products. *Resources Conservation and Recycling* 125: 48–69.

Bigum, M., Brogaard, L., and Christensen, T.H. (2012). Metal recovery from high grade WEEE: a life cycle assessment. *Journal of Hazardous Materials* 207-208: 8–14.

Bimir, M.N. (2020). Revisiting e-waste management in selected African countries. *Journal of the Air and Waste Management Association* 70: 659–669.

Binnemans, K., Jones, P.T., Blanpain, B. et al. (2013). Recycling of rare earths: a critical review. *Journal of Cleaner Production* 51: 1–22.

Bookhagen, B., Bastian, D., Bruchholz, P. et al. (2020). Metal resources in smartphones. *Resources Policy* 68: 1ß1750. (11 pp).

Bovea, M.D., Ibénez-Forés, V., and Pérez-Belis, V. (2020). Repair vs. replacement: selection of best end-of-life scenario for small household electric and electronic equipment based on life cycle assessment. *Journal of Environmental Management* 254: 109679. (14pp).

Boxall, N.J., King, S., Cheng, L.Y. et al. (2018). Urban mining of lithium-ion batteries in Australia: current state and future trends. *Minerals Engineering* 128: 45–55.

van Broekhuizen, P., van Broekhuizen, F., Cornelissen, R., and Reijnders, L. (2012). Workplace exposure to nanoparticles and the application of provisional nanoreference values in times of uncertain risk. *Journal of Nanoparticle Research* 14: 770. (25pp).

Buechler, D.T., Zyaykina, N.N., Spencer, C.A. et al. (2020). Comprehensive element analysis of consumer electronic devices: rare earth, precious and critical elements. *Waste Management* 103: 67–75.

Buekens, A. and Yang, J. (2014). Recycling of WEEE plastics: a review. *Journal of Material Cycles and Waste Management* 16: 415–424.

Cardamone, G.F., Ardolino, F., and Arena, U. (2021). About the environmental sustainability of the European management of WEEE plastics. *Waste Management* 126: 119–132.

Cesaro, A., Marra, A., Belgiorno, V., and Guida, M. (2017). Effectiveness of WEEE mechanical treatment: separation yields and recovered material toxicity. *Journal of Cleaner Production* 142: 2656–2662.

Cesaro, A., Marra, A., Belgiorno, V. et al. (2018a). Chemical characterization and toxicity assessment for the sustainable management of end-of-life cathode ray tubes. *Journal of Material Cycles and Waste Management* 20: 1188–1198.

Cesaro, A., Marra, A., Kuchta, K. et al. (2018b). WEEE management in a circular economy perspective: an overview. *Global NEST Journal* 20: 743–750.

Cesaro, A., Belgiorno; V., Gorrasi, G. et al. (2019). A relative risk assessment of the open burning of WEEE. *Environment Science and Pollution Research* 26: 11042–11052.

Chagnes, A., Cole, G., Ekberg, C. et al. (ed.) (2016). *WEEE Recycling: Research, Development and Policy*, 131–133. Elsevier: Amsterdam.

Chaine, C., Hursthouse, A.S., McLean, B. et al. (2022). Recycling plastics from WEEE: a review of the environmental and human health challenges associated with brominated fire retardants. *International Journal of Environmental Research and Public Health* 19: 776. (27pp).

Chalkidis, A., Jampaiah, D., Aryana, A. et al. (2020). Mercury-bearing wastes: sources, policies and treatment technologies for mercury recovery and safe disposal. *Journal of Environmental Management* 270: 110945. (16pp).

Chancerel, P., Meskers, C.E.M., Hagelüken, C., and Rotter, V.S. (2009). Assessment of precious metal flows during preprocessing of waste electrical and electronic equipment. *Journal of Industrial Ecology* 13: 791–810.

Choi, S., Jung, Y., Kim, Y. et al. (2021). Environmental effect evaluation of refrigerator cycle with life cycle climate performance. *International Journal of Refrigeration* 122: 134–146.

Ciacci, L., Werner, T.T., Vassura, I., and Passarini, F. (2019). Backlighting the European indium recycling potential. *Journal of Industrial Ecology* 23: 426–437.

Ciez, R.E. and Whiteacre, J.E. (2019). Examining different recycling processes for lithium-ion batteries. *Nature Sustainability* 2: 148–156.

Cole, C., Gnanapragasam, A., Cooper, T., and Singh, J. (2019). Assessing barriers to reuse of electrical and electronic equipment, a UK perspective. *Resources Conservation and Recycling* X1: 100004. (10pp).

Das, P., Gabriel, J.P., Tay, C.Y., and Lee, J. (2021). Value added products from thermochemical treatment of contaminated e-waste plastics. *Chemosphere* 269: 128409. (20pp).

Dauwe, T., Altdorfer, F. and Gschrey, B. (2018). *Waste and disposal emissions from F-gas containing refrigerators and potential actions to improve recovery of F gases* (94pp). Available at: www,omgevingvlaanderen.be,

De Farias, C.V., Paulino, J.F., Barcelos, D.A. et al. (2020). Is mercury in fluorescent lamps the only risk to human health? A study of environmental mobility of toxic metals and human health risk assessment. *Chemosphere* 261: 128107. (10pp).

Del Rio, D.D.F., Sovakool, B.F., Foley, A.M. et al. (2022). Decarbonizing the glass industry; a critical and systematic review of developments, sociotechnical system and policy options. *Renewable and Sustainable Energy Reviews* 155: 111885. (29pp).

Deyer, T.D. (2014). Glass recycling. In: *Handbook of Recycling* (ed. E. Worrell and M.A. Reuter), 191–209. London: Elsevier.

Duan, H., Miller, T.R., Liu, G. et al. (2018). Chilling prospect: climate effect om mismanaged refrigerants in China. *Environmental Science and Technology* 52: 6350–6356.

Dutta, D. and Goel, S. (2021). Understanding the gap between formal and informal e-waste recycling facilities in India. *Waste Management* 125: 163–171.

Esposito, L., Cafiero, L., De Angelis, D. et al. (2020). Valorization of the plastic residue from a WEEE treatment plant by pyrolysis. *Waste Management* 112: 1–10.

Fizaine, F. (2020). The economics of the recycling rate. New insights from waste electrical and electronic equipment. *Resources Policy* 67: 101675. (14pp).

Fontana, D., Forte, F., Pietrantonio, M., and Pucciarmati, S. (2021). Recent developments in recycling end-of-life flat panel displays: a comprehensive review focussed on indium. *Critical Reviews in Environmental Science and Technology* 51: 429–456.

Forti, V, Baldé, C.P., Kuehr, R. and Bel, G. (2021). Global e-waste monitor 2020. Global E-waste Statistics Partnership. Available at www.globalwaste.org ()

Gaines, L. (2019). Profitable recycling of low-cobalt lithium-ion batteries will depend on new process developments. *One Earth* 1: 414–415.

Ghosh, B., Ghosh, M.K., Parhi, P. et al. (2015). Waste printed circuit boards recycling: an extensive assessment of current status. *Journal of Cleaner Production* 94: 5–19.

Goosey, E. and Goosey, M. (2019). The materials of waste electrical and electronic equipment. In: *Waste Electrical and Electronic Equipment (WEEE) Handbook* (ed. V. Goodship, A. Stevels, and J. Huisman), 231–262. Amsterdam: Elsevier.

Graedel, T.E., Reck, B.K., Ciacci, L., and Passarini, F. (2019). On the spatial dimension of the circular economy. *Resources* 8 (1): 32. (10pp).

Guevara-Garcia, J.A., Montiel-Corona, V., and Landry, T.D. (2012). Fate and risk evaluation of WEEE including spent batteries and compact fluorescent lamps in Mexico. *International Journal of Environmental Science and Engineering Research* 3: 11–24.

Guo, J., Patton, L., Wang, J., and Xu, Z. (2020). Fate and migration of polybrominated diphenyl ethers in a workshop for waste printed-circuit board de-soldering. *Environmental Science and Pollution Research* 27 (30342): 30351.

Hagelüken, C. (2006a). Improving metal returns and eco-efficiency in electronic recycling. Proceedings of the 2006 IEEE International Symposium on Electronics and the Environment pp 218-223

Hagelüken, C. (2006b). Recycling of electronic scrap at Umicore's integrated metals smelter and refinery. *Erzmetall* 59: 152–161.

Hagelüken, C. (2015). Closing the loop for rare metals used in consumer products: opportunities and challenges. *Natural Resources Management and Policy* 46: 103–119.

Hagelüken, C. and Corti, C.W. (2010). Recycling gold from electronics: cost effective use through 'design for recycling'. *Gold Bulletin* 43: 209–220.

Helbig, C., Thorenz, A., and Tuma, A. (2020). Quantitative assessment of dissipative losses of 18 metals. *Resources Conservation and Recycling* 153: 104537. (8pp).

Heriyanto, P. and F. and Sahajwalla, V. (2018). From waste glass to building materials: an innovative sustaiable solution for waste glass. *Journal of Cleaner Production* 191: 192–206.

Hlavatska, L., Ishchenko, V., Pohrebennyk, V., and Salamon, I. (2021). Material flow analysis of waste electrical and electronic equipment in Ukraine. *Journal of Ecological Engineering* 22: 198–207.

Inamuddin, B.R., Ahamed, M.I., and Asiri, A.M. (ed.) (2020). *Polymers for Light Emitting Displays*. NJ, USA: Wiley Hoboken.

Isimekhai, K.A., Garelick, H., Watt, J., and Purchase, D. (2017). Heavy metals distribution and risk assessment in soil from an informal e-waste recycling site in Lagos State, Nigeria. *Environmental Science and Pollution Research* 24: 17206–17219.

Kajiwara, N., Matsukami, H., Malarvannan, G. et al. (2022). Recycling plastics containing decabromodiphenyl ether in new consumer products including children's toys purchased in Japan and seventeen other countries. *Chemosphere* 289: 133178. (8pp).

Karhu, M., Kotnis, J., Yang, Y. et al. (2018). *Circular Economy and Zero Waste Aspects and Business Models of Production*. European Union, Brussels: Solutions for Critical Raw Materials 206pp.

Kaya, M. (2016). Recovery of metals and nonmetals from electronic waste by physical and chemical recycling processes. *Waste Management* 57: 62–90.

Keri, C. (2019). Recycling cooling and freezing appliances. In: *Waste Electrical and Electronic Equipment (WEEE) Handbook* (ed. V. Goodship, A. Stevels, and J. Huisman), 357–370. Amsterdam: Elsevier.

Khaliq, A., Rhadhani, M.A., Brooks, G., and Masood, S. (2014). Metal extraction processes for electronic waste and existing industrial routes: a review and Australian perspective. *Resources* 3: 153–179.

Kim, K. and Kim, K. (2019). Recycling of end-of-life glass generated from end-of-life LCD devices: a feasibility study using simulated waste glass. *Journal of Cleaner Production* 227: 199–206.

Klose, S. and Pauliuk, S. (2021). Quantifying longevity and circularity of copper for different resource efficiency policies at the material and product levels. *Journal of Industrial Ecology* 25: 979–993.

Lecler, M., Zimmermann, F., Silvente, E. et al. (2015). Exposure to hazardous substances in cathode ray tube (CRT) recycling sites in France. *Waste Management* 39: 226–235.

Lecler, M., Zimmermann, F., Silvente, E. et al. (2018). Improving the work environment in the fluorescent lamp recycling sector by optimizing mercury elimination. *Waste Management* 76: 250–260.

Lee, C., Popuri, S.R., Peng, Y. et al. (2015). Overview on industrial recycling technologies and management strategies of end-of-life fluorescent lamps in Taiwan and other developed countries. *Journal of Material Cycles and Waste Management* 17: 312–323.

Lennartsson, A., Engström, F., Samuelsson, C. et al. (2018). Large-scale WEEE recycling integrated in an ore-based cu extraction system. *Journal of Sustainable Metallurgy* 4: 222–232.

Li, J., Su, G., Letscher, R.J. et al. (2018). Liquid crystal monomers (LCMs): a new generation of persistent, bioaccumulative and toxic (PBT) compounds? *Environmental Science and Technology* .

Li, S., Huang, Y., Zhang, X. et al. (2021). Atmospheric fate and risk investigation of typical liquid crystal monomers. *ACS Sustainable Chemical Engineering* 9: 3600–3607.

Lin, S., Ali, M.U., Zeng, C. et al. (2022). Toxic chemicals from uncontrolled e-waste recycling: exposure, body burden, health impact. *Journal of Hazardous Materials* 426: 127792. (12 pp).

Liu, L. and Keoleian, G.A. (2020). LCA of rare earth and critical metal recovery and replacement decision for commercial lighting management. *Resources Conservation and Recycling* 159: 104845. (12pp).

Liu, L., Dou, Y., Yao, B. et al. (2019). Historical and projected HFC-410A emitted from room air conditioning sector in China. *Atmospheric Environment* 212: 194–200.

Liu, X., Lu, X., Feng, Y. et al. (2022). Recycling WEEE plastics in China: trend and environmental impacts. *Resources Conservation and Recycling* 177: 105978. (10pp).

Lu, B., Song, X., Yang, J., and Yang, D. (2017). Comparison on end-of-life strategies of WEEE in China based on LCA. *Frontiers of Environmental Science & Technology* 11: 7. (12pp).

Luo, D., Chen, Q., Liu, B., and Qiu, Y. (2019). Emergence of flexible white organic light emitting diodes. *Polymers* 11: 384. (28pp).

Maisel, E., Chancerel, P., Dimitrova, G. et al. (2020). Preparing WEEE plastics for recycling—how optimizing particle size in preprocessing can improve separation efficiency. *Resources Conservation and Recycling* 154: 104619. (10pp).

Marra, A., Cesaro, A., and Belgiorno, V. (2018). Separation efficiency of valuable and critical metals in WEEE mechanical treatments. *Journal of Cleaner Production* 186: 490–498.

McMahon, K., Uchendu, C., and Fitzpatrick, C. (2021). Quantifying used electric and electronic equipment exported from Ireland to West Africa in roll-on roll-off vehicles. *Resources Conservation and Recycling* 164: 105177. (12pp).

Meng, W., Wang, X., Yuan, W. et al. (2016). The recycling of leaded glass in cathode ray tubes (CRT). *Procedia Environmental Sciences* 31: 954–960.

Mihai, F., Gnoni, M., Meidiana, C. et al. (2019). Waste, electrical and electronic equipment (WEEE): flows, quantities and management, a global scenario. In: *Electronic Waste Management and Treatment Technology* (ed. M.N.V. Prasad and M. Vithanage), 1–34. Amsterdam: Elsevier.

Mochizuki, T., Takigami, Y., Kondo, T., and Okuzaki, H. (2017). Fabrication of flexible transparent electrodes using PEDOT-PSS and application to resistive touch screen panels. *Journal of Applied Polymer Science* 135: 45972. (6pp).

Mowla, M., Rahman, I., Islam, N., and Aich, N. (2021). Assessment of heavy metal contamination and health risk from indoor dust and air of informal E-waste recycling shops in Dhaka, Bangladesh. *Journal of Hazardous Materials Advances* 4: 1065. (10pp).

Nagajothi, P.G. and Felixkala, I. (2015). Electronic waste management: a review. *International Journal of Applied Engineering Research* 10: 133–138.

Namsheer, K. and Rout, C.K. (2021). Conducting polymers: a comprehensive review on recent advances in synthesis, properties and applications. *RSC Advances* 11: 5659–5697.

Nassar, N.T. (2017). Shift and trends in global anthropogenic stocks of tantalum. *Resources Conservation and Recycling* 125: 233–250.

Navazo, J.M.V., Villalba Mendez, G., and Talens-Peiro, L. (2014). Material flow and energy requirements of mobile phone material recovery processes. *International Journal of Life Cycle Assessment* 19: 567–579.

Norias, R.M., Ascensao, G., Seabra, M.P., and Labrincha, J.A. (2016). Waste glass from end-of-life fluorescent lamps as raw material in geopolymers. *Waste Management* 52: 245–255.

Ohajinwa, C.M., van Bodegom, P.M., Vijver, M.G., and Peijnenburg, W.J.G.M. (2018). Impact of informal electronic waste recycling on metal concentrations in soils and dusts. *Environmental Research* 164: 385–394.

Ohm, T., Myung, S., Jang, W., and Yu, S. (2017). A comparison of refrigerant management practices and suggestions for improvement in South Korea. *Journal of Material Cycles and Waste Management* 19: 631–644.

Okeme, J.O. and Arrandale, V.H. (2019). Electronic waste recycling: occupational exposures and work-related health effects. *Current Environmental Health Reports* 6: 256–268.

Ongondo, F.O., Williams, I.D., and Cherret, T.J. (2011). How are WEEE doing? A global review of the management of electrical and electronic wastes. *Waste Management* 31: 714–730.

Owakowski, P. (2019). Investigating the reasons for storage of WEEE by residents—a potential for removal from households. *Waste Management* 87: 192–203.

Parajuly, K., Thapa, K., Cimpan, C., and Wenzel, H. (2018). Electronic waste and informal recycling in Kathmandu, Nepal: challenges and opportunities. *Journal of Material Cycles and Waste Management* 20: 656–666.

Park, J., Jang, I., Choi, W. et al. (2019). Greenhouse gas emissions offsetting by refrigerant recovery from WEEE: a case study in a WEEE recycling plant in Korea. *Resources Conservation and Recycling* 142: 167–172.

Pascale, A., Sosa, A., Bares, C. et al. (2016). E-waste informal recycling: an emerging source of lead exposure in South America. *Annals of Global Health* 82: 197–201.

Pérez-Belis, V., Bakker, C., Juan, P., and Bovea, M.D. (2017). Environmental performance of end-of-life scenario's for electrical equipment: case study for vacuum cleaners. *Journal of Cleaner Production* 159: 158–170.

Perkins, D.N., Drisse, M.B., Nxele, T., and Sly, P.D. (2014). E-waste: a global hazard. *Annals of Global Health* 80: 286–295.

Pinegar, M. and Smith, Y.R. (2019). Recycling of end-of-life lithium-ion batteries. Part I: commercial processes. *Journal of Sustainable Metallurgy* 5: 402–416.

Pitarch, A.M., Reig, L., Gallardo, A. et al. (2021). Reutilization of hazardous spent fluorescent lamps glass waste as supplementary cementitious material. *Construction and Building Materials* 292: 123424. (13pp).

Pradhan, J.K. and Kumar, S. (2014). Informal e-waste recycling: environmental risk assessment of heavy metal contamination in Mandoli industrial area Delhi India. *Environmental Science and Pollution Research* 21: 7913–79286.

Priya, A. and Hait, S. (2018). Comprehensive characterization of printed circuit boards of various end-of-life electrical and electronic equipment for beneficiation investigation. *Waste Management* 75: 103–123.

Qi, Y., Xiao, X., Lu, Y. et al. (2019). Cathode ray tube glass recycling. *Science of the Total Environment* 650: 2842–2849.

Reijnders, L. (2016). Conserving the functionality of relatively rare metals associated with steel life cycles: a review. *Journal of Cleaner Production* 131: 76–79.

Reijnders, L. (2018). Quality of recycling: conserving functionality of geochemically scarce elements associated with aluminum life cycles. In: *Recycling and Reuse of Materials* (ed. M.A. Holst and T.M. Kjeldsen), 115–180. New York: Nova Science Publishers.

Reijnders, L. (2022). Is near-zero waste production of copper and its geochemically scarce companion elements feasible? *Minerals Processing and Extractive Metallurgy Review* 43: 1021, 1048.

Rey-Raap, N. and Gallardo, A. (2012). Determination of mercury distribution inside spent compact fluorescent lamps by atomic absorption spectrometry. *Waste Management* 32: 944–948.

Rey-Raap, N. and Gallardo, A. (2013). Removal of mercury bonded in residual glass from spent fluorescent lamps. *Journal of Environmental Management* 115: 175–178.

Rieger, F., Oey, J.C., Palchyk, V. et al. (2021). Chemical recycling of WEEE plastics—production of high purity monocyclic aromatic chemicals. *Processes* 9: 530. (16pp).

Rollat, A., Guyonnet, D., Planchon, M., and Tuduri, J. (2016). Prospective analysis of the flow of certain rare earths in Europe at the 2020 horizon. *Waste Management* 49: 427–437.

Roy, Z. and Halder, G. (2020). Replacement of halogenated refrigerants towards sustainable cooling systems: a review. *Chemical Engineering Journal Advances* 3: 100027. (15pp).

Sahajwalla, V. and Gaikwad, V. (2018). The present and future of e-waste plastics recycling. *Current Opinion in Green and Sustainable Chemistry* 13: 102–107.

Salhofer, S., Steuer, B., Ramusch, R., and Beigl, P. (2016). WEEE waste management in Europe and China – a comparison. *Waste Management* 57: 27–35.

Sansotera, M., Navarrini, W., Talaeemashhadi, S., and Venturini, E. (2013). Italian WEEE management system and treatment of end-of-life cooling and freezing equipment for CFCs recovery. *Waste Management* 33: 1491–1498.

Santana, J.J.C., Guerhardt, F., Franzini, C.E. et al. (2021). Refurbishing and recycling of cell phones as a sustainable process of reverse logistics: a case study in Brazil. *Journal of Cleaner Production* 283: 124585. (9pp).

Scheutz, C., Fredenslund, A.M., Nedenskov, J., and Kljeldsen, P. (2010). Release and fate of fluorocarbons in a shredder residue landfill cell—2 field investigation. *Waste Management* 30: 2163–2169.

Shaikh, S., Thomas, K., and Zuhair, S. (2020). An explanatory study of e-waste creation and disposal: upstream considerations. *Resources Conservation and Recycling* 155: 104662. (22pp).

Shittu, O.S., Williams, I.D., and Shaw, P.J. (2021). Global e-waste management: can WEEE make a difference? A review of e-waste trends, legislation, contemporary issues and future challenges. *Waste Management* 120: 549–563.

Shuva, M.A.H., Rhamdhani, M.A., Brooks, G.A. et al. (2016). Thermodynamics data of valuable elements relevant to e-waste processing through primary and secondary copper production. *Journal of Cleaner Production* 131: 795–809.

Signoret, C., Girard, P., Le Guen, A. et al. (2021). Degradation of polystyrene plastics during recycling: accommodation of PP within ABS after WEEE plastics imperfect sorting. *Polymers* 43: 1439. (28pp).

Singh, M., Thind, P.S., and John, S. (2018). Health risk assessment of the workers exposed to the heavy metals on e-waste recycling sites if Chandigarh and Ludhiana, Punjab, India. *Chemosphere* 203: 426–433.

Singh, N., Duan, H., and Tang, Y. (2020). Toxicity evaluation of E-waste plastics and potential repercussions for human health. *Environment International* 137: 105559. (8pp).

Song, Q. and Li, J. (2014). A systematic review of the human body burden of e-waste exposure in China. *Environment International* 63: 82–93.

Soulier, M., Glöser-Chahoud, S., Goldmann, D., and Tercero Espinoza, L.A. (2018). Dynamic analysis of European Coper flows. *Resources Conservation and Recycling* 129: 143–152.

Sovacool, B.K., Griffith, S., Kim, J., and Brazilian, M. (2021). Climate and industrial F-gases: a critical and systematic review of developments, sociotechnical systems and policy options for reducing synthetic greenhouse gas emissions. *Renewable and Sustainable Energy Reviews* 141: 110759. (55pp).

Statista 2021. E-waste generation worldwide. Available at www.statista.com.

Steuer, B., Ramusch, R., and Salhofer, S. (2018). Is there a future for the informal recycling sector in urban China? *Detritus* 4: 189–200.

Terazono, A., Oguchi, M., Yoshida, A. et al. (2017). Material recovery and environmental impact by informal e-waste recycling in the Philippines. In: *Sustainability through Innovation in Product Life Cycle Design* (ed. M. Matsumoto, K. Masui, S. Fukushige, and S. Kondoh), 197–213. Singapore: Springer.

Tran, C.D. and Salhofer, S.P. (2018). Analysis of recycling activities for e-waste in Vietnam. *Journal of Material Cycles and Waste Management* 20: 110–126.

Tran, H.P., Schaubroeck, T., Nguyen, D.Q. et al. (2018). Material flow analysis for management of waste TVs from households in urban areas of Vietnam. *Resources Conservation and Recycling* 139: 78–89.

UNEP (2018). Handbook for the Montreal protocol on substances that deplete the ozone layer (12th edition) available at: https://leap.unep.org/sites/default/files/2020-09/

Vaccari, M., Zambetti, F., Bates, M. et al. (2020). Application of an integrated assessment scheme for sustainable waste management of electric and electronic equipment: the case of Ghana. *Sustainability* 13: 3191. (15pp).

Vats, M.C. and Singh, S.K. (2014). E-waste characteristics and its disposal. *International Journal of Ecological Science and Environmental Engineering* 1: 49–61.

Velásquez-Rodríguez, O.F., Lovik, A.N., and Moreno-Mantilla, C.E. (2021). Evaluation of the environmental impact of end-of-life refrigerators in Colombia by material flow analysis. *Journal of Cleaner Production* 314: 127884. (11 pp).

Wagner, F., Peeters, J.R., De Keyzer, J. et al. (2019). Towards a more circular economy for WEEE plastics—Part B: assessment of the technical feasibility of recycling strategies. *Waste Management* 96: 206–214.

Wang, R. and Xu, Z. (2014). Recycling of non-metallic fractions from waste electric and electronic equipment (WEEE): a review. *Waste Management* 34: 1455–1469.

Welfens, M.J., Nordman, J., and Seibt, A. (2016). Drivers and barriers to return and recycling mobile phones. Case studies of communication and collection campaigns. *Journal of Cleaner Production* 132: 108–121.

Wu, J., Li, T., Wang, J. et al. (2022). Establishment of HCFC-22 national-provincial-gridded emissions inventories in China and the analysis of emission reduction potential. *Environmental Science & Technology* 56: 814–822.

Yang, J., Bertram, J., Schettgen, I. et al. (2020). Arsenic burden in e-waste recycling workers - a cross-sectional study at the Agbogbloshie e-waste recycling site, Ghana. *Chemosphere* 261: 127712. (9pp).

Yazcici, B., Can, Z.S., and Calli, B. (2014). Prediction of future disposal of end-of-life refrigerators containing CFC-11. *Waste Management* 34: 162–166.

van Yken, J., Boxall, N.J., Cheng, K.Y. et al. (2021). E-waste recycling and resource recovery: a review of technologies, barriers and enablers with a focus on Oceania. *Metals* 11: 1313. (40pp).

Yoshida, A., Terazono, A.B., Jr, F.C. et al. (2016). E-waste recycling processes in Indonesia, the Philippines and Vietnam: a case study of cathode ray tube TVs and monitors. *Resources Conservation and Recycling* 106: 48–58.

Zeng, X., Xu, X., Boezen, H.M., and Huo, X. (2016). Children with health impairment by heavy metals in an e-waste recycling area. *Chemosphere* 148: 408–415.

Zhang, L. and Xu, Z. (2016). A review of current progress in recycling technologies for metals from waste electric and electronic equipment. *Journal of Cleaner Production* 127: 19–36.

Zhang, J., Chen, S., Kim, J., and Cheng, S. (2016). Mercury flow analysis and reduction pathways for fluorescent lamps in Mainland China. *Journal of Cleaner Production* 133: 451–458.

Zschiesche, C., Ayhan, M., and Antrelkowitsch, J. (2018). Challenges and opportunities of lead smelting process for complex feed mixture. In: *Extraction 2018. The Minerals, Metals and Materials Series* (ed. B. Davis et al.), 325–338. Cham: Springer.

Zubi, G., Dufo-López, R., Cavalho, M., and Pasaoglu, G. (2018). The lithium-ion battery: state of the art and future perspectives. *Renewable and Sustainable Energy Reviews* 89: 292–308.

3

Generation, Composition, Collection, and Treatment of E-Waste

Monjur Mourshed[1,2], Sharifa Khatun[2], Kaviul Islam[3,4], Nahid Imtiaz Masuk[2], and Mahadi Hasan Masud[1,2]

[1] *Mechanical & Automotive Discipline, School of Engineering, RMIT University, Bundoora Campus, Melbourne, VIC, Australia*
[2] *Department of Mechanical Engineering, Rajshahi University of Engineering and Technology, Rajshahi, Bangladesh*
[3] *Department of Mechanical Engineering, Iowa State University, Ames, IA, USA*
[4] *School of Science and Engineering, Canadian University of Bangladesh, Dhaka, Bangladesh*

Abstract

With the advent of modern technologies, electrical and electronic amenities gained much popularity, which in turn caused a surge in e-waste generation. Moreover, the application of precious materials in electrical and electronic equipment enhances efficiency and service life, which also urges recovery at the end of service life (EoL). Thus, it is high time to closely monitor and regulate e-waste disposal, collection, and treatment strategies and control transborder shipment from developed to underdeveloped countries. This chapter focuses on the factors affecting e-waste generation and collection processes after disposal or at EoL. Also, the composition of e-waste and its respective effects on the environment during the extraction period as raw materials and disposal as waste are now considered serious environmental issues. E-waste treatment procedures are discussed to highlight their contribution to a circular economy rather than only being treated as general or hazardous waste.

Keywords *E-waste; E-waste generation; E-waste management; Treatment*

Abbreviations

WEEE	Waste Electrical and Electronic Equipment
EoL	End of Service Life
EEE	Electrical and Electronic Equipment

Management of Electronic Waste: Resource Recovery, Technology and Regulation, First Edition.
Edited by Anshu Priya.

USA	United States of America
EU	European Union
Mt	Million Tonnes
USD	United States Dollar
TBS	Take-Back System
EPR	Extended Producer Responsibility
GDP	Gross Domestic Product
UN	United Nations
NGO	Non-governmental Organizations
ECR	Extended Consumer Responsibility
PCB	Plastic Circuit Board
ROHS	Restriction of Hazardous Substance

3.1 Introduction

The generation of waste electrical and electronic equipment (WEEE) has increased over the last few decades, which ultimately poses a serious challenge to the safety of the environment. The term e-waste or WEEE generally includes electrical and electronic instruments and consumables that have been utilized in their service life as transferring or generating signals and are disposed of at the EoL (Ranasinghe & Athapattu, 2020). According to StеP, EEEs that are used in offices and households as essential or luxurious gadgets and discarded at the EoL period without any further refabrication or reuse are classified as e-waste (Step Initiative %J Bonn, G. S. 2014). Nowadays, the generation rate of WEEE is higher than municipal solid waste, and about 43.8 million tonnes (Mt) of WEEE generation was recorded only in 2015 (Baldé et al., 2017). Moreover, Baldé et al. (Baldé et al., 2017) reported that in 2016, about 44.7 Mt of e-waste was generated with a per capita consumption of 6.1 kg per year, where Asia contributed the highest portion (~41%) and the United States of America (USA) and European Union (EU) generated ~29% and ~27%, respectively. WEEE generation increases at an alarming rate, and Global E-waste Monitor 2020 reported a total generation of 53.6 Mt in 2020 at a 7.3 kg per year consumption rate (Forti et al., 2020). Following this, it is predicted that around 74.7 Mt of WEEE will be generated by 2030 (Forti et al., 2020). According to an estimation by International Criminal Police Organization, the global e-waste market reaches nearly 20.5–25 billion USD per year (taking 500 dollars per tonne) (Rucevska et al. 2015).

Rapid industrialization and urbanization, technological advancement, the standard of living, etc., insist consumers switch their electronic and electrical gadgets easily. This phenomenon insists producers to design new products with attractive features and update the existing models. On the other hand, level of income, lifestyle, advanced materials, manufacturing technologies, and development in information and technology sectors cause an influx in the per capita consumption of electronic and electrical gadgets in Europe, Oceania, and America (Borthakur & Govind, 2017; Borthakur et al., 2019; Kumar et al., 2017; Li et al., 2015). It is reported that per capita consumption of e-waste is higher in rich countries (per capita 19.6 kg per year) compared to poor countries (per capita 0.06 kg per year) (Borthakur et al., 2019; Srivastava & Pathak, 2020). At the same time, transboundary shipment of WEEE to developing countries from developed countries to take advantage of low-cost recycling and recovery as well as to get rid of the critical management of hazardous WEEE. Along with the population, gross domestic product (GDP) regulates the e-waste generation by controlling material flow, economic elasticity, and purchasing power parity in any country. It is found that a 100 GDP per capita increase in purchasing power standards stimulates 0.27 kg and 0.2 kg of e-waste ready for collection and recycling, respectively (Awasthi et al., 2018).

To cope with consumers' demands, production and consumption of EEE goods as well as WEEE generation are also increasing in a similar pattern. Global e-waste follows a linear flow from North to South as waste to resources, and during this transformation, it causes a serious risk to the environment and human health (Daum et al., 2017; Sugimura & Murakami, 2016). It is quite alarming that about 82.6% or 44.3 Mt of e-waste data is not available (means their destination or treatment process is not recorded); therefore, it is difficult to estimate their adverse effect on environment as most of the e-waste is shipped to developing countries and unfortunately, those countries do not have well-established waste management policy and technology (Baldé et al., 2017; Forti et al., 2020; Heacock et al., 2016; Ilankoon et al., 2018). It is found that around 8% of WEEE is treated as mixed or general waste and disposed of in landfill or passed through incineration, whereas 7–20% of WEEE is transferred to poor countries either as scrap or second-hand goods (Forti et al., 2020). It is also a serious concern that around 70–75% of WEEE in the USA and India is left untreated, and 0.6 Mt of WEEE is dropped as general waste in Europe (Forti et al., 2020). However, the Association of Plastics Manufacturers in Europe classified WEEE as a mixture of ferrous and nonferrous materials, including gold, platinum, lead, copper, cobalt, plastic, and ceramics (Masud et al., 2019; Mourshed et al., 2017; Srivastava & Pathak, 2020). Thus, materials recovery from WEEE is now considered a secondary source for precious materials supply through recovery, along with WEEE management in a planned manner. On the other hand, the recovered materials are not precious, but some of them are also considered as rare materials due to their limited stock in earth crest (Baldé et al., 2017; Forti et al., 2020). Only 15% of WEEE is recovered globally on average (Heacock et al., 2016), but this process has gradually gained more attention. On the other hand, only 17.4% (9.3 Mt) of e-waste was recorded to be stored and recycled through legitimate routes in 2019, whereas the rest of 82.6% (44.4 Mt) was left unrecorded (Forti et al., 2020). However, the remaining part of WEEE that is kept untreated or directly disposed of to the environment (after recovering the valuables) is the growing concern for managing environmental safety and security. As most of the recovery works are conducted by informal sectors, both formal and informal sectors are much more interested in the recovered items rather than the environmentally friendly disposal of remaining parts (Zeng et al., 2016). Therefore, a holistic approach is needed to design a sustainable strategy to emphasize the recovery process and regulate the uncontrolled dumping of WEEE. Figure 3.1 illustrates a general flow chart of WEEE management strategies followed by most countries around the world.

The extraordinary demand for high-tech EEE as well as their dumping through global cross-boundary transportation have turned WEEE into a major concern in the global economic and environmental landscape. Thus, it is obvious to have some global stand-up that can possibly control this high-volume flow of toxic effluent in a responsible and systematic manner. The waste management framework is designed by considering the WEEE collection, transportation, processing (recycle, recover, refurbish, etc.) and disposal factors (Srivastava & Pathak, 2020). However, strengthening the existing rules to control the import–export of e-waste, setting up recovery and recycling plants, providing financial backup to install WEEE processing units, and imposing an efficient strategy to collect e-waste from people are the backbone for the smooth operation of the e-waste management policy (Baldé et al., 2017; Borthakur and Management, 2016; Kumar et al., 2017; Kumar & Rawat, 2018). It is also essential to set up a plan for a sustainable circular economy, and in this very way, researchers strongly support the 3R (reuse, recycle, and reduce) and 4R concepts (reuse, recycle, reduce, and replace) (Garlapati, 2016; Islam et al., 2021; Mourshed et al., 2017).

The last but not the least or the most important aspect of sustainable WEEE management strategy is to raise mass awareness among the people; so that people from all classes get the opportunity to rethink their old electronic and electrical gadgets and find an eco-friendly way to dispose of those items. In light of the above discussion, this chapter aims to focus on the generation pattern of e-waste around the world. At the same time, the composition and dispersion of e-waste components are highlighted to emphasize the fatal effect of e-waste on human beings as well as on environment. This chapter also showcases a comprehensive analysis and respective challenges in the e-waste collection and treatment procedures. Section 2 gives a global e-waste generation scenario, and its composition is given in Section 3. Formal and informal e-waste collection strategies are discussed in Section 4. Post-collection treatment of e-waste is illustrated in Section 5.

3.2 Global E-Waste Generation Scenario

As seen in Figure 3.2, Asia produced the highest amount of e-waste (24.9 Mt) in 2019, followed by America (GESP n.d.; Statistica n.d.; Times n.d.). Asia is

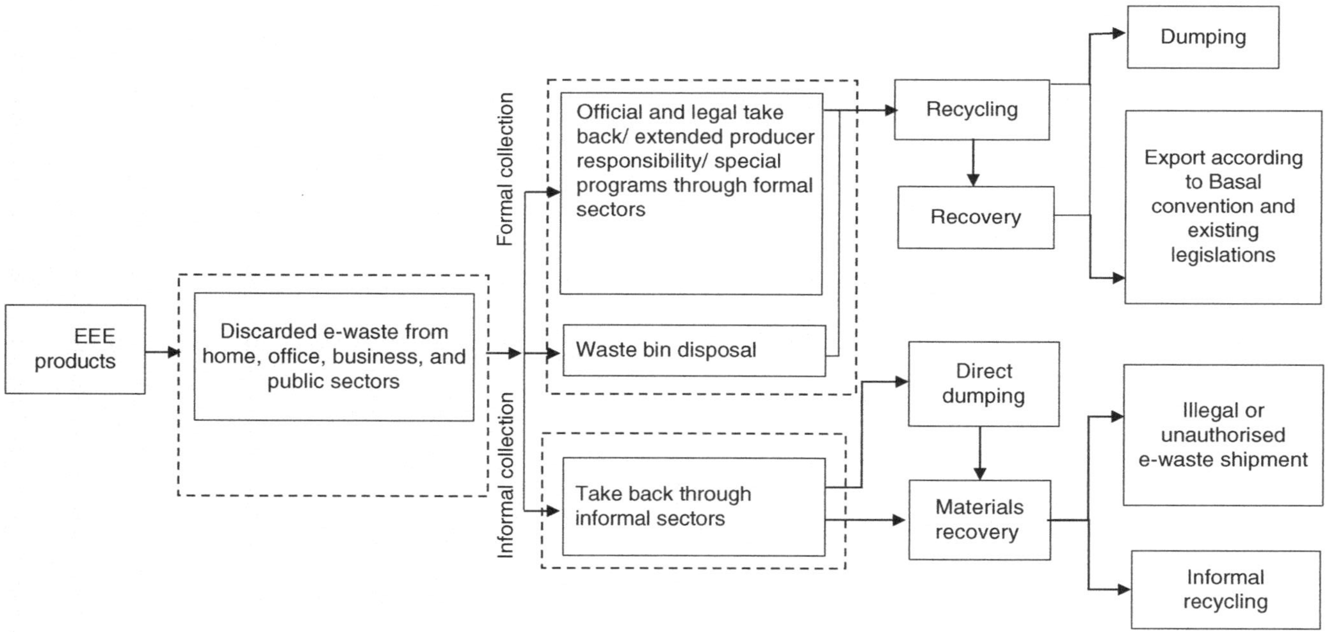

Figure 3.1 Flow diagram of a general WEEE management system (Masud et al. 2022; Pathak et al., 2017; Srivastava & Pathak, 2020; Sthiannopkao et al. 2013).

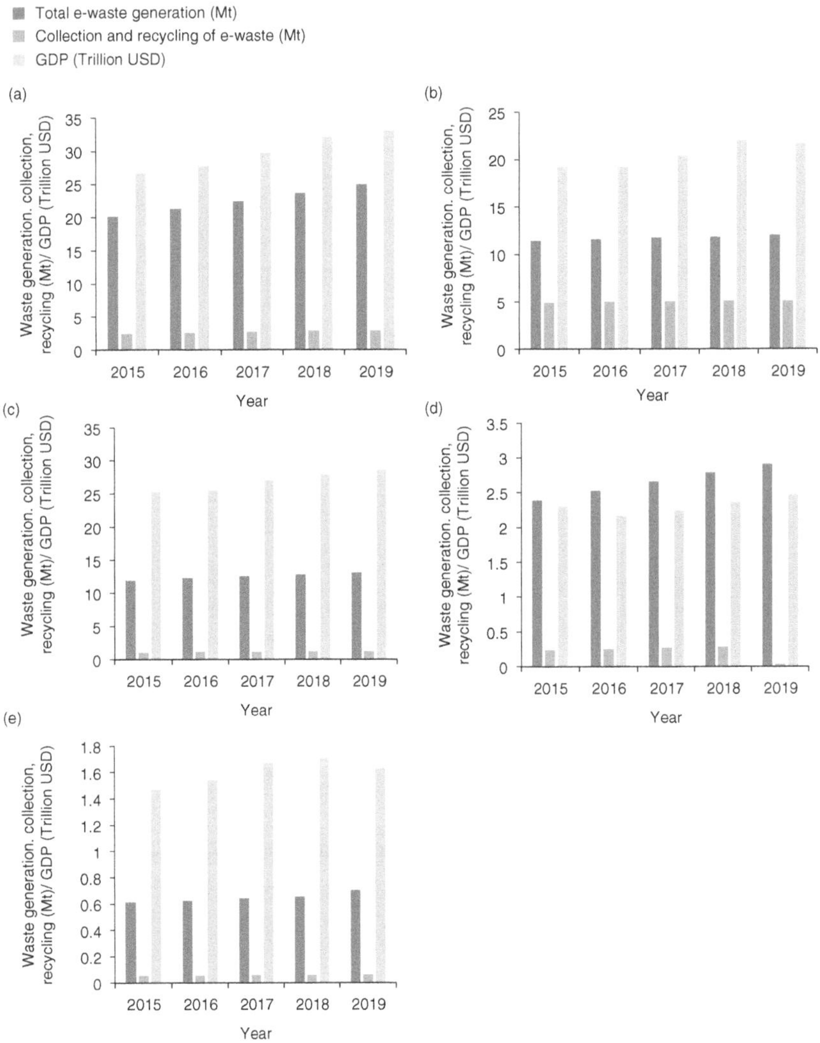

Figure 3.2 E-waste generation scenario in (a) Asia region, (b) Europe region, (c) America region, (d) Africa region, (e) Australia and Oceania region (GESP n.d.; Statistica n.d.; Times n.d.).

a dynamic zone with a wide diversity of growing and established economies, various levels of income, and hence widely diversified grades of electronics consumption, as well as a wide variety of WEEE processing and disposal processes. Baldé et al. (Baldé et al., 2017) reported that the application of EEE is increasing at a faster rate in developing nations than in developed countries. Even though Asia is the world's leading producer of e-waste (globally 46.5%), only

2.9 Mt or 11.7% of this is officially gathered and reused (GESP n.d.; Statistica n.d.; Times n.d.). This recycling and collection rate naturally differs from country to country based on existing national and international legislation. With an annual production of 10.1 Mt (in 2019), China is Asia's largest producer of e-waste; however, this amount is anticipated to reach 27 Mt by 2030 (Baldé et al., 2017; Zeng et al., 2017).

There are several reasons behind this increase in WEEE, and population density is one of the major reasons, which naturally causes a rise in local demand for advanced and newly launched EEE. Moreover, the country's GDP, or per capita income, indicates people's affordability for the latest equipment and appliances such as mobile phones, laptops, tablets and air conditioning systems, which indicates a direct relationship to the generation of WEEE as well. In contrast, those who cannot easily buy new equipment tend to depend on used or second hand or refurbished items (Ongondo & Williams, 2011). In any case, consumption is increasing owing to increased local demand, as illustrated by Figure 3.2, which shows a linear growth in e-waste generation each year.

To fulfill local consumption and global use, Asia, specifically East and South-East Asia, explicitly assembles and manufactures electronic and electrical appliances. South Asian countries, including Bangladesh, India, Sri Lanka, and Pakistan, produce a significant amount of e-waste annually (Partnership, 2019a, 2019b). Apart from production, China and other Asian nations such as Pakistan and India play a significant role in the reuse, refurbishment, and recycling of WEEE, which are either domestically produced and/or imported from developed nations, making them the 4th and 26th greatest producers of e-waste, respectively (Baldé et al. 2017; Ilyas et al. 2020; Srivastava & Pathak 2020). By 2030, emerging nations are expected to dump twice as much electronic trash as developed regions (Sthiannopkao et al. 2013), with e-waste expected to double owing to rising local consumption and the import of EEE (Wu et al., 2015). America is the second-most significant e-waste producer after Asia, and Europe is the third due to the EU's stringent regulations (Baldé et al., 2017); thus, the design, manufacture, collection, disposal, and processing of e-waste are all covered under the EU Directive. It is no wonder that Europe has the highest recycling and collection rates (42.5% or 5.1 Mt) compared to other continents around the world. However, along with the e-waste growth rates, disposal without further processing or open dumping after the recovery of valuable materials is becoming a critical issue to protect the environment.

3.3 General Composition of E-Waste

There are various elements in e-waste that are difficult to present in a comprehensive breakdown of the makeup of the whole waste stream. However, most previous research works focused on five types of materials: plastic, nonferrous metals, ferrous metals, glass, and others. Valuable materials such as silver (Ag), palladium (Pd), gold (Au), and traces of Platinum (Pt), and hazardous materials including lead (Pb), beryllium (Be), mercury (Hg), cadmium (Cd), antimony (Sb), arsenic (As), and indium (In) are both are present in e-waste (Masud et al., 2019).

Overconsumption of copper causes liver damage, causing discomfort feeling for the nasal and throat, while prolonged exposure to cadmium raises the chance of kidney damage, lung cancer, and musculoskeletal diseases (Bhutta et al., 2011; Esteban-Vasallo et al., 2013). Chromium, nickel, lead, and cadmium provide a life-long risk of cancer in human beings. In addition to causing harm to the lungs and kidneys, the reproductive system and neurological system are also getting damaged. Carcinogens and non-carcinogens alike, including brain and liver illnesses, are linked to mercury exposure (Singh et al., 2020). Residents, especially children, are more likely to get cancer and other diseases from the pollution at informal recycling sites (Shaikh, 2021). Figure 3.3 illustrates the effect of e-waste on human health.

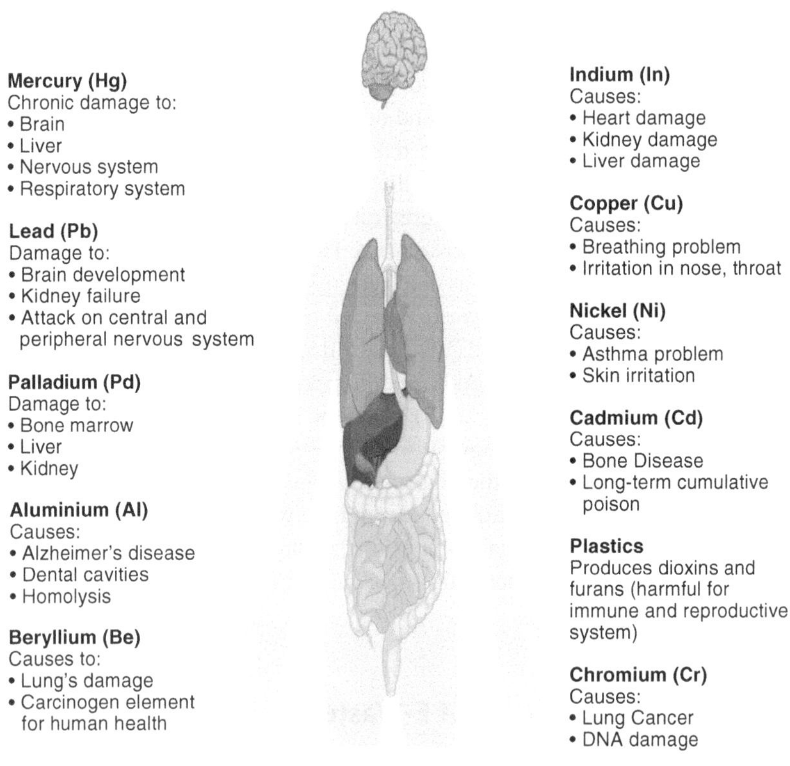

Figure 3.3 Health hazards and risks associated with electronic waste (Jaibee et al., 2015; Ari, 2016; Masud et al., 2019; Masud et al. 2022; Sahajwalla & Gaikwad, 2018; Sheikh et al., 2013; Vats & Singh, 2014; Zhang et al., 2019).

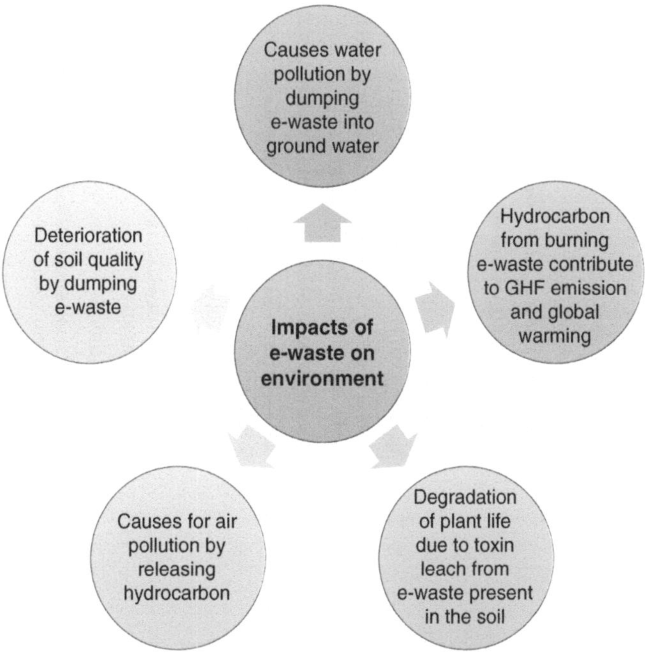

Figure 3.4 Environmental impacts of e-waste (Dopp & Rettenmeier, 2013; Masud et al., 2019; Sahajwalla & Gaikwad, 2018; Shaikh, 2021; Sheikh et al., 2013; Tchounwou et al., 2012; Vats & Singh, 2014).

Besides affecting the human body, as a consequence of improper e-waste recycling and dumping, water, soil, and air are excessively polluted. As a result, there is a risk that chlorinated dioxins, brominated compounds other carcinogens will infiltrated the air and cause cancer in humans and animals. Figure 3.4 illustrates the effect of e-waste on the environment.

E-waste in landfills contains hazardous substances, including lead, mercury, arsenic, and printed circuit boards (PCBs,), which can cause contamination of underlying groundwater or contamination of crops that may be planted nearby or in the area in the future. Large particles from burning, shredding, or dismantling e-waste can remain in the soil for a long period of time and can be harmful to microorganisms in the soil and damage the fertility of the soil (Chatterjee & Kumar, 2009; Kumar et al., 2017; Kumar & Holuszko, 2016). Metals are recovered to some extent, such as gold, copper, aluminum, silver, lead, iron, and tin, and this recovery actually helps to reduce the extraction of virgin materials from nature. On the other hand, many essential metallic substances, such as cobalt, platinum, antimony, germanium, barium, and nickel, remain unrecoverable (Needhidasan et al., 2014). Below, Table 3.1 depicts the relative properties of different e-waste components and their fates after application.

Table 3.1 Fates of different e-waste components after utilization.

Material	Percentage contribution in total e-waste (%)	Heavy material	Toxic material	Health hazard	Environmental hazard	Waste to energy	Waste to product	Direct dumping	Reference
Pb	0.01	√	√	√	√	X	√	X	Jaibee et al. (2015)
Au	0.1	√	X	X	X	X	√	X	Jaibee et al. (2015), Masud et al. (2022)
Hg	0.0022	√	√	√	√	X	√	√	Ari (2016), Vats and Singh (2014)
Cd	0.0094	√	√	√	√	√	√	√	Ari (2016); Vats and Singh (2014)
Cr	0.0063	√	√	√	√	–	√	–	Ari (2016), Masud et al. (2019), Vats and Singh (2014)
Ni	2	√	√	√	√	√	–	√	Masud et al. (2019), Vats and Singh (2014)
Fe	8.11	√	X	X	X	–	√	X	Ari (2016), Masud et al. (2019), Vats and Singh (2014)
Al	2	X	√	√	√	√	√	X	Ari (2016), Masud et al. (2019), Vats and Singh (2014)
Cu	20.12	√	√	√	√	√	√	X	Ari (2016), Masud et al. (2019), Vats and Singh (2014)
Pd	0.01	√	√	√	X	√	√	X	Jaibee et al. (2015), Kaza et al. (2018), Masud et al. (2019), Sheikh et al. (2013), Singh et al. (2020)
Pt	–	√	X	X	X	√	√	X	Masud et al. (2019), Sheikh et al. (2013), Singh et al. (2020)
Ag	0.2	√	X	X	–	√	√	X	Jaibee et al. (2015), Kaza et al. (2018), Masud et al. (2019), Sheikh et al. (2013), Singh et al. (2020)
Plastics	30.23	√	√	√	√	√	√	–	Sahajwalla and Gaikwad (2018), Singh et al. (2020), Vats and Singh (2014)
Zn	1	√	X	X	X	√	√	–	Sahajwalla and Gaikwad (2018), Singh et al. (2020), Vats and Singh (2014)
In	–		√	√	X	√	√	–	Zhang et al. (2019)
Be	–	X	√	√	X	–	√	–	Masud et al. (2019), Sahajwalla and Gaikwad (2018), Singh et al. (2020), Vats and Singh (2014)
Sn	4%	√	X	X √		√	√	–	Masud et al.; Singh et al. (2020), Zhang et al. (2019)

3.4 E-Waste Collection Strategies

3.4.1 Overview

Over 90 jurisdictions, along with well over 2000 sets of legislation, are imposed by different countries around the world as a safeguard to WEEE management and to keep the environment and living beings safe and sound (Veit and Shanghai 2014). Some developed countries (EU, Japan, and South Korea) have implemented both regional (EuP Directives, ROHS Directives) and international (Basel Convention) regulations for the proper management of WEEE, while the USA, Canada, China, and Brazil are still developing control regulations and new laws (Li et al., 2015; Zeng et al., 2016). Moreover, Basel Convention was employed by United Nations (UN) in 1989 to regulate transboundary movement and trafficking of hazardous materials like WEEE, and from 1992, by the Basel Convention agreement, the receiving should have to give consent to the reception of any hazardous materials. Basel Convention emphasis on the transboundary shipment of WEEE products as well as forces the respective authorities to strengthen their existing legislation to control e-waste import and export. Some of the countries have not entered into Basel Convention, whereas the developed countries are looking for dumping sites, and recycling through cheap labor; hence, in 1995, representatives from developed countries passed a ban amendment to completely restrict any kind of hazardous junk shipment to developing and from developed countries (Sthiannopkao et al. 2013; Tansel 2017).

However, most of the e-waste policies that have been followed by the EU and countries outside the EU emerge from the regulations of "Restriction of the Use of Certain Hazardous Substances in EEE (RoHS)" and "Directives on WEEE" (Herat, 2008). These regulations emphasize policy-making and employing government legislation with a view to minimizing e-waste disposal volume by maximizing the reuse, recovery, and recycling approach. In addition to 4R strategies, global and regional legislation also comes into practice and covers up to 66% of the total world population under a legal framework to regulate their discarded e-waste (Baldé et al., 2017). The formal components or organizations to manage e-waste include: (i) European Recycling Forum (ERF) by EuP Directives, (ii) ROHS Directives, (iii) Advance Recycling Fee (ARF) through Take Back System (TBS), (iv) Basal Action Network (BAN) to control trans-boundary transportation, and (v) Extended Producer Responsibility (Pathak & Srivastava, 2019).

Globally, WEEE management is handled using various strategies. Formal and informal e-recycling methods have been implemented for managing e-waste. "Formal e-waste recycling" refers to facilities licensed and authorized to handle electronic waste while maintaining worker protection, industrial hygiene, and environmental pollution controls. The term "Hidden Flow of e-waste" refers to the process by which electronic garbage is managed after it has evaded legitimate

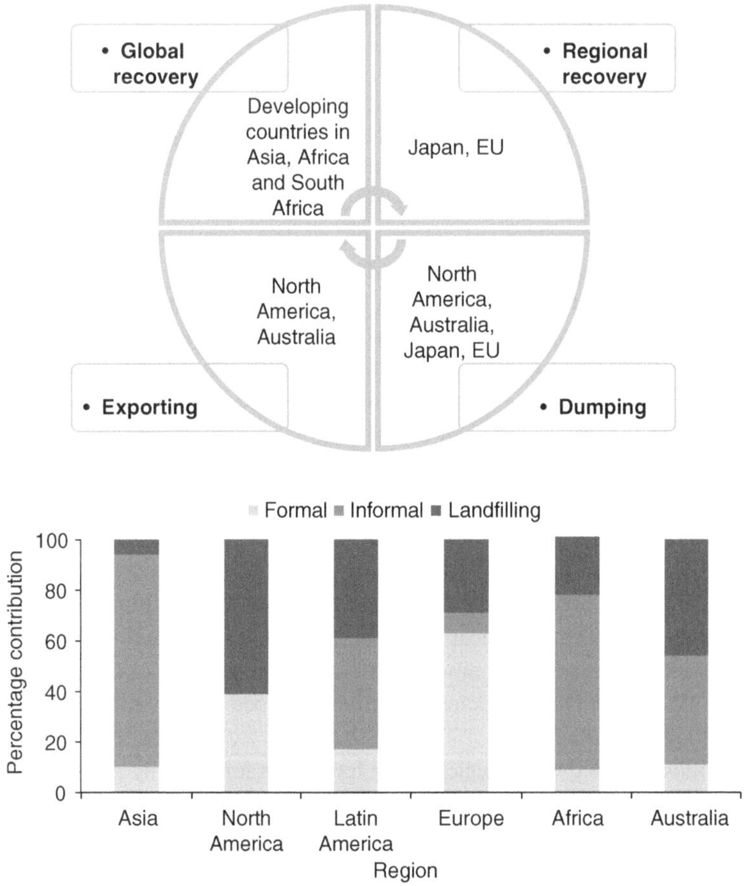

Figure 3.5 Global e-waste management strategies (Borthakur & Govind, 2017; Liebmann, 2015; Lifecycles 2021).

waste collection and administration methods. Total e-waste management is normally classified as formal, informal, landfilling, or open dumping. The following Figure 3.5 illustrates the e-waste management of the different regions, where it can be observed that countries in Asia mostly depend on informal management of e-waste, while developed countries mainly use the formal e-waste management system.

Although developed countries are more focused on the formal collection and management of e-waste through TBS or EPR, or other time-to-time special campaigns from the government, apart from these, small and medium entrepreneurs are also dealing with the e-waste collection, processing, and shipment business. In most cases, developed countries export or dump their e-waste (in its full body or parts) to developing countries in the name of donation or TBS (Kumar et al., 2017), which actually helps developed countries to manage their e-scrap. And it has been found that China, Africa, and other South Asian countries have to manage 50% of total e-waste or in other words, these countries generate the same amount of e-waste (as they receive the e-scrap) (Cucchiella et al., 2015; Golev et al., 2016). However, almost 96% of high-income countries have some set rules for solid waste management, whereas this number is only 60% for low-income countries (Kaza et al., 2018).

EU encourages its 27 member countries to boost the recycling rate of their e-waste (EU directives 2003). Countries in Latin America are not widely accepted and have developed their own policies regarding e-waste management. Africa has also had the same mixed experience. However, more than 85% of EEE stuff in Africa is second-hand, and e-waste management is mainly controlled by informal or unauthorized organizations (Garlapati, 2016). Though half of the states in the USA follow some legislations to assist the WEEE recycling strategy, still the national-level e-waste regulation policy is not in place, whereas Canada has adopted industry-grade WEEE management and recycling system. Australia also has its "National Television and Computer Recycling Scheme" to collect and process e-waste, which is jointly operated by the government and industries. In Asia, Japan, South Korea, and Taiwan have a very good stand on recycling their own e-waste, and it is nearly 75%, 75%, and 82%, respectively; on the other hand, China, India, Bangladesh, and Pakistan have also developed some of their policies to tackle e-waste, but those legislations have not yet been put into practice.

3.5 Formal E-Waste Management

3.5.1 Overview

Due to escalating e-waste amounts and new regulations, formal recyclers are joining the e-waste recycling business to make the total process more flexible. With the implementation of the best available technologies to recycle e-waste in formal management, a notable improvement in material recovery and environmental safety can be achieved. However, procedures associated with formal recycling often begin with sorting, checking, reusing, and repairing collected stuff. The devices that need to be recycled are then disassembled and occasionally crushed, and the materials are extracted either by automated equipment or by human labor. In addition, many formal e-recycling factories serve as suppliers to downstream parties to recover materials such as metal, glass, and plastics (Tsydenova & Bengtsson, 2011). An overview of formal e-waste processing steps is shown in Figure 3.6.

On the other hand, informal recycling mainly relies on local vendors, and the safety measures used in these categories are very poor compared to the formal

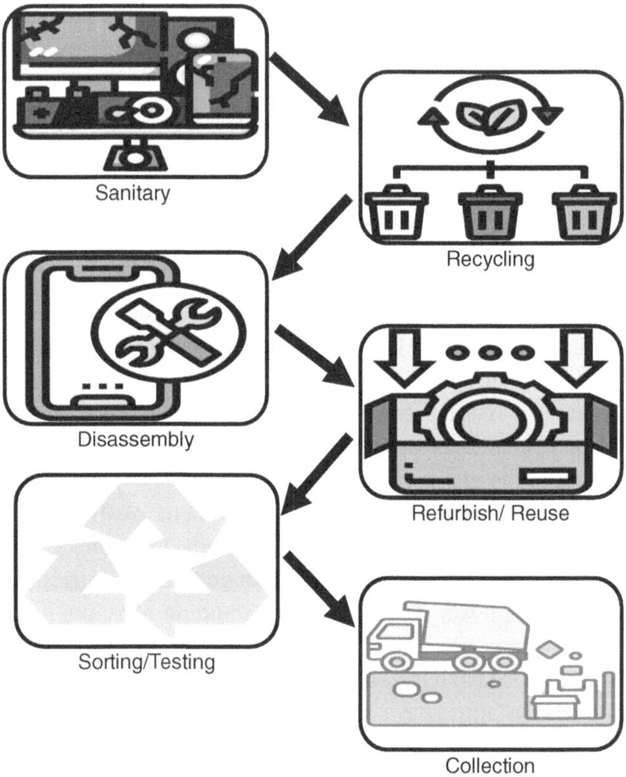

Figure 3.6 Formal E-waste process (Strike, 2021; WA, 2021).

management system. The most common informal recycling methods include checking, dismantling, and collecting valuable materials. Most often, they dump e-waste as general waste in open fields or in water (Matsukami et al., 2015; Song & Li, 2014), and these activities have disproportionately negative health repercussions for underprivileged people (Heacock et al., 2016).

3.5.2 Government Authorities/Municipal Authorities

E-waste is handled in a variety of ways across the world, with legislation and tools implemented by governments to unions. EU formally collects and processes about 25% (2.1 Mt) of its projected 8.7 Mt of annual e-waste generation after maintaining the latest industrial regulations for protecting the respective personnel associated with e-waste management (Mmereki et al. 2016). The EU e-waste regulation explicitly controls the deposition, recovery, and recycling objectives for EU member countries. Thus, it establishes a minimum annual collecting and processing objective of 4 kg per person for all member countries

(Mmereki et al., 2016). A well-developed e-waste management system has been introduced in Switzerland, which recycles 11 kg per person of e-waste generation and processing compared to the EUs' 4 kg per person goal (Mmereki et al., 2016). Another report shows that 33% of e-waste is sorted separately and handled correctly in the EU (Mmereki et al., 2016). According to the ElektroG system, Germany gathered and processed around 0.75 Mt of e-waste in 2006, whereas other EU member states collected approximately 0.019 Mt (Mmereki et al., 2016). However, product stewardship and EPR programs for WEEE management are becoming more popular in the USA and Canada. For instance, in the USA, the state of Maine follows the product stewardship model to manage its e-waste with merchants' active involvement (Wagner, 2009). Non-governmental organizations (NGOs) established in the USA are strongly concerned with the issue of e-waste. For example, one of the largest networks of environmental NGOs in the USA is comprised of the Basel Action Network, the Silicon Valley Toxic Coalition, and the Electronics Take-Back Coalition. All three organizations' shared goal is to advocate for national solutions for the disposal of hazardous waste (Widmer et al., 2005).

In Europe, there are two collection systems, one of which is kerbside collection at multifamily residences, which is a handy choice for inhabitants who live in highly populated regions. A collection experiment in a multifamily home in Vienna revealed that the small appliance collection rate is 0.4–1.1 kg per person per year, while the kerbside collection rate with fixed or colored containers achieved 1.33 kg per person per year in Copenhagen (Salhofer, 2017). Another option is to put a container at collection sites close to public areas. Case studies found that in Germany and Sweden, small appliance collection rates ranged from 0.04 to 0.84 kg/cap/year (Salhofer, 2017). E-waste is collected door-to-door in Japan to prevent it from being contaminated with municipal rubbish. Similar to TBS or EPR policies, retailers and municipal authorities convey the collection and transportation processes and send them back to manufacturers for recycling purposes (Mmereki et al., 2016).

3.5.3 Extended Producer Responsibility

Switzerland is the first country to introduce EPR to manage e-waste, and other countries in the EU, some states of the USA, and Canada also followed this footstep to attain a sustainable circular economy through successful e-waste management. However, most developing and underdeveloped countries are still lagging behind in setting up sound e-waste management guidelines. Nowadays, many countries implement EPR policies to enable waste management, irrespective of the type of waste. One of the most significant things that is regulated under the EPR plan is e-waste management. The EPR system was implemented in 2003 by the EU to encourage recycling habits (Masud et al. 2022). A key component of EPR is

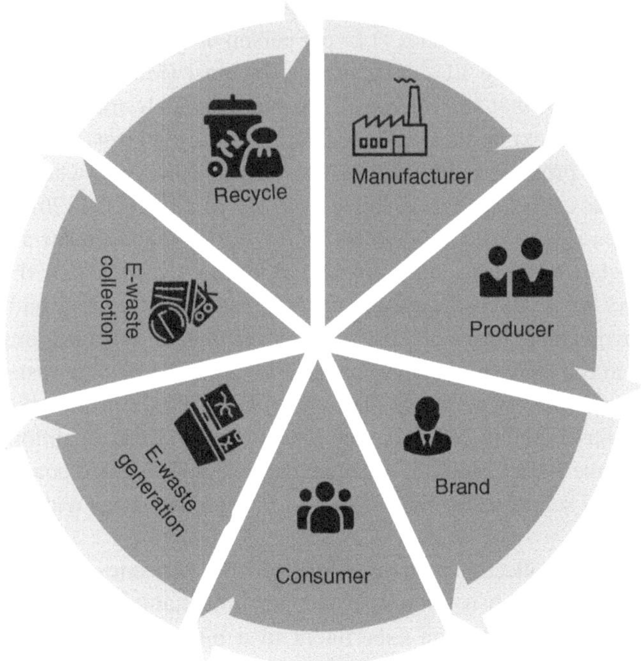

Figure 3.7 Flow process of EPR policy.

the increased involvement of manufacturers in the long-term management of their manufactured goods as well as in solving environmental issues (Jang 2010). Thus, producers are similarly responsible for their goods' entire life cycle under the EPR strategy. Figure 3.7 highlights the EPR flow process. In EPR, producers include not just manufacturers and suppliers but also importers and retailers. The responsibility is distributed via a variety of economic incentives, including financial assistance for the collection and sustainable recycling of WEEE, as well as the promotion of greener production. The Polluter Pays Principle or 3P is the foundation of the EPR strategy. Due to industrial operations that generate e-waste, manufacturers are considered the main contributors to producing e-waste under the EPR. According to the EPR model, EEE manufacturers are responsible for paying for contamination caused by the e-waste (ESCAP 2021).

EU first launched this program, and the EU legal system is primarily dependent on it, as manufacturers are obliged to use both individual and collective TBS to manage the products service life, along with the post-consumer stage (Jinghua et al., 2020; Ogushi and Kandlikar, 2007; Toyasaki et al., 2011). In Japan, the EPR concept is well-known and widely used where manufacturers are obliged by the existing laws to collect e-waste from authorized collection points and proceed with recovery steps (Masud et al., 2019). The implementation of the EPR plan in 2003 was the most significant development in South Korea, with manufacturers collecting around 70% of e-waste (Manish & Chakraborty, 2019). Other countries, including Canada and USA, are working to implement EPR-based systems (Thompson & Oh, 2006). Three methods exist in Brazil for disposing of and recycling e-scrap: (i) collection by retailers, (ii) collection by manufacturers, and (iii) collection by NGOs and social organizations.

3.5.4 Extended Consumer Responsibility

Consumers are also responsible for the safe disposal of their products, and in this way, TBS, extended consumer responsibility (ECR), and other specific programs help to smooth this collection and successive processing system. For example, some countries provide particular collection bins (like Japan) for the collection of wastes like hazardous, municipal, recyclable, etc. In contrast, some countries apply deposition fees so that customers can get back their money after returning their used EEE goods through authorized collectors or dealers (Wang et al., 2016). In Japan, residential users pay a charge to cover a percentage of the recycling and transport expenses. Thus, in ECR, charges for the processing of e-waste are added in advance to the original price of the goods. In addition, applying higher charges and taxation to luxurious EEE goods can be another approach to reduce environmental hazards with a bit negative impact on the market-based economy (Borthakur et al., 2019; Borthakur and Management, 2016).

3.5.5 Take Back Policy

The core content of EPR is the identification of active TBS or design for the environment (DfE) components for safe e-waste disposal (Palmeira et al., 2018). Conversely, due to this strict environmental legislation, first-world countries use developing and under-developed countries like India, China, and Africa as dumping sites, simulating donation or take-back strategies (Kumar et al., 2017; Yu et al., 2010). Figure 3.8 shows the whole take-back process in a flowchart. The role of the manufacturer to handle the obsolete EEE is the prime concern in TBS. In this scenario, handling the obsolete EEE items through manufacturers also ensures more safety and reliability in the management and processing of e-waste than through other retailers or vendors.

The EU directive has by far the most comprehensive national regulatory system since it covers the whole life cycle of an EEE. The most recent update in the EU directive has included new recycling and take-back collecting targets that comprise the addition of "one-to-zero", or wholesalers must accept old products without purchasing a new one from them—as a collection option for low WEEE. Japan has also established a legal structure comparable to the EU to take care of the obsolete EEE. In the European region, most domestic e-waste collection programs

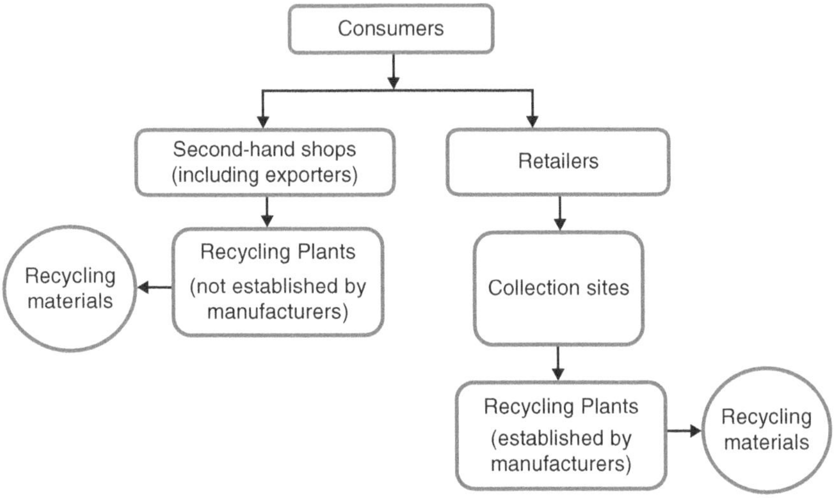

Figure 3.8 Flow diagram of a take-back policy.

are integrated with existing municipal recycling and hazardous household trash collection programs, as well as retailer-sponsored take-back programs. Whereas some countries like France and Belgium, place a high value on waste take-back via recycling plants (Salhofer, 2017). Electronic and electrical items that are no longer in use can be returned or exchanged for free by several organizations. For example, Lenovo and Nokia were among the first Chinese corporations to do so (Perkins et al., 2014).

Commercial take-back programs for items have been implemented in Singapore. For example, the cost of refurbished or old cell phones is getting fixed by the sellers considering the physical condition of that item. Then contact is done between seller and customer, and during the contract duration (e.g. 2 or 3 years), the cell phones are leased at a lower price than their purchase price. Consequently, nearly 95% of second-hand cell phones are taken back to Singapore (Mmereki et al., 2016). Customers in the USA can return their old electronics for free to a number of manufacturers (Toshiba, Apple, Samsung, Sony, HP, Lenovo, LG, Panasonic) at designated drop-off locations or by sending them by postage (Masud et al. 2022; Mmereki et al., 2016).

3.6 Informal E-Waste Management

3.6.1 Overview

Informal sectors mainly control the e-waste collection and processing in developing and under-developing countries, which is also the major source of income for many people to lead their livelihood (Heacock et al., 2016). "Tokai" in Bangladesh or "Kawariwala" in India, or local vendors in some other countries are moving from door-to-door to collect e-waste to run their businesses, which not only helps to collect e-waste but also ensures the consumers have some positive pricing for their obsolete EEE products (Borthakur et al., 2019). E-waste workers rely on the informal sector for a living, yet they face serious health risks (Dutta et al., 2018; Rautela et al., 2021).

Lack of adequate government legislation and safety measures in the informal sectors that are handling e-waste in developing and transitional countries results in a small amount of material recovery (Perkins et al., 2014). E-waste is deconstructed and processed crudely in the informal process, and its hazardous elements may cause severe damage to the human body. When toxic substances (like acid from lead-acid batteries) are handled without proper safety and disposed of without following existing laws and regulations; this will result in serious damage to the environment and human beings.

3.6.2 Local Vendors

Informal sectors are growing up due to the lack of adequate arrangements and infrastructure to collect and process e-waste by responsible authorities like municipalities. The informal sector is unacquainted with scientific processes to manage e-waste in a proper way, as well as they are not abided by the existing laws and regulations. The informal sector is mainly responsible for gathering, segregating, dismantling, and discarding the collected items. Waste pickers and small scrap merchants conduct the majority of the door-to-door collection. They frequently collect e-waste from customers after reasonable compensation. Local waste pickers are one of the best media to collect e-waste from every corner, which in turn makes it easier for the responsible authorities to proceed with the next steps (Chatterjee & Kumar, 2009). A flow chart of informal e-waste processing is shown in Figure 3.9.

Diverse sources of e-waste are sorted into numerous categories, including plastic, modules, glass, components, and metal, based on their marketability for maximum financial benefits (Chatterjee & Kumar, 2009). After collection by the waste pickers, sorting and in some cases, dismantling or disassembling of e-waste items are mainly carried out by the scrap merchants.

The scrap dealers sell all the disassembled and separated plastic, glass, and metal pieces to metal or glass smelters, plastic re-processors, and glass or metal smelters who specialize in transforming these scraps of plastic, iron, copper, glass, and aluminum into new materials. For non-formal entities, the process of smelting and recycling metals is unfamiliar; thus, they opt to sell the material. It is interesting to note that the informal sector plays a dominating role in handling e-waste in South Asian countries, and only in India do informal recyclers handle 95%

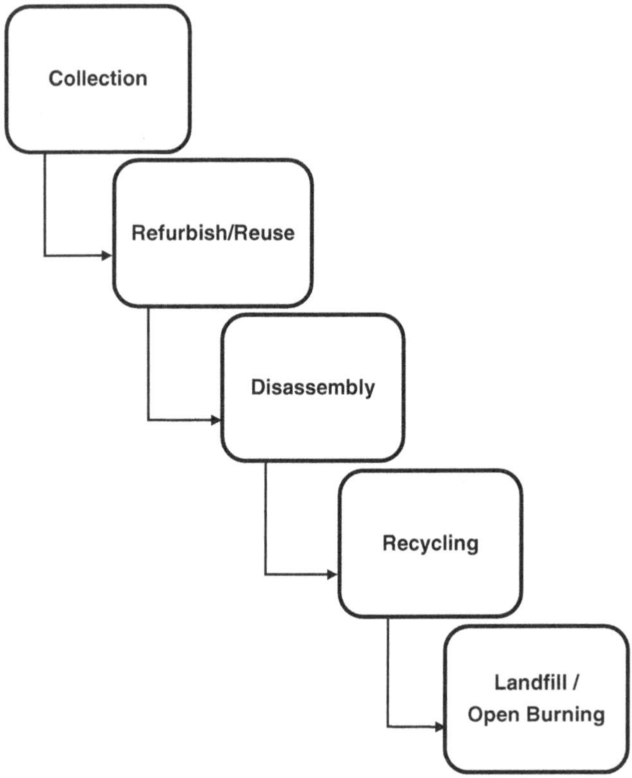

Figure 3.9 Informal E-waste Process (Barker, 2020; Record, 2019; Starr, 2019).

of e-waste (Borthakur & Sinha, 2013). South Africa dismantles almost a tenth of its collected e-waste and sends the recovered parts to be reused by manufacturers (partnership). Ghana, on the other hand, handles 40–60% of its domestically produced e-waste, with the informal sector managing 95% of it (Maes & Preston-Whyte, 2022).

On the other hand, household rubbish, especially e-waste, is often seen as a valuable product that is generally swapped for cash. Unsurprisingly, inhabitants prefer informal methods of collecting e-waste because the collection service takes place conveniently at the customers' doorstep and at the same time provides monetary refunds. In many major Chinese cities and South Asian countries, unofficial household e-waste collection exceeds formally collected or received quantities (Masud et al. 2022; Salhofer, 2017). The practice of selling e-waste to private or local collectors also supports the establishment of a robust informal recycling industry as well as refueling the circular economy of the country (Chi et al., 2014). E-waste disassembly is a big business in China that relies heavily on primitive recycling processes like acid bathing, melting, and burning to recover precious minerals and metals for industry and to provide a steady revenue, despite the considerable health dangers involved.

3.6.3 Others

Along with the collection and processing steps (like recycling, recovery, refurbishment, etc.), it is also important to ensure the proper dumping of scrap parts, such as plastic cases, wire covers, and plastic circuit boards (Wang et al., 2016; Widmer et al., 2005). E-waste disposal sites close to residential or agricultural areas are considered environmental hot spots. Such sites are responsible for spreading diseases and odors, releasing leachates and poisonous liquids into rivers, ponds, and groundwater. According to a report in 2019, about 50 tonnes of mercury, 71-k tonnes of brominated flame retardants plastics, and 98 Mt. of greenhouse gases (CO_2 equivalent) were generated from the discarded WEEE (Forti et al., 2020). Thus, after material recovery, it is essential to dump the e-scrap in a well-managed way to minimize the spread of ecotoxicology.

About 82.3% of e-waste was burned and discarded into landfills in the USA in 2009, whereas only 17.7% was recycled (Partnership). It is reported that two-thirds of EU's waste stream is still possibly heading to landfills and substandard treatment facilities inside or outside the EU (Lucier & Gareau, 2019; Mmereki et al., 2016). In this scenario, UAE and Singapore allegedly act as pre-distribution centers for transferring e-waste from the EU and USA to South Asian countries, where India and Pakistan serve as the main destination countries (Mmereki et al., 2016). A significant portion of the world's e-waste is collected in open dumpsites in various African countries, most notably Ghana and Nigeria (Maes & Preston-Whyte, 2022). In Australia, 43% of all electronic waste is recycled using low-value methods, such as shredding items to extract metals. Following this step, any remaining items are sent to the landfill (Bontinck et al., 2021).

Therefore, proper consideration should be taken to minimize regional and global dumping (through transboundary shipment). It is reported that Asia imported about 93% of e-waste, whereas China (covers Korea, Japan, the Western part of the USA, Australia, and South-East Asia) and India (covers the Eastern part of the USA, EU, Central Asia, and part of South-East Asia) are lead importers. Basel law was introduced to regulate this transboundary shipment of e-waste as poor countries always preferred to dump or manage e-waste in rich countries.

3.7 Treatment of E-Waste

3.7.1 Overview

The above discussion gives an idea about the volume of e-waste collection with respect to e-waste generation throughout the world. If the collected wastes are dumped without any treatment, then it will not only add volume to the existing waste but also cause a great loss in terms of waste to product or material recovery strategies. However, the treatment or processing of e-waste is always challenging as it comprises numerous elements where some of which can be recycled, reused, recovered, refurbished, etc., and the remaining parts should be dumped in a proper way by following the existing rules and regulations. In this way, controlled treatment of e-waste adds value to our economy instead of polluting the surrounding environment and ecosystem (Soetrisno & Delgado-Saborit, 2020). When the recovered materials are used in new devices or the obsolete products are managed to be reduced, the material life cycle is increased, and at the same time, the extraction of new materials from nature is suppressed. Therefore, further treatment of e-waste offers a circular approach to the material flow in place of traditional linear movement and ultimately extends the materials' useful life.

3.8 Reuse and Refurbish

Reusing and refurbishing e-waste have been highly encouraged at both national and international levels. Direct reuse or waste avoidance practices involve the handling of something along to a subsequent owner or selling it outright to someone else (Ongondo et al., 2011; Read et al., 2009). However, once the item enters the waste stream, it should be then passed through proper cleaning, repair, safety testing, etc., or a combination of all these steps to meet the standard for being used as a refurbished item. On the other hand, depending on the complexity and economic aspects, sometimes recycling or recovery of items is more preferable once the EEEs are tagged as e-waste; although, reuse is at a higher position than recycling in the waste hierarchy (Williams, 2015). Therefore, reusing and refurbishing actually prolongs the product's useful life and saves the environment by reducing the e-waste generation rate. The potential challenges to the reuse and refurbishing approach are discussed in Figure 3.10.

3.9 Recycle

Recycling is considered one of the most common and oldest techniques for processing e-waste. Recycling adds more economic and environmental value than reusing or refurbishing; thus, most of the research work focuses on recycling. EEE items are the result of a combination of different types of components that are manufactured from a wide range of materials. Hence, a contribution from different parties is required to recycle specific parts or components with the required technologies. On the other hand, all the materials in e-waste are not recyclable or not economically feasible to pass through a recycling system. Therefore, it is

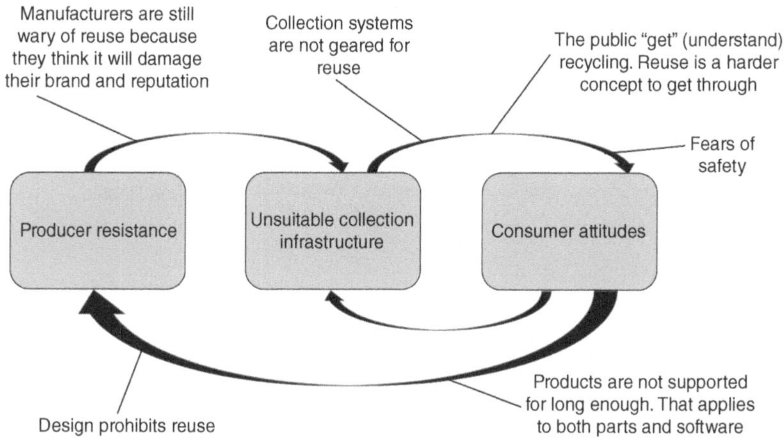

Figure 3.10 Illustrates the intricate interdependencies between reuse barriers.

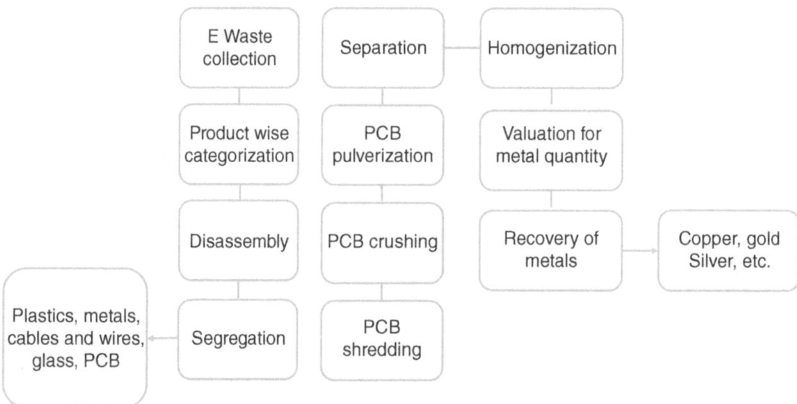

Figure 3.11 Recycling of e-waste and recovery of valuable materials.

important to sort out the essential and recyclable materials and to build a supply chain network of trace materials from both domestic and international points of view. According to a report by the Association of Electronics Recyclers, around 7000 people were engaged in the electronics recycling business in the USA in 2019, and their annual contribution to the economy is more than USD 700 million (Kang & Schoenung, 2005). Figure 3.11 presents a conceptual approach to efficient e-waste material recycling (Akram et al., 2019).

The nature and composition of e-waste are totally different from other types of solid waste; thus, it requires special attention while passing through the recycling system. The efficient recycling process depends on the arrangement of recycling infrastructure to handle a wide variety of e-waste. Infrastructure development is

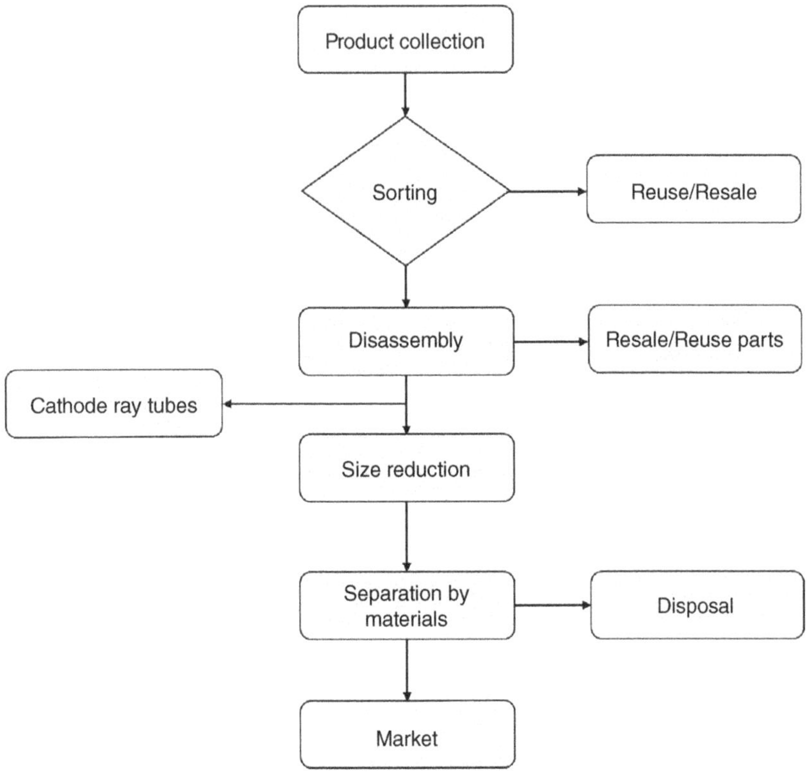

Figure 3.12 A simplified flowchart of the electronic goods recycling process.

one of the most crucial parts of recycling, as it determines the method applied for recycling and the volume of waste that can be handled by the arrangement. Moreover, infrastructure design must have to consider the available and feasible technologies to adopt in the recycling system, plan the business strategies to resell the recycled items, follow the specific guidelines for the recycling process applied by the government, etc. Figure 3.12 depicts a flow diagram to represent the corresponding roles of collecting, sorting, recovering, recovery, recycling, and disposing of e-waste in the recycling process.

Disassembly begins with the recovery of high-value parts and ends with low-value parts to optimize economic value. Unwanted substances that are dangerous and harmful should be removed carefully during disassembly. For example, printer ink is dangerous, and it must be removed before recycling. On the other hand, materials that are laborious and economically not feasible to recover are dispatched as scrap, such as laminated metals, and are discarded as removing laminated covering is laborious. Therefore, dismantling and recovery orders are crucial to optimize output efficiency.

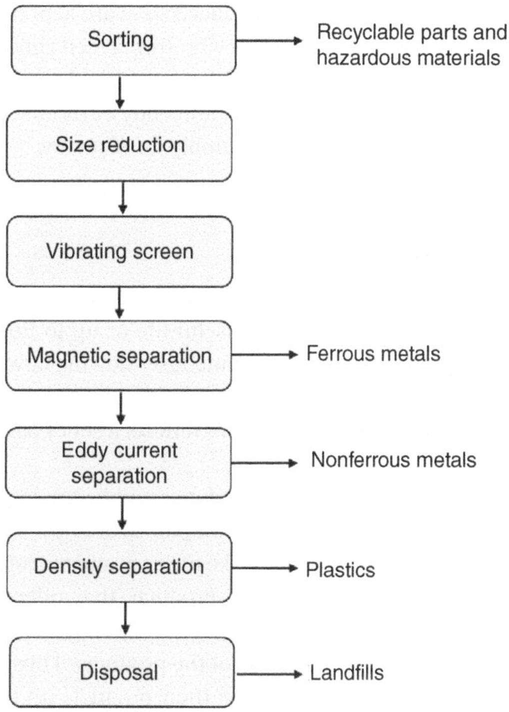

Figure 3.13 A simplified diagram of the process steps at a material recovery.

3.10 Recovery

At the materials recovery facility, the ultimate fate of collected e-waste is determined by categorizing it as either reusable or recyclable or dumped as waste. Thus, the reusable items are separated, and recovery of valuable materials and parts is carried out before sending them for recycling or dumping as scrap. However, the age of the parts and their mechanical condition are important factors at this stage in order to maximize the economic value of recovered items. Figure 3.13 depicts a schematic diagram of the process involved in material recovery.

Although disassembling and sorting are important parts of e-waste treatment, which is also a sophisticated job and requires enough experience. Moreover, it helps to classify recoverable items and store them for future use. After arriving at the recovery section, obsolete e-wastes are tested through the plug-and-play to check their operating conditions. If any items are reusable after some repair work, then they are separated and resold the necessary repairs. Those items that totally obsolete and fail the plug-and-play tests are sent to the disassembly section.

Many particles still have complex shapes that must be reduced to optimal sizes for recovery. E-waste, in particular, is made up of various valuable materials that are linked together by electrical circuits. These large assemblies must be reduced in size in the next process. Reduced sizes are kept on a vibrating screen to be processed further. The motor in a vibrating screen generates centrifugal force, which vibrates the screen box. Materials move quickly to other separations like magnetic separation, density separation, and eddy current separation after falling into the screen from the inlet to disassembly for recovery.

3.11 Reduce

Using the EEEs at their full useful life or up to EoL enables to reduce the need for new products while simultaneously slowing down the e-waste generation rate. Thus, it is time to plan the consumption pattern of EEEs and, at the same time, manage the e-waste. Figure 3.14 reflects a set of pathways for the possible means of reducing e-waste.

Due to the advancement of science and technology, electronic gadgets are regularly updated, and electronic companies are now competing to attract customers with numerous features to make life easy and smooth. In this very way, customers are also inclined to the new products that offer higher efficiency and better functionality. Thus, customers are switching their products to get the updated ones without waiting for EoL of the products. Thus, proper steps should be taken to encourage customers to use their products up to EoL, and at the same time, programs like TBS or swapping the gadgets should be launched; so that at least product usage up to EoL can be managed through reselling.

In possible cases, repairing the EEEs rather than throwing them away economically helpful for both customers and repair shop owners. Customers benefit from repairing as they will not buy a new item if they get the repair service; on the other hand, repair shops indirectly boost the circular economy by using the recovered items from obsolete items. Thus, recovered items can be used in repair shops, and this form of repair shop is getting popular in Europe (like Fairphone) and Australia.

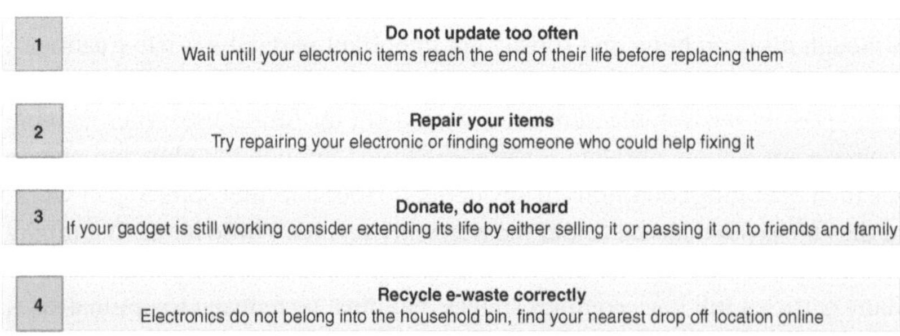

Figure 3.14 Four ways to reduce e-waste carbon footprint.

Considering reselling the electronic item if it is old enough or when someone switches to new electronic gadgets is a good idea to hold the e-waste generation. To promote this idea, several online business media (like Facebook marketplace) are working, and nowadays, local traders are also engaged in reselling of EEEs as second-hand items. Therefore, people from every class in society get the opportunity to use electronic gadgets (like smartphones, laptops, tabs, etc.).

When the electronic products are totally obsolete and dead, i.e. they reach EoL, they should be collected as e-waste and processed in a special manner (to ensure dumping of the product is done after the possible recovery and recycling steps). To facilitate these processes, programs like EPR and TBS (which are renamed under some other names in different countries) are working.

3.12 Rethinking

Now-a-days, electronic gadgets are making life easier and becoming part and parcel of our daily lives. This large-scale dependence on EEEs is also responsible for adding a huge volume of waste to our existing waste stream, which consequently impacts our environmental safety and harmony. Rethinking does not involve any modern technology to manage e-waste but gives underlined ideas to use and manage our electronic gadgets in a more sensible manner. Rethinking emphasis on the use of EEEs up to their EoL, considering repairing or swapping before switching to a new product or item, and also planning for recycling or recovering rather than dumping as trash. Rethinking influences the e-waste generation, collection, and disposal by linking up producers, customers, and parties involved in e-waste processing through the following three main categories:

1. Customer variables: include age, gender, standard of living, lifestyle, social awareness, and perception of customers from all categories of life.
2. Product variables: include product features, price, warranty period, and recyclable or recovery options to attract people.
3. TBS or EPR or ECR variables: this includes provision for repairing or collecting the device and promotions or incentives to encourage sustainable e-waste management.

3.13 Conclusion

The chapter presents a global e-waste generation, collection, and treatment process scenario. The collection and treatment activities of e-waste need to be upgraded to cope with the generation pattern of various forms of e-waste, and at the same time, sustainable e-waste management strategies should be introduced.

To ensure proper e-waste management, the collection system for e-waste should be expanded and developed in both urban and rural areas, and proper initiatives should be taken to make people aware of the recycling of it e-waste rather than treating e-waste as general waste or municipal waste. This chapter also sheds light on the composition of e-waste and analyzes its relative impact in terms of economic and environmental issues. As EEEs have precious materials in their composition that need to be recovered; on the other hand, those EEEs also possess dangerous and toxic elements that need to be disposed of safely. Thus, characterization and segregation of e-waste should be done with expertise and following adequate safety measures. This chapter presents a comprehensive discussion on the contribution of formal and informal sectors to managing e-waste and reveals that a combined approach from government authorities along with manufacturers, and consumers plays a crucial role in stimulating the process of collection and treatment of e-waste. In summary, strengthening existing rules and legislation to handle e-waste, establishing adequate infrastructure to ease the collection and treatment of e-waste through special campaigns, providing special bins or drop-in centers, setting up certified treatment facilities, training up staff to deal with e-waste and emphasizing on recycle–reduce–recover to facilitate circular economy are the key components to face the potential challenges with e-waste.

References

Akram, R., Fahad, S., Hashmi, M.Z. et al. (2019). Trends of electronic waste pollution and its impact on the global environment and ecosystem. *Environmental Science and Pollution Research* 26 (17): 16923–16938.

Ari, V. (2016). A review of technology of metal recovery from electronic waste. E-Waste in transition—From pollution to resource.

Awasthi, A.K., Cucchiella, F., D'Adamo, I. et al. (2018). Modelling the correlations of e-waste quantity with economic increase. *Science of The Total Environment* 613–614: 46–53. https://doi.org/https://doi.org/10.1016/j.scitotenv.2017.08.288.

Baldé, C., Forti, V., Gray, V., Kuehr, R., & Stegmann, P. J. I. E. V. (2017). The global E-waste monitor–2017, United Nations University (UNU), international telecommunication union (ITU) & international solid waste association (ISWA), Bonn/Geneva/Vienna. 978–992.

Barker, K. (2020). Canada-wide initiative launched to improve e-waste reuse and recycling. Retrieved March, 2022, from https://www.recyclingproductnews.com/article/34577/canada-wide-initiative-launched-to-improve-e-waste-reuse-and-recycling

Bhutta, M.K.S., Omar, A., and Yang, X. (2011). Electronic waste: a growing concern in today's environment. *Economics Research International* 1: 34–44.

Bontinck, P., Bricout, J., Grant, T., & Legoe, G. (2021E-Product Stewardship in Australia: Evidence Report. A report prepared for the Commonwealth Department of Agriculture, Water and the Environment by Iceni Group and Lifecycles, Sydney, Australia.

Borthakur, A. and Govind, M. (2017). Emerging trends in consumers' E-waste disposal behaviour and awareness: a worldwide overview with special focus on India. *Resources, Conservation and Recycling* 117: 102–113. 10.1016/j.resconrec .2016.11.011.

Borthakur, A. J. I. J. Management, W. (2016). Policy implications of e-waste in India: a review 17(3–4), 301–317.

Borthakur, A. and Sinha, K. (2013). Electronic waste management in India: a stakeholder's perspective. *Electronic Green Journal* 1 (36): 1–21.

Borthakur, A., Govind, M., and Singh, P. (2019). Chapter 2—inventorization of E-waste and its disposal practices with benchmarks for depollution: the global scenario. In: *Electronic Waste Management and Treatment Technology* (ed. M.N.V. Prasad and M. Vithanage), 35–52. Butterworth-Heinemann 10.1016/B978-0-12-816190-6.00002-9.

Chatterjee, S. and Kumar, K. (2009). Effective electronic waste management and recycling process involving formal and non-formal sectors. *International Journal of Physical Sciences* 4 (13): 893–905.

Chi, X., Wang, M.Y., and Reuter, M.A.J.J. (2014). E-waste collection channels and household recycling behaviors in Taizhou of China. *Journal of Cleaner Prodution* 80: 87–95.

Cucchiella, F., D'Adamo, I., Koh, S.L. et al. (2015). Recycling of WEEEs: an economic assessment of present and future e-waste streams. *Renewable and Sustainable Energy Reviews* 51: 263–272.

Daum, K., Stoler, J., and Grant, R.J.J.I. (2017). Toward a more sustainable trajectory for e-waste policy: a review of a decade of e-waste research in Accra. *Ghana* 14 (2): 135.

Dopp, E. and Rettenmeier, A. (2013). Tin toxicity. In: *Encyclopedia of Metalloproteins* (ed. Krestinger et al.), 2235–2239. Amsterdam: Elsevier.

Dutta, D., Kumari, A., Panda, R. et al. (2018). Close loop separation process for the recovery of Co, Cu, Mn, Fe and Li from spent lithium-ion batteries. *Separation and Purification Technology* 200: 327–334.

ESCAP, U (2021). Toward sustainable e-waste management in Asia and the Pacific.

Esteban-Vasallo, M.D., Aragonés, N., Pollan, M. et al. (2013). Human placenta and markers of heavy metals exposure. In: *Respond. Environmental Health Perspectives*, vol. 121(1) (ed. Esteban-Vasallo et al.), a10–a11.

Forti, V., Balde, C. P., Kuehr, R., & Bel, G. (2020). The Global E-waste Monitor 2020: Quantities, flows and the circular economy potential.

Garlapati, V.K. (2016). E-waste in India and developed countries: management, recycling, business and biotechnological initiatives. *Renewable and Sustainable Energy Reviews* 54: 874–881. 10.1016/j.rser.2015.10.106.

GESP. The Global E-waste Statistics Partnership https://globalewaste.org/country-sheets

Golev, A., Schmeda-Lopez, D.R., Smart, S.K. et al. (2016). Where next on e-waste in Australia? *Waste Management* 58: 348–358. 10.1016/j.wasman.2016.09.025.

Heacock, M., Kelly, C.B., Asante, K.A. et al. (2016). E-waste and harm to vulnerable populations: a growing global problem. *Environmental Health Perspectives* 124 (5): 550–555.

Herat, S. (2008). Recycling of cathode ray tubes (CRTs) in electronic waste. *Clean – Soil, Air, Water* 36 (1): 19–24. https://doi.org/10.1002/clen.200700082.

Ilankoon, I., Ghorbani, Y., Chong, M.N. et al. (2018). E-waste in the international context—a review of trade flows, regulations, hazards, waste management strategies and technologies for value recovery. *Waste Management* 82: 258–275.

Islam, M.K., Khatun, M.S., Arefin, M.A. et al. (2021). Waste to energy: an experimental study of utilizing the agricultural residue, MSW, and e-waste available in Bangladesh for pyrolysis conversion. *Heliyon* 7 (12): e08530.

Jaibee, S., Abd Rahim, A.K., Mohamad, F. et al. (2015). Review on current status of waste electric and electronic product in Malaysia. *Applied Mechanics and Materials.*

Jang, Y.C. (2010). Waste electrical and electronic equipment (WEEE) management in Korea: generation, collection, and recycling systems. *Journal of Material Cycles and Waste Management* 12: 283–294.

Jinghua, X., Xueqing, F., Jun, C., and Shuyou, Z. (2020). Precision forward design for 3D printing using kinematic sensitivity via Jacobian matrix considering uncertainty. *The International Journal of Advanced Manufacturing Technology* 110 (11): 3257–3271.

Kang, H.-Y. and Schoenung, J.M. (2005). Electronic waste recycling: a review of US infrastructure and technology options. *Resources, Conservation and Recycling* 45 (4): 368–400.

Kaza, S., Yao, L., Bhada-Tata, P., and Van Woerden, F. (2018). *What a Waste 2.0: a Global Snapshot of Solid Waste Management to 2050*. The World Bank.

Kumar, A. and Holuszko, M.J.R. (2016). Electronic waste and existing processing routes: a Canadian perspective. *Resources* 5 (4): 35.

Kumar, S. and Rawat, S.J.G.I.Q. (2018). Future e-Waste: standardisation for reliable assessment. *Government Information Quarterly* 35 (4): S33–S42.

Kumar, A., Holuszko, M., and Espinosa, D.C.R. (2017). E-waste: an overview on generation, collection, legislation and recycling practices. *Resources, Conservation and Recycling* 122: 32–42. 10.1016/j.resconrec.2017.01.018.

Li, J., Zeng, X., Chen, M. et al. (2015). "Control-Alt-Delete": rebooting solutions for the e-waste problem. *Environmental Science Technology* 49 (12): 7095–7108.

Liebmann, A. (2015). ICT Waste Handling: Regional and Global End-of-Life Treatment Scenarios for ICT Equipment.

Lifecycles, I. G. a (2021). E-product stewardship in Australia EXECUTIVE REPORT. www.awe.gov.au

Lucier, C.A. and Gareau, B.J. (2019). Electronic waste recycling and disposal: an overview. *Assessment and Management of Radioactive and Electronic Wastes* 1–12.

Maes, T. and Preston-Whyte, F. (2022). E-waste it wisely: lessons from Africa. *SN Applied Sciences* 4 (3): 1–12.

Manish, M. A., & Chakraborty, P. (2019). E-Waste Management in India: Challenges and Opportunities. Retrieved February, 2022, from https://www.teriin.org/article/e-waste-management-india-challenges-and-opportunities#:~:text=E%2Dwaste%20recyclers%20use%20processes,managed%20by%20an%20unorganized%20sector.

Masud, M.H., Akram, W., Ahmed, A. et al. (2019). Towards the effective E-waste management in Bangladesh: a review. *Sustainability* 26 (2): 1250–1276.

Masud, M.H., Mourshed, M., Mahjabeen, M. et al. (2022). Global electronic waste management: current status and way forward. *Paradigm Shift in E-waste Management* 9–47.

Matsukami, H., Tue, N.M., Suzuki, G. et al. (2015). Flame retardant emission from e-waste recycling operation in northern Vietnam: environmental occurrence of emerging organophosphorus esters used as alternatives for PBDEs. *Science of The Total Environment* 514: 492–499.

Mmereki, D., Li, B., Baldwin, A., and Hong, L. (2016). The generation, composition, collection, treatment and disposal system, and impact of E-waste. *E-waste in Transition-from Pollution to Resource* 65–93.

Mourshed, M., Masud, M.H., Rashid, F., and Joardder, M.U.H. (2017). Towards the effective plastic waste management in Bangladesh: a review. *Environmental Science and Pollution Research* 24 (35): 27021–27046. https://doi.org/10.1007/s11356-017-0429-9.

Needhidasan, S., Samuel, M., and Chidambaram, R. (2014). Electronic waste–an emerging threat to the environment of urban India. *Journal of Environmental Health Science and Engineering* 12 (1): 1–9.

Ogushi, Y. and Kandlikar, M. (2007). *Assessing Extended Producer Responsibility Laws in Japan*, 4502–4508.

Ongondo, F. and Williams, I. (2011). Mobile phone collection, reuse and recycling in the UK. *Waste Management* 31 (6): 1307–1315.

Ongondo, F.O., Williams, I.D., and Cherrett, T.J. (2011). How are WEEE doing? A global review of the management of electrical and electronic wastes. *Waste Management* 31 (4): 714–730.

Palmeira, V. N., Guarda, G. F., & Kitajima, L. F. W. J. D. (2018). Illegal international trade of e-waste-Europe. 1(1), 48.

Partnership, T. G. E.-w. S (2019a). Pakistan-2019-E-waste statistics https://globalewaste.org/statistics/country/pakistan/2019

Partnership, T. G. E.-w. S (2019b). Sri-lanka-2019-E-waste statistics https://globalewaste.org/statistics/country/sri-lanka/2019

Pathak, P. and Srivastava, R.R. (2019). Environmental management of e-waste. In: *Electronic Waste Management and Treatment Technology*, 103–132. Elsevier.

Pathak, P., Srivastava, R.R., and Ojasvi. (2017). Assessment of legislation and practices for the sustainable management of waste electrical and electronic equipment in India. *Renewable and Sustainable Energy Reviews* 78: 220–232. https://doi.org/https://doi.org/10.1016/j.rser.2017.04.062.

Perkins, D.N., Drisse, M.-N.B., Nxele, T., and Sly, P.D. (2014). E-waste: a global hazard. *Annals of Global Health* 80 (4): 286–295.

Ranasinghe, W.W. and Athapattu, B.C. (2020). Challenges in E-waste management in Sri Lanka. In: *Handbook of Electronic Waste Management*, 283–322. Elsevier.

Rautela, R., Arya, S., Vishwakarma, S. et al. (2021). E-waste management and its effects on the environment and human health. *Science of The Total Environment* 773: 145623.

Read, M., Gregory, M.K., and Phillips, P.S. (2009). An evaluation of four key methods for monitoring household waste prevention campaigns in the UK. *Resources, Conservation and Recycling* 54 (1): 9–20.

Record, T. I. (2019). Kalasin's burning e-waste dump needs a legal fix. Retrieved March, 2022, from https://theisaanrecord.co/2019/09/23/kalasin-e-waste-burning-legal-fix

Rucevska, I. N. C., Isarin, N., Yang, W., Liu, N., Yu, K., Sandnæs, S., Olley, K., McCann, H., Devia, L., Bisschop, L., Soesilo, D., Schoolmeester, T., Henriksen, R., Nilsen, R. (2015). Waste crime—waste risks: gaps in meeting the global waste challenge A UNEP Rapid Response Assessment, United Nations Environment Programme and GRID-Arendal, Nairobi and Arendal, Issue. www.grida.no ISBN: 978-82-7701-148-6.

Sahajwalla, V. and Gaikwad, V. (2018). The present and future of e-waste plastics recycling. *Current Opinion in Green and Sustainable Chemistry* 13: 102–107.

Salhofer, S. (2017). E-waste collection and treatment options: a comparison of approaches in Europe, China and Vietnam. In: *Source Separation and Recycling*, 227–227. Springer.

Shaikh, S. (2021). A critical analysis of e-waste management and recycling in Pakistan: a life cycle assessment Victoria University.

Sheikh, A., Correa, P., da Silva, E.L. et al. (2013). Energy conversion using Pd-based catalysts in direct ethanol fuel cell. *Renewable Energy and Power Quality Journal* 1 (11): 342–345.

Singh, N., Duan, H., and Tang, Y. (2020). Toxicity evaluation of E-waste plastics and potential repercussions for human health. *Environment International* 137: 105559.

Soetrisno, F.N. and Delgado-Saborit, J.M. (2020). Chronic exposure to heavy metals from informal e-waste recycling plants and children's attention, executive function and academic performance. *Science of the Total Environment* 717: 137099.

Song, Q. and Li, J. (2014). A systematic review of the human body burden of e-waste exposure in China. *Environment International* 68: 82–93.

Srivastava, R.R. and Pathak, P. (2020). Policy issues for efficient management of E-waste in developing countries. In: *Handbook of Electronic Waste Management*, 81–99. Elsevier.

Starr, R. (2019). Top E-Waste Recycling Tips for Small Business. Retrieved March, 2022, from https://smallbiztrends.com/2019/04/e-waste-recycling.html

Statistica. Electronic waste worldwide https://www.statista.com/study/56028/global-electronic-waste

Step Initiative %J Bonn, G. S (2014). Solving the E-Waste Problem (Step) White Paper: One Global Definition of E-waste. 3576.

Sthiannopkao, S., Wong, M.H.J.S., and o. t. T. E. (2013). Handling e-waste in developed and developing countries. *Initiatives, Practices, and Consequences* 463: 1147–1153.

Strike, T. N. (2021). Refurbish E-waste to protect environment. https://thenewsstrike.com/refurbish-e-waste-to-protect-environment

Sugimura, Y. and Murakami, S. (2016). Problems in Japan's governance system related to end-of-life electrical and electronic equipment trade. *Resources, Conservation and Recycling* 112: 93–106. https://doi.org/https://doi.org/10.1016/j.resconrec.2016.04.009.

Tansel, B.J.E. (2017). From electronic consumer products to e-wastes: global outlook, waste quantities, recycling challenges. *Environment International* 98: 35–45.

Thompson, S., & Oh, S. (2006). Do sustainable computers result from design for environment and extended producer responsibility? Analyzing e-waste programs in Europe and Canada. Proceedings of the International Solid Waste Association's 2006 Annual Congress, Copenhagen, Denmark

Times, S. List of continents by GDP https://statisticstimes.com/economy/continents-by-gdp.php

Toyasaki, F., Boyacı, T., and Verter, V. (2011). An analysis of monopolistic and competitive take-back schemes for WEEE recycling. *Production and Operations Management* 20 (6): 805–823.

Tsydenova, O. and Bengtsson, M. (2011). Chemical hazards associated with treatment of waste electrical and electronic equipment. *Waste Management* 31 (1): 45–58.

Vats, M. and Singh, S. (2014). E-waste characteristic and its disposal. *International Journal of ecological science and environmental engineering* 1 (2): 49–61.

Veit, R. J. S. E. E.-W.A. Shanghai. (2014). Evolution of producer responsibility and product stewardship.

WA, L. R. (2021). E-Waste Recycling. https://lightrecyclewa.org/e-waste-recycling

Wagner, T.P. (2009). Shared responsibility for managing electronic waste: a case study of Maine, USA. *Waste Management* 29 (12): 3014–3021.

Wang, Z., Zhang, B., and Guan, D.J.N. (2016). Take responsibility for electronic-waste disposal. *Nature* 536 (7614): 23–25.

Widmer, R., Oswald-Krapf, H., Sinha-Khetriwal, D. et al. (2005). Global perspectives on e-waste. *Environmental Impact Assessment Review* 25 (5): 436–458.

Williams, I. (2015). Forty years of the waste hierarchy. *Waste management (New York, NY)* 40: 1–2.

Wu, Q., Leung, J.Y., Geng, X. et al. (2015). Heavy metal contamination of soil and water in the vicinity of an abandoned e-waste recycling site: implications for dissemination of heavy metals. *Science of The Total Environment* 506: 217–225.

Yu, J., Williams, E., Ju, M., and Shao, C. (2010). Managing e-waste in China: policies, pilot projects and alternative approaches. *Resources, Conservation and Recycling* 54 (11): 991–999.

Zeng, X., Xu, X., Boezen, H.M., and Huo, X.J.C. (2016). Children with health impairments by heavy metals in an e-waste recycling area. *Chemosphere* 148: 408–415.

Zeng, X., Duan, H., Wang, F., and Li, J. (2017). Examining environmental management of e-waste: China's experience and lessons. *Renewable and Sustainable Energy Reviews* 72: 1076–1082.

Zhang, J., Wang, H., Yuan, X. et al. (2019). Tailored indium sulfide-based materials for solar-energy conversion and utilization. *Journal of Photochemistry and Photobiology C: Photochemistry Reviews* 38: 1–26.

4

Toxicity Characterization and Environmental Impact of E-Waste Processing

Shahriar Shams, Pg Rusydina Idris, and Ismawi Yusof

Civil Engineering Programme Area, Universiti Teknologi Brunei, Jalan Tungku Link, Gadong, BE 1410, Brunei Darussalam

Abstract

Advancement of technologies, industrialization, and increased standard of living have led to significant growth of electrical and electronic products in the last decade. Electronic product sales have grown in developing countries due to their affordability, and the disposal of electronic waste (e-waste) affects the ecosystem directly and indirectly. E-metals, waste plastics, polymers, ceramics, and toxic compounds threaten soil, water, and air quality. The situation is further aggravated by inappropriate dismantling, and release of toxic and hazardous compounds such as lead, tin, antimony, mercury, chromium, copper, nickel, and cadmium, acids, and volatile compounds during various processing techniques. The hazardous and toxic compounds from in situ treatment processes like hydrometallurgy or pyrometallurgy exposure can release highly toxic emissions. This chapter highlights the state of the environment with details of the contamination path for hazardous and toxic compounds getting into soil, water, and air with possible impacts on vegetation, ecosystem, and human health.

Keywords *e-waste; hazardous; toxic; environment; health; ecosystem; disposal*

4.1 Introduction

The use of electrical and electronic equipment (EEE) has increased significantly because of affordability and has become part and parcel of everyday life, starting with mobile phones, laptops, PCs, and other computer-related products or devices. Advancement of technologies, industrialization, and increased standard of living have led to significant growth of electrical and electronic products in the last decade. The products' short lives with new updates and models are

Management of Electronic Waste: Resource Recovery, Technology and Regulation, First Edition.
Edited by Anshu Priya.

making electrical and electronic products more obsolete at the end of their life cycle. The e-waste is growing at a rate of 3.2% per year, and it was estimated that 52.2 million tonnes of e-waste will be produced globally in 2021 (Abdelbasir et al. 2018; Alghazo et al. 2019; Baldé et al. 2017), and 70–80% of e-waste has been exported for recycling and disposal to developing countries in Asia and Africa (Shamim and Murshea 2015). The most challenging part of e-waste is the accumulation of e-waste and the lack of recycling facilities. Many countries, both developed and developing, are concerned about the increased volume of e-waste produced each year. Unfortunately, the majority of e-waste ends up in dumpsites or in landfills due to a lack of recycling options and proper collection strategies. This improper disposal of e-waste not only has an environmental impact but also poses a serious threat to the health and well-being of people. Some of the impacts of e-waste are short-term, while most can have secondary or tertiary long-term impacts, which include groundwater contamination from toxic hazardous substances leached from the improper dismantling process and poor handling and management of e-waste. The aim of this chapter is to understand the pathways, by which the e-waste possessing toxic elements reaches soil, water, and air during its dismantling process and leads to health hazards.

The e-waste consists of three phases of operations: (i) decontamination; (ii) dismantling; and (iii) segregation, as shown in Figure 4.1. Decontamination is done to make the product nonhazardous by removing liquids and gases. However, during the extraction of liquids and gases, accidents may happen and contaminate the surrounding environment. Dismantling is the second phase where the product is broken down to extract precious or valuable and recyclable materials, either

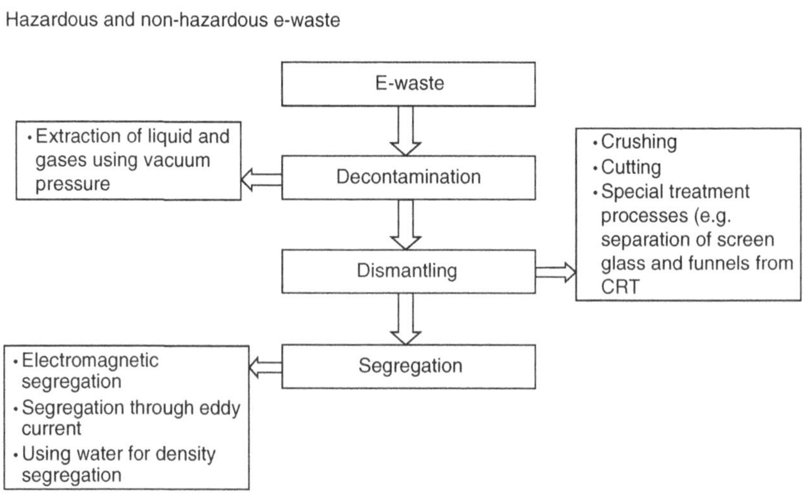

Figure 4.1 E-waste processing.

mechanically or manually. In particular, manual methods of dismantling pose a considerable risk if adequate safety measures are not taken. The third phase of the e-waste process includes the segregation of hazardous and nonhazardous e-waste.

Waste-printed circuit boards (PCB) serve as the principal component of electronic gadgets, which contain a variety of chemicals. Some outdated electronic equipment can include more than 1000 distinct chemicals with a combination of metals, polymers, and ceramics. This diverse composition may contain poisonous and dangerous substances. Currently, no standardized, entrenched technique to handle a broad range of e-waste is available because of the inherent complexity of these gadgets. The recovery of precious metals is the major focus of current processing methods since it is economically advantageous and the materials are found in substantial amounts in e-waste in comparison to the concentrations of equivalent ores. Crude burning, physical dismantling, mechanical segregation and treatment, pyrometallurgy, hydrometallurgy, and other techniques are some of the ones employed to accomplish these aims. Wherever the functioning of these processes is insufficient or dangerous, it may result in the emission of extremely toxic pollutants that can contaminate the air as well as the formation of new hazardous substances. Currently, a sizeable amount of e-waste is reprocessed utilizing pyrometallurgy or hydrometallurgy. To increase the recovery of materials, disintegration and mechanical segregation are typically done before the metallurgical procedures. Materials from outdated EEE can also be recovered through mechanical processing alone. Hammering, crushing, dismantling, electrostatics, gravity, density-based, and magnetic segregation are several methods used to segregate metals and nonmetals (Cui and Forssberg 2003). In order to recover metals, hydrometallurgical recycling methods typically involve the following steps: disbanding of the material, cleansing/refinement, electrowinning, chemical cutback, or crystallization. Methods including solidification, cementation, and extraction of solvent have been used in the concentration step. The pyrometallurgical method of recovering metals from e-waste involves heating the material either alone or in combination with other materials to promote the production of slag and concentrate and purify the desired metals. Smelting, converting, refining, and electrorefining are the processes employed. Bio-metallurgy, which makes use of microorganisms to enhance the leaching of metals, is one of the most recent methods being utilized to recycle e-waste. This strategy has, however, only been applied on a research scale thus far.

4.2 Impact of E-Waste

The impact of e-waste disposal and extraction processes can contaminate soil, water, and air, either directly or indirectly. In order to characterize the hazardous compounds contained in electronic equipment, a thorough examination of the combination of various forms of e-waste is required for various types of home appliances, heavy and light IT equipment, and accessories.

4.2.1 Direct Impact

The direct impact of e-waste is usually present as a primary constituent that possesses a hazardous and/or a toxic component and is released directly into water, soil, and air during dismantling or disposal of e-waste. Some of these hazardous constituents can be metallic or organic pollutants. Metallic constituents like lead, which is available in cathode ray tubes (CRTs), fluorescent tubes, solder in PCBs, and liquid crystal displays (LCDs), can pollute soil with toxicity. Tin and Antimony used in PCBs are toxic and volatile (Williams 2010).

The majority of nickel, which exists in Ni–Cd batteries as hydroxide and PCBs at trace levels, is found in alloys with numerous metals like copper, chromium, and cadmium. In tiny homes, electrical and electronic products, such as cordless phones, mobile phones, music systems, and calculators, use different shapes of Ni–Cd batteries, which can directly cause soil and water pollution. Persistent organic pollutants (POPs), such as Polychlorinated Biphenyls (PCBs), are poisonous, nondegradable, and bioaccumulative in organisms. PCBs exist in capacitors and transformers in the form of coolants and dielectric fluids. These substances can be transferred directly by air, water, and migrating animals. These may build up in animal, terrestrial, and aquatic environments in addition to human bodies.

Particles of polybrominated biphenyls (PBBs) are mostly airborne but can also be ingested by the soil and sediments during combustion processes. Thermoplastics typically use the flame-retardant ingredient hexabromocyclododecane (HBCD), which can easily volatilize and leak since it is not chemically bound to the polymer. Chlorofluorocarbons (CFCs) and hydrochlorofluorocarbons (HCFCs), which are mostly found in refrigerators, air conditioners, and freezers, are dangerous refrigerant gases (McCulloch et al. 2003). These are mostly emitted into the air and blend well since they are gaseous at ambient temperature and have poor solubility in water. The most dangerous substances emitted are CFC-12, HCFC-22, and HFC-134a because they increase the greenhouse effect on a worldwide scale.

4.2.2 Indirect Impact

The leftovers or residues produced during e-waste handling and processing for the extraction of important materials are known as secondary pollutants. These are often created indirectly during the pyrometallurgical or hydrometallurgical processes used to remediate e-waste.

E-waste handling, physical disassembly, or shredding in processing workshops can produce a lot of dust (Lau et al. 2014). Even loading and unloading equipment has the potential to release tiny dust particles that may contain ceramics, polymers, and even heavy metals. High quantities of Ni, Cd, Zn, Pb, Cr, Cu, and Hg have been shown to leak during shredding and dismantling operations. These metals are released through improper processing techniques involving high temperatures due to crude burning, de-soldering, and metal melting, in addition to preprocessing operations.

POPs include polychlorinated dibenzo-p-dioxins and dibenzofurans (PCDD/Fs), which are extremely durable and less degradable, may move great distances, and build up in the cells of living animals (Li et al. 2008). Byproducts of manufacturing and combustion processes include PCDD/Fs. Large quantities of these compounds are released during the crude burning of e-waste (Leung et al. 2007).

4.3 Environmental Impact

Due to the toxic materials and substances present in e-waste, it poses a significant challenge to the environment (Ashfaq and Khatoon 2014). Heavy metals as well as flame retardants can leak straight from the e-waste and contaminate both the soil and groundwater if incorrectly disposed of in landfills or illegally dumped. Common e-waste materials that are present in large quantities are carbon, lead, tin, copper, silicon, beryllium, iron, and aluminum. Sometimes, cadmium, mercury, and thallium are also present in small quantities. These substances are potentially hazardous. The most common substances present in e-waste that have an impact on the environment are shown in Table 4.1.

According to Devi et al. (2004), when PVC cable insulation is burned to remove copper from the wire, it emits very dangerous dioxins and furans. E-waste is known to cause irreversible impacts on both humans and the environment. E-waste releases inorganic and organic materials when it is disposed of in a non-environmentally friendly manner. Even abandoned e-waste recycling facilities might continue to be a source of POPs, plastics, and trace metals, endangering the environment and people's health. Three decades after the banning of recycling, polychlorinated biphenyls (PCBs) and a number of heavy metals are still present at reasonably high levels in e-waste recycling zones (Wu et al. 2019).

4.3.1 Impact on Soil

Heavy metals and organic substances frequently pollute the soil in the e-waste regions. Rice grown on contaminated soil affected by e-waste has impacted the food chain (Zhang et al. 2012). Any electrical equipment, if disposed of improperly after its end of life, will contaminate the soil with heavy metals such as Ni, Cd, Pb, Zn, Cu, Cr, and Hg (Dharini et al. 2017; Khan et al. 2021). Particularly, POPs, which have toxic effects and are released as a result of dismantling, repairing, and burning at the recycling center, can reduce the microbial community, diversity of

Table 4.1 Environmental impact of common substances present in e-waste.

Substances	Process	Impact
Mercury	Detected in fluorescent tubes, tilt switches, and flat-screen displays, and released into the environment when ECBs and switches were broken.	Methylated mercury, which bioaccumulates in living things and concentrates through human food and aquatic life, is created when inorganic mercury blows out in waterbodies.
Lead	Breaking and removing yoke (CRT, Monitors, ATM, Video Cameras, etc.), then dumping	Harmful phosphor is released into the environment and leaches into groundwater, affecting aquatic life.
Cadmium	Computer chip de-soldering and removal, followed by acid baths and crude burning to eliminate any remaining metals.	Air emissions and soil and water discharge
PAHs	Copper from computer lines is removed via open burning and stripping.	Emitted into the soil, water, and air.
Hydrocarbons	Shredding and low-temperature melting of plastics from printers, keyboards, monitors, etc. to be reused	Air emissions and discharge into the soil
Sulfur	Found in lead acid batteries	Produces sulfuric acid when released into the environment

their species, basic processes, and enzyme activities with significant changes to soil ecosystem function and impacting "soil-crop-food pathway" (Salam and Varma 2019; Kaushik 2018; Colombo et al. 2016). Soils tested at an abandoned e-waste recycling site in Bangalore had up to 2850 mg/kg Pb, 39 mg/kg Cd, 180 mg/kg Sb, 957 mg/kg Sn, 49 mg/kg Hg, and 2.7 mg/kg Bi, which is one-hundredfold larger compared to the other places close to the same area (Ha et al. 2009). In a study conducted to determine the concentrations of metals close to a discarded e-waste recycling facility, Wu et al. (2019) discovered concentrations of heavy metals in the burnt site's field were significantly higher than those in samples from the rice field and stream, which were either significantly polluted with Ag, Sb, and Sn at the crude burnt site or greatly contaminated with Ni, Cd, Zn, Pb, Cu, and Ba.

4.3.2 Impacts on Water

Water is contaminated by acid washing of e-waste and metals leaching from discarded e-waste recycling facilities. The "ecological balance" of the aquatic environment may be impacted by the heavy metal poisoning of a river, which will also reduce the diversity of the aquatic system. Well, waters have been polluted by heavy metals leaked from e-waste disposal facilities in Nigeria (Alabi and Bakare 2014). The samples collected from the surface water and sediment from the Lianjiang River indicated water pollution due to a nearby e-waste recycling facility (Xu et al. 2015). The electronic devices, dead cells, and batteries disposed of improperly with other household waste contain lead and pollute underground water. When CRTs from old computer monitors, video cameras, and televisions are disassembled to remove the yoke, the water becomes contaminated. One cell phone battery has enough cadmium to contaminate 600 m^3 of water (Trick 2002).

4.3.3 Impact on Air

Improper recycling sites have high mean concentrations of particulate matter (PM_{10}), POPs, and heavy metals, which pollute the surrounding air (Qin et al. 2019). The air is contaminated by hydrocarbons from the open-air burning of copper wires to get the precious copper inside the lines and the removal of gold plating from computer chips (Kaushik 2018). E-waste can pollute the air when it is transferred to countries with poorly regulated recycling systems, which is frequent in developing countries. E-waste is frequently disassembled and shredded in underdeveloped nations, releasing dust or big particles into the local environment. Low-value e-waste is frequently burnt, and it typically contains a lot of plastic. Uncontrolled or improperly controlled burning frequently occurs at lower temperatures and emits pollutants like dioxins that are harmful to both humans and animals. Burning also releases fine particles ($PM_{2.5}$ and PM_{10}), which can travel hundreds of kilometers and increase the risk for a variety of chronic illnesses and malignancies, have detrimental effects on lung health, and get past the body's defenses. Finally, when recycling is not adequately controlled, higher-value elements like gold and silver are frequently removed from compacted electronics and e-waste using acids, de-soldering, and other chemicals and procedures that emit harmful gases to nearby areas. Besides, broken and burnt CRT releases poisonous gases into the atmosphere (Ramachandra and Saira 2004). By 2040, it was predicted that the combined manufacture of plastics from e-waste and the burning of plastic garbage would result in around 400 million tonnes of CO_2 emissions worldwide (Raudaskoski et al. 2019).

4.4 Health Impact

The main concern about e-waste is the deleterious effect of the materials on humans. This can have either a direct or indirect effect. One of the materials with a significant impact on human health is lead. Up to 6.3 percent of an average PC is made up of it. Every computer has an average of 2 kg of lead, including the display, which makes up about 20% of the total weight. In the monitor glass, lead serves as a radiation barrier. Lead is present in significant amounts in the tin-lead solder used to join computer chips to PCBs. Cd, Cr, Cr, Hg, Sb, and Br flame retardants are other e-waste components that raise environmental health issues. E-waste is often processed utilizing backyard recycling in poor nations. Environmental and health studies conducted close to recycling operations have revealed higher PCDD/Fs ratios. Chlorinated dioxins and furans are known to have hazardous effects on the nervous system, skin, liver, gastrointestinal tract, and immune system in both people and animals. Some of the components and their health impacts are discussed in Table 4.2.

Furthermore, substantial levels of PBDEs were discovered in the blood samples of employees working at the dismantling plant. When IT and computer accessories are recycled, harmful flame retardants are released into the atmosphere. One type of flame retardant was present in workers at deconstruction sites at levels 70 times higher than those in hospital cleaners. Due to their frequent exposure to flame retardants in the air, full-time computer screen workers had somewhat higher blood levels of flame retardants than cleaners did. It was found that the fine dust particles extracted from the Copper Smelter Recycling facility can easily penetrate the human respiratory system and make someone ill. Higher blood levels of PBDEs were subsequently observed in human tissues (including the placenta, serum, breast milk, and hair) in the e-waste recycling area with increased reactive oxygen species (ROS) and interleukin-8 (IL-8), and both may cause DNA damage that may result in cardiovascular issues or even cancer.

4.5 Ecological Impact

The impact of electronic waste leads to an alarming ecological problem. Hazardous and toxic materials from e-waste can easily infiltrate soil and waterbody through leaching, impacting soil microorganisms and aquatic species in a variety of ways, including their growth and reproduction. A study done by Liu et al. (2015) observed indigenous microbial assemblages on e-waste-affected soils obtained from Guiyu, China. Proteobacteria, Acidobacteria, Bacteroidetes, and Firmicutes predominated in the e-waste-affected communities, which showed a substantial difference in the microbial structural makeup between the polluted and uncontaminated soils. The quantity of fungus is inversely proportionate to the amount of contaminated e-waste, indicating a one-to-one relationship between pollution and organisms (Sanusi 2015). Huang et al. (2013) pointed out that while direct impacts are seen and toxins have the capacity to negatively affect many aquatic creatures' characteristics, composition, metabolism, and growth, indirect impacts are created on food chain organisms and through ecological stress. Fish and other aquatic species have been observed to be exposed to a variety of hazardous materials. Carp fish in China's Nanyang River were shown to bioaccumulate PBDEs (Polybrominated diphenyl ethers) up to 766 ng/g in fresh

Table 4.2 Health impact of common substances found in e-waste.

Elements	Occurrence	Health impact
Lead	Glass, panels, gaskets, solders, and monitors.	Humans' central and peripheral nerve systems, circulation of blood, kidneys, and reproductive systems are all harmed by it.
Cadmium	Infrared detectors, semiconductors, resistors, etc.	Very poisonous that can have a permanent and bioaccumulative adverse impact on human health.
Chromium	Hexavalent chromium is utilized to harden steel housings and prevent untreated and galvanized steel plates from corrosion.	It quickly penetrates cell membranes, damages DNA, and also results in asthmatic bronchitis.
Copper	Due to its great conductivity, copper is one of the most often utilized metals in EEE. Copper is recovered by open burning and acid leaching. Mobile phone ecotoxicity hazards are mainly by copper components at 45,818–77,938 PAF m^3/kg (Singh et al. 2019).	Low-temperature combustion raises the possibility of dioxin generation and copper particulate matter emissions, directly contaminating the air and subjecting humans to high levels of oxidative stress and neurological effects (Gaetke and Chow 2003).
Mercury	Used in fluorescent lights, batteries, LCDs, switches, thermostats, and sensors and is exceedingly toxic to people, is susceptible to leaking, particularly from fluorescent lamps, which are more likely to shatter when handled or disposed of inappropriately.	Mercury has the potential to be harmful to the kidneys and the brain. It bioaccumulates and is harmful to the skin and respiratory systems.
Barium	Barium is mainly present in CRTs as barium oxide (12%) and strontium oxide (12%) (Herat 2008) can pollute the air as it produces deadly oxides while mixing with air in an unstable form.	Brief contact with Barium might cause severe health problems.
Beryllium	Usually used in motherboards.	Due to the inhalation of fumes and dust, Beryllium is extremely carcinogenic and mostly responsible for lung cancer. Additionally, it contributes to skin conditions including warts and chronic beryllium illness, or berylliosis.

weight, while e-waste-polluted rivers indicated that PBDE concentrations were up to 16,000 ng/g (Luo et al. 2009). Near an e-waste recycling facility, the top predator water snake had fresh weight averages of 16,512 ng/g PCBs (Polychlorinated biphenyls), highly carcinogenic chemical compounds, which are 16 times higher than the usual value (Wu et al. 2008). The composition of the soil is altered in unpredictable and complex ways by heavy metals (from improper e-waste disposal and incomplete recycling activities), PBDEs (from burning, shredding, and dismantling), dioxins/furans (from incomplete burning), and acidification (from recycling practices that involve leaching). These modifications may be extremely detrimental to soil microorganisms, plants, animals, and other creatures that depend on these plants for existence. In polluted soils, plants frequently have harmed cell structure, changed metabolism, and decreased development. Additionally, e-waste can have a double impact on some plant species by contaminating the soil underneath them as well as by coming into direct contact with the toxins. For instance, lead may coat the surface of leaves, slowing down photosynthesis and harming or killing plants.

4.6 Impact from Processing E-Waste

The hunt for novel processing techniques that might sort and handle a wide variety of outdated electronic waste without losing a lot of valuable metals in an environmentally responsible way is currently ongoing. Some methodologies have previously been used in practice, while others are still in the research phase. Pyrometallurgy, hydrometallurgy, and electrochemistry are a few methods that have been suggested for recovering metal values. However, some of the methods were well known for their impact, and some are still under investigation.

The most typical first stage in processing e-waste is the mechanical process to segregate the material fraction and minimize heavier particles. Before mechanical processing, a number of unit operations are used, such as hand disassembly, hammering, crushing, magnetic segregation, and vibration. There are several methods used when precious metal from the circuit board has to be extracted.

4.6.1 Smelting Method

Due to the high concentration of copper on the IC board, it may be processed alongside regular ore as high-grade copper ore. The resin is burned off, and the fiberglass is transformed into slag during the smelting process. During the refining process, the precious metal is recovered using the conventional technique. In addition to causing air pollution and a loss of noble metals, the technique cannot be used to recover materials with low metal values. Other metals, including aluminum, tin, and lead, cannot be recovered using this method.

With relatively poor yields, this technology uses crude burning, leaching with cyanide, and heating and melting to extract mostly gold, copper, and silver. Waste incineration facilities provide a substantial contribution to cadmium and mercury emissions each year. Additionally, heavier particles that are not emitted into the air are formed into slag and might be disposed of and then re-enter the environment. Therefore, if no mitigating actions are taken, the burning of e-waste will result in increased emissions.

Pyrometallurgy has the advantages of simple operation and high efficiency. However, it requires high energy consumption, and this method will cause brominated flame retardants in PCBs to produce dioxins.

Groundwater contamination from landfills is mostly brought on by Brominated Flame Retardants (BFRs) and heavy metals, while the burning of PCBs can also result in the production of polybrominated dibenzodioxins and dibenzofurans (PBDD/Fs). Another effect would come from open burning, which emits dust and harmful gases that might be readily ingested and disrupt the local food and water sources.

4.6.2 Hydrometallurgical Method

Dissolving a metallic fraction by the lixiviant during dissolution, requires efficient discharge and segregation of metallic from nonmetallic substances. The extraction and purification operations are simplified by the mechanical process for the segregation of metallic substances. The nonmetallic fraction is concentrated as small particles during air separation because the precious metal is present as a thin coating on the surface of other metals. It is considered the best option when the hydrometallurgical method is compared with the pyrometallurgical method due to its low pollution, cost-effectiveness, and simpler application, and it is a widely used technique. However, there are some downsides to this method. Solvent extraction technology is a widely used technique, but its employment has been hampered by the application of excessive amounts of hazardous and organic solvents, processing time, high costs, and undesirable chemical waste discharge. Significant expenditure is often needed for the downstream processing, in addition to the usage of strong acids where bases are mostly present and the treatment of created acidic waste. Despite their effectiveness, several of these systems have high energy requirements and cause new issues with waste management. Hydrometallurgy poses serious environmental problems since it uses a lot of leaching chemicals and releases waste liquid.

4.6.3 Physical Separation Method

In order to maximize the importance of recyclable e-waste with a greater metal concentration, this procedure eliminates the nonreusable material from the IC board. This kind of separation plant often includes a number of physical processing tools, including hammering, grinding, screens, magnetic segregation, and electrical conductivity segregation, among others. A number of metal pieces of varied shapes and compositions are produced using this technique. Economic return and minimum environmental pollution should be considered when using this method. The Physical separator involves high operating costs and energy. Efficient handling and disposal of e-waste have influenced the labor cost and the storage cost. This method also produces smell, dust, and black fumes during the recycling of e-waste, which are considered as health hazards from e-waste.

4.6.4 Scrapping Method

This technique involves melting solder and manually scraping off the installed electrical components so that the clean board and the removed material may be handled separately. Because resin and metallic pieces are processed separately, recycling efficiency is improved.

4.7 Conclusions

The disposal of e-waste at the end of its lifespan has become a major challenge globally. The handling, storage, dismantling, and disposal of e-waste have both direct and indirect impacts on the environment. The e-waste comprising plastics, metals, polymers, and ceramics, along with a number of hazardous and toxic compounds, poses a serious threat to soil, water, and air quality. The situation is further aggravated by inappropriate dismantling and release of toxic, hazardous, and volatile compounds during various processing techniques such as hydrometallurgy or pyrometallurgy. One of the most modern techniques for recycling e-waste is bio-metallurgy, which uses microorganisms to assist with the leaching of metals. To develop recycling methods to extract precious materials in an environmentally friendly way, a thorough understanding of their behavior is required. Additionally, it's critical to avoid the use of risky processing methods and processes that might pollute the environment.

References

Abdelbasir, S.M., Hassan, S.S.M., Kamel, A.H., and El-Nasr, R.S. (2018). Status of electronic waste recycling techniques: a review. *Environmental Science and Pollution Research* 25 (17): 16533–16547.

Alabi, O.A. and Bakare, A.A. (2014). Cytogenotoxic effects and reproductive abnormalities induced by E-waste contaminated underground water in mice. *The Japan Mende Society. Cytologia* 79 (3): 331–340.

Alghazo, J., Ouda, O., Alanezi, F. et al. (2019). Potential of electronic waste recycling in Gulf Cooperation Council states: an environmental and economic analysis. *Environmental Science and Pollution Research* 26 (35): 35610–35619.

Ashfaq, A. and Khatoon, A. (2014). Environmental impacts and assessment of electronic waste management. *International Journal of Current Microbiology and Applied Sciences* 3 (7): 772–779.

Baldé, C.P., Forti, V., Gray, V. et al. (2017). *The Global e-Waste Monitor 2017: Quantities, Flows, and Resources*. Bonn/Geneva/Vienna: United Nations University (UNU), International Telecommunication Union (ITU) & International Solid Waste Association (ISWA).

Colombo, F., Macdonald, C.A., Jeffries, T.C. et al. (2016). Impact of forest management practices on soil bacterial diversity and consequences for soil processes. *Soil Biology and Biochemistry* 94: 200–210.

Cui, J. and Forssberg, E. (2003). Mechanical recycling of waste electric and electronic equipment: a review. *Journal of Hazardous Materials* 2003 (99): 243–263. https://doi.org/10.1016/s0304-3894(03)00061-x.

Devi, B.S., Shobha, S.V., and Kamble, R.K. (2004). E-Waste: the hidden harm of technology revolution. *Journal of Indian Association for Environmental Management* 31: 196–205.

Dharini, K., Cynthia, J.B., Kamalambikai, B. et al. (2017). Hazardous E-waste and its impact on soil structure. *IOP Conference Series: Earth and Environmental Science* 80 (012057).

Gaetke, L.M. and Chow, C.K. (2003). Copper toxicity, oxidative stress, and antioxidant nutrients. *Toxicology* 2003 (189): 147–163. https://doi.org/10.1016/S0300-483X(03)00159-8.

Ha, N.N., Agusa, T., Ramu, K. et al. (2009). Contamination by trace elements at e-waste recycling sites in Bangalore, India. *Chemosphere* 76 (1): 9–15.

Herat, S. (2008, 2008). Recycling of cathode ray tubes (CRTs) in electronic waste. *CLEAN – Soil, Air, Water* 36: 19–24. https://doi.org/10.1002/clen.200700082.

Huang, J., Nkrumah, P.N., Anim, D.O., Mensah, E. 2013. *E-Waste Disposal Effects on the Aquatic Environment*: Accra, Ghana.

Kaushik, V. (2018). E-waste and its management. *International Journal for Research in Applied Science & Engineering Technology* 6 (VI): https://doi.org/10.22214/ijraset.2018.6031.

Khan, J., Kumar, A., Giri, A. et al. (2021). Impact of electronic waste pollutants on underground water. *Groundwater Geochemistry* 265–281: https://doi.org/10.1002/9781119709732.ch13.

Lau, W.K.Y., Liang, P., Man, Y.B., and Chung, S.S. (2014). Wong M.H. Human health risk assessment based on trace metals in suspended air particulates, surface dust, and floor dust from e-waste recycling workshops in Hong Kong, China. *Environmental Science and Pollution Research* 21: 3813–3825. https://doi.org/10.1007/s11356-013-2372-8.

Leung, A., Luksemburg, W.J., Wong, A.S., and Wong, M.H. (2007). Spatial distribution of polybrominated diphenyl ethers and polychlorinated dibenzo-p-dioxins and dibenzofurans in soil and combusted residue at Guiyu, an electronic waste recycling site in Southeast China. *Environmental Science & Technology* 2007 (41): 2730–2737. https://doi.org/10.1021/es0625935.

Li, Y., Jiang, G., Wang, Y. et al. (2008). Concentrations, profiles and gas-particle partitioning of PCDD/Fs, PCBs and PBDEs in the ambient air of an E-waste dismantling area, southeast China. *Chinese Science Bulletin* 2008 (53): 521–528. https://doi.org/10.1007/s11434-008-0125-8.

Liu, J., He, X.X., Lin, X.R. et al. (2015). Ecological effects of combined pollution associated with e-waste recycling on the composition and diversity of soil microbial communities. *Environmental Science & Technology* 49 (11): 6438–6447.

Luo, Y., Luo, X.J., Lin, Z. et al. (2009). Polybrominated diphenyl ethers in road and farmland soils from an e-waste recycling region in Southern China: concentrations, source profiles, and potential dispersion and deposition. *Science of the Total Environment* 407 (3): 1105–1113.

McCulloch, A., Midgley, P.M., and Ashford, P. (2003). Releases of refrigerant gases (CFC-12, HCFC-22 and HFC-134a) to the atmosphere. *Atmospheric Environment* 37: 889–902. https://doi.org/10.1016/S1352-2310(02)00975.

Qin, Q., Xu, X., Dai, Q. et al. (2019). Air pollution and body burden of persistent organic pollutants at an electronic waste recycling area of China. *Environmental Geochemistry and Health* 41 (1): 93–123.

Ramachandra, T.V. and Saira, V.K. (2004). Environmentally sound options for waste management. *Environmental Journal of Human Settlement* 1: 3–11.

Raudaskoski, A., Lenau, T., Jokinen, T., Gisslen, A.V., Metze, A.-L., 2019. Designing plastics circulation: electrical and electronic products, eds., Nordic Council of Ministers.

Salam, M.D. and Varma, A. (2019). A review on impact of E-waste on soil microbial community and ecosystem function. *Pollution* 5 (4): 761–774.

Sanusi, A.I. (2015). Impact of burning e-waste on soil physicochemical properties and soil microorganisms. *British Microbiology Research Journal* 8 (2): 434–443.

Shamim, A.K. and Murshea, R.I. (2015). E-waste, trading impact on public health and ecosystem services in developing countries. *International Journal of Waste Resources* 5: 01–12.

Singh, N., Duan, H., Ogunseitan, O.A. et al. (2019). Toxicity trends in E-Waste: a comparative analysis of metals in discarded mobile phones. *Journal of Hazardous Materials* 380: 120898.

Trick, J. (2002). A mobile is not just for Christmas, Tuesday, 24th December 2002. In: . http://news.bbc.co.uk (accessed 19 August 2021).

Williams, P.T. (2010). Valorization of printed circuit boards from waste electrical and electronic equipment by pyrolysis. *Waste and Biomass Valorization* 2010 (1): 107–120. https://doi.org/10.1007/s12649-009-9003-0.

Wu, J.P., Luo, X.J., Zhang, Y. et al. (2008). Bioaccumulation of polybrominated diphenyl ethers (PBDEs) and polychlorinated biphenyls (PCBs) in wild aquatic species from an electronic waste (e-waste) recycling site in South China. *Environment International* 34 (8): 1109–1113.

Wu, Z., Gao, G., and Wang, Y. (2019). Effects of soil properties, heavy metals, and PBDEs on microbial community of e-waste contaminated soil. *Ecotoxicology and Environmental Safety* 180: 705–714.

Xu, X., Zeng, X., Boezen, H.M., and Huo, X. (2015). E-waste environmental contamination and harm to public health in China. *Frontiers of Medicine* 9 (2): 220–228.

Zhang, W., Wu, Y., and Simonnot, M.O. (2012). Soil contamination due to e-waste disposal and recycling activities: a review with special focus on China. *Pedosphere* 22 (4): 434–455.

5

Exposure to E-Wastes and Health Risk Assessment

Atul Kumar[1], Abhishek Sharma[1], and Anshu Priya[2]

[1] *Department of Veterinary Public Health and Epidemiology, CSK HP Agricultural University, Palampur, Himachal Pradesh, India*
[2] *School of Energy and Environment, City University of Hong Kong, Kowloon, Hong Kong*

Abstract

Rapid development in technology has always benefited mankind, but this also engenders an undesirable aspect related to its handling and management. Due to their unsustainable ways of production, distribution, usage, and disposal, only 1/8th to 1/4th of generated e-waste undergoes formal recycling, and even lesser quantum is being salvaged. The bulk of e-waste is processed informally. Since e-wastes contain elements possessing potential environmental contamination attributes, their ignorant disposal leads to the leaching of toxic substances. Eventually, these toxicants find their way into the food chain and influence life expectancy of consumers in several ways. Populations in proximity to e-waste dumping/recycling sites are more predisposed to serious health implications, especially children, pregnant women, and old/ill people. Hence, the goal of this chapter is to highlight the exposure pathways of different types of e-waste, health risk assessments, and formulation of mitigation strategies to minimize their influence on human and animal health.

Keywords *e-waste; exposure; human health; risk assessment*

5.1 Introduction

In today's swiftly progressing era of technology, Electrical and Electronic Equipment (EEE) has become indispensable and are irreplaceable moieties in modern society, impacting human lives both positively and negatively. Electronic devices have become part and parcel of our lives and are enhancing living standards. Because of the rising dependence of humans on electrical items for an affluent lifestyle, the intensified demand and consumption have led to scaled-up

Management of Electronic Waste: Resource Recovery, Technology and Regulation, First Edition.
Edited by Anshu Priya.

opportunities for the national and international trade of EEEs. These include devices ranging from small household items to hefty medical care equipment. One cannot imagine their work getting done without the use of electronic devices. EEE includes items that either have a power cord or battery. EEE, at the end of their use cycles, when disposed of, generates a waste stream known as Electronic Waste (e-waste) or Waste Electrical and Electronic Equipment (WEEE).

Different agencies and researchers have defined EEE and e-waste in several ways. As per the "Solving the E-Waste Problem" (StEP) Initiative, EEE is "Any household or business item with circuitry or electrical components with power or battery supply". "E-waste" or "Electronic Waste" often referred to as "e-scrap" or "end-of-life electronics" are the devices reaching the end of their useful life, including all components, subassemblies, and consumables, that no longer satisfy their original purpose and are discarded (Environment Protection Agency).

Owing to rapid socio-economic development, advancement in research and development in information technologies, and across-the-border movement of electronics, e-waste management issues (including collection, recycling, and treatment) have drawn considerable attention worldwide. This pressure is more on rapidly modernizing and developing economies of the world. The ever-changing customer's requirements, reduced lifespan of electronic items, rapid advancement in technology, high obsolescence rate, low recycling rate, and the advent of modern gadgets in the market have led to frequent and unnecessary purchases, rooting to increased e-waste volumes (Gagliardi and Mirabile n.d., Perkins et al. 2014). As per recent studies, the standard life duration of a computer has plummeted by 50%, from a duration of 4 years down to 2 years (Byster et al. 2002; Duffert et al. 2009).

However, due to the possibility of recovering precious elements such as iron, aluminum, gold, silver, platinum, and rare earth metals, e-waste is now being recognized as a valuable resource on a worldwide scale. Although, it is one of the most rapidly expanding types of waste simultaneously it is also a major environmental concern (Widmer et al. 2005; Duffert et al. 2009; Lundgren 2012; Noel-Brune et al. 2013).

Recycling of this waste is a very challenging issue. When compared to other types of municipal or industrial waste, e-waste is physically and chemically different. It contains both valuable and harmful components (Figure 5.1). For example, gold (from computer processors), silver, palladium, platinum (from catalytic converters), copper, aluminum, iron, etc. have economic value if recycled but also lay out a seat to potentially toxic substances like dioxins, polychlorinated biphenyls, lead, mercury, cadmium, and nickel, which if released into the environment can have detrimental effects on human, animal, and environmental health. E-waste also contains nondegradable items like polycarbonate (PC), acrylonitrile butadiene styrene (ABS), etc., which persist in the environment for hundreds of years. Therefore, e-wastes need distinctive processing and recycling

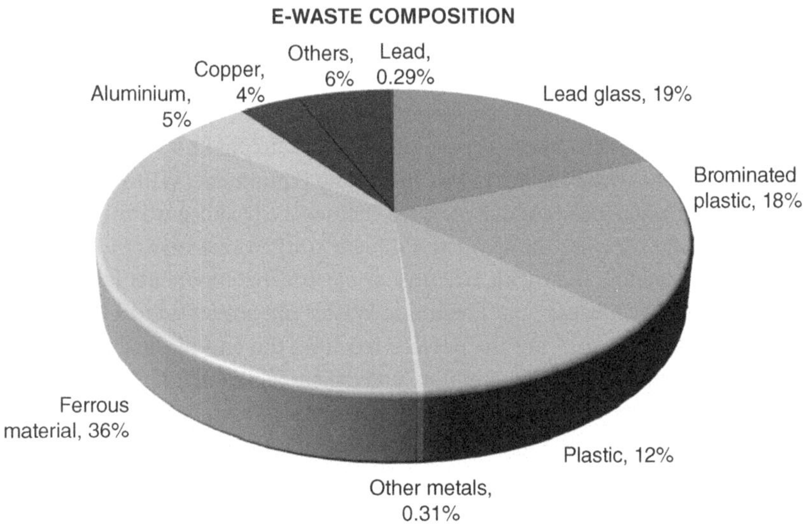

Figure 5.1 Composition of e-waste.

techniques to prevent environmental pollution and adverse impacts on "One Health" (Kumar 2022). But due to strict environmental regulatory regimes in developed countries, the bulk of the world's e-waste is sent to small dealers in developing nations, wherein a substantial amount gets recycled informally.

More than 70% of e-waste produced worldwide is transported to Asian and African nations (Ongondo et al. 2011; Diaz-Barriga 2013). China, being the world's biggest producer and importer of WEEE, with around 70% of global e-waste being dumped in China alone. In 2019, 53.6 million metric tonnes (Mt) of e-waste were produced worldwide, and Asia has been the leading generator of e-waste (24.9 Mt), followed by the Americas (13.1 Mt), Europe (12 Mt), Africa (2.9 Mt), and Oceania (0.7 Mt). Formal collection and recycling were only done with 17.4% of the total amount of e-waste produced around the globe in 2019. Given the present rate of production, the total e-waste production globally is anticipated to reach 74.7 Mt by the year 2030 and double in the next 16 years, making it a quite significant concern (Figure 5.2). From 2016 to 2050, the amount of WEEE around the world is expected to grow by 70%. The impact that Bitcoin mining could have is not taken into account in this projection. There has been little study on the environmental consequences of crypto-currencies like bitcoin that has taken into account the increasing quantity of short-lived hardware used by bitcoin miners, which might accelerate the increase in global electronic trash. Bitcoin miners have progressively resorted to specialized gear with more computational power over time, leading to the progressive increase in computational e-waste generation as the mining equipment becomes obsolete after every 1.5 years.

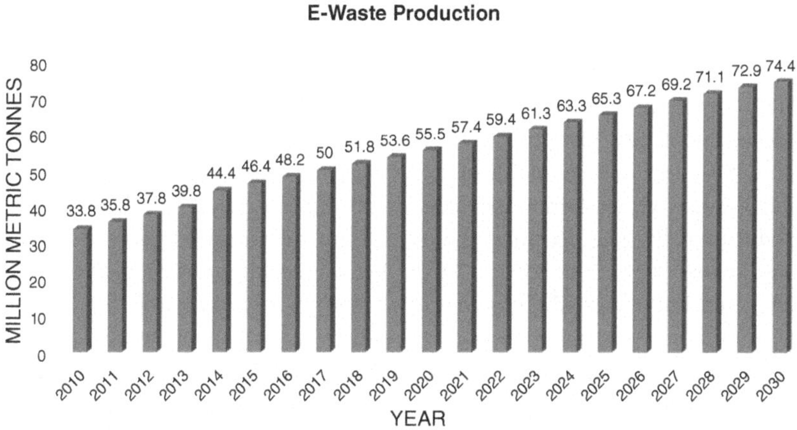

Figure 5.2 Global e-waste production.

The massive output of e-waste has generated multidimensional challenges for present treatment solutions, with adverse effects on ecosystems, human health, and animal health (Kumar and Gupta 2021). High levels of direct occupational exposure occur among workers who recycle e-waste without adequate protection; nonetheless, the more general populace is exposed to the pollutants via the air, food, water, and dust (Duffert et al. 2009; Chen et al. 2011; Chan and Wong 2013). Therefore, keeping in view this emerging global health issue, the present chapter focuses on the exposure pathways of different types of e-waste to vulnerable populations, health risk assessments, and formulation of mitigation strategies to minimize their influence on human and animal health.

5.2 E-Waste Categorization and Vulnerable Population

Products assorted as WEEE include both "white goods" like freezers, washing machines, dryers, etc. as well as "brown goods" like televisions, radios, computers, game consoles, digital media players, and home entertainment systems, etc. There is a wide variety of devices that fall under the umbrella of "WEEE," but for statistical purposes, EEE is categorized based on homology in purpose, constituent materials, typical heft, and similarity in features typical toward the end of life as per WEEE European Community Directive 2012/19/EU Annex. III (Forti et al. 2020). Category 1 includes temperature exchange equipment, e.g. freezers, air conditioners, etc. Category 2 includes screens and monitors, e.g. TV, laptops, etc. Category 3 has lamps such as fluorescent lamps and LEDs. Categories 4 and 5 include large and small equipment like washing machines, copying equipment, electronic tools and toys, and sports equipment, and last Category 6 includes small IT and telecommunication equipment such as mobile phones and GPS.

The population most vulnerable to numerous health risks as a result of environmental contamination is the people living in close proximity to an e-waste disposal and recycling facility. Inappropriate e-waste practices have a disproportionately detrimental effect on the health of marginalized populace. Many people who recycle e-waste, either formal recyclers or informal recyclers, are less wealthy and educationally qualified than the average for their respective communities (Oteng-Ababio 2012; Diaz-Barriga 2013). A considerable proportion of e-waste recyclers are impoverished children and women, mostly in metropolitan areas. Many developing and middle-income countries have thriving recycling businesses, such as Malaysia, Thailand, China, India, Pakistan, Philippines, Vietnam, and Ghana, where e-waste stream plays an important employment opportunity for unemployed people (Modak 2011). Workers often execute the uncontrolled and inexplicable gathering, processing, and redistribution of e-waste at temporary locations, homes, shoddy workshops, and open public areas. Informal sector recycling of e-waste is optimized for maximum stack and output and concurrently environmental and occupational safety are not at all emphasized. Children are regarded as the ideal workers for the disposal of e-waste because of their little, nimble hands, which allow them to effortlessly disassemble, used EEE. They are more likely to be put in situations where they are exposed to potentially harmful compounds at e-waste sites because of their less-developed sense of vulnerability. As children are in their growing phase, their daily consumption of water, food, and air is substantially greater in proportion to their weight and height than that of adults. Additionally, in comparison to adults, children have a substantially high surface area to body weight ratio, which results in a much significant likelihood for dermal absorption (Duffert et al. 2009). Furthermore, children have a diminished capacity to detoxify substances, which may put them at greater risk of adverse effects from exposure to e-waste (Caravanos et al. 2013). According to studies, numerous pollutants included in e-waste were detected in children's and pregnant women's serum. Typically, young children display hand-to-mouth activity and crawl on the ground, which foreseeably results in direct exposure to potentially hazardous chemicals through the oral route. Where informal recycling operations have been in operation for more than ten years, cumulative exposures are foreseeable to be substantial. Major concerns regarding neurodevelopmental deficits are raised for children living in e-waste recycling communities who may have been exposed to high levels of toxicant combinations throughout their lives. Children and developing fetuses are in crucial stages of synaptogenesis, myelination, migration, differentiation, and development of neurons. There may be negative consequences if these precise mechanisms in human neurodevelopment are disturbed (Dietrich et al. 2010). E-waste incorporates a variety of recognized developmental neurotoxicants that can cause behavioral and motor skill problems across the lifetime as well as irreversible cognitive deficiencies in children (Chen et al. 2011). Poor individuals with few career opportunities who are anxious to feed themselves and their families tend to live in communities with rudimentary, informal recycling processes; health and safety concerns are put second to their primary concerns. A case study of how e-waste poisoning may affect practically all citizens' everyday lives is the Agbogbloshie region of Ghana, which is home to around 40,000 people. Approximately 215 000 tonnes of used consumer electric and electronic items, mostly from Western Europe, are shipped into this region each year, as a result, it has become one of Africa's major informal e-waste dumping and processing sites. Pure Earth has listed Agbogbloshie among the top 10 most hazardous dangers in the world due to the area's significant overlap between residential, commercial, and residential areas.

5.3 Exposure Pathways and Health Implications of E-Waste

When e-waste is poorly recycled, it is possible for people there to be directly or indirectly exposed, as a consequence, to a number of hazardous compounds, either present in WEEE or created and then discharged (Figure 5.3). Direct exposure to e-waste includes making physical contact with toxic materials through the skin or via inhalation and ingestion. The air, dust, soil, water, and food may all be ecological sources of exposure to a variety of hazardous compounds, which are released throughout the collecting process as well as informal and official recycling processes such as atmospheric deposition (dry/wet), leaching, chemical/biological degradation, and volatilization (in air/soil) (Li and Achal 2020). The risk of exposure to hazardous substances is substantially increased if un-recommended ways of recycling are practiced to salvage precious materials. Some of the old-fashioned and obsolete ways to recover valuable minerals from e-waste include (Chi et al. 2011):

(i) Physical disassembly of electronic garbage using hammer, chisel, screwdriver, and bare hands.
(ii) Open-air combustion of e-waste, such as wire stacks for metal recovery.
(iii) Heating computer plastic casings to remove flammable polymers and segregate metals.
(iv) Using acid baths to extract gold from processors and other materials.
(v) To liberate precious chips, printed circuit boards are desoldered over charcoal grills.
(vi) Disposal or combustion of waste plastics, for example, in the open, in rivers, and in irrigation canals.

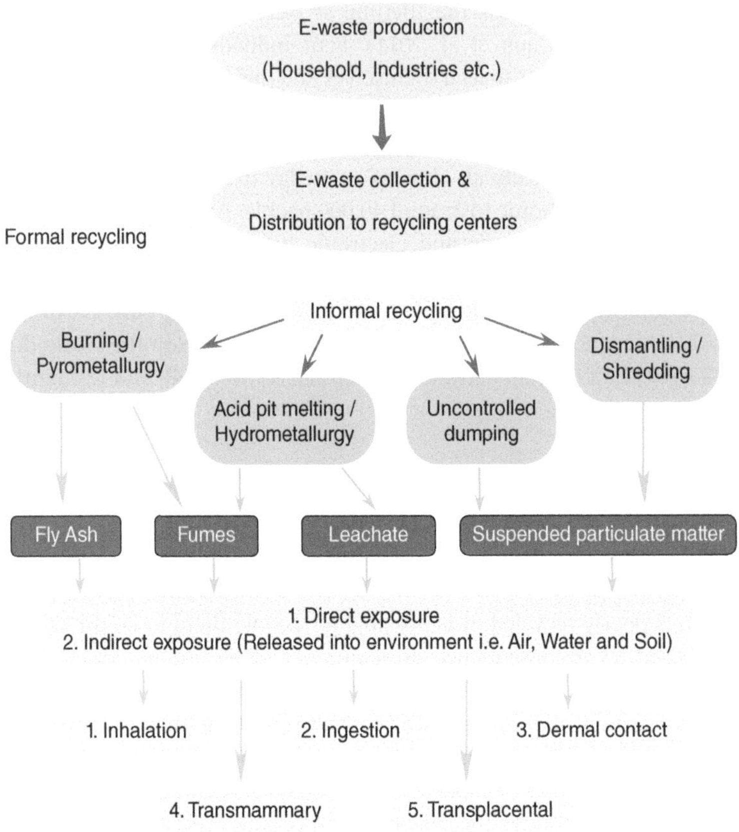

Figure 5.3 Exposure pathways of e-waste.

These primitive techniques generate toxic fumes and leachates that typically comprise polyhalogenated dioxins, furans, and heavy metals (like arsenic, cadmium, mercury, lead, chromium, etc.), which are released by incomplete combustion at low temperatures, pyrometallurgical, and hydrometallurgical processes.

When e-waste is incinerated, heavy metals are also released into the atmosphere, like lead, cadmium, and mercury (Duffert et al. 2009; Frazzoli et al. 2010; Chan and Wong 2013). Incinerators of waste that are not under proper control (hospital waste, or backyard burning of home garbage) often discharge the most dioxins of all the sources of dioxin emissions. Polychlorinated dibenzo-p-dioxins (PCDD)/Fs are generally emitted as airborne pollutants and oftentimes pollute the air, soil, and food.

Such obsolete methods may seem effective to recyclers who lack training and equipment; nonetheless, they do not guarantee workplace safety or environmental protection. Further, the environment-to-food chain is another potential target of e-waste scenarios, in addition to direct human exposure at work and in the community, exposing the general public to a plethora of toxicants, primarily harmful chemical substances, poly aromatic hydrocarbons (PAHs), and persistent organic pollutants (POPs). There are five major routes via, which people are exposed to potentially hazardous substances:

(i) Ingestion
(ii) Inhalation
(iii) Dermal
(iv) Transmammary
(v) Transplacental/Transgenerational

"Take-home exposure" is another form of indirect exposure to toxic chemicals that might emerge from inefficient practices regarding the recycling of electronic debris. This route of exposure refers to the second-hand exposure to hazardous compounds that may occur, especially in the case of youngsters, when the toxicant is carried into the house on contaminated clothes, or other things that have been left over following the recycling of electronic trash. Take-home exposure may result in low-level, chronic, and long-term exposure. Long-term contact with hazardous compounds derived from e-waste (PCDD/Fs, polybrominated diphenyl ethers (PBDEs), polychlorinated biphenyls (PCBs), and PAHs) harms physiological systems, including the neurological, reproductive, and endocrine systems. These chemical components have been shown to cause cancer (e.g. PCDD/Fs, PBDEs, PAHs) and neurotoxic (e.g. PCDD/Fs, PBDEs, PCBs) (Frazzoli et al. 2010; Priyadharshini and Meenambal 2011). Not only can e-waste recycling materials travel great distances, but they also have exceptional environmental persistence. POPs, the predominant toxicants in e-waste, are inert to biological breakdown and have a high propensity to bioaccumulate in the food chain. These compounds have the potential to continue to be present in the environment over extended periods of time due to their long half-lives. Some compounds may have synergistic or modifying effects in a mixture of chemicals that might be very hazardous.

E-waste impacts lands and rivers through particle fallout and water that runs off of contaminated soil, as well as the downstream transfer of contaminated sediments in aquatic systems (Yu et al. 2006). Therefore, agriculture and animal husbandry practices within and directly next to waste dumps, or downstream of polluted streams, probably expose the general populace to hazardous substances through contaminated water and soil. Continuous exposure to toxic chemicals also increases the amount of chemicals that are loaded into pastures, potentially leading to high levels of toxic chemical absorption in cereals (for example, rice readily absorbs arsenic) and other vegetable foods. For instance, PBDEs are often readily transferred from soils/sediments to plant roots. In addition, it is possible that contamination of pastures will lead to severe contamination of various animal-sourced foods as well. Consumption of toxicants over an extended period of time by dairy animals may happen when pastures are contaminated (by water, soil, or vegetation). Generally, the compounds found in e-waste bioaccumulate in tissues and are excreted in animal-sourced foods like eggs and milk (Kierkegaard et al. 2007). Studies also indicate the presence of brominated flame retardants (BFRs) in seafood (Isosaari et al. 2005).

Since toxic elements like POPs from e-waste may be transferred via breastfeeding and infrequently by transplacental route, therefore transgenerational exposures also have long-term impacts on the generations to come (Frazzoli et al. 2010).

5.4 Chemical Composition of E-Waste and Health Risks Associated with Their Exposure

Various chemical components of e-waste and their potential impact on health are summarized in Table 5.1.

5.4.1 Persistent Organic Pollutants (POPs)

They are organic chemical substances. They possess a particular combination of physical and chemical properties such that, once released into the environment, they stay unaltered for many years, show biomagnification at trophic levels in the food chain, and are hazardous to the health of humans and animals (UNEP n.d., Stockholm Convention). POPs released as a result of processing techniques for e-waste include BFRs, PBDEs, and PCBs. It was determined that uncontrolled open combustion was the primary source of POPs (Batterman et al. 2009).

5.4.2 Polycyclic Aromatic Hydrocarbons (PAHs)

PAHs are produced during the combustion of coal, oil, gas, wood, garbage, and tobacco. They are released as combustion by-products of e-waste.

5.4.3 Dioxins

Dioxins are released due to open-air burning of wires and plastics (Wen et al. 2008) and are found in higher concentrations at open burning sites. Routes of exposure to dioxins are air, water, soil, food, dust, and vapors.

5.4.4 Heavy Metals

Dismantling and pit-acid procedures to extract precious metals like gold cause heavy metals to contaminate land and surrounding waters.

Table 5.1 Routes of exposure to various chemical components of e-waste and health impacts.

Route of exposure	Compound	Adverse health effect	Reference
POPs			
Ingestion, inhalation, or transplacental	BFRs	Cancer, implications on reproductive health, neurobehavioral and developmental disorders, and alterations in thyroid function	Kim et al. (2014)
	PBDEs	Thyroid function, neurobehavioral development, and reproductive development. hormone impact	Sanders et al. (2005), Peters et al. (2006), Pacyniak et al. (2007), Darnerud (2008),
	PBBs	Cancer, Thyroid dysfunction, Lymphoma, Breast cancer	Bahn et al. (1980), Hoque et al. (1998)
Ingestion, inhalation, dermal contact, or transplacental	PCBs	Multiple sites for carcinogenesis include the liver, thyroid, immune system, reproductive system, and neurobehavioral development. Act as tumor promoters	Tharappel et al. (2002), Negri et al. (2003), Shi et al. (2019),
PAHs			
Ingestion, inhalation, or dermal contact	Anthracene, Acenapthylene, Acenapthene, Benzopyrene, Benzofluoranthene, Benzopyrelene, Chrysene, Dibenzoanthracene, Fluoranthene, and Pyrene	Teratogenic, carcinogenic, and mutagenic.	IARC (1983)
Dioxins			
Ingestion, inhalation, dermal contact, or transplacental	Perfluroalkyls	Thyroid abnormalities, Breast cancer, Hepatic damage, Renal cancer, Ulcerative colitis, Testicular carcinoma, and Pre-eclampsia	Fenton et al. (2009), White et al. (2011), Barry et al. (2013)
	PCDDs and polychlorinated dibenzofurans (PCDFs)	Neurobehavioral impairment, Reproductive abnormalities, Carcinogenicity, Endocrine disruption, and Immunity development	Larsen (2006), Eskenazi et al. (2018), Vinceti et al. (2018), Ames et al. (2019)
Ingestion, inhalation, or dermal contact	Dioxin-like polychlorinated biphenyls	Diabetes, cancer, porphyria, endometriosis, early menopause, decreased thyroid and testosterone levels, cardiovascular disease, altered immune response, abnormalities of the skin, teeth, and nails, altered hearing, psychomotor, and gender-related neurobehavioral patterns, as well as impaired cognitive	White and Birnbaum (2009)

Elements			
Ingestion, inhalation, dermal contact, or transplacental	Nickel	Lung fibrosis, kidney and cardiovascular diseases, and cancer of the respiratory tract	McGregor et al. (2000), Zambelli, and Ciurli (2013)
Ingestion, inhalation, or transplacental	Beryllium	Lung cancer, Berylliosis	Sanderson et al. (2001)
Ingestion or inhalation	Cadmium	Kidney damage, Inhibition of progesterone and testosterone synthesis, Bone abnormalities, and lung emphysema	Piasek and Laskey (1999), Barbier et al. (2005), Kumar and Singh (2014)
	Chromium (VI)	Carcinogenic, Reproductive, and Hormonal function disruption	Banu et al. (2008), Quinteros et al. (2008)
	Zinc	Copper deficiency, Anemia	Magee and Matrone (1960)
Ingestion, inhalation, or dermal contact	Aluminum	Pulmonary lesions, Myocardial disorders, Neurodegenerative diseases, Fetal toxicity, and Skeletal development	El Hangouche et al. (2017), Iijima et al. (2017), Mold et al. (2018),
	Arsenic	Carcinogenicity, Neurobehavioral changes, Dermal alterations, Possibility of developing diabetes	Tsai et al. (2003), Centeno et al. (2006), Gamble et al. (2018)
	Antimony	Irritation to skin, eyes, and lungs, damage to heart, liver, and kidney	Roper (1992), Kumar, and Singh (2014)
	Barium	Heart failure, Kidney failure, Lung edema, and Gastric hemorrhages	Barnes and Eltherington (1966), Milanov and Stoyanov (1982)
	Bismuth	Renal damage, Eczema	İşlek et al. (2001), Kumar and Singh (2014),
	Cobalt	Neurological, Cardiovascular, and Endocrine functions disruption.	Tower (2010)
	Copper	Abnormal reproduction and Developmental toxicity	Keen et al. (1996), Hanna et al. (1997)
	Gallium	Renal toxicity and anemia	Chitambar (2010)

(Continued)

Table 5.1 (Continued)

Route of exposure	Compound	Adverse health effect	Reference
	Germanium	Renal failure, Anemia, Neuropathy, and Bone marrow defects	Tao and Bolger (1997)
	Indium	Renal, cardiac, and hepatic damage and carcinogenicity	IARC (2018)
	Iron	Hepatic damage	EFSA (2004)
	Lithium	Ataxia, Verbal impairment, Tremors,	Bauer and Gitlin (2016)
	Mercury	Neurodegenerative disorders, Cardiotoxicity, Neurobehavioral developmental abnormalities in children	Bridges and Zalups (2010), Shamim et al. (2015)
	Molybdenum	Male reproductive abnormalities, Increased blood uric acid, Risk of stroke	Albin and Oskarsson (2022)
	Lead	Encephalopathy, Renal toxicity, and Anemia in children, Reproductive and Early Developmental toxicities	National Research Council (1993)
	Selenium	Disruption of endocrine and thyroid functions, Hepatotoxicity, and Gastrointestinal disturbances	Vinceti et al. (2001)
	Silver	Argyria and Argyriosis	Lansdown (2006)
	Tin	Dermal and ocular irritation, cholangitis, liver, and renal failures	Winship (1988)
	Vanadium	Gastrointestinal disorders, Green-black tongue, Rhinitis, Conjunctivitis, Pneumonia, and Depression	USEPA (2009), Assem, and Levy (2012)

5.5 Health Risk Assessments

The process of determining the chance of occurrence of an event (in this case, exposure to heavy metals) and the likely size of its detrimental impact over a particular time period is known as health risk assessment. E-waste contains a plethora of toxicants, i.e. heavy metals, plastics, POPs, etc. The nonscientific and un-recommended disposal of such wastes has a significant impact on both the health of humans and the environment, especially among children and pregnant women during their sensitive windows of development. E-waste has the potential to cause severe health risks among people living in contaminated surroundings. Therefore, their exposure risk assessment is of utmost importance to safeguard human and animal health. Since exposure may take place via various routes, viz. ingestion, inhalation, and dermal, therefore, due consideration should be given to the exposure pathway while evaluating the risk. There are different ways to assess human health risks, but the hazard index method (HI) and carcinogenic risk (CR)-based estimations are most widely applied. The slope factor (SF) and reference dose (RfD) are other primary toxicity risk variables that are used for risk assessment. The SF is used for CR assessment and represents the risk of acquiring cancer based on the daily dose of mg per kg of body weight (mg/kg/bw), while the RfD is used for noncarcinogenic risk assessment.

5.5.1 Noncarcinogenic Risk Assessment

The following equations are used to determine the noncarcinogenic hazard quotient (HQ) for each heavy metal (USEPA, 2007):

$$\mathrm{HQ} = \frac{\mathrm{ADD} * \mathrm{RBA}}{\mathrm{RfD}}$$

wherein ADD is the average daily dose of heavy metals (mg/kg/day); RBA is the relative bioavailability of the metal; and RfD is the reference dose (μg/kg/day).

Exposure to an element with an $HQ > 1$ may have health consequences that are not carcinogenic, whereas an $HQ < 1$ indicates that there is no risk of potential harm from that element. Exposure to two or more substances may have synergistic or accumulative effects. Several simultaneous exposures to sub-threshold levels can result in adverse health effects. The cumulative HQ is HI, which may be calculated as

$$HI = \Sigma\, HQ_1 + HQ_2 + HQ_3 + \ldots + HQ_n$$

Since exposure from two or three pathways may have synergistic or accumulative effects, the total exposure hazard index (HIt) may also be calculated as:

$$HIt = \Sigma\, HQ_i \text{ i.e.}, HQ_{ing} + HQ_{inh} + HQ_{derm}$$

Table 5.2 Noncarcinogenic and carcinogenic risk assessment scales.

Hazard quotient/ hazard index	Noncarcinogenic risk	Cancer cases per number of inhabitants	Risk level/ cancer risk
< 0.1	Negligible	<1 per 10^6	1/Very low
$\geq 0.1 < 1$	Low	>1 per 10^6 <1 per 10^5	2/Low
$\geq 1 < 4$	Medium	>1 per 10^5 <1 per 10^4	3/Medium
≥ 4	High	>1 per 10^4 <1 per 10^3	4/High

Source: Adapted from USEPA, (1999).

Values of $HIt < 1$ indicate no substantial risk of noncarcinogenic effects, whereas values of $HIt > 1$ indicate the possibility of noncarcinogenic effects, with the likelihood increasing with the HI value.

Pregnant women and lactating mothers residing in proximity to formal and informal recycling facilities impose fetal and post-fetal health risks via their trans-placental and trans-mammary routes. Despite the fact that the placenta may function as a selective transporter that inhibits the flow of potentially hazardous compounds to the growing baby, some environmental pollutants may easily or partly breach the placental barrier. Under one such circumstance, breastfed babies' daily heavy metal intake may be determined with the use of the following formula (Anderson et al. 1994):

$$\mathrm{DI} = \frac{\mathrm{IR} * C}{\mathrm{BW}}$$

where DI is Daily Intake in μg/kg-bw/day; IR is Ingestion rate in L/day; and C is concentration in μg/L.

HQ for infants may be calculated as: $HQ = \frac{DI}{RfD}$

5.5.2 Carcinogenic Risk Assessment

E-waste contains a variety of chemicals and components, some of which are potentially carcinogenic. The parameter to assess health risk is CR, or Incremental lifetime cancer risk (ILCR). The likelihood that a person will acquire cancer over the course of their lifetime at a certain exposure level is how CR is represented. The cancer risk probability is determined from the SF (Table 5.2).

$$\mathrm{CR\ or\ ILCR} = (\mathrm{ADD} \times \mathrm{RBA} \times \mathrm{SF}),$$

$\mathrm{CR} < 0.0001$ is considered to be acceptable; however, $\mathrm{CR} > 0.0001$ is viewed as unsatisfactory, and it is recommended to undertake some form of remediation (Moya et al. 2011).

Recently, it has been found that the number of heavy metals can pass through the placental barrier, providing substantial health risks to fetuses. It has been observed that blood samples taken from umbilical cords and neonates as young as one day old contained considerable concentrations of heavy metals like lead, putting them at risk for lifelong physiological, neurological, and developmental disorders.

Since routes of exposure may vary depending upon the person's occupation, his/her inhabitation, or any other factor. Therefore, it becomes pertinent to evaluate health risks due to exposure to e-waste through various routes. Hence, the ADD through oral intake (ADD_{ing}), respiratory route (ADD_{inh}), and dermis (ADD_{derm}) should always be calculated by using the following equations:

$$ADD_{ing} = \frac{C * \text{Ing}\,R * \text{EF} * \text{ED} * \text{CF}}{\text{BW} * \text{AT}}$$

$$ADD_{inh} = \frac{C * \text{Inh}\,R * \text{EF} * \text{ED}}{\text{BW} * \text{AT}}$$

$$ADD_{derm} = \frac{C * \text{SA} * \text{SL} * \text{ABS} * \text{EF} * \text{ED} * \text{CF}}{\text{BW} * \text{AT}}$$

where C refers to the concentration of the pollutant present in the air (mg/kg); Ing R refers to ingestion rate; For inhalation, Inh R is inhalation rate, which is about 15.2 m^3/day of soil (USEPA 1997); EF refers to exposure frequency (days/year); ED refers to exposure duration, which varies according to occupation/category; AT is the average time of exposure. In case of noncarcinogenic compounds, EF = AT, whereas in case of carcinogens, AT varies situation-wise. For noncarcinogens: AT = ED × 365 days per year, and this kind of intake is referred to as chronic daily intake. For carcinogens: AT = Lifetime (70 years) × 365 days per year, and intake is called lifetime average daily dose; BW refers to average body weight, which is 60 kg for adults and 15 kg for children (Kumar et al. 2018; Sharma et al. 2022). SA refers to skin area in case of dermal contact, and the exposed skin area is =2253 cm^2; SL refers to skin adherence factor, which is =0.2 mg/cm^2/day; ABS refers to dermal absorption factor, which is 0.001 (Chang et al. 2009). The unit conversion factor is denoted by $CF = 10^{-6}$.

Children who live in the vicinity of e-waste recycling sites, ingest contaminated soil and dust, as a result of which they are exposed to various health hazards and potential carcinogens during their lifetime. Therefore, in that case, ADD_{ing} should take into account the ingestion rate of soil and dust, which may be calculated as follows (Yang et al. 2022):

$$\text{Ing}\,R = \frac{(W_{\text{faeces}} * C_{\text{faeces}} + V_{\text{urine}} * C_{\text{urine}}) - (W_{\text{food}} * C_{\text{food}})}{C_{\text{soil/dust}}}$$

where W_{faeces} and W_{food} denote dry weight of feces and weight of food consumed (kg/day), respectively; C_{faeces}, C_{food}, and $C_{\text{soil/dust}}$ refer to the concentration of tracer elements in feces, food, and soil/dust (mg/kg), respectively; and Volume of urine is V_{urine} (mL/day).

5.6 E-Waste Management

A vast range of e-waste from households, business enterprises, industries, and the public sector has been added to the waste stream as a result of technological advancements. Cradle-to-cradle policies and strategies are required for effective management of e-waste. There have been many international legislation and efforts on the management of e-waste so far. Some of them are (Ilankoon et al. 2018):

The Basel Convention: It focuses on decreased production of harmful waste, advancement of eco-friendly waste management practices at disposal sites, restriction of hazardous waste transboundary movements, and formation of legal frameworks that permit transboundary movements.

The European Union's WEEE Directive: It includes no-cost take-back programs for customers, recycling and reuse initiatives, and the segregation of WEEE from other trash types during the collection process. It encourages "eco-design" – the design of products that decrease e-waste and make it easier to recover. According to it, producer is responsible for product end-of-life treatment.

StEP (Solving the E-waste Problem): StEP is founded on studies and pilot projects examining the processing of e-waste, formulation of a plan and objectives to reduce the WEEE issue, improvement of member and industry-wide communication and education around e-waste issues.

G-8 3Rs Initiative: Under this initiative, 3Rs are Reduce, Reuse, and Recycle.

National Strategy for Electronics Stewardship (NSES), USA: According to it, there should be improvements in electronic product design. The United States government will establish guidelines that will lead to better handling and management of worn or abandoned electronic equipment here at home and less harm to developing countries as a result of this kind of technology being sent there.

GeSI (Global e-Sustainability Initiative): This initiative primarily deals with bringing together ICT firms, industry groups, and nonprofit organizations. E-waste is one of the primary emphasis areas of this global initiative. It also aims to utilize innovative technology to achieve sustainable aims and control ICT-related hazards.

IETC (International Environmental Technology Centre) – UNEP: Waste management using ecologically sustainable technologies is discussed, with special attention to countries in developing phases and with transitional economies.

As evident, developed countries have strict laws and regulations to properly manage and recycle WEEE. However, in developing nations, e-waste is non-judiciously transported and recycled in small towns and villages using crude technology, despite the Basel Convention's restrictions on the transportation of hazardous waste (LaDou and Lovegrove 2008). By developing two major directives – the WEEE directive and the Restriction of the Use of Certain Hazardous Substances (RoHS) in EEE Regulations – the European Union has taken the initiative in protecting the environment in its region. But such regulations are either lacking or not enforced strictly in most parts of the world.

To effectively implement the e-waste disposal system, waste management should always make use of deposit-refund schemes such as the advance recycling fee (ARF) and the advance disposal charge (ADC), where e-waste producers should be given a compensatory amount. Extended producer responsibility (EPR) is an environmental protection policy that places the producer in charge of the whole product lifespan, including take-back, recycling, and ultimate disposal. This means that the responsibility of the producer continues even after the product has been purchased. Including this in the legal framework is crucial if EPR is to become an essential step in the production of electronic and electrical equipment. Also, the usage of hazardous compounds such as lead, cadmium, mercury, PCBs, and other toxic substances for which suitable alternatives have been identified is decreasing. Numerous nations have implemented the RoHS standards for the production of electrical and electronic devices (Wath et al. 2011).

Hydrometallurgy, pyrometallurgy, and informal practices related to e-waste disposal should be addressed using sustainable and cost-effective techniques. Heavy metal pollution can be minimized through the use of biohydrometallurgy, which is an excellent alternative method. Countries like China, India, Nigeria, and Vietnam recycle e-waste in a primitive manner (Chen et al. 2011). The aim of recycling is to profitably recover valuable metals such as gold, silver, platinum, palladium, and tin. Toxicants in e-waste produce serious contamination of nearby air, dust, soil, and water. This is due to a lack of strong environmental regulations and worker safety. Therefore, formal recycling is of utmost importance, keeping in view the adverse and cumulative health risks associated with e-waste and their disposal. Formal recycling reduces occupational health risks for recyclers as well as its environmental impact. Formal recycling is often accomplished via a three-stage process. The outcome of the first level of treatment serves as input for the second level, and so on. WEEE collected and segregated is directly subjected to first-level treatment.

First-level treatment includes:

- Decontamination of solid e-waste items
- Disassembling
- Segregation

Second-level treatment includes:

- Use of hammers to separate metal parts from plastic parts.

- Shredding to break large items into smaller pieces.
- Certain treatment processes like
  - Cathode-ray tube treatment
  - Electromagnetic segregation or Density separation using water

Third-level treatment includes:

- Recycling
  - Chemical treatment to recover plastic or metal parts
  - Treatments like shredding and breaking of plastic and glass items
  - Thermal treatment to recover plastic
- Recovery
  - Use of reverberator or electrolytic techniques to recover metals like lead and copper.
  - Using leaching, smelting, and electrolytic processes to recover precious metals like Gold, Platinum, etc.

Recycled metals and polymers may be utilized to manufacture new equipment, eliminating the need to mine virgin earth metals and the environmental effect of mining these metals.

5.7 Conclusion

E-waste is a serious challenge, but without a universally accepted definition, its associated environmental and health risks management seems to be a herculean task for the international community. Very likely, concerns about e-waste's impact on ecosystems and human health pollution are grossly underestimated. The ever-increasing production of e-waste poses a serious matter to those emerging nations, where the native population along with the environment are at risk. Therefore, organizations such as the United Nations stress the need to take a proactive and effective approach to the management of this issue on a global scale in a manner that is sustainable. The processing methods used in this context are typically rudimentary due to the nonenforcement of regulations and the paucity of resources. This leads to the generation of waste that is frequently more dangerous than its original materials. The detrimental effects of electronic trash may be mitigated by the development of eco-friendly and effective clean-up methods and the regulation of contaminated sources. Most e-waste collection, dismantling, and recycling tasks are completed by unorganized sectors because they lack the necessary infrastructure and technical expertise to handle such waste. The people engaged in the informal sector are also devoid of knowledge regarding the adverse harmful effects of e-waste on environment and public health. Concrete action must be taken to stop the release of these toxic compounds into the environment, decrease resource waste by recovering important metals, and safeguard locals and employees from toxic exposure. The vast volume of garbage produced, together with sloppy recycling and disposal, remains the major source of harmful contaminants. Legislative frameworks are needed to enhance waste management techniques. Using these rules in conjunction with locally tailored EPR frameworks, it is possible to develop waste management strategies that benefit the widest possible range of constituents. As a result, it would have little effect on the ecosystems. Even if it would be beneficial, it is unrealistic to expect all EEE to be cleansed of all traces of harmful materials. While further study is obviously needed, it's also important to get the word out about the dangers of recycling electronic trash. Further research is needed to determine the exposure source and the possible chronic detrimental effects on mothers and their neonates, notwithstanding the conclusion that health risk assessment revealed no significant concern for developmental toxicity in infants. Diseases may be prevented, or their progression slowed if early childhood exposures are identified sooner. Birth cohorts might be used to identify and track lifestyle variables affecting health from fetus through adulthood, such as food and chemical, an exotoxin, and endotoxin exposures. For a full understanding of the major impacts, routes, and mechanisms of environmental hazards on human health, further research is needed to examine multipollutant exposure and use multivariate statistical methods. Toxic metals and compounds found in electronic trash, such as lead, mercury, cadmium, chromium, PCBs, PBDEs, and PAHs, have been linked to damage in a wide variety of body systems. Partnerships between the public and commercial sectors are essential to the creation and implementation of effective E-waste management plans. Therefore, global collaborations involving producers, people, policymakers, etc. are of utmost significance in curbing the impact of e-waste on human and environmental health.

References

Albin, M. and Oskarsson, A. (2022). *Molybdenum: Handbook on the Toxicology of Metals*, 601–614. Academic Press.

Ames, J., Warner, M., Siracusa, C. et al. (2019). Prenatal dioxin exposure and neuropsychological functioning in the Seveso Second Generation Health Study. *International Journal of Hygiene and Environmental Health* 222 (3): 425–433.

Anderson, M.A., Dewey, K.G., Frongillo, E. et al. (1994). *WHO Working Group on Infant Growth: An Evaluation of Infant Growth*. Geneva, Switzerland: World Health Organization.

Assem, F.L. and Levy, L.S. (2012). *Inhalation Toxicity of Vanadium: Vanadium*, 209–224. Dordrecht: Springer.

Bahn, A.K., Mills, J.L., Snyder, P.J. et al. (1980). Hypothyroidism in workers exposed to polybrominated biphenyls. *New England Journal of Medicine* 302 (1): 31–33.

Banu, S.K., Samuel, J.B., Arosh, J.A. et al. (2008). Lactational exposure to hexavalent chromium delays puberty by impairing ovarian development, steroidogenesis and pituitary hormone synthesis in developing Wistar rats. *Toxicology and Applied Pharmacology* 232 (2): 180–189.

Barbier, O., Jacquillet, G., Tauc, M. et al. (2005). Effect of heavy metals on, and handling by, the kidney. *Nephron Physiology* 99 (4): 105–110.

Barnes, C.D. and Eltherington, L.G. (1966). *Drug Dosage in Laboratory Animals: A Handbook*. Oakland, California: University of California Press.

Barry, V., Winquist, A., and Steenland, K. (2013). Perfluorooctanoic acid (PFOA) exposures and incident cancers among adults living near a chemical plant. *Environmental Health Perspectives* 121 (11–12): 1313–1318.

Batterman, S., Chernyak, S., Gouden, Y. et al. (2009). PCBs in air, soil and milk in industrialized and urban areas of KwaZulu-Natal, South Africa. *Environmental Pollution* 157 (2): 654–663.

Bauer, M. and Gitlin, M. (2016). *Lithium and its History: The Essential Guide to Lithium Treatment*, 25–31. Cham: Springer.

Bridges, C.C. and Zalups, R.K. (2010). Transport of inorganic mercury and methylmercury in target tissues and organs. *Journal of Toxicology and Environmental Health B* 13 (5): 385–410.

Byster, L., Westervelt, S., Gutierrez, R. et al. (2002). *Exporting Harm: The High-tech Trashing of Asia*. Seattle: Basel Action Network.

Caravanos, J., Clarke, E.E., Osei, C.S. et al. (2013). Exploratory health assessment of chemical exposures at e-waste recycling and scrapyard facility in Ghana. *Journal of Health and Pollution* 3 (4): 11–22.

Centeno, J.A., Tchounwou, P.B., Patlolla, A.K. et al. (2006). Environmental pathology and health effects of arsenic poisoning. In: *Managing arsenic in the environment: from soil to human health*, 311–327. Australia: CSIRO Publishing.

Chan, J.K.Y. and Wong, M.H. (2013). A review of environmental fate, body burdens, and human health risk assessment of PCDD/Fs at two typical electronic waste recycling sites in China. *Science of the Total Environment* 463: 1111–1123.

Chang, J., Liu, M., Li, X. et al. (2009). Primary research on health risk assessment of heavy metals in road dust of Shanghai. *China Environmental Science* 29 (5): 548–554.

Chen, A., Dietrich, K.N., Huo, X. et al. (2011). Developmental neurotoxicants in e-waste: an emerging health concern. *Environmental Health Perspectives* 119 (4): 431–438.

Chi, X., Streicher-Porte, M., Wang, M.Y. et al. (2011). Informal electronic waste recycling: A sector review with special focus on China. *Waste Management* 31 (4): 731–742.

Chitambar, C.R. (2010). Medical applications and toxicities of gallium compounds. *International Journal of Environmental Research and Public Health* 7 (5): 2337–2361.

Darnerud, P.O. (2008). Brominated flame retardants as possible endocrine disrupters. *International Journal of Andrology* 31 (2): 152–160.

Diaz-Barriga, F. (2013). Evidence-based intervention programs to reduce children's exposure to chemicals in e-waste sites: Discussion Paper for WHO Working Meeting on E-waste and Children's Health, 1-90

Dietrich, K.N., Yeates, K.O., Ris, M.D. et al. (2010). *Environmental Toxicants. Pediatric Neuropsychology: Research, Theory, and Practice*, 211–264. New York: Guilford Press.

Duffert, C., Brune, M.N., and Prout, K. (2009). *Background Document on Exposures to E-Waste*. Geneva, Switzerland: World Health Organization.

El Hangouche, A.J., Fennich, H., Alaika, O. et al. (2017). Reversible myocardial injury and intraventricular thrombus associated with aluminium phosphide poisoning. *Case Reports in Cardiology* 2017: 6287015.

Eskenazi, B., Warner, M., Brambilla. et al. (2018). The Seveso accident: A look at 40 years of health research and beyond. *Environment International* 121: 71–84.

European Food Safety Authority, EFSA (2004). Opinion of the Scientific Panel on Dietetic products, nutrition, and allergies [NDA] related to the Tolerable Upper Intake Level of Iron.

Fenton, S.E., Reiner, J.L., Nakayama, S.F. et al. (2009). Analysis of PFOA in dosed CD-1 mice. Part 2: Disposition of PFOA in tissues and fluids from pregnant and lactating mice and their pups. *Reproductive Toxicology* 27 (3–4): 365–372.

Forti V., Baldé C.P., Kuehr R., Bel G. The Global E-waste Monitor (2020). Quantities, flows and the circular economy potential. United Nations University (UNU)/United Nations Institute for Training and Research (UNITAR) – co-hosted SCYCLE Programme, International Telecommunication Union (ITU) & International Solid Waste Association (ISWA), Bonn/Geneva/Rotterdam. Available at: https://ewastemonitor.info/wp-content/uploads/2020/11/GEM_2020_def_july1_low.pdf.

Frazzoli, C., Orisakwe, O.E., Dragone, R. et al. (2010). Diagnostic health risk assessment of electronic waste on the general population in developing countries' scenarios. *Environmental Impact Assessment Review* 30 (6): 388–399.

Gagliardi, D. and Mirabile, M. (n.d.). *Overview of OSH Issues Related to the e-waste Management*. Geneva, Switzerland: World Health Organization (WHO).

Gamble, A.V., Givens, A.K., and Sparks, D.L. (2018). Arsenic speciation and availability in orchard soils historically contaminated with lead arsenate. *Journal of Environmental Quality* 47 (1): 121–128.

Hanna, L.A., Peters, J.M., Wiley, L.M. et al. (1997). Comparative effects of essential and nonessential metals on preimplantation mouse embryo development in vitro. *Toxicology* 116 (1–3): 123–131.

Hoque, A., Sigurdson, A.J., Burau, K.D. et al. (1998). Cancer among a Michigan cohort exposed to polybrominated biphenyls in 1973. *Epidemiology* 9 (4): 373–378.

IARC (1983). Polynuclear aromatic compounds, part 1, chemical, environmental, and experimental data: IARC Monographs on the Evaluation of the Carcinogenic Risk of Chemicals to Man. *IARC Scientific Publications* 32: 33–451.

IARC (2018). Welding, Molybdenum Trioxide, and Indium Tin Oxide. IARC Monographs on the Evaluation of Carcinogenic Risks to Humans Volume 118. ISBN: 978-92-832-0156-4.

Iijima, Y., Bando, M., Yamasawa, H. et al. (2017). A case of mixed dust pneumoconiosis with desquamative interstitial pneumonia-like reaction in an aluminum welder. *Respiratory Medicine Case Reports* 20: 150–153.

Ilankoon, I.M.S.K., Ghorbani, Y., Chong, M.N. et al. (2018). E-waste in the international context – A review of trade flows, regulations, hazards, waste management strategies and technologies for value recovery. *Waste Management* 82: 258–275.

İşlek, İ., Uysal, S., Gök, F. et al. (2001). Reversible nephrotoxicity after overdose of colloidal bismuth subcitrate. *Pediatric Nephrology* 16 (6): 510–514.

Isosaari, P., Lundebye, A.K., Ritchie, G. et al. (2005). Dietary accumulation efficiencies and biotransformation of polybrominated diphenyl ethers in farmed Atlantic salmon (Salmo salar). *Food Additives and Contaminants* 22 (9): 829–837.

Keen, C.L., Chang, L.W., Magos, L. et al. (1996). *Teratogenic effects of essential trace metals: Deficiencies and excesses: Toxicology of metals*, 977–1001. London: CRC Lewis Publishers.

Kierkegaard, A., Asplund, L., de Wit, C.A. et al. (2007). Fate of higher brominated PBDEs in lactating cows. *Environmental Science & Technology* 41 (2): 417–423.

Kim, Y.R., Harden, F.A., Leisa-Maree, L. et al. (2014). Health consequences of exposure to brominated flame retardants: a systematic review. *Chemosphere* 106: 1–19.

Kumar, A. (2022). Disaster response under "One Health" Approach: Contribution of Veterinary Public Health. In: *Management of Animals in Disasters* (ed. S. Verma and H.T. Prem), 245–261. Singapore: Springer https://doi.org/10.1007/978-981-16-9392-2_21.

Kumar, A. and Gupta, V. (2021). E-waste: an emerging threat to "one health". In: *Environmental Management of Waste Electrical and Electronic Equipment* (ed. M.H. Chaudhery), 49–61. Elsevier Inc. https://doi.org/10.1016/B978-0-12-822474-8.00003-9.

Kumar, U. and Singh, D.N. (2014). Electronic waste: Concerns & hazardous threats. *International Journal of Current Engineering and Technology* 4: 802–811.

Kumar, A., Gill, J.P.S., and Bedi, J.S. (2018). Multiresidue determination of pesticides in market honey from northern India using QuEChERS approach and assessment of potential risks to consumers. *Current Science* 115 (2): 283–291.

Ladou, J. and Lovegrove, S. (2008). Export of electronics equipment waste. *International Journal of Occupational and Environmental Health* 14 (1): 1–10.

Lansdown, A.B. (2006). Silver in health care: antimicrobial effects and safety in use. *Biofunctional Textiles and the Skin* 33: 17–34.

Larsen, J.C. (2006). Risk assessments of polychlorinated dibenzo-p-dioxins, polychlorinated dibenzofurans, and dioxin-like polychlorinated biphenyls in food. *Molecular Nutrition & Food Research* 50 (10): 885–896.

Li, W. and Achal, V. (2020). Environmental and health impacts due to e-waste disposal in China—A review. *Science of the Total Environment* 737: 139745.

Lundgren, K. (2012). *The Global Impact of E-Waste: Addressing the Challenge*. International Labour Office. Programme on Safety and Health at Work and the Environment (SafeWork).

Magee, A.C. and Matrone, G. (1960). Studies on growth, copper metabolism and iron metabolism of rats fed high levels of zinc. *The Journal of Nutrition* 72 (2): 233–242.

McGregor, D.B., Baan, R.A., Partensky, C. et al. (2000). Evaluation of the carcinogenic risks to humans associated with surgical implants and other foreign bodies – a report of an IARC Monographs Programme Meeting. *European Journal of Cancer* 36 (3): 307–313.

Milanov, M.P. and Stoyanov, I.N. (1982). A possibility of substituting Ca^{2+} by Ba^{2+} or Mn^{2+} during the contractile processes of complex stomach smooth muscles. *General Pharmacology: The Vascular System* 13 (6): 511–513.

Modak, P. (2011). Waste: investing in energy and resource efficiency. In: *Towards a Green Economy: Pathways to Sustainable Development and Poverty Eradication*. United Nations Environment Programme Available at: http://www.unep.org/greeneconomy/Portals/88/documents/ger/o.o CoverFrontmatter.pdf.

Mold, M., Umar, D., King, A. et al. (2018). Aluminium in brain tissue in autism. *Journal of Trace Elements in Medicine and Biology* 46: 76–82.

Moya, J., Phillips, L., Schuda, L. et al. (2011). *Exposure Factors Handbook*, 2011e, vol. 52. Washington: US Environmental Protection Agency.

National Research Council (1993). *Measuring Lead Exposure in Infants, Children, and Other Sensitive Populations*. Washington (DC): National Academies Press (US) Adverse Health Effects of Exposure to Lead. Available from: https://www.ncbi.nlm.nih.gov/books/NBK236465/ (accessed 9 August 2022).

Negri, E., Bosetti, C., Fattore, E. et al. (2003). Environmental exposure to polychlorinated biphenyls (PCBs) and breast cancer: a systematic review of the epidemiological evidence. *European Journal of Cancer Prevention* 12 (6): 509–516.

Noel-Brune, M., Goldizen, F.C., Neira, M. et al. (2013). Health effects of exposure to e-waste. *The Lancet Global Health* 1 (2): 70.

Ongondo, F.O., Williams, I.D., and Cherrett, T.J. (2011). How are WEEE doing? A global review of the management of electrical and electronic wastes. *Waste Management* 31 (4): 714–730.

Oteng-Ababio, M. (2012). When necessity begets ingenuity: e-waste scavenging as a livelihood strategy in Accra, Ghana. *African Studies Quarterly* 13: 1–21.

Pacyniak, E.K., Cheng, X., Cunningham, M.L. et al. (2007). The flame retardants, polybrominated diphenyl ethers, are pregnane X receptor activators. *Toxicological Sciences* 97 (1): 94–102.

Perkins, D.N., Drisse, M.N.B., Nxele, T. et al. (2014). E-waste: a global hazard. *Annals of Global Health* 80 (4): 286–295.

Peters, A.K., Nijmeijer, S., Gradin, K. et al. (2006). Interactions of polybrominated diphenyl ethers with the aryl hydrocarbon receptor pathway. *Toxicological Sciences* 92 (1): 133–142.

Piasek, M. and Laskey, J.W. (1999). Effects of in vitro cadmium exposure on ovarian steroidogenesis in rats. *Journal of Applied Toxicology* 19 (3): 211–217.

Priyadharshini, S. and Meenambal, T. (2011). A survey on electronic waste management in Coimbatore. *International Journal of Engineering, Science and Technology* 3: 2099–2104.

Quinteros, F.A., Machiavelli, L.I., Miler, E.A. et al. (2008). Mechanisms of chromium (VI)-induced apoptosis in anterior pituitary cells. *Toxicology* 249 (2–3): 109–115.

Roper, W.L. (1992). *Toxicological profile for antimony and compounds*, 1–5. Atlanta: Agency for Toxic Substances and Disease Registry. Public Health Statement.

Sanders, J.M., Burka, L.T., Smith, C.S. et al. (2005). Differential expression of CYP1A, 2B, and 3A genes in the F344 rat following exposure to a polybrominated diphenyl ether mixture or individual components. *Toxicological Sciences* 88 (1): 127–133.

Sanderson, W.T., Ward, E.M., Steenland, K. et al. (2001). Lung cancer case–control study of beryllium workers. *American Journal of Industrial Medicine* 39 (2): 133–144.

Shamim, A., Mursheda, A.K., and Rafiq, I. (2015). E-waste trading impact on public health and ecosystem services in developing countries. *International Journal of Waste Resources* 5 (4): 1–18.

Sharma, N., Kumar, A., Singh, S. et al. (2022). Multi-residue determination of pesticides in vegetables and assessment of human health risks in Western Himalayan region of India. *Environmental Monitoring and Assessment* 194: 332. https://doi.org/10.1007/s10661-022-09992-9.

Shi, J., Xiang, L., Luan, H. et al. (2019). The health concern of polychlorinated biphenyls (PCBs) in a notorious e-waste recycling site. *Ecotoxicology and Environmental Safety* 186: 109817.

Tao, S.H. and Bolger, P.M. (1997). Hazard assessment of germanium supplements. *Regulatory Toxicology and Pharmacology* 25 (3): 211–219.

Tharappel, J.C., Lee, E.Y., Robertson, L.W. et al. (2002). Regulation of cell proliferation, apoptosis, and transcription factor activities during the promotion of liver carcinogenesis by polychlorinated biphenyls. *Toxicology and Applied Pharmacology* 179 (3): 172–184.

Tower, S.S. (2010). Arthroprosthetic cobaltism: Neurological and cardiac manifestations in two patients with metal-on-metal arthroplasty: A case report. *The Journal of Bone & Joint Surgery* 92 (17): 2847–2851.

Tsai, S.Y., Chou, H.Y., The, H.W. et al. (2003). The effects of chronic arsenic exposure from drinking water on the neurobehavioral development in adolescence. *Neurotoxicology* 24 (4–5): 747–753.

United Nations Environment Programme (UNEP), n.d. Stockholm convention, http://chm.pops.int/TheConvention/ThePOPs/tabid/673/Default.aspx

United States Environmental Protection Agency (1997). Exposure factors handbook. In: *EPA/600/P95/002F*. Washington DC: Office of Research and Development.

United States Environmental Protection Agency (1999). *A Risk Assessment Multiway Exposure Spreadsheet Calculation Tool*. Washington DC: United States Environmental Protection Agency.

United States Environmental Protection Agency. (2009). *Provisional Peer-reviewed Toxicity Values for Vanadium and Its Soluble Inorganic Compounds Other Than Vanadium Pentoxide (CASRN 7440-62-2- and Others) EPA/690/R-09/070F*.

USEPA (2007). Appendix F: The Risk Quotient Method and Levels of Concern. Available at: https://www3.epa.gov/pesticides/endanger/litstatus/effects/appendix_f_rq_method_and_locs.pdf

Vinceti, M., Wei, E.T., Malagoli, C. et al. (2001). Adverse health effects of selenium in humans. *Reviews on Environmental Health* 16 (4): 233–252.

Vinceti, M., Malagoli, C., Werler, M.M. et al. (2018). Adverse pregnancy outcomes in women with changing patterns of exposure to the emissions of a municipal waste incinerator. *Environmental Research* 164: 444–451.

Wath, S.B., Dutt, P.S., and Chakrabarti, T. (2011). E-waste scenario in India, its management and implications. *Environmental Monitoring and Assessment* 172 (1): 249–262.

Wen, S., Yang, F.X., Gong, Y. et al. (2008). Elevated levels of urinary 8-hydroxy-2′-deoxyguanosine in male electrical and electronic equipment dismantling workers exposed to high concentrations of polychlorinated dibenzo-p-dioxins and dibenzofurans, polybrominated diphenyl ethers, and polychlorinated biphenyls. *Environmental Science and Technology* 42 (11): 4202–4207.

White, S.S. and Birnbaum, L.S. (2009). An overview of the effects of dioxins and dioxin-like compounds on vertebrates, as documented in human and ecological epidemiology. *Journal of Environmental Science and Health. Part C* 27 (4): 197–211.

White, S.S., Stanko, J.P., Kato, K. et al. (2011). Gestational and chronic low-dose PFOA exposures and mammary gland growth and differentiation in three generations of CD-1 mice. *Environmental Health Perspectives* 119 (8): 1070–1076.

Widmer, R., Oswald-Krapf, H., Sinha-Khetriwal, D. et al. (2005). Global perspectives on e-waste. *Environmental Impact Assessment Review* 25 (5): 436–458.

Winship, K.A. (1988). Toxicity of tin and its compounds. *Adverse Drug Reactions and Acute Poisoning Reviews* 7 (1): 19–38.

Yang, Y., Zhang, M., Chen, H. et al. (2022). Estimation of Children's soil and dust ingestion rates and health risk at e-waste dismantling area. *International Journal of Environmental Research and Public Health* 19 (12): 7332.
Yu, X.Z., Gao, Y., Wu, S.C. et al. (2006). Distribution of polycyclic aromatic hydrocarbons in soils at Guiyu area of China, affected by recycling of electronic waste using primitive technologies. *Chemosphere* 65 (9): 1500–1509.
Zambelli, B. and Ciurli, S. (2013). *Nickel and Human Health: Interrelations Between Essential Metal Ions and Human Diseases*, 321–357. Dordrecht: Springer.

6

Metal Resources in Electronics: Trends, Opportunities and Challenges

Marcelo P. Cenci, Daniel D. Munchen, José C. Mengue Model, and Hugo M. Veit

Materials Engineering Department, Federal University of Rio Grande do Sul, Porto Alegre/RS, Brazil

Abstract

Throughout the years, technological advances have imposed a variable content of metals in electronics. New functionalities and manufacturing methods require the use of a wide range of metals; some of them which have high strategic and economic relevance. Since the use of electronics is increasing, the major concerns rely on the metals' shortage supply and waste of resources. To close the materials' cycle – the return to the production chain – recycling appears as an alternative, which brings opportunities for a sustainable lifecycle. However, the challenges faced are economic and know-how barriers, yet to be overcome. The present chapter analyzes the use of metals in electronics, embracing the composition timeline, future tendencies, and opportunities. In addition, it will discuss how composition can impact the use of resources and the further recycling process. The analysis is mainly based on the composition of printed circuit boards, screens, batteries, LED lamps, and magnets.

Keywords *E-waste; Recycling; Metals; Cell phone; Computer; PCB; Battery; LED lamp; Magnet; Urban mining*

6.1 Introduction

The amount of metals in electric and electronic equipment (EEE) has decreased worldwide, mostly substituted by polymers and composites, which are lighter and cheaper. For instance, in a 10-year gap, the content of gold and silver in personal computers' motherboards has declined by 40%. Even so, the content of both these metals in e-waste was estimated to be 50 times higher than in mineral resources (Izatt and Hagelüken 2016). According to Forti et al. (2020), the value

Management of Electronic Waste: Resource Recovery, Technology and Regulation, First Edition. Edited by Anshu Priya.

of raw materials that e-waste generated in 2019 was approximately $57 billion USD, mostly regarding metals, such as copper, iron, and gold. Additionally, the recycling rate was reported at only 17.4% in the same year, meaning that only $10 billion USD was recovered from e-waste. EEE characterization has been conducted for over three decades. But the consensus on how to divide the quantification is somewhat confusing. Most researchers categorize not by chemical classes, such as "metals," "polymers" "ceramics," but by parts, such as "cables," "printed circuit board," and so on, which in fact contain mixtures of metals, polymers, and composites. Consequently, metals in EEE are difficult to quantify due to the variability of the parts in which they are contained. This leads to an undersized quantification that impairs recycling.

The three important reasons for recovering metals from e-waste are economic, environmental, and public health. From an economic perspective, metals have market value, as pure or as alloys, depending on the new application; from a recycling perspective, at a certain scale, it eases the burden on the environment because it minimizes the extraction of raw materials from mineral deposits, as well as contributes to the security of raw materials supply and to a more circular economy; and finally from the public health perspective, the hazardous substances that represent risks for human health and/or the environment – brominated flame retardants and toxic metals like mercury and lead – are avoided (Arduin et al. 2020).

The EEE addressed in this chapter has undergone a great variability in metal composition due to the changing technologies over years. In addition, aside from the base metals (Fe, Cu, Al, and Ni), other metals (rare earths, Ta, Ga, and Ge, for instance) have gained attention in electronic devices, demanding higher volumes of exploration and refinement, hence, requiring future recycling.

6.2 Composition of Different EEE Components: Past, Present, and Tendencies

6.2.1 Printed Circuit Boards (PCBs)

Printed circuit boards (PCBs) (or printed wiring boards – PWBs) are components commonly found in various types of electronic equipment, from the simplest to the most technological. Basically, PCBs are produced in the form of a sandwich of conductive layers (or conductive tracks) and insulating layers, usually produced with copper, phenolic and epoxy resins, and fiberglass (LaDou 2006; Rocchetti et al. 2018). The main function of PCBs is to connect different types of electronic components/circuits in the same structure. The first developments, for what we now call PCBs, began in the early twentieth century, with the German Albert Hanson, in 1903, who described the idea of a conductor (copper or bronze) in

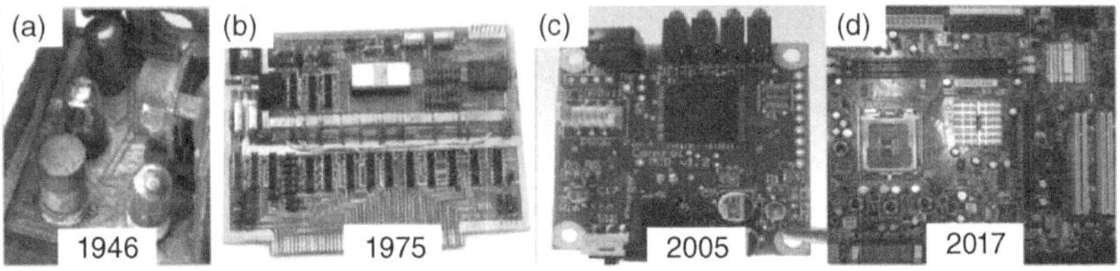

Figure 6.1 Evolution of PCBs over the years. Source: Martins et al. (2021)/Reproduced with permission from Elsevier.

the form of a flat sheet interspersed with an insulating sheet in multiple layers. In 1094, Thomas Edison performed chemical experiments for the deposition of conductors on linen paper (Harper 2003). In the following years, there was a great development in the design of PCBs; however, until the 1950s the use of PCBs was very restricted and especially aimed at military applications. (Harper 2003). In 1952, Motorola announced the use of "plated circuits" in household radio equipment, and from then on, the use of PCBs became increasingly common. Figure 6.1 presents different PCBs throughout their history.

The production of PCBs on a global level has changed in recent decades. In the 1980s, the United States of America accounted for 42% of global production, dropping to less than 10% in 2006. However, in 2006, three-quarters of the global production of PCBs was already carried out in Asia, especially in China, with more than 1000 producers (LaDou 2006). The amount of PCBs produced globally is difficult to estimate, but it is considered that about 3–5% of the electronic waste generated is composed of PCBs, which would currently be equivalent to about 2 million metric tons (Faraji et al. 2022).

In terms of chemical composition, PCBs have also changed throughout their history. In addition to the time issue, the type of PCBs, their application, and their manufacturer also affected the variation of chemical composition. Among the metals, copper stands out, having been used since the first generations of PCBs for the production of conductive tracks/paths (Hadi et al. 2015). Other metals also play a key role, such as lead, tin, and silver, used as metals/alloys for soldering (George and Pecht 2016), nickel, gold, platinum, and palladium, often used as coating metals (corrosion-resistant and electrical conductors) (Huang et al. 2022), in addition to aluminum, iron, and zinc used in more structural applications. The insulating layer also had changes in the materials used throughout history, starting with wood, paper, phenolic resin, epoxy resin, and finally epoxy resin reinforced with fiberglass (Kaya 2016; Williams 2005). Table 6.1 presents some examples of PCB composition over a period of 15 years (1991–2006). It can be clearly seen that base metals such as copper, nickel, tin, lead, zinc, and aluminum did not show a significant variation trend during this period. With regard to iron and gold, there is a tendency for the levels to decrease over the years. Silver showed a trend toward stability.

Table 6.1 Metallic composition of PCBs found by different authors in a period of 15 years (1991–2006).

%	1991	1997	1997	2004	2004	2006	2006
Cu	20	10	22	26.8	15.6	17.85	23.47
Al	2	7	–	4.7	–	4.78	1.33
Pb	2	1.2	1.55	–	1.35	4.19	0.99
Zn	1	1.6	–	1.5	0.16	2.17	1.51
Ni	2	0.85	0.32	0.47	0.28	1.63	2.35
Fe	8	–	3.6	5.3	1.4	2.0	1.22
Sn	4	–	2.6	1.0	3.24	5.28	1.54
Sb	0.4	–	–	0.06	–	–	–
Au/ppm	1000	280	350	80	420	350	570
Pt/ppm	–	–	–	–	–	4.6	30
Ag/ppm	2000	110	–	3300	1240	1300	3301
Pd/ppm	50	–	–	–	10	250	294

Adapted from Ogunniyi et al. (2009).

Another possible comparison is in relation to the type of PCB. Table 6.2 shows the differences between a computer motherboard, memory board, and smartphone board.

In Table 6.2, it is possible to verify that the largest amount of metals in the composition of PCBs occurs in cell phones, with emphasis on the high content of copper, silver, gold, nickel, and tin. The development of smartphones, with new features for each new model launched, makes it possible to foresee the use of increasingly lighter and more technological PCBs.

Regarding the new trends in the PCB industry, we can basically mention two lines of action: the first involving the search for more sustainable manufacturing

Table 6.2 Composition differences for different types of PCBs

PCB type	Content (wt%)									Total
	Ag	Al	Au	Cu	Fe	Ni	Sn	Zn	Pb	
Motherboard	0.045	2.660	0.002	19.650	8.340	0.270	1.990	2.550	NR	**35.51**
Memory board	0.053	1.403	0.069	13.830	1.470	0.320	1.700	0.043	NR	**18.91**
Mobile phone	0.2	NR	0.12	41.8	0.66	1.93	4.57	NR	0.19	**49.47**

NR = not reported.

Adapted from Martins et al. (2021) and Tan et al. (2017).

processes from an environmental point of view (Ochoa et al. 2019) and the second line in the area of flexible electronics (chemical/bio sensors, human–computer interfaces, flexible optoelectronics and displays, electronics skins, and many others) (Khalifeh 2020; Wang et al. 2022).

6.2.2 LED Lamps

By characteristics of eco-friendliness, energy-efficiency, and reliability, the development of white LEDs and LED lamps was considered a revolution in the market of general lighting, which has allowed the spread of a more sustainable alternative to light sources (Montoya et al. 2017; Thejokalyani and Dhoble 2014). The consumption of LED lamps has rapidly grown in the last decade, being expected to reach 76% of the total market share in 2025 (Placek 2021; Zissis and Bertoldi 2018). As LEDs are based on solid-state lighting, the possibility of design variations is also a factor that increases the acceptance of LED lamps by consumers. A wide range of formats, colors, and materials are available for decorative lamps and aesthetic applications (Gassmann et al. 2016). For these reasons, some consumers of LED lamps expect modern and unique designs, and this has implications for the material composition of devices. It is also important to note that the LED industry is still in its growth stage, and a wide spectrum of design possibilities is especially expected (Dzombak et al. 2017). Figure 6.2(a) demonstrates the variability of structures and designs, and Figure 6.2(b) demonstrates the current typical components that LED lamps can be divided into. Macro components are mainly polymeric and metallic pieces of carcass, a PCB, and a module where the LEDs are mounted.

Due to this alternance of designs, some studies have explored the material composition of LED lamps from a recycling and life cycle assessment perspective.

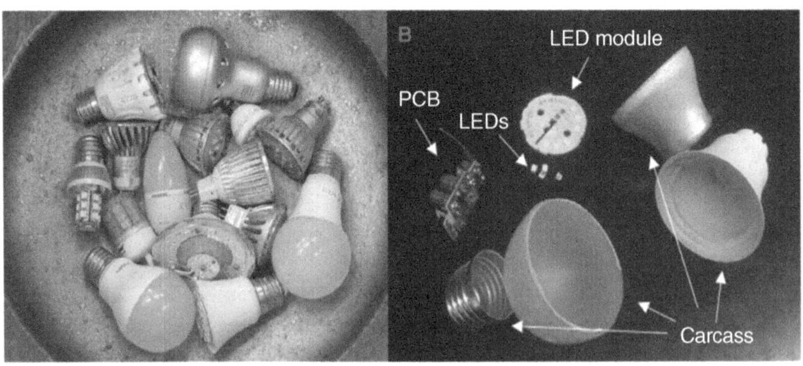

Figure 6.2 Design variability of LED lamps (a) and typical components (b). Sources: Gassmann et al. (2016)/LUGER RESEARCH e.U.Institute for Innovation & Technology/ https://www.led-professional.com/resources-1/articles/led-lamps-recycling-technology-for-a-circular-economy/last accessed June 20, 2023 and Cenci et al. (2020)/Reproduced with permission from Elsevier.

Table 6.3 Mass fraction of LED lamps' macro components over time.

Components	2009[a)]	2012[b)]	2014[c)]	2016[d)]	2019[e)]	2020[f)]	2020[g)]	2022[h)]
Glass (%)	12.1	0	6	15	0	0	0	0
Carcass metals (%)	40.3	44.8	20.1	44.2	21.6	20.97	18.6	16
Carcass polymers (%)	23.4	9.4	52.5	21.3	48	50.07	53.2	55
PCB and other electronic parts (%)	21.7	45.8	21.3	19.5	30.4	28.96	28.2	29
Total lamp (g)	NA	178	75.84	NA	58.24	39.21	42.89	NA

NA – not analyzed.

a) European Lamp Companies Federation (2009).
b) Scholand and Dillon (2012).
c) Sangwan et al. (2014).
d) Gassmann et al. (2016).
e) Dillon et al. (2019).
f) Martins et al. (2020).
g) Cenci et al. (2020).
h) Benmamas et al. (2022).

Table 6.3 explores the evolution of the lamps' composition by macro components since the year 2009, when LED lamps became consolidated in the market (Cho et al. 2017). All cited studies evaluated several brands and models of lamps, with the exception of the study of Scholand and Dillon (2012), which used only one model.

From Table 6.3, it is possible to visualize some structural and compositional tendencies for the lamps. In general terms, glass and carcass metals are being minimized, while carcass polymers are increasing in concentration. The total lamp weight is also decreasing, which may reflect improvements in the design and material selection of the lamps. For example, the mentioned minimization in the use of glass and metals can lead to a decrease in weight.

Glass is mostly applied as a structural material where the light can pass with minimum interferences. It was widely applied in incandescent and fluorescent (FL) lamps, but LED lamps can easily substitute it with polymers. In many lamps, the application of polymers as a medium where the light passes through is useful to promote a better dispersion of it (acting as a diffuser) (Dzombak et al. 2017). Cenci et al. (2020) found the use of polycarbonate as diffuser in all six models of LED lamps examined. In Figure 6.2(b), a polycarbonate bulb can be seen (identified as one of the carcass components).

Metals in the carcass typically refer to aluminum pieces applied as heat exchangers and Edison screws and pins for electricity connection (commonly made of aluminum under a protective layer of nickel) (Cenci et al. 2020). Early lamps had their mass dominated by heat sinks (components to promote heat exchange). However, their use is being minimized as result of efforts toward thermal performance of the devices (Dzombak et al. 2017). LED products are severely affected by the devices' thermal management, in which a poor thermal performance can cause degradation of light over time. Thus, despite the tendencies to minimize the use of metals, their use is essential and tend to continue to be relevant in the composition of lamps.

It is worth mentioning that some studies evaluated the macro-design evolution of LED lamps over time, and their results reinforce the aforementioned tendencies. Dzombak et al. (2017) found a decrease in the size and use of aluminum heat sinkers over time (with potential substitution by polymers in some cases), and the authors did not mention glass as diffusers. Dillon et al. (2019) also found a decrease in metals, the absence of glass, an increase in the use of polymers, and pointed out that the total mass of the lamps is significantly decreasing.

About the electronic parts (PCBs, LED modules, and LEDs) as macro-components, Wehbie and Semetey (2022) and Dillon et al. (2019) pointed out that the size of the LED modules is decreasing over time due to the enhancements in the luminous efficiency and heat transfer properties of the LEDs. Authors found out that, in modern designs, it is common to assemble the PCB and the LED module as only one component, potentially saving materials and minimizing weight. The PCBs of LED lamps have the main function of supplying appropriate voltage and current (Górecki 2013; Uddin et al. 2013), are typically single-sided and constructed with capacitors and transformers as main components, and are considered low-grade PCBs.

LED pieces are the most valuable and complex component of LED lamps, containing, in relevant concentrations, elements such as Ga, In, Cu, Ag, Au, and rare earths (Cenci et al. 2021a). Basically, LEDs are structurally composed of a diode (or chip, which emits light), two electrical contacts (anode and cathode and wire bonds), a base made of copper or aluminum that serves as heat sink, a polymeric case for protection, and lens and encapsulator (mostly a phosphor material embedded in epoxy and silicone resins) (Nardelli et al. 2017). Over time, three types of LEDs were developed to emit white light in lamps. The first approach consists in mounting the LEDs with three chips of primary colors (red, green, and blue) to generate white light. The second approach is similar to FL lamps, in which the light from an ultraviolet chip passes through a phosphor (added into the encapsulator material) for conversion to a broadband white light. The third approach applies a blue chip and a phosphor-conversing material that absorbs part of the blue light and converts it to yellow (red and green), composing the white light (Cho et al. 2017; Feezell and Nakamura 2018). Although, these approaches do not configure an evolution over time, but rather different options, the market is dominated by the third approach, applying a GaN-based diode as chip and yttrium–aluminum garnet doped with cerium as phosphor ($Y_3Al_5O_{12}$:Ce, also called yellow phosphor) (Cho et al. 2017; Nair et al. 2020). In addition, a wide range of different

Table 6.4 Composition of waste white LEDs

Elements (%)	Characterization of waste LEDs			
	Cenci et al. (2020)	Cenci et al. (2021a)	Mir et al. (2022)	Oliveira et al. (2020)
Cu	56.84	65.03	15.01	50.45
Sn	8.89	4.26	3.09	10.3
Al	0.22	NA	32.38	0.29
Fe	0.61	NA	27.31	NA
Mn	NA	NA	0.94	NA
Zn	NA	NA	0.85	NA
Ni	0.13	NA	0.18	0.14
Pb	0.51	NA	0.03	8.36
Ag	0.6	0.12	0.06	0.17
Au	0.05	0.06	0.03	NA
Ga	0.23	0.17	0.05	0.07
In	ND	ND	<0.01	NA
As	NA	NA	0.02	NA
Y	0.49	0.36	0.12	0.4
Gd	NA	NA	NA	0.01
Eu	NA	NA	0.01	NA
Ce	0.01	0.01	0.01	0.01

NA – not analyzed; ND – not detected.

diode materials and phosphors can generate special light features for different applications. The review studies of Alim et al. (2021) and Nair et al. (2020) are recommended as an overview of the different materials that can be applied to LED chips and phosphors. Table 6.4 gathers some characterization results of white LEDs collected from different brands and models of lamps.

Alim et al. (2021) have reviewed some material tendencies of LEDs. The authors pointed out that Au wire bonding has been the primary option for LED packaging since the 90s, and it is being substituted for Cu and Ag due to lower costs and similar efficiency. Ag wire is gaining the market, and Ag paste is also used as chip-attachment material, justifying the high concentrations of Ag in Table 6.4. The authors also pointed out a tendency of minimizing the thickness of the phosphor layer embedded in a matrix, possibly leading to a reduction in the concentration of critical materials associated with the phosphors. The study of Gaffuri et al. (2021) presented some alternative structures for white LEDs, in order to promote a reduction in the use of critical materials. However, their application in lamps is still uncertain. The authors highlighted the optimization of nanotechnology in the field, mainly in the proposition of nanoscale blue chips and phosphor grains, which are also expected to reduce the number of light defects. There are also possibilities for the entire substitution of GaN-based chips with rare earth element-based phosphors. Quantum-dots LEDs (QLEDs), and organic LEDs (OLEDs) are some alternatives currently under research for general lighting applications (Arquer et al. 2021; Pode 2020). Carbon dots, and sulfide-nitride-silicate matrices (with Mn as a doping agent) are some alternatives for phosphors. These advances tend to reduce the use of rare earth metals and increase the importance of polymers in structures. Additional insights about QLEDs and OLEDs are given in the screen section.

6.2.3 Screens

It has been more than 70 years since cathode ray tubes (CRTs) were introduced to consumers and became consolidated in the market (Yu-Gong et al. 2016). CRTs were an essential part of all monitor or television devices for a long time until they started to be continuously replaced by thinner and lighter devices in the early 2000s (Qi et al. 2019; Singh et al. 2016). Some new technologies entered and conquered the market, and the current waste stream of obsolete screens is composed of a variety of technologies, namely, CRTs (which are still relevant from a waste management point of view), liquid crystal displays (LCDs) with FL and LED backlights, plasma display panels (PDPs), and OLED displays (Qi et al. 2019; Ryan et al. 2011; Shama et al. 2016). About their technological life cycle, Jia (2018) claims that, while CRTs, PDPs, and FL LCDs are in their elimination phase, LED LCDs are in their maturity phase, and OLEDs and QLEDs are in their growth phase.

Although this section is focused on display technology, it is important to mention that screens are composed of several structural and electronic parts. Polymeric, ferrous, aluminum, and other pieces compose the carcass, and PCBs, speakers, and cables can be a source of valuable elements, such as gold, copper, palladium, and silver (Buchert et al. 2012; Herat 2008; Ryan et al. 2011; Wong 2004). It is well known that these parts are constantly evolving to possibly lighter and thinner designs. For example, a typical CRT device can easily reach more than 20 kg due to the thick layer of glass (Yu-Gong et al. 2016). Back in 2003, according to Aoe et al. (2003), a 32-in. television weighed 59 kg as CRT, 28 kg as PDP, and 21 kg as LCD. Nowadays, a 32-in. LED LCD screen weights around 5 kg (Lo 2022), and OLED devices are even thinner and lighter (Shama et al. 2016).

Before evaluating the evolution of the composition of the different types of displays, a brief description of their operation and technology is useful to comprehend the basic designs. Firstly, the designs of CRT, PDP, and LCD screens are explored. Then, the modern OLED and QLED designs are explained in more detail. Examples of typical designs of these screens are shown in Figure 6.3.

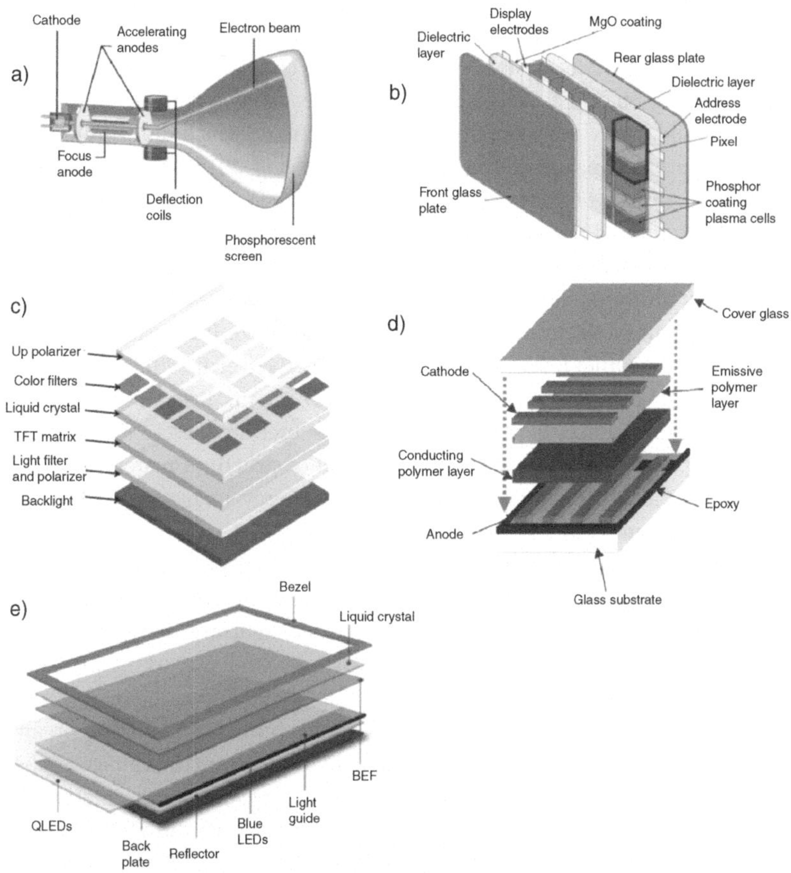

Figure 6.3 Screen design evolution. (a) CRT display design. Source: Adapted from Dhamija (2020); (b) PDP display design. Source: Kalyani and Dhoble (2015)/ with permission of Elsevier; (c) LCD display design. Source: Dai et al. (2017)/ John Wiley & Sons; (d) OLED display design. Source: Electronics Hub (2019)/Electronicshub; (e) photoluminescent QLED display design. Source: Tassone (2021)/ARC Publication.

A CRT is composed of a conical tube and a panel made of three types of glasses: an electron gun (cathode, a heated filament), focusing and deflecting coils, anodes to accelerate the electron beam, and a FL screen including a metallic shadow mask and coated phosphors (Herat 2008; Singh et al. 2016; Yu-Gong et al. 2016). Electrons are directed to hit the FL dots on the screen panel, forming an image. To provide a necessary shielding, the glasses of the neck and tube are impregnated with lead oxide, and the panel glass contains barium and strontium oxides (Nnorom et al. 2011). The working principle of PDP panels is based on the UV emission of noble gases (xenon, neon, argon, and helium) when a voltage is applied. The UV emission in each cell of the panel (pixels) reaches coated Red-Green-Blue phosphors (RGB oxides containing rare earth elements Y, Gd, Eu, and Tb) that generate visible colors (Buchert et al. 2012; Días et al. 2015; Hischier and Baudin 2010). Cells are contained between two metal oxides-coated glass plates and other components such as electrodes (including indium tin oxide [ITO]), protective, and dielectric layers (Chen et al. 2015; Manganello 2015; Tarafder et al. 2012). LCD panels need an FL or LED backlight source. Basically, the light emitted passes through polarizing and diffuser filters (polymeric sheets) and is directed to an array of liquid crystal pixels (biphenyl, azomethine, pyrimidine, or difluoro phenylene) placed between two glasses and electronically arranged by a pair of electrodes (ITO) (Gabriel et al. 2020; Juchneski et al. 2013; Shama et al. 2016). Other layers are also present, and Ryan et al. (2011) pointed out that more than 25 components can be found in LCD screens. Besides the backlight itself, the main differences between a FL or LED source are that a LED backlight allows lighter and thinner devices (minimization of structural materials) (Chen 2018; Peeters et al. 2012), and is more efficient in energy, image-generation, and environmental terms (Chen 2018; Savvilotidou et al. 2014).

OLED and QLED are currently growing screen technologies. OLED panels are flat sources producing homogeneous light over an area, without the need for a backlight unit (Pode 2020). The absence of a backlight unit permits the construction of lighter and thinner screens. OLEDs have similar working-principles to inorganic LEDs, in which a voltage applied to anode and cathode electrodes promotes the recombination of electrons and holes (into a sandwich of organic layers), releasing photon energy. The emitted color depends on the voltage applied and the composition of the various organic layers (particularly the material of the emitting layer) (Azrain et al. 2018; Pode 2020). Kalyani and Dhoble (2015) described many materials and designs applied to OLED panels. The construction of the panels relies on an active-matrix array for most complex displays (active matrices need a thin film transistor layer like LCD screens), with the cells containing an RGB structure (OLEDs for red, green, and blue emission into a pixel) (Yin et al. 2019). Cells are basically mounted on a glass substrate, an anode layer is positioned above the substrate (commonly an ITO layer and the thin film transistor serves as anode in active matrices), and organic active layers are sandwiched between the anode and the cathode, in which the last is commonly made of Mg:Ag, LiF, and Mg: Al-based materials.

Although QLEDs can be both photoluminescent and electroluminescent materials, current QLED screens are quite similar to LCD screens (applied as photoluminescent material), in which QLEDs replace the backlight unit, drastically improving the efficiency in energy, lifetime, and image-quality terms (Arquer et al. 2021; Dai et al. 2017; Jia 2018; Tassone 2021). QLEDs are crystals ranging from 2 to 10 nm, mostly made of cadmium selenide-based materials (Arquer et al. 2021), and due to their small size, they influence the energy gap and the color emitted by the electrons (the color emitted depends on the material and size of the crystals, for example, nearer to 2 nm tends to blue, and nearer to 10 nm tends to red) (Tassone 2021). A QLED array can be positioned above a UV or blue LED backlight, absorbing the radiation and emitting RGB colors. In practice, QLEDs substitute the phosphor materials of the LED LCD backlight (which contains Y, Eu, and others), and also present structures with substrate, cathode, anode, and emitting layers (Dai et al. 2017; Jia 2018; Tassone 2021).

At this point, only by analyzing the evolution of technology, it is possible to infer some tendencies in the composition of obsolete screens. CRT devices relied on a high portion of glass in their structures. Other types of screens also apply layers of glass into the panels, but its application, in quantity, is less and less needed. The emerging of newer screens based on LED technology is leading to an increase in the importance of semiconductor materials, both as electrodes or diodes (a layer of ITO as anode is still present in most types of screens). The increasing importance of polymeric layers as functional materials can also be noted, mainly considering the advances from CRT screens, in which polymers had no application, until OLED screens, in which polymers are the central materials. A reduction in the use of phosphors is also notable, mainly considering newer OLED and QLED devices, with particular importance for the reduction in the use of critical rare earth metals. In environmental terms, advances in screen technologies eliminated the use of lead glasses in CRTs and the use of mercury in FL LCDs. Table 6.5 provides composition data of different types of screens, in which tendencies can be analyzed. For OLED and QLED screens, characterization studies from a waste management perspective are still rare, as they are in a growth stage in the market.

Table 6.5 confirmed some of the tendencies aforementioned. Data show that the mass fraction of the display compared with the total device is being reduced over time. This may be due to the minimization of the use of glass and the increasing use of polymers. The concentration of phosphors is also reducing (for example, the use of Y and Eu), and newer OLEDs and QLEDs additionally replace these materials. In, which is a traditional target in recycling studies of screens, is keeping their relevance over time (concentrations are maintained from PDP panels to LED LCD), and OLED and QLED screens also require ITO layers. In LED LCD screens, the element Ga emerged in importance, as it is a constituent of the LED semiconductor materials, and may be relevant for recycling as critical material. Current photoluminescent QLED screens still need a backlight of blue LEDs. However, modern QLED screens apply blue-OLEDs as backlight, overcoming the use of Ga (Pode 2020). Even being minimized in comparison with older screens, OLED screens still contain some rare earth metals in their composition. In literature, W, Mo, V, In, Eu, Tb, Ir, Li, and others are cited as possible constituents of the organic emissive layers, electrodes, and other components (Pode 2020; Wang et al. 2020).

With the advances in OLED and QLED screens, these technologies still have a great potential for improvements, and a prediction for their material compositions

Table 6.5 Composition of screens according to different authors

Materials	Composition[e]			
	CRT[a]	PDP[b]	FL LCD[c]	LED LCD[d]
Display fraction weight (%)	67	26	21	24
Glass (%)	85	64	45	45
BaO	*7.8*	*11.25*	*NF*	*NF*
SrO	*7.8*	*9*	*NF*	*NF*
PbO	*9.5*	*0*	*NF*	*NF*
Phosphors	*13*	*8.16*	*NF*	*NF*
Polymers (%)	NF	NF	53	53
Al (ppm)	NF	NF	5	399
Cu (ppm)	NF	NF	14	11
Zn (ppm)	NF	NF	1305	112
Sn (ppm)	NF	NF	255	508
In (ppm)	NF	326	263	273
Y (ppm)	NF	1661	53	11
Ga (ppm)	NF	NF	0	11
Eu (ppm)	NF	152	4	0.2
La (ppm)	NF	NF	3	0
Ce (ppm)	NF	NF	2	1
Gd (ppm)	NF	455	0.4	5
Tb (ppm)	NF	NF	1	0
Others	15 (polymers, copper, steel)	Structural materials	~1	~1

a) Herat (2008), Matsuto et al. (2004), Nnorom et al. (2011), Yu-Gong et al. (2016).
b) Chen et al. (2015), Hischier and Baudin (2010), Manganello (2015).
c) Amato and Beolchini (2018), Buchert et al. (2012), Juchneski (2013), Savvilotidou et al. (2014).
d) Dhiman and Gupta (2020), Juchneski (2013), Lahtela et al. (2019), and Peeters et al. (2012).
e) Considering CRT and PDP TVs, and LCD monitors; NF – reliable data not found in the literature.

and recycling targets is difficult to provide as there is no consolidated design yet. According to Jia (2018), the future of the screen market may be strongly dictated by the developments in flexible panels. This has some implications for recycling because some materials have to be replaced by flexible ones. The most relevant adaptation is the replacement of the ITO electrode layer (too brittle for this new application), and some candidates for substitution are based on silver nanowires, graphene, PEDOT (poly [3,4-ethylenedioxythiophene] polystyrene sulfonate), thin metal layers, and carbon nanotubes (Pode 2020). The replacement of the ITO layer is also the subject of studies, in which SnO, SbO, ZnO, graphene, carbon nanotubes, copper nanowires, and others are candidates (Kalyani and Dhoble 2015; U.S. Geological Survey 2022). In addition, advances toward pure organic OLEDs are reported in the review of Anand et al. (2021) and toward white OLEDs in the reviews of Chiu et al. (2022) and Yin et al. (2019).

In comparison with OLEDs, electroluminescent QLEDs promise a generation of cost-effective, energy-saving, ultrathin, and simplified-structural displays (Dai et al. 2017). The main advance from photoluminescent to electroluminescent QLEDs is the absence of a backlight and LCD unit, which allows a reduction in the use of materials, size, and weight. Materials for the fabrication of electroluminescent QLEDs are still evolving. Currently, literature mostly reports RGB QLEDs based on Cd–Se–Zn materials, including a structure with a thin film transistor layer, anode electrode of ITO, and other layers that contain organic materials and metal oxides (Arquer et al. 2021; Dai et al. 2017). Efforts are being made to eliminate Cd in the structure (due to environmental issues), and InP, ZnSe, silicon, and carbon-based materials are candidates (Arquer et al. 2021). Similarly to the current advances in OLEDs toward white OLEDs and flexible displays (requiring material substitutions), future advances in QLED screens may also pass through these stages, and new applicable materials will be uncovered.

6.2.4 Batteries

Since the electrochemical study began, in 1600, the perception of "animal electricity" by Galvani, in 1780, and the first battery created by Volta, in 1800, great advances have been made in the development of the very efficient and modern batteries that are available today. In Figure 6.4, a brief timeline of battery evolution is presented.

The studies about lithium-ion batteries (LIBs) began in 1976 by Stanley Whittingham (Whittingham 1976); however, some problems related to the dendrite (lithium crystals) formation during the discharge process made them unsafe. After years of development, in 1991, the Sony Corporation inserted into the market the first safety LIB, which was produced according to Akira Yoshino's work (Nishi 2001). Since then, with continuous improvement, they are the most employed resource energy in portable devices, such as smartphones, laptops, electric vehicles, and energy storage systems (noncontinuous/intermittent generation, i.e. photovoltaic solar energy) because of their high density of energy, no memory effect, and long cycle of life compared to the other kinds of batteries (Broussely and Pistoia 2007; Lavoie et al. 2017). The lithium ores have been explored for LIBs manufacturing for over 40 years. Due to this being a well-established technology whose use has grown steadily, it is extremely important that obsolete LIBs are used as a secondary source of lithium.

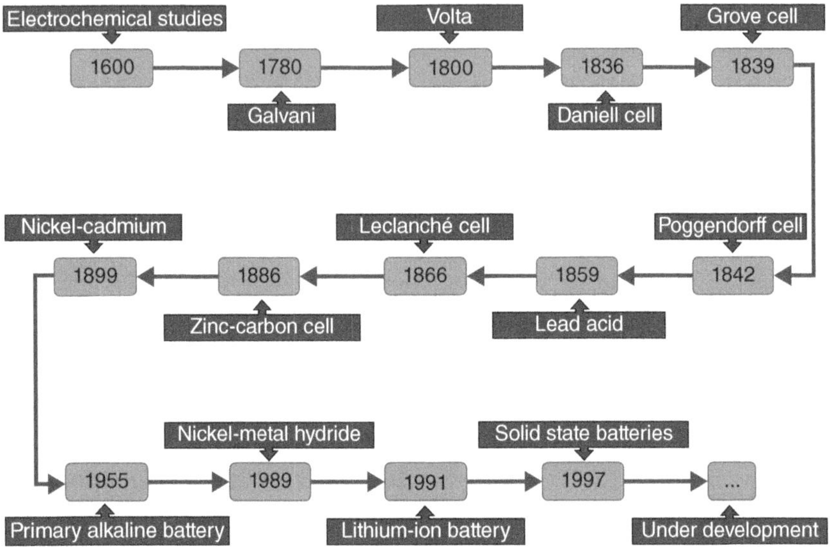

Figure 6.4 Timeline of battery evolution. Source: Adapted from Buchmann (2011).

The market has shown continuous growth, and it is expected that this projection will remain positive for LIBs; as an effort to attend this great and significant demand, many investments have been made in order to increase their stability and robustness. A report with 18 scenarios, as a result of 8 demands created from 4 different supply conditions, demonstrates that the amount of lithium is critical for the process of transition to more sustainable energies that the world needs (Greim et al. 2020).

Batteries are the main application of lithium consumption, which has been increasing in the last few years. According to the United States Geological Survey (USGS) report, 65% was used in batteries in 2020 (U.S. Geological Survey 2020), while in 2018 it represented 48% (Ober 2018). The projection for 2022 shows a growing tendency, reaching 74% (U.S. Geological Survey 2022). The same increase is observed for cobalt, another important metal present in LIBs. The German Mineral Resources Agency (DERA) reported that the cobalt demand worldwide would increase by 103% by 2026 due to the progressive scenario. In 2017, 43.7% of the cobalt demand was attributed to LIB production, of which 8.2% was for electric vehicles, pointing to a new application boom (Al Barazi et al. 2018). This continuous increase can be justified by politics engaged in the decarbonization of transport and mobility as a goal to reduce emissions of greenhouse gases (Ruhnau et al. 2019).

About 25 industries in North America and Europe are engaged in LIB recycling. Some partnerships between battery recyclers and automobile companies have been made to ensure material sources are used in new LIBs. Those initiatives are crucial because the high demand for lithium around the world, made the price increase from $ 7000 per ton in January 2021 to about $ 26,200 in November 2021 (U.S. Geological Survey 2022).

According to Liu et al. (2021), in 2018, 4.2% (400 tons) of lithium needed to manufacture new LIBs was provided from batteries at end-of-life (EOL). In a study conducted by Barik et al. (2016), the average percentage of cobalt and lithium present in a smartphone LIB is 19.25% (3.63 g) and 2.36% (0.44 g), respectively.

The LIBs' life cycle is approximately two years, when they achieve their EOL and are considered e-waste; hence, it is really necessary to embrace initiatives that claim to be eco-friendly destinations. According to The Global E-Waste Monitor (Forti et al. 2020), it is estimated that the world will generate 11 million tons of LIBs by 2030. On the other hand, around just 5% of them are recycled due to inefficient collection, after their useful lives, and the costs inherent to the recycling processes, including the labor involved in reverse logistics. Aware of these difficulties, recycling policies need to be intensified (Natarajan et al. 2018).

6.2.5 Magnets

In 1969, the uses of permanent magnets (PM) – materials that retain large magnetic moment in the absence of magnetizing force – were described worldwide as: loudspeakers 30%, motors and generators 20%, measuring instruments 15%, communications and electronics 15%, latching and holding devices 10%, separation equipment 4%, and toys, novelties, and miscellaneous 6% (Gould 1969). According to the chronology of PM discoveries, the only 2 PMs in use at that time were AlNiCo and barium/strontium ferrites. But in this same year, samarium–cobalt alloys were reported for the first time with improved magnetic properties in comparison with the precursors (Das 1971). Later, in the 1980s, another important type of magnet was developed: Neodymium–Iron–Boron (NdFeB), with the highest magnetic strength ever reported, which broadened the applications' diversity.

These four main PM subgroups differ greatly regarding the characteristics that mostly influence applications: the magnetic strength, the operating temperature, and the cost. Table 6.6 depicts these features according to different authors. For instance, for motors demanding high temperatures, SmCo is more suitable since NdFeB magnets only stand up to an average of 100 °C. Conversely, generators used in wind turbines – providing green energy – contain NdFeB magnets due to their high magnetic strength demand combined with their rather low cost compared to SmCo. The continuous change in composition, and hence in microstructure and chemical phase, remains the key factor in magnetic improvements throughout the years. Moreover, the constant pursuit of reduced device sizes combined with increased efficiency prompted a fast development of different PMs. For instance, it is reported that to achieve the same 1000 G (magnetic force), an NdFeB magnet

Table 6.6 Permanent magnets' characteristics relevant for applications

	Ferrites	Alnico	SmCo	NdFeB
Magnetic strength [$(BH)_{max}$] (MGOe)[a)]	0.7–5.2	1.5–5.1	16–34	28–56
Operating temperature (°C)[a)]	350	450–550	250–350	70–220
Cost ($/kg)[b)]	6.4	56	95	70

Note: all PMs were fabricated according to the sintered method.
a) Wang et al. (2022).
b) Cui et al. (2022).

developed in 1995 needed a piece volume of 0.22 cm^3, while an AlNiCo magnet produced in 1950 needed 11.9 cm^3 (Lewis and Jiménez-Villacorta 2012).

The Sr/Ba-ferrites are ceramic magnets that, since their conception, have been mostly composed of iron, barium, strontium, and lead oxides. Although a rather low magnetic product $(BH)_{max}$, the cost was – and still remains – inexpressive and is chemically inert (Lewis and Jiménez-Villacorta 2012). Meanwhile, AlNiCo magnets, besides aluminum, nickel, and cobalt, also maintain iron, copper, titanium, and silicon in their composition throughout the years.

Back in the late 1960s, the ferromagnetic intermetallic phase of the type $RE(TM)_5$ was discovered, in which RE could be yttrium, cerium, praseodymium, or samarium, and TM a transition metal, mostly cobalt. The continuous exploration of the alloys led to the discovery of $RE_2(TM)_{17}$. Instabilities in co-production, mostly in Democratic Republic of Congo in the late 1970s, led the PM industry to achieve new developments, which resulted in the $RE_2Fe_{14}B$ class of materials, where RE could be neodymium, praseodymium, dysprosium, or even a mixture. The dysprosium addition (terbium and holmium also, but dysprosium preferably due to better efficiency/cost ratio) increases coercivity and temperature dependence: 1 at.% of neodymium substituted by dysprosium increased anisotropy ($H_{c,j}$) in 6 kOe (Tenaud et al. 1991). The neodymium content of a magnet can be somewhat reduced with the substitution of praseodymium, which has a minor impact on magnetic performance. This practice began in the early 2000s and was hastened by the price increases in 2010 and 2011 (Smith and Eggert 2016).

In terms of NdFeB magnets, comparing the years 2000 and 2007 in Japan, the share of applications increased, especially for motors and generators, from nearly 27% to 46%, meanwhile HDD/communications, MRI machines, and acoustic devices have withered (Sugimoto 2011). In the meantime, in China, from 2005 to 2015, the REPM application shares also showed an increase in motors, electric/hybrid vehicles, and wind generators (Dong et al. 2017). NdFeB production represented 75% of the market (of magnets) in 1991; in 1990, it was reported at US$ 285 million in sales (Tenaud et al. 1991).

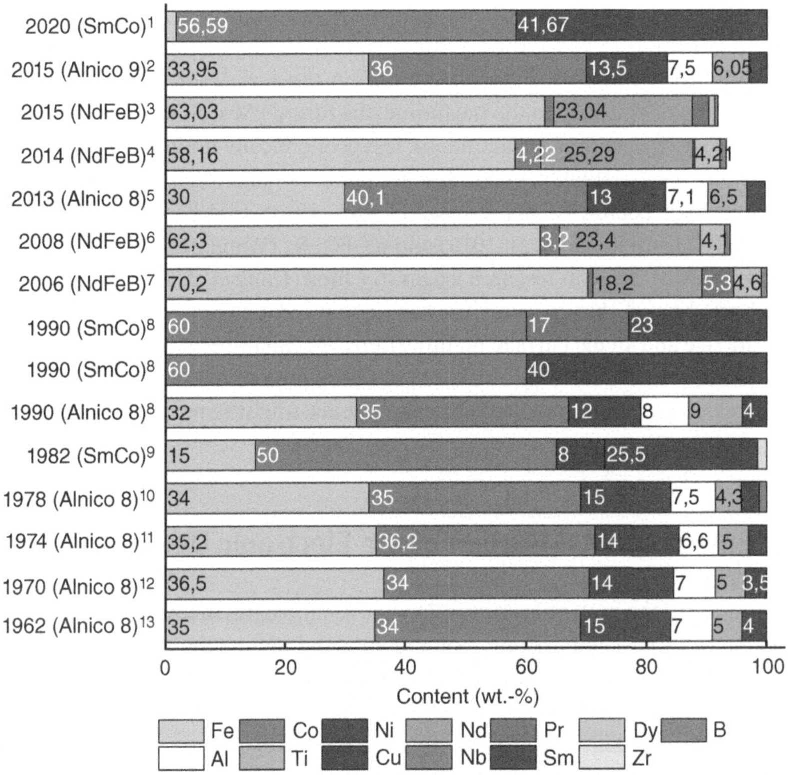

Figure 6.5 Elemental composition variation (wt.-%) of PM in accordance with the year of production. Note: ferrites were not included since compositions are mostly expressed in oxide form. Sources: [1]Prodius et al. (2020), [2]Sun et al. (2015), [3]Ueberschaar and Rotter (2015), [4]Hoogerstraete et al. (2014), [5]Xing et al. (2013), [6]Miura et al. (2008), [7]Saguchi et al. (2006), [8]Strnat (1990), [9]Rabenberg et al. (1998), [10]Durand-Charre et al. (1978), [11]Iwama and Takeuchi (1974), [12]Hoffmann and Stäblein (1970), and [13]Cronk (1966).

A few PM compositions throughout the years are depicted in Figure 6.5. It can be observed that iron is the main element, present in most of the magnets due to supply availability in addition to the lower price. Cobalt is also found in most NdFeB magnets, although in low amounts. This fact highlights the magnetic importance of this metal over the years. Another significant aspect is the sum of rare earths, which can reach up to 30 wt% in NdFeB. Regarding the year's analysis, the changes are subtle within the same type, mostly relying on the specific application that the industry demands. In economic terms, chemical element substitutability is affected by the production and consumption patterns in response to market changes (for example, price changes) (Smith and Eggert 2016).

Software programs based on property calculations are a modern tool used nowadays to search for modified PM for future purposes (McCallum et al. 2014).

In the case of NdFeB magnets, the enhancements concentrate on increasing the operating temperature by forming a thin, magnetically harder shell (nanometric scale) enriched with dysprosium or terbium on the surface of $Nd_2Fe_{14}B$ phase (Sepehri-Amin et al. 2016). In the meantime, the future PM scenario also develops without rare earths, with more abundant, less economically limited, and environmentally restricted elements, in addition to higher remanent magnetization with reasonable high coercivity. Some research focuses on iron-based compounds, such as FeNi (Kurichenko et al. 2019) and α''-$Fe_{16}N_2$ (Wang 2020). Others, such as MnBi and MnGa, are in research stages in China (Dong et al. 2017). Moreover, researchers showed that hexaferrites have enhanced magnetic properties with nanoscale size and shape particle modifications (de Julián Fernández et al. 2021), which means that not even the oldest PM has been forgotten. But from a practical point of view, the limitations to fabricate in bulk might suppress expectations (Wang et al. 2022).

6.3 Environmental Burden of the Electronic Devices

The evolution of the material composition and technologies modifies the environmental burden associated with electronic devices throughout their entire life cycle. This section aims to briefly discuss the environmental effects of the technological changes over time mentioned in the previous sections.

The positive or negative environmental contribution of a material/technological change can be analyzed through several parameters. Some important and commonly employed are toxicity, cumulative energy demand, depletion of natural resources, and global warming potential (Andrae and Andersen 2010; Kosai and Yamasue 2019). The contributions must be analyzed from the point of view of the equipment's entire life cycle, also considering characteristics of recyclability, durability, repairability, and others (Cenci et al. 2021b).

Regarding the toxicity of some materials employed in the electronic industry, a milestone in this scenario was the implementation of the RoHS directive in 2003 (updated in 2011), which restricts the use of hazardous substances in electrical and electronic equipment. Restricted substances and their maximum concentrations are Pb (0.1%), Hg (0.1%), Cd (0.01%), Cr^{+6} (0.1%), polybrominated biphenyls (0.1%), and polybrominated diphenyl ethers (0.1%) (George and Pecht 2016; Horn 2016; Wienold et al. 2011). This led to many changes in the composition of electronic components, for example, the substitution of Pb in soldering and emergence of halogen-free flame retardants.

Life cycle analysis (LCA) is an important tool to evaluate the environmental impact of products and services and can be used to compare different technologies and materials. Some studies point out that most of the environmental impact of the life cycle of electronics is from the premanufacturing phase (processing of raw materials), evidencing the importance of the materials' choice and evolution (Cenci et al. 2021b). Nuss and Eckelman (2014) assembled information about the cradle-to-gate environmental burdens of 63 metals. Gold and the platinum group displayed the highest environmental impacts for all categories analyzed (global warming potential, energy demand, eutrophication, acidification, and human toxicity). Rare earths and some other metals are also relevant, such as Eu, Sc, Ru, Dy, Ta, Ga, Tl, and others (Kosai and Yamasue 2019; Nuss and Eckelman 2014). Their substitution, also considering the ones mentioned in the RoHS directive, could potentially lead to significant environmental savings.

Analyzing LCA studies for electronic products, the PCBs play a major role in terms of environmental impacts (due to the acquisition of raw materials and the manufacturing phase). Andrae and Andersen (2010) highlighted the relevance of the PCBs' environmental impacts for the laptop and mobile phone scenarios. For example, the acquisition of raw materials and manufacturing of the PCBs consumed 60% of the total energy required in the life cycle and 48% of the global warming potential. Another good example is for screens, in which Bhakar et al. (2015) reported that PCBs were the most impacting component for both CRT, FL-LCD, and LED–LCD types in all impact categories analyzed (between 60% and 80% of total impact). In the same study, the author reported that from CRT screens to LED–LCD screens, there were significant improvements in environmental performance, highlighting the reduction in energy consumption. More efforts are needed to evaluate the environmental performance and compare it with the emerging OLED screens. In a cradle-to-gate approach, Amasawa et al. (2016) pointed out that OLED screens had a better environmental performance than LED–LCD screens in terms of global warming potential. However, Yeom et al. (2018) pointed out that OLEDs have higher toxicity and resource depletion potential than LED–LCDs.

For lamps, the main source of environmental impacts is phase use (Tähkämö et al. 2012). Due to their less energy consumption, LED lamps perform better on environmental parameters than other types of lamps. In the case of LED lamps, their impact during the raw material acquisition and manufacturing phases is higher than that of other types. However, it is compensated by less energy consumption, and this tends to improve as long as the performance of LED devices is advancing (Sangwan et al. 2014; Tähkämö et al. 2012).

Magnets are essential to clean energy alternatives, as electric vehicles and wind energy (Jin et al. 2016). In a cradle-to-gate LCA analysis approach, Bailey et al. (2021) compared the environmental impacts of Nd and SmCo magnets. Most of the environmental impacts are due to the electroplating step (Nd magnets) and co-production (SmCo magnets), and Nd magnets performed better in all impact categories. Karal et al. (2021) also reported that most of the environmental impacts of the Nd magnets' life cycle are from the Nd raw material exploitation and refining phases. Additionally, according to Jin et al. (2016), recycling Nd from wasted magnets significantly reduces the environmental burden in all impact categories.

Batteries, namely Li-ion NCM batteries, had most of their environmental impact from the preparation of the raw material phase, mainly from the cathode active material, wrought aluminum, and electrolyte (Sun et al. 2020). The authors stated that the recycling phase is significant for minimizing negative impacts. Comparing different types of batteries, the evolution from lead-acid batteries to modern ones was beneficial in environmental aspects (Hiremath et al. 2015; Wang et al. 2018). Wang et al. (2018) reported that LFP batteries have the smallest environmental impacts among lead-acid and lithium manganese types. Additionally, Hiremath et al. (2015) compared Li-ion batteries (mixture of nickel-cobalt-manganese [NCM], lithium-iron-phosphate [LFP], and lithiun-manganese oxide batteries [LMO]) with lead–acid, sodium–sulfur, and vanadium-redox batteries and found the first to be the most beneficial for the environment. The authors claimed that the results did not show significant dependency on the application and recommended improvements in their round-trip efficiency.

6.4 Recycling and Metal Recovery

The recovery of metals from WEEE is divided into preprocessing – activities include dismantling, sorting, and comminution – , physical processing, such as electrostatic separation, which is based on different physical properties of metals, and chemical processing – higher cost and energy demand – focusing mainly on metals refining (Li et al. 2019). Among the different final processing techniques, the most used consist of hydro- and pyrometallurgical (or combined). Electrometallurgical recovery is also nowadays gaining attention.

6.4.1 PCBs

In recent decades, the amount of electronic waste, including PCBs, has increased significantly, with generation rates higher than domestic solid waste. However, recycling of PCBs (which represent between 3% and 5% of total WEEE) remains a challenge worldwide (Faraji et al. 2022). The presence of different materials such as polymers, ceramics, and a wide range of metals (e.g. copper, aluminum, iron, nickel, lead, tin, and zinc), in addition to precious metals (gold, silver, platinum, and palladium) increases the complexity for any type of recycling process (Awasthi et al. 2016). The most technically enabled processes are based on pyrometallurgy, hydrometallurgy, and electrometallurgy, however, all of them have advantages and disadvantages. This makes it difficult to develop an efficient and universal process for dealing with PCB waste. Among the metals, the nonferrous Ag, Au, and Cu are the most economically interesting and the most cited in studies on PCB recycling (Yang et al. 2021).

6.4.2 LED Lamps

The challenge of recycling LED lamps starts in the pretreatment phase for the dismantling of the macro components, in which some approaches are under research to efficiently deliver separated flows of aluminum pieces, PCBs and electronic parts, and LEDs. Martins et al. (2020) proposed mechanical methods to separate the lamps into flows of polymers, metallic housing parts, PCBs, and LED modules. Benmamas et al. (2022) proposed electrodynamic fragmentation as an initial tool for dismantling.

An efficient separation of the LEDs from the LED module is also a crucial step for recycling routes, and keeping the metallic aluminum piece recyclable is favorable. Balinski et al. (2022) tested a chemical route (organic solvent) and a thermal route for the liberation of LEDs. Other authors have proposed pyrolysis for this step, such as Zhan et al. (2015) and Zhou et al. (2019). An interesting physical approach for LEDs could be electrostatic separation, as in the study of Cenci et al. (2021a), which efficiently separated a conductive fraction rich in copper, tin from gold, and a nonconductive fraction rich in gallium, yttrium, and cerium.

The focus of the material recovery from LEDs is gallium, rare-earth elements, and gold. The yttrium fixed into a polymeric matrix and the refractory nature of gallium in white LEDs (GaN) require special attention for their extraction, usually performed by thermal methods, such as in the study of Nagy et al. (2017), which employed sodium carbonate for an oxidative roasting at 1200 °C. Downstream processes usually apply hydrometallurgy solutions to complete the recycling route.

6.4.3 Screens

In the primary dismantling of screens, plastic casings, PCBs, and metallic parts have consolidated routes for recycling. The focus of research is normally on the recovery of the light source materials and the panel containing indium (Fontana et al. 2020). As can be seen in Section 6.2.3, most modern designs of screens contain critical and valuable elements in the light sources (LEDs and OLEDs, for example) and layers of ITO.

According to Fontana et al. (2020) and Amato and Beolchini (2018), the main challenge is the recovery of indium, still mostly performed at lab scale with very few pilot plant solutions, being not sustainable in economic and environmental terms. The main global use of indium is for ITO layers, and less than 1% of the waste indium is currently recycled (Akcil et al. 2019). In the patent review of Amato and Beolchini (2018), the authors stated that most of the recycling routes start with a mechanical process of panels in order to increase the contact area for the extracting phase, which is mostly performed by hydrometallurgical techniques. Leaching can be performed by acid and base solutions, and organic alternatives are a recent hotspot in research. Chloride is a common extracting agent of indium that can be applied both for leaching and volatilization processes. For example, in the study of Flerus et al. (2018) and Takahashi et al. (2009), some parameters involved in the volatilization of indium in the form of chlorides, and its posterior recovery, are assessed.

6.4.4 Batteries

Pyro and hydrometallurgical processes are a consolidated way of recovering valuable metals from LIBs, but they present high operating costs, as well as the generation of secondary waste. Several researchers have been looking for alternatives to recycling that are more economical and/or have a lower environmental impact (Mohanty et al. 2021; Thompson et al. 2021). Biometallurgy has been studied as an alternative to obtain more economically and environmentally favorable recovery routes (Roy et al. 2021).

Concerning the types of processes that have been more abundantly studied/used in the recycling of LIBs, the following are mentioned: pretreatment, hydrometallurgical, and pyrometallurgical processes (Pražanová et al. 2022); in addition to the direct recycling method (Harper et al. 2019).

No consolidated pretreatment exists in LIB recovery, because it depends on the number of steps, which the process has, and the way it used to be carried out. Kim et al. (2021) presented a possible route that consists of discharge, dismantling, comminution, classification, and separation; if the following step will be "dissolution" than the obtained product is able to be used in new LIBs by direct recycling, but in case of thermal treatment after separation, the product needs to be performed by hydrometallurgy.

Hydrometallurgy consists of obtaining the metal from the waste by dissolving it in an aqueous medium or molten salt with a subsequent concentration/recovery process, such as cementation, precipitation, and solvent extraction. The leaching solution, used in the ions dissolution, can be acidic or basic, and the types of acids can be organic or inorganic (Hayes 2021). Well-established hydrometallurgy methodologies have been generally carried out by inorganic acids – not so environmentally friendly – such as HNO_3, H_2SO_4, and HCl. Therefore, efforts have been made to develop more ecological routes with the application of organic acids such as citric and DL-malic (Musariri et al. 2019).

Pyrometallurgical processes use high temperatures and have been widely used to obtain metals and other types of materials. This type of processing allows the change of physical and chemical structures, conferring different characteristics and properties, which are not possible to achieve at room temperature. In pyrometallurgy, furnaces at high temperatures are used in order to reduce the oxides to a metallic alloy; in addition to this alloy, slag, and gases are the result (Hayes 2021). A large amount of BILs can be processed in a single batch – a significant advantage of pyrometallurgy, and no need to carry out pretreatment to concentrate the material, as is usually done in hydrometallurgical processes (Meng et al. 2018; Zhang et al. 2018).

6.4.5 Magnets

The main processes for the hydrometallurgical recovery of REEs consist of the leaching of magnetic scrap using solvents with high digestion efficiency, followed by the separation of REEs (Nd, Pr, Dy, and Tb) using mostly solvent extraction, ionic liquids, or precipitation. The challenges for the hydrometallurgical process are the selective dissolution, the concentration and separation of REEs from the main species, and the total recovery of the metals of interest (Firdaus et al. 2016). The purity of the recovered REE compounds depends on the composition of the leachate, i.e. the type and amount of impurities present.

On the other hand, pyrometallurgical recovery techniques of REE from PM use remelting, in which slag – concentrated in REE – and metallic phases are separated. The slag consequently follows a refinement in the electroslag process (Jowitt et al. 2018). Challenges for the recovery of REEs from the magnetic scrap are reported as: (1) the mixture of different wastes produced (complexity of the magnetic waste composition); (2) the effects of contaminants on the recycling process; (3) optimization for mutual separation of REEs; and (4) the economic feasibility of the recycling and recovery process and the life cycle of products containing magnets (availability of scrap metal containing magnets) (Yang et al. 2017).

6.5 Major Challenges in Management

The rising tendency in EEE consumption will continue in the next few years since advances in technology engage accessibility and penetration in new markets. Consequently, after a period, the EEE obsolescence will demand a precise management system due to the form of waste that is chemically and physically different from others. However, the three main issues regarding the metal recovery from EEE consist of (1) collecting and general logistics; (2) heterogeneous content and complexity; and (3) economic and environmental scale-up.

National legislation on e-waste in developed countries requires formal collecting – from households directly or in specific collecting points – whereas in countries with no legislation, e-waste is disposed of in regular bins or in dumpsites, which engages informal workers in pursuing unsafe recycling (Forti et al. 2020; Shittu et al. 2021). In addition, transboundary movement of WEEE, in spite of reallocating and intensifying economic activity, might also generate illegal movements in less strictly regulated countries (Chauhan et al. 2018).

Another aspect that difficult recycling deals with is the mixture of components in a WEEE. As already mentioned, metals are between layers of polymers or ceramics, in cables, in PCBs, and in other complex structures, which require the same proportional complex separation processes. Moreover, the most valuable metal contents are generally low, which might impose larger volumes of material with more steps of treatment (Ambaye et al. 2020; Chauhan et al. 2018). Besides being technically possible, the recovery of all resources in WEEE is not economically feasible yet (Parajuly et al. 2019), but might enhance considerably with manual separation, which is the main issue when scaling up due to labor cost (Forti et al. 2020). According to a survey, most WEEE handlers considered low-profit margins and the value of recycled products the main challenge (Osmani et al. 2021). Another factor is the complex product designs. Despite the concept of "design for recycling" exists for more than two decades, its implementation is not fully implemented.

6.6 Concluding Remarks and Perspectives

Considering the information and studies cited in this chapter, we can conclude that the WEEE theme is still far from a worldwide consensus on the best way to manage this type of solid waste. There are several challenges to be overcome in the technical, economic, environmental, tax, and logistical areas (internal and external of the countries) and even in the population's awareness about the importance of the correct destination of these materials.

This chapter has focused on some types of equipment/components (PCBs, LED lamps, screens, magnets, and batteries); however, there are many other types of electronics that are part of the problem and increase the complexity of managing them. Each of these equipment/components has a different chemical composition and naturally their recycling is focused on different materials. In a uniform way, it can be said that metals in general are the main target materials; however, each type of equipment/component will have in its composition some specific metals of greater interest. In general, we can say that WEEE recycling will always focus, to a greater or lesser extent, on nonferrous metals (such as copper, nickel, tin, cobalt, lithium, indium, and aluminum), precious metals (such as gold, silver, and palladium), rare earths, and those recently considered critical, not yet mentioned in the previous groups (e.g. magnesium, titanium, niobium, and graphite).

The perspectives show that the technological development of the electronics sector tends to be maintained for the future, and this, naturally, leads to a constant generation of this type of solid waste. We can also predict variations in terms of chemical composition for this equipment (depending on different manufacturers, technologies, and years of manufacture); however, it is not expected that the target materials for recycling will be different from those already mentioned.

References

Akcil, A., Agcasulu, I., and Swain, B. (2019). Valorization of waste LCD and recovery of critical raw material for circular economy: a review. *Resources, Conservation and Recycling* 149: 622–637. https://doi.org/10.1016/J.RESCONREC.2019.06.031.

Al Barazi, S., Brandenburg, T., Kuhn, T., Schmidt, M., Vetter, S., 2018. *Rohstoffrisikobewertung – Kobalt. Berlin.*

Alim, M.A., Abdullah, M.Z., Aziz, M.S.A., and Kamarudin, R. (2021). Die attachment, wire bonding, and encapsulation process in LED packaging: a review. *Sensors and Actuators A: Physical* 329: 112817. https://doi.org/10.1016/J.SNA.2021.112817.

Amasawa, E., Ihara, T., Ohta, T., and Hanaki, K. (2016). Life cycle assessment of organic light emitting diode display as emerging materials and technology. *Journal of Cleaner Production* https://doi.org/10.1016/j.jclepro.2016.07.025.

Amato, A. and Beolchini, F. (2018). End of life liquid crystal displays recycling: a patent review. *Journal of Environmental Management* 225: 1–9. https://doi.org/10.1016/j.jenvman.2018.07.035.

Ambaye, T.G., Vaccari, M., Castro, F.D. et al. (2020). Emerging technologies for the recovery of rare earth elements (REEs) from the end-of-life electronic wastes: a review on progress, challenges, and perspectives. *Environmental Science and Pollution Research* 2729 (27): 36052–36074. https://doi.org/10.1007/S11356-020-09630-2.

Anand, V., Mishra, R., and Barot, Y. (2021). Recent advances in the development of pure organic white light emitters. *Dyes and Pigments* 191: 109390. https://doi.org/10.1016/J.DYEPIG.2021.109390.

Andrae, A.S.G. and Andersen, O. (2010). Life cycle assessments of consumer electronics – are they consistent? *International Journal of Life Cycle Assessment* 15: 827–836. https://doi.org/10.1007/S11367-010-0206-1/FIGURES/3.

Aoe, T., Michiyasu, T., Matsouka, Y., and Shikata, N. (2003). Case study for calculation of factor X (eco-efficiency) – comparing CRT TV, PDP TV and LCD TV. In: *Proceedings of EcoDesign 2003: Third International Symposium on Environmentally Conscious Design and Inverse Manufacturing. Institute of Electrical and Electronics Engineers (IEEE), Tokyo*, 650–655. https://doi.org/10.1109/ECODIM.2003.1322749.

Arduin, R.H., Mathieux, F., Huisman, J. et al. (2020). Novel indicators to better monitor the collection and recovery of (critical) raw materials in WEEE: focus on screens. *Resources, Conservation and Recycling* 157: 104772. https://doi.org/10.1016/J.RESCONREC.2020.104772.

Arquer, F.P.G., Talapin, D.V., Klimov, V.I. et al. (2021). Semiconductor quantum dots: technological progress and future challenges. *Science (80-)* 373: eaaz8541. https://doi.org/10.1126/SCIENCE.AAZ8541.

Awasthi, A.K., Zeng, X., and Li, J. (2016). Integrated bioleaching of copper metal from waste printed circuit board—a comprehensive review of approaches and challenges. *Environmental Science and Pollution Research* 2321 (23): 21141–21156. https://doi.org/10.1007/S11356-016-7529-9.

Azrain, M.M., Mansor, M.R., Fadzullah, S.H.S.M. et al. (2018). Analysis of mechanisms responsible for the formation of dark spots in organic light emitting diodes (OLEDs): a review. *Synthetic Metals* 235: 160–175. https://doi.org/10.1016/J.SYNTHMET.2017.12.011.

Bailey, G., Orefice, M., Sprecher, B. et al. (2021). Life cycle inventory of samarium-cobalt permanent magnets, compared to neodymium-iron-boron as used in electric vehicles. *Journal of Cleaner Production* 286: https://doi.org/10.1016/J.JCLEPRO.2020.125294.

Balinski, A., Recksiek, V., Stoll, M. et al. (2022). Liberation and separation of valuable components from LED modules: presentation of two innovative approaches. *Recycling* 7: 26. https://doi.org/10.3390/RECYCLING7030026.

Barik, S.P., Prabaharan, G., and Kumar, B. (2016). An innovative approach to recover the metal values from spent lithium-ion batteries. *Waste Management* 51: 222–226. https://doi.org/10.1016/J.WASMAN.2015.11.004.

Benmamas, L., Bouzidi, Y., Houset, G. et al. (2022). Selective separation of plastic LED lamp components using electrodynamic fragmentation for material recovery. *Waste Management* 144: 210–220. https://doi.org/10.1016/J.WASMAN.2022.03.024.

Bhakar, V., Agur, A., Digalwar, A.K., and Sangwan, K.S. (2015). Life cycle assessment of CRT, LCD and LED monitors. *Procedia CIRP* 29: 432–437. https://doi.org/10.1016/J.PROCIR.2015.02.003.

Broussely, M. and Pistoia, G. (2007). *Industrial Applications of Batteries: From Cars to Aerospace and Energy Storage*, 1ee. Elsevier.

Buchert, M., Manhart, A., Bleher, D., Pingel, D., 2012. *Recycling Critical Raw Materials from Waste Electronic Equipment. Darmstadt.*

Buchmann, I. (2011). *Batteries in a Portable World: A Handbook on Rechargeable Batteries for Non-engineers*, 3ee. Cadex Electronics Inc.

Cenci, M.P., Dal Berto, F.C., Schneider, E.L., and Veit, H.M. (2020). Assessment of LED lamps components and materials for a recycling perspective. *Waste Management* 107: 285–293. https://doi.org/10.1016/j.wasman.2020.04.028.

Cenci, M.P., Dal Berto, F.C., Camargo, P.S.S., and Veit, H.M. (2021a). Separation and concentration of valuable and critical materials from wasted LEDs by physical processes. *Waste Management* 120: 136–145. https://doi.org/10.1016/j.wasman.2020.11.023.

Cenci, M.P., Scarazzato, T., Munchen, D.D. et al. (2021b). Eco-friendly electronics—a comprehensive review. *Advanced Materials Technology* 7: 2001263. https://doi.org/10.1002/admt.202001263.

Chauhan, G., Jadhao, P.R., Pant, K.K., and Nigam, K.D.P. (2018). Novel technologies and conventional processes for recovery of metals from waste electrical and electronic equipment: challenges & opportunities – a review. *Journal of*

Environmental Chemical Engineering 6: 1288–1304. https://doi.org/10.1016/j.jece.2018.01.032.

Chen, C.F. (2018). LEDs for liquid crystal display (LCD) backlighting. In: *Nitride Semiconductor Light-Emitting Diodes (LEDs): Materials, Technologies, and Applications* (ed. J. Huang, H.-C. Kuo, and S.-C. Shen), 619–646. Woodhead Publishing https://doi.org/10.1016/B978-0-08-101942-9.00019-8.

Chen, M., Jiang, P., Chen, H. et al. (2015). Leaching assessments of toxic metals in waste plasma display panel glass. *Journal of the Air & Waste Management Association (1995)* 65: 743–750. https://doi.org/10.1080/10962247.2015.1016634/SUPPL_FILE/UAWM_A_1016634_SM4072.DOCX.

Chiu, N.C., Smith, K.T., and Stylianou, K.C. (2022). Metal-organic frameworks for white light emission: from synthesis to device fabrication. *Coordination Chemistry Reviews* 459: 214441. https://doi.org/10.1016/J.CCR.2022.214441.

Cho, J., Park, J.H., Kim, J.K., and Schubert, E.F. (2017). White light-emitting diodes: history, progress, and future. *Laser & Photonics Reviews* 11: 1600147. https://doi.org/10.1002/LPOR.201600147.

Cronk, E.R. (1966). Recent developments in high-energy alnico alloys. *Journal of Applied Physics* 37: 1100. https://doi.org/10.1063/1.1708350.

Cui, J., Ormerod, J., Parker, D. et al. (2022). Manufacturing processes for permanent magnets: part I—sintering and casting. *JOM* 74: 1279–1295. https://doi.org/10.1007/S11837-022-05156-9/FIGURES/14.

Dai, X., Deng, Y., Peng, X. et al. (2017). Quantum-dot light-emitting diodes for large-area displays: towards the dawn of commercialization. *Advanced Materials* 29: 1607022. https://doi.org/10.1002/ADMA.201607022.

Das, D.K. (1971). Influence of sintering temperature on magnetic properties of samarium-cobalt magnets. *IEEE Transactions on Magnetics* 7: 432–435. https://doi.org/10.1109/TMAG.1971.1067144.

de Julián Fernández, C., Sangregorio, C., de la Figuera, J. et al. (2021). Progress and prospects of hard hexaferrites for permanent magnet applications. *Journal of Physics D: Applied Physics* 54: 153001. https://doi.org/10.1088/1361-6463/ABD272.

Dhamija, A., 2020. *Computer Graphics |Cathode Ray Oscilloscope| Cathode Ray Tube (Video Display Technology) [WWW Document]*. https://www.geeksforgeeks.org/computer-graphics-cathode-ray-tube-video-display-device ().

Dhiman, S. and Gupta, B. (2020). Cyphos IL 104 assisted extraction of indium and recycling of indium, tin and zinc from discarded LCD screen. *Separation and Purification Technology* 237: 116407. https://doi.org/10.1016/j.seppur.2019.116407.

Días, G.C., Alvarado-Rivera, J., Zayas, M.E. et al. (2015). Glass-ceramic, using TV screen and galvanic wastes as raw materials. *International Journal of Engineering and Innovative Technology* 4: 168–172.

Dillon, H.E., Ross, C., and Dzombak, R. (2019). Environmental and energy improvements of LED lamps over time: a comparative life cycle assessment. *LEUKOS* 16: 229–237. https://doi.org/10.1080/15502724.2018.1541748.

Dong, S., Li, W., Chen, H., and Han, R. (2017). The status of Chinese permanent magnet industry and R&D activities. *AIP Advances* 7: 056237. https://doi.org/10.1063/1.4978699.

Durand-Charre, M., Bronner, C., and Lagarde, J.P. (1978). Relation between magnetic properties and crystallographic texture of columnar alnico 8 permanent magnets. *IEEE Transactions on Magnetics* 14: 797–799. https://doi.org/10.1109/TMAG.1978.1060013.

Dzombak, R., Padon, J., Salsbury, J., and Dillon, H. (2017). Assessment of end-of-life design in solid-state lighting. *Environmental Research Letters* 12: 084013. https://doi.org/10.1088/1748-9326/AA7AB1.

Electronics Hub, (2019). *OLED Display Technology [WWW Document]. Electronics Hub*. https://www.electronicshub.org/oled-display-technology (Accessed 28 April 2022).

European Lamp Companies Federation, 2009. *Material Composition [WWW Document]*.

Faraji, F., Golmohammadzadeh, R., and Pickles, C.A. (2022). Potential and current practices of recycling waste printed circuit boards: a review of the recent progress in pyrometallurgy. *Journal of Environmental Management* 316: 115242. https://doi.org/10.1016/J.JENVMAN.2022.115242.

Feezell, D. and Nakamura, S. (2018). Invention, development, and status of the blue light-emitting diode, the enabler of solid-state lighting. *Comptes Rendus Physique* 19: 113–133. https://doi.org/10.1016/j.crhy.2017.12.001.

Firdaus, M., Rhamdhani, M.A., Durandet, Y. et al. (2016). Review of high-temperature recovery of rare earth (Nd/Dy) from magnet waste. *Journal of Sustainable Metallurgy* 2: 276–295. https://doi.org/10.1007/S40831-016-0045-9/FIGURES/16.

Flerus, B., Swiontek, T., Bokelmann, K. et al. (2018). Thermochemical modelling and experimental validation of in situ indium volatilization by released halides during pyrolysis of smartphone displays. *Metals (Basel)* 8: 1040. https://doi.org/10.3390/met8121040.

Fontana, D., Forte, F., Pietrantonio, M., and Pucciarmati, S. (2020). Recent developments on recycling end-of-life flat panel displays: a comprehensive review focused on indium. *Critical Reviews in Environmental Science and Technology* https://doi.org/10.1080/10643389.2020.1729073.

Forti, V., Baldé, C.P., Kuehr, R., and Bel, G. (2020). *The Global E-Waste Monitor 2020: Quantities, Flows, and the Circular Economy Potential*. Bonn/Geneva/Rotterdam: UNU/UNITAR/ITU/ISWA.

Gabriel, A.P., Kasper, A.C., and Veit, H.M. (2020). Acid leaching of indium from the screens of obsolete LCD monitors. *Journal of Environmental Chemical Engineering* 8: 103758. https://doi.org/10.1016/J.JECE.2020.103758.

Gaffuri, P., Stolyarova, E., Llerena, D. et al. (2021). Potential substitutes for critical materials in white LEDs: technological challenges and market opportunities.

Renewable and Sustainable Energy Reviews 143: 110869. https://doi.org/10.1016/J.RSER.2021.110869.

Gassmann, A., Zimmermann, J., Gauß, R. et al. (2016). *LED Lamps Recycling Technology for a Circular Economy [WWW Document]*. LED Prof https://www.led-professional.com/resources-1/articles/led-lamps-recycling-technology-for-a-circular-economy (

George, E. and Pecht, M. (2016). RoHS compliance in safety and reliability critical electronics. *Microelectronics and Reliability* 65: 1–7. https://doi.org/10.1016/J.MICROREL.2016.07.150.

Górecki, K. (2013). The influence of power supply voltage on exploitive parameters of the selected lamps. *Informacije MIDEM* 43: 193–198.

Gould, J.E. (1969). Permanent magnet applications. *IEEE Transactions on Magnetics* 5: 812–821. https://doi.org/10.1109/TMAG.1969.1066646.

Greim, P., Solomon, A.A., and Breyer, C. (2020). Assessment of lithium criticality in the global energy transition and addressing policy gaps in transportation. *Nature Communications* 11: 1–11. https://doi.org/10.1038/s41467-020-18402-y.

Hadi, P., Xu, M., Lin, C.S.K. et al. (2015). Waste printed circuit board recycling techniques and product utilization. *Journal of Hazardous Materials* 283: 234–243. https://doi.org/10.1016/j.jhazmat.2014.09.032.

Harper, C. (2003). *Electronic Materials and Processes Handbook*, 3ee. McGraw-Hill.

Harper, G., Sommerville, R., Kendrick, E. et al. (2019). Recycling lithium-ion batteries from electric vehicles. *Nature* 2019 (575): 75–86. https://doi.org/10.1038/s41586-019-1682-5.

Hayes, P. (2021). *Process Principles in Minerals and Materials Production: With a Focus on Metals Production and Recycling*, 4ee. Deborah Fay.

Herat, S. (2008). Recycling of cathode ray tubes (CRTs) in electronic waste. *CLEAN* 36: 19–24. https://doi.org/10.1002/CLEN.200700082.

Hiremath, M., Derendorf, K., and Vogt, T. (2015). Comparative life cycle assessment of battery storage systems for stationary applications. *Environmental Science & Technology* 49: 4825–4833. https://doi.org/10.1021/ES504572Q/SUPPL_FILE/ES504572Q_SI_001.PDF.

Hischier, R. and Baudin, I. (2010). LCA study of a plasma television device. *International Journal of Life Cycle Assessment* 15: 428–438. https://doi.org/10.1007/S11367-010-0169-2/FIGURES/6.

Hoffmann, A. and Stäblein, H. (1970). Investigations of high-coercivity alnico alloys. *IEEE Transactions on Magnetics* 6: 225–230. https://doi.org/10.1109/TMAG.1970.1066731.

Hoogerstraete, T.V., Blanpain, B., Van Gerven, T., and Binnemans, K. (2014). From NdFeB magnets towards the rare-earth oxides: a recycling process consuming only oxalic acid. *RSC Advances* 4: 64099–64111. https://doi.org/10.1039/C4RA13787F.

Horn, D.A. (2016). The international development of RoHS. In: *IEEE Conference on Technologies for Sustainability*, 34–38. Phoenix, AZ, USA: IEEE https://doi.org/10.1109/SusTech.2016.7897139.

Huang, T., Zhu, J., Huang, X. et al. (2022). Assessment of precious metals positioning in waste printed circuit boards and the economic benefits of recycling. *Waste Management* 139: 105–115. https://doi.org/10.1016/J.WASMAN.2021.12.030.

Iwama, Y. and Takeuchi, M. (1974). Spinodal Decomposition in Alnico 8 Magnet Alloy. *Transactions of the Japan Institute of Metals* 15: 371–377. https://doi.org/10.2320/MATERTRANS1960.15.371.

Izatt, R.M. and Hagelüken, C. (2016). Recycling and sustainable utilization of precious and specialty metals. In: *Metal Sustainability: Global Challenges, Consequences, and Prospects* (ed. R.M. Izatt), 1–22. John Wiley & Sons, Ltd https://doi.org/10.1002/9781119009115.CH1.

Jia, H. (2018). Who will win the future of display technologies? *National Science Review* 5: 427–431. https://doi.org/10.1093/NSR/NWY050.

Jin, H., Afiuny, P., McIntyre, T. et al. (2016). Comparative life cycle assessment of NdFeB magnets: virgin production versus magnet-to-magnet recycling. *Procedia CIRP* 48: 45–50. https://doi.org/10.1016/J.PROCIR.2016.03.013.

Jowitt, S.M., Werner, T.T., Weng, Z., and Mudd, G.M. (2018). Recycling of the rare earth elements. *Current Research in Green and Sustainable Chemistry* 13: 1–7. https://doi.org/10.1016/J.COGSC.2018.02.008.

Juchneski, N.C.F. (2013). *Monitores de LCD: Caracterização dos materiais e processamento mecânico das placas de circuito impresso*. UFRGS.

Juchneski, N.C.F., Scherer, J., Grochau, I.H., and Veit, H.M. (2013). Disassembly and characterization of liquid crystal screens. *Waste Management & Research* 31: 549–558. https://doi.org/10.1177/0734242X13485795.

Kalyani, N.T. and Dhoble, S.J. (2015). Novel materials for fabrication and encapsulation of OLEDs. *Renewable and Sustainable Energy Reviews* 44: 319–347. https://doi.org/10.1016/J.RSER.2014.11.070.

Karal, E., Kucuker, M.A., Demirel, B. et al. (2021). Hydrometallurgical recovery of neodymium from spent hard disk magnets: a life cycle perspective. *Journal of Cleaner Production* 288: 125087. https://doi.org/10.1016/j.jclepro.2020.125087.

Kaya, M. (2016). Recovery of metals and nonmetals from electronic waste by physical and chemical recycling processes. *Waste Management* 57: 64–90. https://doi.org/10.1016/j.wasman.2016.08.004.

Khalifeh, S. (2020). Optimization of polymeric structures of organic printed circuit boards. In: *Polymers in Organic Electronics* (ed. S. Khalifeh), 203–249. ChemTec Publishing https://doi.org/10.1016/B978-1-927885-67-3.50010-9.

Kim, S., Bang, J., Yoo, J. et al. (2021). A comprehensive review on the pretreatment process in lithium-ion battery recycling. *Journal of Cleaner Production* 294: 126329. https://doi.org/10.1016/J.JCLEPRO.2021.126329.

Kosai, S. and Yamasue, E. (2019). Global warming potential and total material requirement in metal production: identification of changes in environmental impact through metal substitution. *Science of The Total Environment* 651: 1764–1775. https://doi.org/10.1016/J.SCITOTENV.2018.10.085.

Kurichenko, V.L., Karpenkov, D.Y., Karpenkov, A.Y. et al. (2019). Synthesis of FeNi tetrataenite phase by means of chemical precipitation. *Journal of Magnetism and Magnetic Materials* 470: 33–37. https://doi.org/10.1016/J.JMMM.2017.11.040.

LaDou, J. (2006). Printed circuit board industry. *International Journal of Hygiene and Environmental Health* 209: 211–219. https://doi.org/10.1016/J.IJHEH .2006.02.001.

Lahtela, V., Virolainen, S., Uwaoma, A. et al. (2019). Novel mechanical pre-treatment methods for effective indium recovery from end-of-life liquid-crystal display panels. *Journal of Cleaner Production* 230: 580–591. https://doi.org/10.1016/J .JCLEPRO.2019.05.163.

Lavoie, Y., Danet, F., and Lombard, B. (2017). Lithium-ion batteries for industrial applications. In: *Petroleum and Chemical Industry Technical Conference*, 283–290. IEEE https://doi.org/10.1109/PCICON.2017.8188747.

Lewis, L.H. and Jiménez-Villacorta, F. (2012). Perspectives on permanent magnetic materials for energy conversion and power generation. *Metallurgical and Materials Transactions A* 44: 2–20. https://doi.org/10.1007/S11661-012-1278-2.

Li, Z., Diaz, L.A., Yang, Z. et al. (2019). Comparative life cycle analysis for value recovery of precious metals and rare earth elements from electronic waste. *Resources, Conservation and Recycling* 149: 20–30. https://doi.org/10.1016/j .resconrec.2019.05.025.

Liu, W., Liu, W., Li, X. et al. (2021). Dynamic material flow analysis of critical metals for lithium-ion battery system in China from 2000–2018. *Resources, Conservation and Recycling* 164: 105122. https://doi.org/10.1016/J.RESCONREC .2020.105122.

Lo, E., 2022. *How Much Does a TV Weigh? (Complete Guide) [WWW Document].* https://voltfixer.com/how-much-does-a-tv-weigh ().

Manganello, M. (2015). *Fundamental Study of Recycling Plasma Display Panels.* Colorado School of Mines.

Martins, T.R., Tanabe, E.H., and Bertuol, D.A. (2020). Innovative method for the recycling of end-of-life LED bulbs by mechanical processing. *Resources, Conservation and Recycling* 161: 104875. https://doi.org/10.1016/J.RESCONREC .2020.104875.

Martins, T.A.G., Caldas, M.P.K., de Moraes, V.T. et al. (2021). Recovering metals from motherboard and memory board waste through sulfuric leaching. *Journal of Environmental Chemical Engineering* 9: 106789. https://doi.org/10.1016/J.JECE .2021.106789.

Matsuto, T., Jung, C.H., and Tanaka, N. (2004). Material and heavy metal balance in a recycling facility for home electrical appliances. *Waste Management* 24: 425–436. https://doi.org/10.1016/J.WASMAN.2003.12.002.

McCallum, R.W., Lewis, L., Skomski, R. et al. (2014). Practical aspects of modern and future permanent magnets. *Annual Review of Materials Research* 44: 451–477. https://doi.org/10.1146/ANNUREV-MATSCI-070813-113457.

Meng, Q., Zhang, Y., and Dong, P. (2018). Use of electrochemical cathode-reduction method for leaching of cobalt from spent lithium-ion batteries. *Journal of Cleaner Production* 180: 64–70. https://doi.org/10.1016/J.JCLEPRO.2018.01.101.

Mir, S., Vaishampayan, A., and Dhawan, N. (2022). A review on recycling of end-of-life light-emitting diodes for metal recovery. *JOM* 2021: 1–13. https://doi.org/10.1007/S11837-021-05043-9.

Miura, K., Itoh, M., and Machida, K.I. (2008). Extraction and recovery characteristics of Fe element from Nd–Fe–B sintered magnet powder scrap by carbonylation. *Journal of Alloys and Compounds* 466: 228–232. https://doi.org/10.1016/J.JALLCOM.2007.11.013.

Mohanty, A., Sahu, S., Sukla, L.B., and Devi, N. (2021). Application of various processes to recycle lithium-ion batteries (LIBs): a brief review. *Materials Today: Proceedings* 47: 1203–1212. https://doi.org/10.1016/J.MATPR.2021.03.645.

Montoya, F.G., Peña-García, A., Juaidi, A., and Manzano-Agugliaro, F. (2017). Indoor lighting techniques: an overview of evolution and new trends for energy saving. *Energy and Buildings* 140: 50–60. https://doi.org/10.1016/J.ENBUILD.2017.01.028.

Musariri, B., Akdogan, G., Dorfling, C., and Bradshaw, S. (2019). Evaluating organic acids as alternative leaching reagents for metal recovery from lithium ion batteries. *Minerals Engineering* 137: 108–117. https://doi.org/10.1016/J.MINENG.2019.03.027.

Nagy, S., Bokányi, L., Gombköto, I., and Magyar, T. (2017). Recycling of gallium from end-of-life light emitting diodes. *Archives of Metallurgy and Materials* 62: 1161–1166. https://doi.org/10.1515/AMM-2017-0170.

Nair, G.B., Swart, H.C., and Dhoble, S.J. (2020). A review on the advancements in phosphor-converted light emitting diodes (pc-LEDs): phosphor synthesis, device fabrication and characterization. *Progress in Materials Science* 109: 100622. https://doi.org/10.1016/J.PMATSCI.2019.100622.

Nardelli, A., Deuschle, E., de Azevedo, L.D. et al. (2017). Assessment of light emitting diodes technology for general lighting: a critical review. *Renewable and Sustainable Energy Reviews* 75: 368–379. https://doi.org/10.1016/J.RSER.2016.11.002.

Natarajan, S., Boricha, A.B., and Bajaj, H.C. (2018). Recovery of value-added products from cathode and anode material of spent lithium-ion batteries. *Waste Management* 77: 455–465. https://doi.org/10.1016/J.WASMAN.2018.04.032.

Nishi, Y. (2001). The development of lithium ion secondary batteries. *Chemical Record* 1: 406–413. https://doi.org/10.1002/TCR.1024.

Nnorom, I.C., Osibanjo, O., and Ogwuegbu, M.O.C. (2011). Global disposal strategies for waste cathode ray tubes. *Resources, Conservation and Recycling* 55: 275–290. https://doi.org/10.1016/J.RESCONREC.2010.10.007.

Nuss, P. and Eckelman, M.J. (2014). Life cycle assessment of metals: a scientific synthesis. *PLoS ONE* 9: e101298. https://doi.org/10.1371/JOURNAL.PONE.0101298.

Ober, J.A. (2018). Mineral commodity summaries 2018. *Mineral Commodity Summaries* https://doi.org/10.3133/70194932.

Ochoa, M.L.A., He, H., Schoenung, J.M. et al. (2019). Design parameters and environmental impact of printed wiring board manufacture. *Journal of Cleaner Production* 238: 117807. https://doi.org/10.1016/J.JCLEPRO.2019.117807.

Ogunniyi, I.O., Vermaak, M.K.G., and Groot, D.R. (2009). Chemical composition and liberation characterization of printed circuit board comminution fines for beneficiation investigations. *Waste Management* 29: 2140–2146. https://doi.org/10.1016/J.WASMAN.2009.03.004.

Oliveira, R.P., Botelho Junior, A.B., and Espinosa, D.C.R. (2020). Characterization of wasted LEDs from tubular lamps focused on recycling process by hydrometallurgy. In: *Energy Technology 2020: Recycling, Carbon Dioxide Management, and Other Technologies* (ed. X. Chen, Y. Zhong, L. Zhang, et al.), 317–325. Cham: Springer https://doi.org/10.1007/978-3-030-36830-2_30.

Osmani, M., Pollard, J., Forde, J. et al. (2021). Circular economy business model opportunities, challenges, and enablers in the electrical and electronic equipment sector: stakeholders' perspectives. In: *Proceedin gs SARDINIA 2021*. CISA.

Parajuly, K., Kuehr, R., Awasthi, A.K., Fitzpatrick, C., Lepawsky, J., Smith, E., Widmer, R., Zeng, X., 2019. *Future E-Waste Scenarios. Bonn/Osaka*.

Peeters, J.R., Vanegas, P., Dewulf, W., and Duflou, J.R. (2012). Active disassembly for the end-of-life treatment of flat screen televisions: challenges and opportunities. In: *Design for Innovative Value Towards a Sustainable Society*, 535–540. Dordrecht: Springer https://doi.org/10.1007/978-94-007-3010-6_103.

Placek, M. (2021). *Global Lighting Market: LED Penetration Rate [WWW Document]*. Statista https://www.statista.com/statistics/246030/estimated-led-penetration-of-the-global-lighting-market (

Pode, R. (2020). Organic light emitting diode devices: an energy efficient solid state lighting for applications. *Renewable and Sustainable Energy Reviews* 133: 110043. https://doi.org/10.1016/J.RSER.2020.110043.

Pražanová, A., Knap, V., and Stroe, D.I. (2022). Literature review, recycling of lithium-ion batteries from electric vehicles. *Part I: Recycling Technology. Energies* 15: 1086. https://doi.org/10.3390/EN15031086.

Prodius, D., Gandha, K., Mudring, A.V., and Nlebedim, I.C. (2020). Sustainable Urban Mining of critical elements from magnet and electronic wastes. *ACS Sustainable Chemistry & Engineering* 8: 1455–1463. https://doi.org/10.1021/ACSSUSCHEMENG.9B05741/SUPPL_FILE/SC9B05741_SI_001.PDF.

Qi, Y., Xiao, X., Lu, Y. et al. (2019). Cathode ray tubes glass recycling: a review. *Science of The Total Environment* 650: 2842–2849. https://doi.org/10.1016/J.SCITOTENV.2018.09.383.

Rabenberg, L., Mishra, R.K., and Thomas, G. (1998). Microstructures of precipitation-hardened SmCo permanent magnets. *Journal of Applied Physics* 53: 2389. https://doi.org/10.1063/1.330867.

Rocchetti, L., Amato, A., and Beolchini, F. (2018). Printed circuit board recycling: a patent review. *Journal of Cleaner Production* https://doi.org/10.1016/j.jclepro.2018.01.076.

Roy, J.J., Cao, B., and Madhavi, S. (2021). A review on the recycling of spent lithium-ion batteries (LIBs) by the bioleaching approach. *Chemosphere* 282: 130944. https://doi.org/10.1016/J.CHEMOSPHERE.2021.130944.

Ruhnau, O., Bannik, S., Otten, S. et al. (2019). Direct or indirect electrification? A review of heat generation and road transport decarbonisation scenarios for Germany 2050. *Energy* 166: 989–999. https://doi.org/10.1016/J.ENERGY.2018.10.114.

Ryan, A., O'Donoghue, L., and Lewis, H. (2011). Characterising components of liquid crystal displays to facilitate disassembly. *Journal of Cleaner Production* 19: 1066–1071. https://doi.org/10.1016/J.JCLEPRO.2011.01.009.

Saguchi, A., Asabe, K., Fukuda, T. et al. (2006). Recycling of rare earth magnet scraps: carbon and oxygen removal from Nd magnet scraps. *Journal of Alloys and Compounds* 408–412: 1377–1381. https://doi.org/10.1016/J.JALLCOM.2005.04.178.

Sangwan, K.S., Bhakar, V., Naik, S., and Andrat, S.N. (2014). Life cycle assessment of incandescent, fluorescent, compact fluorescent and light emitting diode lamps in an Indian scenario. *Procedia CIRP* 15: 467–472. https://doi.org/10.1016/J.PROCIR.2014.06.017.

Savvilotidou, V., Hahladakis, J.N., and Gidarakos, E. (2014). Determination of toxic metals in discarded liquid crystal displays (LCDs). *Resources, Conservation and Recycling* 92: 108–115. https://doi.org/10.1016/J.RESCONREC.2014.09.002.

Scholand, M. and Dillon, H.E. (2012). *Life-Cycle Assessment of Energy and Environmental Impacts of LED Lighting Products Part 2: LED Manufacturing and Performance*. Richland, WA: (United States). https://doi.org/10.2172/1044508.

Sepehri-Amin, H., Ohkubo, T., and Hono, K. (2016). Micromagnetic simulations of magnetization reversals in Nd-Fe-B based permanent magnets. *Materials Transactions* 57: 1221–1229. https://doi.org/10.2320/MATERTRANS.M2015457.

Shama, S., Jindal, U., Goyal, M. et al. (2016). A review-on different types of displays. *International Journal of Multimedia and Ubiquitous Engineering* 11: 327–332. https://doi.org/10.14257/ijmue.2016.11.8.33.

Shittu, O.S., Williams, I.D., and Shaw, P.J. (2021). Global E-waste management: can WEEE make a difference? A review of e-waste trends, legislation, contemporary issues and future challenges. *Waste Management* 120: 549–563. https://doi.org/10.1016/J.WASMAN.2020.10.016.

Singh, N., Li, J., and Zeng, X. (2016). Global responses for recycling waste CRTs in e-waste. *Waste Management* 57: 187–197. https://doi.org/10.1016/J.WASMAN.2016.03.013.

Smith, B.J. and Eggert, R.G. (2016). Multifaceted material substitution: the case of NdFeB magnets, 2010–2015. *JOM* 68: 1964–1971. https://doi.org/10.1007/S11837-016-1913-2.

Strnat, K.J. (1990). Modem permanent magnets for applications in electro-technology. *Proceedings of the IEEE* 78: 923–946. https://doi.org/10.1109/5.56908.

Sugimoto, S. (2011). Current status and recent topics of rare-earth permanent magnets. *Journal of Physics D: Applied Physics* 44: 064001. https://doi.org/10.1088/0022-3727/44/6/064001.

Sun, Y.L., Zhao, J.T., Liu, Z. et al. (2015). The phase and microstructure analysis of Alnico magnets with high coercivity. *Journal of Magnetism and Magnetic Materials* 379: 58–62. https://doi.org/10.1016/J.JMMM.2014.12.003.

Sun, X., Luo, X., Zhang, Z. et al. (2020). Life cycle assessment of lithium nickel cobalt manganese oxide (NCM) batteries for electric passenger vehicles. *Journal of Cleaner Production* 273: 123006. https://doi.org/10.1016/J.JCLEPRO.2020.123006.

Tähkämö, L., Puolakka, M., Halonen, L., and Zissis, G. (2012). Comparison of life cycle assessments of LED light sources. *Journal of Light & Visual Environment* 36.

Takahashi, K., Sasaki, A., Dodbiba, G. et al. (2009). Recovering indium from the liquid crystal display of discarded cellular phones by means of chloride-induced vaporization at relatively low temperature. *Metallurgical and Materials Transactions A: Physical Metallurgy and Materials Science* 40: 891–900. https://doi.org/10.1007/s11661-009-9786-4.

Tan, Q., Dong, Q., Liu, L. et al. (2017). Potential recycling availability and capacity assessment on typical metals in waste mobile phones: a current research study in China. *Journal of Cleaner Production* 148: 509–517. https://doi.org/10.1016/j.jclepro.2017.02.036.

Tarafder, A., Singh, S.P., Karmakar, B., and Karmakar, B. (2012). Environmentally friendly and new generation glasses for plasma TV. *Kanch* 4: 42–48.

Tassone, G. (2021). *Display Technology: From LCD to QLED*. Politecnico di Torino.

Tenaud, P., Lemaire, H., and Vial, F. (1991). Recent improvements in NdFeB sintered magnets. *Journal of Magnetism and Magnetic Materials* 101: 328–332. https://doi.org/10.1016/0304-8853(91)90767-5.

Thejokalyani, N. and Dhoble, S.J. (2014). Novel approaches for energy efficient solid state lighting by RGB organic light emitting diodes – a review. *Renewable and Sustainable Energy Reviews* 32: 448–467. https://doi.org/10.1016/J.RSER.2014.01.013.

Thompson, D., Hyde, C., Hartley, J.M. et al. (2021). To shred or not to shred: a comparative techno-economic assessment of lithium ion battery hydrometallurgical recycling retaining value and improving circularity in LIB supply chains. *Resources, Conservation and Recycling* 175: 105741. https://doi.org/10.1016/J.RESCONREC.2021.105741.

U.S. Geological Survey (2020). *Mineral Commodity Summaries 2020*. Reston, VA, USA: U.S. Geological Survey https://doi.org/10.3133/mcs2020.

U.S. Geological Survey (2022). Mineral commodity summaries 2022. *Mineral Commodity Summaries* https://doi.org/10.3133/MCS2022.

Uddin, S., Shareef, H., and Mohamed, A. (2013). Power quality performance of energy-efficient low-wattage LED lamps. *Measurement* 46: 3783–3795. https://doi.org/10.1016/J.MEASUREMENT.2013.07.022.

Ueberschaar, M. and Rotter, V.S. (2015). Enabling the recycling of rare earth elements through product design and trend analyses of hard disk drives. *Journal of Material Cycles and Waste Management* 17: 266–281. https://doi.org/10.1007/S10163-014-0347-6/FIGURES/9.

Wang, J.P. (2020). Environment-friendly bulk Fe16N2 permanent magnet: review and prospective. *Journal of Magnetism and Magnetic Materials* 497: 165962. https://doi.org/10.1016/J.JMMM.2019.165962.

Wang, Q., Liu, W., Yuan, X. et al. (2018). Environmental impact analysis and process optimization of batteries based on life cycle assessment. *Journal of Cleaner Production* 174: 1262–1273. https://doi.org/10.1016/J.JCLEPRO.2017.11.059.

Wang, S., Zhang, H., Zhang, B. et al. (2020). Towards high-power-efficiency solution-processed OLEDs: material and device perspectives. *Materials Science & Engineering R: Reports* 140: 100547. https://doi.org/10.1016/J.MSER.2020.100547.

Wang, H., Lamichhane, T.N., and Paranthaman, M.P. (2022). Review of additive manufacturing of permanent magnets for electrical machines: a prospective on wind turbine. *Materials Today Physics* 24: 100675. https://doi.org/10.1016/J.MTPHYS.2022.100675.

Wang, Y., Xu, C., Yu, X. et al. (2022). Multilayer flexible electronics: manufacturing approaches and applications. *Materials Today Physics* 23: 100647. https://doi.org/10.1016/J.MTPHYS.2022.100647.

Wehbie, M. and Semetey, V. (2022). Characterization of end-of-life LED lamps: evaluation of reuse, repair and recycling potential. *Waste Management* 141: 202–207. https://doi.org/10.1016/J.WASMAN.2022.01.037.

Whittingham, M.S. (1976). Electrical energy storage and intercalation chemistry. *Science (80-)* 192: 1126–1127. https://doi.org/10.1126/SCIENCE.192.4244.1126.

Wienold, J., Recknagel, S., Scharf, H. et al. (2011). Elemental analysis of printed circuit boards considering the ROHS regulations. *Waste Management* 31: 530–535. https://doi.org/10.1016/j.wasman.2010.10.002.

Williams, T. (2005). Printed circuits. In: *The Circuit Designer's Companion* (ed. T. Williams), 40–69. Elsevier https://doi.org/10.1016/B978-075066370-0/50003-1.

Wong, H.Y.K. (2004). Electronic display reuse and recycling. *UNSW* https://doi.org/10.26190/UNSWORKS/5111.

Xing, Q., Miller, M.K., Zhou, L. et al. (2013). Phase and elemental distributions in alnico magnetic materials. *IEEE Transactions on Magnetics* 49: 3314–3317. https://doi.org/10.1109/TMAG.2013.2252155.

Yang, Y., Walton, A., Sheridan, R. et al. (2017). REE recovery from end-of-life NdFeB permanent magnet scrap: a critical review. *Journal of Sustainable Metallurgy* 3: 122–149. https://doi.org/10.1007/S40831-016-0090-4/FIGURES/5.

Yang, M., Liu, H., Ye, B., and Qian, W. (2021). Recycling of printed circuit boards by abrasive waterjet cutting. *Process Safety and Environmental Protection* 148: 805–812. https://doi.org/10.1016/J.PSEP.2021.01.052.

Yeom, J.M., Jung, H.J., Choi, S.Y. et al. (2018). Environmental effects of the technology transition from liquid–crystal display (LCD) to organic light-emitting diode (OLED) display from an e-waste management perspective. *International Journal of Environmental Research* 12: 479–488. https://doi.org/10.1007/S41742-018-0106-Y/FIGURES/5.

Yin, Y., Ali, M.U., Xie, W. et al. (2019). Evolution of white organic light-emitting devices: from academic research to lighting and display applications. *Materials Chemistry Frontiers* 3: 970–1031. https://doi.org/10.1039/C9QM00042A.

Yu-Gong, Tian, X.M., Wu, Y.F. et al. (2016). Recent development of recycling lead from scrap CRTs: a technological review. *Waste Management* 57: 176–186. https://doi.org/10.1016/J.WASMAN.2015.09.004.

Zhan, L., Xia, F., Ye, Q. et al. (2015). Novel recycle technology for recovering rare metals (Ga, In) from waste light-emitting diodes. *Journal of Hazardous Materials* 299: 388–394. https://doi.org/10.1016/J.JHAZMAT.2015.06.029.

Zhang, X., Li, L., Fan, E. et al. (2018). Toward sustainable and systematic recycling of spent rechargeable batteries. *Chemical Society Reviews* 47: 7239–7302. https://doi.org/10.1039/C8CS00297E.

Zhou, J., Zhu, N., Liu, H. et al. (2019). Recovery of gallium from waste light emitting diodes by oxalic acidic leaching. *Resources, Conservation and Recycling* 146: 366–372. https://doi.org/10.1016/J.RESCONREC.2019.04.002.

Zissis, G. and Bertoldi, P. (2018). *Status of LED-Lighting World Market in 2017*. European Commission. Ispra, European Commission, Ispra.

7

Urban Mining of e-Waste: Conversion of Waste to Wealth

Piotr Nowakowski

Silesian University of Technology, Faculty of Transport and Aviation Engineering, Katowice, Poland

Abstract

Urban mining is a part of a circular economy. One of the categories of products widely used in households is electrical and electronic equipment. There is a big variety of materials required to manufacture any electronic or electrical product. This chapter discusses the recycling potential of various categories of e-waste. The secondary raw materials for recovery are mainly ferrous, nonferrous metals, and plastics. There are also precious and rare earth metals. Efficient collection should be provided using all available methods, including stationary collection points, municipal waste collection centers, collections in containers, and on-demand collection. Properly conducted waste collections contribute to lowering emissions in transportation and saving the cost of vehicles and staff. However, it also requires education and information campaigns addressed to individuals on how to properly dispose of end-of-life equipment. In this context, the education curricula should focus on environmental hazards and health issues of improperly disposed e-waste.

Keywords *urban mining; e-waste; recycling; circular economy*

7.1 The Principles of Urban Mining and the Life Cycle of Electrical and Electronic Equipment

Urban mining is a concept similar to waste management, or closing the loops, or cradle to cradle (Krook and Baas 2013). However, some more precise meaning should be applied to urban locations. This term takes into consideration the area inside city borders and all stocks included in the urban settlement (Krook 2010). Therefore, the secondary raw materials from the resources are easily accessible. Moreover, urban mining is similar to mineral mining, and various

Management of Electronic Waste: Resource Recovery, Technology and Regulation, First Edition.
Edited by Anshu Priya.

processes like identification of resources, collection, transportation, processing, and refining resemble the classic meaning of mining (Ottoni et al. 2020). The most significant resources regarding metal contents, purity of materials, and other valuable components are deposited in waste electrical and electronic equipment (WEEE), commonly named e-waste (Cossu et al. 2012). All stages of urban mining are included in a broader concept of circular economy (Andersen 2007; Korhonen et al. 2018). Therefore, they cannot be investigated separately, but several assumptions need to be considered within the closed loop of a product's design until the end of life (Islam and Huda 2018).

The concern to protect the natural environment and humans exposed to the negative impact of waste generated from used equipment has increased over the past 50 years. This forced various institutions to take important steps to counteract the negative effects of environmental pollution and prepare the legal and organizational framework for handling used and disposed WEEE. This resulted in the adoption of appropriate legal regulations in several countries, initially in Switzerland, South Korea, Japan, and Taiwan (Chung and Murakami-Suzuki 2008; Hischier et al. 2005), which served as an example in building the foundations of the waste equipment management system. The obtained results initiated similar actions in other countries, including the European Union (European Commission 2012; European Union 2003a), and then in other countries, including the two countries with the highest population, India and China (Chung and Zhang 2011; Heeks et al. 2014). Two key Directives concerning WEEE and the restriction of the use of hazardous substances in the production of such devices (Restriction of Hazardous Substances – ROHS) (Bogaert et al. 2008; European Union 2003b) have become models of e-waste management for other countries (Ongondo et al. 2010).

Handling and managing WEEE are interdisciplinary. It is related not only to environmental protection and waste management, but also to legal, financial, and social issues of waste management (Hester and Harrison 2009). However, technical aspects are of particular importance when choosing construction materials for new products, and later – at the end of their life cycle – regarding the collection, transportation, and recycling of WEEE (Nowakowski et al. 2017; Parajuly and Wenzel 2017). The concept of a new electrical or electronic product, its dimensions, and operational features are carried out at the design stage. When the electrical or electronic equipment reaches the end of life the collection, transportation, and processing of equipment, as well as related problems with the disassembly of used equipment, need to be taken into consideration (Wang and Gupta 2011). This is a part of the circular economy where urban mining can be applied (Figure 7.1).

The end of life of the equipment and the decision to dispose of depends on the attitude of individuals and social behavior (Nowakowski 2016; Saphores et al. 2012). The questions from individuals like when, where, and how to dispose of WEEE overlap with the technical and managerial approach: what kind of

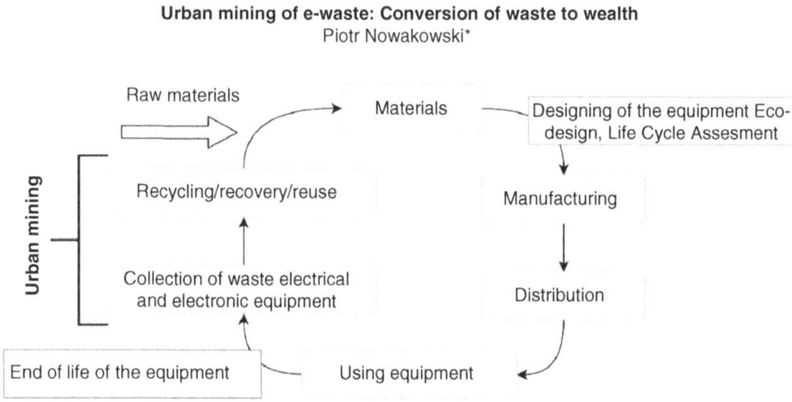

Figure 7.1 Urban mining tasks in the circular economy.

collection, what type of collection vehicles are required, and how to recycle each category of e-waste (Borthakur and Govind 2017; Edumadze et al. 2013). The collected e-waste can undergo various methods of treatment (European Union 2003a):

- recovery – activities that do not pose a threat to life, human health, or the environment, consisting in the use of waste in whole or in part and leading to the recovery of substances, materials, or energy from waste and their use;
- reuse of waste equipment or its components for the same purpose for which they were designed and manufactured;
- processing – activities that are undertaken after the waste equipment has been handed over to the entrepreneur operating the treatment plant for disassembly, which includes the removal of hazardous components and materials, cutting or shredding, and preparation for recovery or disposal;
- recycling – a form of recovery that consists in reprocessing substances or materials contained in waste in production processes, aimed at obtaining substances or materials for primary or similar purposes, except energy recovery;
- waste neutralization – subjecting waste to biological, physical, or chemical transformations to produce final material neutral for the environment.

The functioning of households has been supported by the extensive use of electrical and electronic equipment. The number of types of appliances used in the kitchen (for preparing dishes, grinding products, and thermal processing), information and communication equipment (routers, entertainment centers, and components of the so-called smart home), as well as toys, has increased significantly (Forti et al. 2020). A natural consequence of the extensive use of electronic and electrical equipment is the significant weight and number of waste items in the household waste stream. Eurostat statistics show that in 2019, an average of 10 kg of WEEE waste per capita was collected in the European Union (Eurostat 2022).

A design of new equipment takes into consideration not only the dimensions and materials to be used but also what is going to happen after the end of life of a product. Legislators demand that the design and production of electrical and electronic equipment must take into account the last phase of the life cycle, including disassembling, mechanical processing, and recycling (Charter 2001). Disassembly should be relatively simple and allow materials to be recycled. At the design stage of electrical and electronic equipment, it is planned to use appropriate raw materials for the production of components (Bogue 2007). The methodology used in many manufacturing companies is called Design for Environment (Ferrão and Amaral 2006). In the European Union, various programs for environmentally friendly design and product cycle management have been developed in recent years (European Commission 2009). Taking into account environmental and economic aspects, eco-design concept has emerged with some modifications and improvements that occurred later (Brezet et al. 1997; Charter 2001). The main factors having an impact on the environment and environmental impact costs are highlighted in Figure 7.2. It shows the importance of the design stage for the entire life cycle in terms of costs and environmental impact of the designed electrical and electronic equipment. The design stage contributes to approximately 80% of environmental impact costs (Lee and Park 2005; Van Nes et al. 1999).

Another tool for estimating the environmental impact of electrical and electronic equipment in its life cycle is the Life Cycle Assessment (LCA) method. It analyzes the impact on the environment and human health. Its purpose is to identify and quantify environmental loads, i.e. materials, energy, emissions,

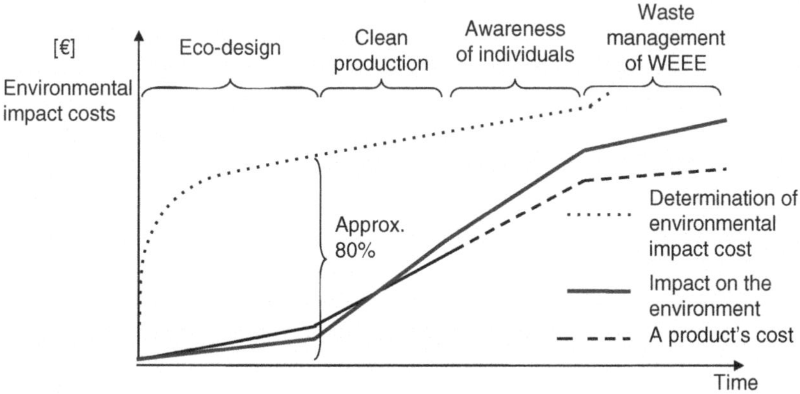

Figure 7.2 Environmental impact costs in the life cycle of electrical and electronic equipment. Source: Adapted from Lee and Park (2005).

and waste released into the environment, and to estimate the available options to reduce the burden (Ismail and Hanafiah 2019). Producers of electrical and electronic equipment use LCA to assess the negative effects of production, use, and possible mishandling of waste.

7.2 Materials for Recovery from Electrical and Electronic Equipment

WEEE is a source of valuable materials. Depending on the type of equipment, different categories of materials are used (Rosa et al. 2019). The most important materials used in the construction of electrical and electronic devices include iron alloys and nonferrous metal alloys (Islam et al. 2020). The second large group includes plastics (Chandrasekaran et al. 2018). In addition, there are additives in the form of precious metals, rare metals, and rare earth metals. Hazardous substances, although occurring in small amounts, are of particular importance for the degradation of the natural environment (Zeng et al. 2017). Ferrous metals are the common materials in electrical and electronic equipment that ensure the appropriate strength for the fastening of individual parts and assembling modules of the entire electrical or electronic product. Elements made of steel are characterized by strength, durability, and stability of the structure, which facilitate the achievement of the assumed mechanical features of a product. Commonly, metal sheets are used to make cabinets (refrigerators, washing machines, or dishwashers) and assemblies supporting printed boards (personal computers, TV sets, and audio equipments). Figure 7.3 shows the average material content for six common categories of equipment used in households (Buekens and Yang 2014; Cui and Forssberg 2003; Cui and Zhang 2008; Huisman and Magalini 2007).

Raw materials obtained after disassembling WEEE, after their further processing and recycling, can be used in the new supply chain. This is a significant contribution to reducing the exploitation of raw materials and meeting the demands of a circular economy. It significantly reduces energy consumption compared to processing ores or crude oil or from deposits acquired from urban mining Table 7.1 (Kang and Schoenung 2005; Kirschen et al. 2002).

The proportion of nonferrous metals in electrical or electronic equipment is significantly lower, but they are irreplaceable due to their electrical (copper) or mechanical properties (lightweight – aluminum). Another major group of materials found in electrical and electronic equipment is plastics. Thermoplastics (i.e. polyethylene – PE, polypropylene – PP, acrylonitrile-butadiene-styrene copolymer – ABS, polyvinyl chloride – PVC) and thermosetting, as well as glass-epoxy laminates for the production of printed circuit boards, are common in a broad range of electrical and electronic equipment (Buekens and Yang 2014). They are mainly used for the production of housing elements, keyboards,

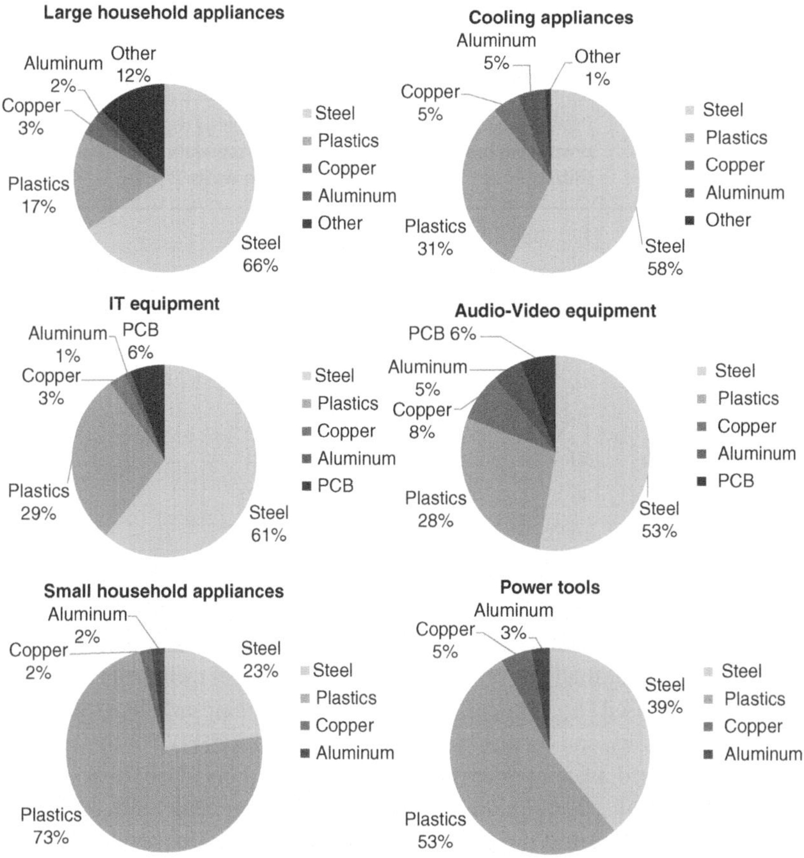

Figure 7.3 Average material content in six categories of household e-waste. Source: Compiled results from reports by Buekens and Yang (2014), Cui and Forssberg (2003), Cui and Zhang (2008), and Huisman and Magalini (2007).

insulating materials (cables), as carriers for mounting plates, and airtight closures of components. Plastics make up variable contents of the weight of collected electronic waste, and the polymers used differ depending on the category of equipment (Maris et al. 2015). Most plastics are found in toys (65% of the equipment weight) and sports equipment, as well as control and supervision instruments (60% of the equipment weight). Relatively small amounts of them are in medical devices and lighting equipment (<10% by weight of the equipment). The application of the metals and common plastics in electrical and electronic equipment is presented in Table 7.2.

Many products belonging to different groups of equipment contain hazardous materials. Dismantling suggestions for both groups of technical measures indicate the mandatory removal of such substances or components (Ahirwar and

Table 7.1 Comparing primary energy required for the extraction of materials from ores or crude oil with the energy required for recovering materials from secondary raw materials.

Material	Primary energy required for processing ores or crude oil [MJ/kg]	Energy required for recycling secondary materials from e-waste [MJ/kg]
Aluminum	170–300	17
Copper	30–125	10–20
Brass	80	10–20
Carbon steel	20–40	10
Stainless steel	46	10
Polyethylene	70–110	15
Polystyrene	60–100	15
Polypropylene	80	15
Glass	10–35	5–7

Tripathi 2021). This should be considered when selecting a disassembly strategy. Precious metals are a small but very important component of technical measures. They are mainly used in electrical engineering due to their excellent electrical conductivity, ductility, and corrosion resistance (Chancerel 2009). For this reason, they have been used as components of solders, connections of microprocessors and integrated circuits, coatings for contacts, cable plugs, etc. Platinum is of great importance in the electrical industry, as is palladium. They are used in the production of hard drives to increase the recording capacity of: capacitors, integrated circuits, and LCD panels (Hagelüken 2012). The content of precious metals varies depending on the group of electronic equipment. They appear on printed circuit boards with elements; their content is: for silver 200–400 mg (landline phone, laptop, video camera, game consoles); for gold, about 100 mg (monitor, game console), 350–480 mg (plasma TV, laptop) (Bigum et al. 2011; Lin et al. 2009). The content of palladium in electronic equipment is about 50 mg (copier, LCD TV) and 140 mg (laptops) (Lim et al. 2011). Metals belonging to the group of rare or rare earth metals constitute a small alloy addition, which, however, significantly affects the properties of the alloy. Thanks to these additives, alloys with specific mechanical or magnetic properties are obtained. Rare or rare earth metals allow for the production of elements such as displays, diodes, and magnets. (Binnemans et al. 2013; Schüler et al. 2011). Rare earth metals are additives found in various components in electronic equipment. Their content is, for example, gallium up to 5 mg (mobile phones, laptops), strontium up to 1400 mg (game consoles, plasma TV), tantalum – 90–400 mg (mobile phones, video cameras) up to 2200 mg

Table 7.2 Brief characteristics of the main materials in manufacturing electrical and electronic equipment.

Material	Brief description of the material
Carbon steel	The cheapest construction material from the metal group, at the same time, guarantees the appropriate strength of the elements. Easy to process, and as a ferromagnetic material, it is recovered in the process of magnetic separation.
Stainless steel	The material is minimally corrosive and therefore long-lasting in product. The elements are resistant to external factors, e.g. water and steam.
Aluminum	Light material, does not corrode but is more expensive than steel components; used in housings and elements to maintain the stability of the structure.
Copper	A necessary metal for all electrical connections due to its very low specific resistance. Corrosion resistant. Relatively expensive, therefore, it is advisable to obtain a high recovery rate of this metal in the recycling.
Brass	An alloy of mainly copper, zinc, and other minor components. Corrosion resistant.
Polystyrene Polypropylene Acrylonitrile butadiene styrene (ABS)	Common materials for construction elements of almost all electronic devices. Light, cheap, and easy to process, as well as having aesthetic value. They do not conduct electricity and are easy to process and recycle; they are also used as refuse-derived fuel (energy recovery).
Polyurethane	Necessary in a foamed form, as a filler and insulator in cooling equipment.
Polyvinyl chloride	Plastic that occurs most often in cables. The problem is the addition of chlorine and phthalates, which make energy recovery more difficult.

(laptops), neodymium, and praseodymium. 28–35% of the mass of permanent magnets (Binnemans et al. 2013; Cui and Forssberg 2003). Neodymium is an essential component of the strongest permanent magnets. It has been used in the manufacturing of microphones, speakers, and hard drives (Menad et al. 2016). Gallium is used in the production of LEDs, optical sensors, and solar cells. In electronic equipment, it is found in advanced communication devices, wireless cards, and high-performance computers. In modern smartphones, the content of gallium is 10 times greater than in a standard mobile phone (de Oliveira et al. 2021; Zhou et al. 2019). Indium is used in the production of various coatings in the production of LCD screens, other panels, and small electrical components. Tantalum is mainly used in small electronic components, and most often it is a component of capacitors (60% of this metal's production). Owing to their properties, it was possible to largely miniaturize the capacitors for use in modern IT and telecommunications equipment (Peelman et al. 2018; Zhang et al. 2017).

WEEE has high recycling potential, depending on the mass share of individual materials, the ease of their separation, and the purchase prices of secondary raw materials. In practice, the recovery rate should be close to 100%. Large construction elements that are manually disassembled are easy to identify and sort. The biggest problem is with small components and small devices. External dimensions of the parts used make identification difficult, and manual separation becomes unprofitable. The same is with printed circuit boards and components integrated with them (Goodship and Stevels 2012). Disassembly of WEEE uses destructive disassembly (mechanical methods) and nondestructive disassembly. The nondestructive method aims to recover parts or components for reuse or regeneration. Most often, it is carried out manually. Then, parts are obtained that can be reused after testing their functionality. In WEEE processing plants, mainly mechanical disassembly is used to obtain process efficiency and economic efficiency. As a rule, there are several separate disassembly lines in the plant. This is mainly due to the characteristics of waste equipment and separate requirements related to the safety of disassembly and the effectiveness of the entire process. For example, for the two most common household appliances – washing machines and refrigerators – the disassembly lines must be different (Nowakowski 2020). Refrigerators and other refrigeration devices go through a special refrigerant removal section due to the danger of pentane explosions when grinding or shredding such equipment. The disassembly of IT devices (computers, laptops), or small kitchen devices, is carried out differently. For the first-mentioned group, it is extremely important to separate elements containing valuable materials such as gold, silver, and palladium (processors, gold-plated connectors, and capacitors) from elements containing hazardous substances (e.g. electrolytic capacitors, batteries, or accumulators). The group of small kitchen appliances is virtually free of precious metals and hazardous substances, so they can be directly processed mechanically (Bovea et al. 2016). The shredded parts are separated on several specialized separators using the basic physical properties of the materials (e.g. susceptibility to the magnetic field or gravity). The magnetic separator can separate iron and steel from other metals (aluminum and copper). In most cases, the fractions after separation have a high degree of purity (Kirschen et al. 2002).

7.3 The Collections and Social Attitude Toward Disposal of E-Waste

7.3.1 Methods of WEEE Collections

E-waste should be disposed of legally and responsibly by households. Companies responsible for the collection of such waste, fulfilling the requirements of the

WEEE Directive, have many collection options (European Commission 2012). The most important ones include stationary collection at the Municipal Waste Collection Points or the Municipal Selective Waste Collection Points. Large stores selling new electrical and electronic equipment also collect waste equipment. In many cases, it is an offer of a free collection of used equipment upon delivery of new equipment. This applies in particular to the supply of large equipment such as washing machines, refrigerators, and dishwashers. In many frequently visited places, for example, near supermarkets, in the alleys of shopping malls, and also in housing estates, bins and containers for the collection of small electronic equipment have been placed (Nowakowski et al. 2021). Often, it is also possible to leave used batteries in them.

Another type of collection is the mobile collection. In a given area (e.g. several villages or districts), residents leave used equipment in front of the property, and employees of the collecting company take the used equipment away. The wide use of IT systems and mobile applications has allowed the introduction of a new method of mobile collection – on-demand waste collection (Król et al. 2016). It consists of reporting to the collecting company the willingness to dispose of waste equipment via a website, mobile application, or telephone notification (Nowakowski et al. 2018; "ZeroNet," 2022). This is a very convenient method because a precise pick-up time can be arranged and an individual can get help with the removal of heavy equipment. Having data on the location and time of collection allows you to develop an accurate collection plan, and thanks to IT systems, it is possible to optimize routes, which reduces costs. Another way of collecting WEEE is in spontaneously organized forms that accompany various events, actions, and events of a pro-ecological nature. An example is Earth Days, promotions in stores when buying new equipment (discount for returning old equipment, etc.). Apart from the above-mentioned legal methods of WEEE waste disposal, there are also illegal and inappropriate methods. The purpose of such activities is to acquire used devices and then disassemble and sell the most valuable components and raw materials (Chi et al. 2011; Li and Tee 2012).

Some residents dispose of used small equipment together in municipal waste with a mixed fraction, or sometimes in a selective collection together with plastic or metal. This is not the right way, as it wastes valuable raw materials and risks contaminating the mixed waste stream (Nowakowski 2019; Parajuly et al. 2016). Companies collecting municipal waste, although they have municipal waste separation stations, are not adapted to the proper processing of used electronic equipment. Often, fragments of such a device, after passing through the separation line, may end up in a landfill or in a composting plant (this is the case, for example, in the case of button-type batteries – they are small and end up in the smallest fraction for composting during separation on sieves). The worst way to dispose of waste equipment is to throw it in a forest, ditch, or other place outside the buildings.

7.3.2 The Awareness of the Inhabitants When Choosing the Method of Waste Disposal

The basic factor influencing the level of collection of waste equipment from households is the awareness of residents regarding the appropriate methods of handling this type of waste. On the one hand, the WEEE Directive or other legislation similar to EU Directive specifies the prohibition of improper disposal of waste equipment with municipal waste, and on the other hand, it provides numerous methods and options for legal and recommended ways to remove these appliances from households. However, it often happens that the inhabitants are not fully aware of what to do, and sometimes it may be due to the ease of disposal of, for example, small equipment by placing it in containers or containers with mixed municipal waste (Huisman et al. 2012; Magalini et al. 2012; Milovantseva and Saphores 2013).

Figure 7.4 shows the residents' options how to choose handling waste electrical and electronic equipment.

There are many factors influencing the decision of residents to end their use and their will to get rid of WEEE. It can be full or partial wear of the equipment or damage that is difficult to repair or expensive (Nowakowski 2019; Wilson et al. 2017). Often, however, residents wonder about energy-saving equipment and cheaper operation (e.g. less water consumption in washing machines or dishwashers). Smaller and cheaper equipment is considered redundant and suitable to be disposed of if corrosion, cracks, or discoloration appear on it. There

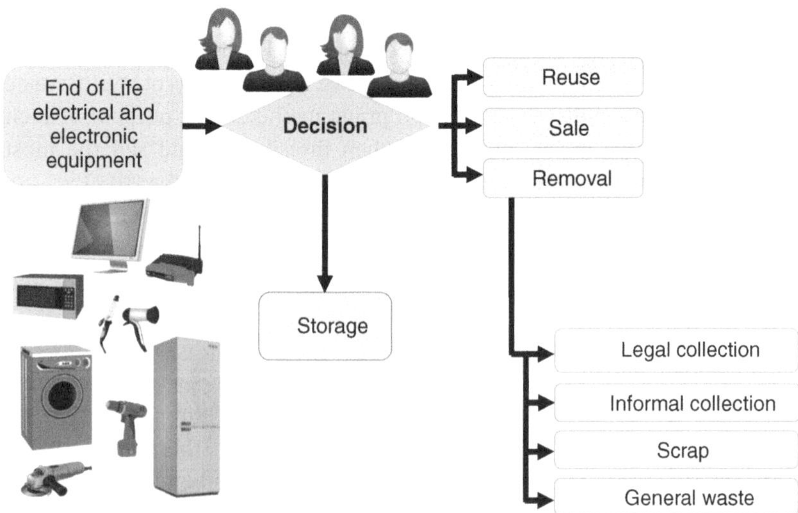

Figure 7.4 Fate of end of life electrical and electronic equipment – Decision making by individuals.

is no doubt that educational and promotional campaigns are an important step in improving the residents' awareness of the proper handling of WEEE waste, as well as their willingness to go to the trouble of handing over waste equipment to companies dealing with legal collection (Borthakur and Govind 2017; Islam et al. 2016). This is one of the purposes for which the WEEE Directive has established WEEE Recovery Organizations in the European Union. The main goal of information and educational campaigns is to make residents aware of:

- the risks of environmental pollution due to improper disposal of equipment, especially when placing it in mixed municipal waste, or abandoning it in green areas,
- benefits related to the recyclability of raw materials used in the construction of the equipment, which will be possible after the equipment is handed over to the appropriate dismantling and processing plant,
- a possibility of donating the equipment at specific points or at the place of residence under the collection schedule for a given area.

The activities aimed at improving the efficiency of WEEE collection directed at residents can be provided as a general rule on a national level but also as a micro initiative in local communities – e-waste collection at schools, local events, or small shops (Sivathanu 2016; Wang et al. 2018).

7.4 Discussion and Conclusion

The circular economy approach gives a great opportunity to manage all resources in supply chains. One of the categories of waste generated in households is WEEE. Our society is progressively dependent on electrical and electronic equipment, and more waste products would have to be collected (Cucchiella et al. 2015). The high recycling potential of disposed e-waste of by households allows for saving energy and recovering valuable raw materials. The deposits of secondary raw materials of significant importance are in urban communities and can therefore be explored as urban, mining taking into consideration all sustainable goals (Xavier et al. 2019).

The effectiveness of urban mining of WEEE depends on three key pillars:

- legislation that establishes a framework for WEEE collection, treatment, and recycling;
- organizational and technical resources needed for the collection and recycling of the waste equipment;
- social attitude in increasing awareness of the high recycling potential of e-waste by individuals and participating in effective collections of waste electrical and electronic items from the households.

The legislation regarding e-waste collection and recycling systems is already present in many countries (Patil and Ramakrishna 2020; Thakur and Kumar 2021). After the introduction of the WEEE legislation, we can observe an increase in the level of waste equipment collection from households (Forti et al. 2020). The effort made by WEEE Recovery Organizations, collecting companies, and numerous local governments brings measurable benefits for the acquisition of valuable raw materials suitable for recycling and the neutralization of substances that threaten the natural environment (Patil and Ramakrishna 2020; Shahabuddin et al. 2022). Therefore, the experience of the countries that had already introduced the WEEE legislation and the benefits for the environment and the local population should encourage other countries to implement similar laws. The technical and organizational parts of the collection and recycling of e-waste must provide all necessary resources for effective and safe e-waste collection, disassembling, and recycling. This process consists of three important steps. The first step is the selection of the appropriate waste collection type. At the same time, it is necessary to improve by informing residents about the ways of legal disposing of waste equipment. There are relatively few containers and bins for the collection of small equipment. At the same time, sets of containers found in shopping malls, shopping avenues, and supermarkets should be marked (Nowakowski et al. 2021). The on-demand mobile collection has a great potential for development, especially when route optimization software is employed (Nowakowski et al. 2018). Another issue to be solved is informal collection. This is especially important in developing countries. Therefore, various programs and proposals have been issued to encourage the informal sector to participate in the official recycling network (Gu et al. 2016; Xue et al. 2019). Additional solutions allow using information technologies in supporting e-waste collections to improve effectiveness and lower collection costs (Alves et al. 2021; Friege et al. 2015; Nowakowski et al. 2018). The recycling of various categories requires a preliminary selection of equipment with similar features or dimensions. Therefore, there are separate disassembling lines for the individual categories of e-waste like cooling equipment, information technology equipment, and large and small household appliances (Alam et al. 2022; Arduin et al. 2020; Xu et al. 2018). Also, new techniques and image recognition methods have been applied to identify materials after disassembling to achieve pure fractions for recycling (Andrade et al. 2022; Charles et al. 2020; Grigorescu et al. 2019).

The collection rate of e-waste directly depends on the attitude and behavior of individuals in society. In this case, proper information campaigns should be applied to each age group. Several studies indicate a specific approach depending on the individual country (Otto et al. 2018; Singh et al. 2022; Uhunamure et al. 2021). Collecting companies and e-waste recovery organizations provide information campaigns to educate individuals about available legal methods of WEEE disposal. Educating individuals is an important factor in improving collection rates of WEEE. Properly conducted education campaigns encourage individuals for legal disposal of e-waste and discourage them to stockpile old equipment in households or dispose of in landfills (Gu et al. 2016; Oliveira et al. 2012).

This chapter indicates that the awareness of individuals and social behavior are important for legal e-waste collection in urban mining. Available collection methods are flexible for residents, and WEEE can be disposed of conveniently. The disassembly and recycling technologies allow for the efficient recovery of materials deposited in the components of electrical and electronic equipment. It was highlighted that the recycling potential of individual categories of e-waste is significant. Besides common metals and plastics, some alloys include rare earth metals. Additional improvement is required for the development of new recycling methods for rare earth metals and for lowering the energy use in the collection and treatment processes. Conversion of waste to wealth is possible when all partners in the reverse supply chain cooperate. The benefits for the environment and society include the minimization of hazardous substances and the maximization of resources for energy recovery and recycling.

References

Ahirwar, R. and Tripathi, A.K. (2021). E-waste management: a review of recycling process, environmental and occupational health hazards, and potential solutions. *Environmental Nanotechnology, Monitoring and Management* 15: 100409.

Alam, T., Golmohammadzadeh, R., Faraji, F., and Shahabuddin, M. (2022). E-waste recycling technologies: an overview, challenges and future perspectives. In: *Paradigm Shift E-waste Management*, 143–176.

Alves, R., Ferreira, K.L.A., Lima, R.d.S., and Moraes, F.T.F. (2021). An action research study for elaborating and implementing an electronic waste collection program in Brazil. *Systemic Practice and Action Research* 34: 91–108.

Andersen, M.S. (2007). An introductory note on the environmental economics of the circular economy. *Sustainability Science* 2: 133–140.

Andrade, D.F., Castro, J.P., Garcia, J.A. et al. (2022). Analytical and reclamation technologies for identification and recycling of precious materials from waste computer and mobile phones. *Chemosphere* 286: 131739.

Arduin, R.H., Mathieux, F., Huisman, J. et al. (2020). Novel indicators to better monitor the collection and recovery of (critical) raw materials in WEEE: focus on screens. *Resources, Conservation and Recycling* 157: 104772.

Bigum, M., Brogaard, L., and Christensen, T.H. (2011). Metal recovery from high-grade WEEE: a life cycle assessment. *Journal of Hazardous Materials* 207: 8–14.

Binnemans, K., Jones, P.T., Blanpain, B. et al. (2013). Recycling of rare earths: a critical review. *Journal of Cleaner Production* 51: 1–22. https://doi.org/10.1016/j.jclepro.2012.12.037.

Bogaert, S., Van Acoleyen, M., Van Tomme, I., De Smet, L., Fleet, D., Salado, R., 2008. *Study on RoHS and WEEE Directives*. ARCADIS RPA DG ENTR Eur. Comm.

Bogue, R. (2007). Design for disassembly: a critical twenty-first century discipline. *Assembly Automation* 27: 285–289. https://doi.org/10.1108/01445150710827069.

Borthakur, A. and Govind, M. (2017). Emerging trends in consumers' E-waste disposal behaviour and awareness: a worldwide overview with special focus on India. *Resources, Conservation and Recycling* 117: 102–113. https://doi.org/10.1016/j.resconrec.2016.11.011.

Bovea, M.D., Pérez-Belis, V., Ibáñez-Forés, V., and Quemades-Beltrán, P. (2016). Disassembly properties and material characterisation of household small waste electric and electronic equipment. *Waste Management* 53: 225–236.

Brezet, H., Van Hemel, C., Böttcher, H. et al. (1997). *Ecodesign: A Promising Approach to Sustainable Production and Consumption*. UNEP.

Buekens, A. and Yang, J. (2014). Recycling of WEEE plastics: a review. *Journal of Material Cycles and Waste Management* 16: 415–434. https://doi.org/10.1007/s10163-014-0241-2.

Chancerel, P. (2009). *Substance Flow Analysis of the Recycling of Small Waste Electrical and Electronic Equipment – An Assessment of the Recovery of Gold and Palladium*. Universitätsbibliothek, Technische Universität Berlin.

Chandrasekaran, S.R., Avasarala, S., Murali, D. et al. (2018). Materials and energy recovery from e-waste plastics. *ACS Sustainable Chemistry & Engineering* 6: 4594–4602.

Charles, R.G., Douglas, P., Dowling, M. et al. (2020). Towards increased recovery of critical raw materials from WEEE – evaluation of CRMs at a component level and pre-processing methods for interface optimisation with recovery processes. *Resources, Conservation and Recycling* 161: 104923. https://doi.org/10.1016/j.resconrec.2020.104923.

Charter, M. (2001). Integrated product policy (IPP) and eco-product development (EPD). *Environ. Conscious Des. Inverse Manuf. 2001 Proc. EcoDesign 2001 Second Int. Symp. On* 672–677. https://doi.org/10.1109/.2001.992445.

Chi, X., Streicher-Porte, M., Wang, M.Y.L., and Reuter, M.A. (2011). Informal electronic waste recycling: a sector review with special focus on China. *Waste Management* 31: 731–742. https://doi.org/10.1016/j.wasman.2010.11.006.

Chung, S.-W., Murakami-Suzuki, R., 2008. *A Comparative Study of e-Waste Recycling Systems in Japan, South Korea and Taiwan from the EPR Perspective: Implications for Developing Countries [WWW Document]*. http://www.ide.go.jp/English/Publish/Download/Spot/pdf/30/007.pdf (accessed 19 October 2011).

Chung, S.-S. and Zhang, C. (2011). An evaluation of legislative measures on electrical and electronic waste in the People's Republic of China. *Waste Management* 31: 2638–2646. https://doi.org/10.1016/j.wasman.2011.07.010.

Cossu, R., Salieri, V., and Bisinella, V. (2012). *Urban Mining: A Global Cycle Approach to Resource Recovery from Solid Waste*. CISA Publ.

Cucchiella, F., D'Adamo, I., Lenny Koh, S.C., and Rosa, P. (2015). Recycling of WEEEs: an economic assessment of present and future e-waste streams. *Renewable and Sustainable Energy Reviews* 51: 263–272. https://doi.org/10.1016/j.rser.2015.06.010.

Cui, J. and Forssberg, E. (2003). Mechanical recycling of waste electric and electronic equipment: a review. *Journal of Hazardous Materials* 99: 243–263.

Cui, J. and Zhang, L. (2008). Metallurgical recovery of metals from electronic waste: a review. *Journal of Hazardous Materials* 158: 228–256. https://doi.org/10.1016/j.jhazmat.2008.02.001.

Edumadze, J.K.E., Tenkorang, E.Y., Armah, F.A. et al. (2013). Electronic waste is a mess: awareness and proenvironmental behavior among university students in Ghana. *Applied Environmental Education and Communication* 12: 224–234. https://doi.org/10.1080/1533015X.2013.876250.

European Commission, 2009. *Directive 2009/125/EC of the European Parliament and of the Council of 21 October 2009 Establishing a Framework for the Setting of Ecodesign Requirements for Energy.*

European Commission, 2012. *Directive 2012/19/EU of the European Parliament and of the Council of 4 July 2012 on Waste Electrical and Electronic Equipment (WEEE) Text with EEA Relevance – LexUriServ.do.*

European Union, 2003a. *Directive 2002/96/EC of the European Parliament and of the Council of 27 January 2003 on Waste Electrical and Electronic Equipment (WEEE) [WWW Document].* http://eur-lex.europa.eu/LexUriServ/LexUriServ.do?uri=OJ:L:2003:037:0024:0038:en:PDF (accessed 28 October 2019).

European Union, 2003b. *Directive 2002/95/EC of the European Parliament and of the Council of 27 January 2003 on the Restriction of the Use of Certain Hazardous Substances in Electrical and Electronic Equipment [WWW Document].* http://eur-lex.europa.eu/LexUriServ/LexUriServ.do?uri=OJ:L:2003:037:0019:0023:en:PDF (accessed 17 October 2019).

Eurostat, 2022. *Waste statistics – Electrical and Electronic Equipment [WWW Document].* https://ec.europa.eu/eurostat/statistics-explained/index.php?title=Waste_statistics_-_electrical_and_electronic_equipment (accessed 10 August 2022).

Ferrão, P. and Amaral, J. (2006). Design for recycling in the automobile industry: new approaches and new tools. *Journal of Engineering Design* 17: 447–462. https://doi.org/10.1080/09544820600648039.

Forti, V., Baldé, C.P., Kuehr, R., and Bel, G. (2020). *The Global E-waste Monitor 2020: Quantities, Flows and the Circular Economy Potential*, 1–119.

Friege, H., Oberdörfer, M., and Günther, M. (2015). Optimising waste from electric and electronic equipment collection systems: a comparison of approaches in European countries. *Waste Management & Research* 33: 223–231.

Goodship, V. and Stevels, A. (2012). *Waste Electrical and Electronic Equipment (WEEE) Handbook*. Elsevier Science.

Grigorescu, R.M., Grigore, M.E., Iancu, L. et al. (2019). Waste electrical and electronic equipment: a review on the identification methods for polymeric materials. *Recycling* 4: 32.

Gu, Y., Wu, Y., Xu, M. et al. (2016). The stability and profitability of the informal WEEE collector in developing countries: a case study of China. *Resources, Conservation and Recycling* 107: 18–26. https://doi.org/10.1016/j.resconrec.2015.12.004.

Hagelüken, C. (2012). Recycling the platinum group metals: a European perspective. *Platinum Metals Review* 56: 29–35. https://doi.org/10.1595/147106712X611733.

Heeks, R., Subramanian, L., and Jones, C. (2014). Understanding e-waste management in developing countries: strategies, determinants, and policy implications in the Indian ICT sector. *Information Technology for Development* 1, 15. https://doi.org/10.1080/02681102.2014.886547.

Hester, R.E. and Harrison, R.M. (2009). *Electronic Waste Management*. Royal Society of Chemistry.

Hischier, R., Wager, P., and Gauglhofer, J. (2005). Does WEEE recycling make sense from an environmental perspective?: the environmental impacts of the Swiss take-back and recycling systems for waste electrical and electronic equipment (WEEE). *Environmental Impact Assessment Review* 25: 525–539. https://doi.org/10.1016/j.eiar.2005.04.003.

Huisman, J., Magalini, F., 2007. *2008 Review of Directive 2002/96 on Waste Electrical and Electronic Equipment (WEEE).*

Huisman, J., Van der Maesen, M., Eijsbouts, R.J.J., Wang, F., Baldé, C.P., Wielenga, C.A., 2012. *The Dutch WEEE Flows.*

Islam, M.T. and Huda, N. (2018). Reverse logistics and closed-loop supply chain of Waste Electrical and Electronic Equipment (WEEE)/E-waste: a comprehensive literature review. *Resources, Conservation and Recycling* 137: 48–75.

Islam, M.T., Abdullah, A.B., Shahir, S.A. et al. (2016). A public survey on knowledge, awareness, attitude and willingness to pay for WEEE management: case study in Bangladesh. *Journal of Cleaner Production* 137: 728–740. https://doi.org/10.1016/j.jclepro.2016.07.111.

Islam, A., Ahmed, T., Awual, M.R. et al. (2020). Advances in sustainable approaches to recover metals from e-waste – a review. *Journal of Cleaner Production* 244: 118815.

Ismail, H. and Hanafiah, M.M. (2019). An overview of LCA application in WEEE management: current practices, progress and challenges. *Journal of Cleaner Production* 232: 79–93.

Kang, H.Y. and Schoenung, J.M. (2005). Electronic waste recycling: a review of US infrastructure and technology options. *Resources, Conservation and Recycling* 45: 368–400.

Kirschen, M., Pfeifer, H., and Wahlers, F.J. (2002). Mass and energy balances of stainless steel Electric Arc Furnace (EAF). In: *AIM Conference Proceedings. Seventh European Electric Steelmaking Conference, Venice, Italy*, 2–3.

Korhonen, J., Honkasalo, A., and Seppälä, J. (2018). Circular economy: the concept and its limitations. *Ecological Economics* 143: 37–46.

Król, A., Nowakowski, P., and Mrówczyńska, B. (2016). How to improve WEEE management? Novel approach in mobile collection with application of artificial intelligence. *Waste Management* 50: 222–233. https://doi.org/10.1016/j.wasman.2016.02.033.

Krook, J. (2010). Urban and landfill mining: emerging global perspectives and approaches. *Journal of Cleaner Production* 16: 1772–1773.

Krook, J. and Baas, L. (2013). Getting serious about mining the technosphere: a review of recent landfill mining and urban mining research. *Journal of Cleaner Production* 55: 1–9. https://doi.org/10.1016/j.jclepro.2013.04.043.

Lee, K.-M., Park, P.-J., 2005. *ECODESIGN Best Practice of ISO/TR 14062.*

Li, R.C. and Tee, T.J.C. (2012). A reverse logistics model for recovery options of e-waste considering the integration of the formal and informal waste sectors. *Procedia – Social and Behavioral Sciences* 40: 788–816.

Lim, S.-R., Kang, D., Ogunseitan, O.A., and Schoenung, J.M. (2011). Potential environmental impacts of light-emitting diodes (LEDs): metallic resources, toxicity, and hazardous waste classification. *Environmental Science & Technology* 45: 320–327. https://doi.org/10.1021/es101052q.

Lin, K.-L., Chang, W.-K., Chang, T.-C. et al. (2009). Recycling thin film transistor liquid crystal display (TFT-LCD) waste glass produced as glass–ceramics. *Journal of Cleaner Production* 17: 1499–1503. https://doi.org/10.1016/j.jclepro.2009.05.012.

Magalini, F., Huisman, J., Wang, F. et al. (2012). *Household WEEE Generated in Italy (WAIT), Analysis on Volumes & Consumer Disposal Behavior for Waste Electric and Electronic Equipment (Technical Report).* Italy: ECODOM.

Maris, E., Botané, P., Wavrer, P., and Froelich, D. (2015). Characterizing plastics originating from WEEE: a case study in France. *Minerals Engineering* 76: 28–37. https://doi.org/10.1016/j.mineng.2014.12.034.

Menad, N., Seron, A., Save, M. et al. (2016). Recovery of permanent magnets type NdFeB from WEEE. In: *6th International Conference on Engineering for Waste and Biomass Valorisation.*

Milovantseva, N. and Saphores, J.-D. (2013). E-waste bans and U.S. households' preferences for disposing of their e-waste. *Journal of Environmental Management* 124: 8–16. https://doi.org/10.1016/j.jenvman.2013.03.019.

Nowakowski, P. (2016). The influence of residents' behaviour on waste electrical and electronic equipment collection effectiveness. *Waste Management & Research* 34: 1126–1135. https://doi.org/10.1177/0734242X16669997.

Nowakowski, P. (2019). Investigating the reasons for storage of WEEE by residents – a potential for removal from households. *Waste Management* 87: 192–203. https://doi.org/10.1016/j.wasman.2019.02.008.

Nowakowski, P. (2020). Reconfigurable recycling systems of E-waste. In: *E-waste Recycling and Management*, 19–38. Springer.

Nowakowski, P., Król, A., and Mrówczyńska, B. (2017). Supporting mobile WEEE collection on demand: a method for multi-criteria vehicle routing, loading and cost optimisation. *Waste Management* 69: 377–392. https://doi.org/10.1016/j.wasman.2017.07.045.

Nowakowski, P., Szwarc, K., and Boryczka, U. (2018). Vehicle route planning in e-waste mobile collection on demand supported by artificial intelligence algorithms. *Transportation Research Part D: Transport and Environment* 63: 1–22. https://doi.org/10.1016/j.trd.2018.04.007.

Nowakowski, P., Kuśnierz, S., Płoszaj, J., and Sosna, P. (2021). Collecting small-waste electrical and electronic equipment in Poland—how can containers help in disposal of e-waste by individuals? *Sustainability* 13: 12422. https://doi.org/10.3390/su132212422.

Oliveira, C.R.d., Bernardes, A.M., and Gerbase, A.E. (2012). Collection and recycling of electronic scrap: a worldwide overview and comparison with the Brazilian situation. *Waste Management* 32: 1592–1610. https://doi.org/10.1016/j.wasman.2012.04.003.

de Oliveira, R.P., Benvenuti, J., and Espinosa, D.C.R. (2021). A review of the current progress in recycling technologies for gallium and rare earth elements from light-emitting diodes. *Renewable and Sustainable Energy Reviews* 145: 111090.

Ongondo, F., Williams, I., and Cherrett, T. (2010). How are WEEE doing? A global review of the management of electrical and electronic wastes. *Waste Management* 31: 714–730. https://doi.org/10.1016/j.wasman.2010.10.023.

Otto, S., Kibbe, A., Henn, L. et al. (2018). The economy of E-waste collection at the individual level: a practice oriented approach of categorizing determinants of E-waste collection into behavioral costs and motivation. *Journal of Cleaner Production* 204: 33–40. https://doi.org/10.1016/j.jclepro.2018.08.293.

Ottoni, M., Dias, P., and Xavier, L.H. (2020). A circular approach to the e-waste valorization through urban mining in Rio de Janeiro, Brazil. *Journal of Cleaner Production* 261: 120990. https://doi.org/10.1016/j.jclepro.2020.120990.

Parajuly, K. and Wenzel, H. (2017). Potential for circular economy in household WEEE management. *Journal of Cleaner Production* 151: 272–285. https://doi.org/10.1016/j.jclepro.2017.03.045.

Parajuly, K., Habib, K., and Liu, G. (2016). Waste electrical and electronic equipment (WEEE) in Denmark: flows, quantities and management. *Resources, Conservation and Recycling* https://doi.org/10.1016/j.resconrec.2016.08.004.

Patil, R.A. and Ramakrishna, S. (2020). A comprehensive analysis of e-waste legislation worldwide. *Environmental Science and Pollution Research* 27: 14412–14431.

Peelman, S., Kooijman, D., Sietsma, J., and Yang, Y. (2018). Hydrometallurgical recovery of rare earth elements from mine tailings and WEEE. *Journal of Sustainable Metallurgy* 4: 367–377.

Rosa, P., Sassanelli, C., and Terzi, S. (2019). Circular business models versus circular benefits: an assessment in the waste from electrical and electronic equipments sector. *Journal of Cleaner Production* 231: 940–952.

Saphores, J.-D.M., Ogunseitan, O.A., and Shapiro, A.A. (2012). Willingness to engage in a pro-environmental behavior: an analysis of e-waste recycling based on a national survey of U.S. households. *Resources, Conservation and Recycling* 60: 49–63. https://doi.org/10.1016/j.resconrec.2011.12.003.

Schüler, D., Buchert, M., Liu, R. et al. (2011). *Study on Rare Earths and Their Recycling*. Öko-Inst. EV Darmstadt.

Shahabuddin, M., Uddin, M.N., Chowdhury, J.I. et al. (2022). A review of the recent development, challenges, and opportunities of electronic waste (e-waste). *International journal of Environmental Science and Technology* 1–8.

Singh, S., Dasgupta, M.S., and Routroy, S. (2022). Evaluation of sustainable e-waste collection method for urban and rural region of India. *Waste Management & Research* 40: 545–555.

Sivathanu, B. (2016). User's perspective: knowledge and attitude towards e-waste. *International Journal of Applied Environmental Sciences* 11: 413–423.

Thakur, P. and Kumar, S. (2021). Evaluation of e-waste status, management strategies, and legislations. *International journal of Environmental Science and Technology* 1–10.

Uhunamure, S.E., Nethengwe, N.S., Shale, K. et al. (2021). Appraisal of households' knowledge and perception towards e-waste management in Limpopo Province, South Africa. *Recycling* 6: https://doi.org/10.3390/recycling6020039.

Van Nes, N., Cramer, J., and Stevels, A. (1999). A practical approach to the ecological lifetime optimization of electronic products. In: *Environmentally Conscious Design and Inverse Manufacturing, 1999. Proceedings. EcoDesign'99: First International Symposium On*, 108–111. IEEE.

Wang, H.-F. and Gupta, S.F. (2011). End-of-life management: disassembly and reuse. In: *Green Supply Chain Management: Product Life Cycle Approach*. McGraw Hill Professional, Access Engineering.

Wang, Z., Guo, D., Wang, X. et al. (2018). How does information publicity influence residents' behaviour intentions around e-waste recycling? *Resources, Conservation and Recycling* 133: 1–9. https://doi.org/10.1016/j.resconrec.2018.01.014.

Wilson, G.T., Smalley, G., Suckling, J.R. et al. (2017). The hibernating mobile phone: dead storage as a barrier to efficient electronic waste recovery. *Waste Management* https://doi.org/10.1016/j.wasman.2016.12.023.

Xavier, L.H., Giese, E.C., Ribeiro-Duthie, A.C., and Lins, F.A.F. (2019). Sustainability and the circular economy: a theoretical approach focused on e-waste urban mining. *Resources Policy* 101467. https://doi.org/10.1016/j.resourpol.2019.101467.

Xu, Y., Zhang, L., Yeh, C.-H., and Liu, Y. (2018). Evaluating WEEE recycling innovation strategies with interacting sustainability-related criteria. *Journal of Cleaner Production* 190: 618–629.

Xue, Y., Wen, Z., Bressers, H., and Ai, N. (2019). Can intelligent collection integrate informal sector for urban resource recycling in China? *Journal of Cleaner Production* 208: 307–315. https://doi.org/10.1016/j.jclepro.2018.10.155.

Zeng, X., Yang, C., Chiang, J.F., and Li, J. (2017). Innovating e-waste management: from macroscopic to microscopic scales. *Science of the Total Environment* 575: 1–5.

ZeroNet [WWW Document], 2022. *ZeroNet*. www.thezeronet.com (accessed 11 August 2022).

Zhang, S., Ding, Y., Liu, B., and Chang, C. (2017). Supply and demand of some critical metals and present status of their recycling in WEEE. *Waste Management* 65: 113–127. https://doi.org/10.1016/j.wasman.2017.04.003.

Zhou, J., Zhu, N., Liu, H. et al. (2019). Recovery of gallium from waste light emitting diodes by oxalic acidic leaching. *Resources, Conservation and Recycling* 146: 366–372.

8

Life Cycle Assessment and Techno-Economic of E-waste Recycling

Deblina Dutta[1], *Rahul Rautela*[2,3], *Pankaj Meena*[2], *Venkata Ravi Sankar Cheela*[4], *Pranav Prashant Dagwar*[1,2], *and Sunil Kumar*[2]

[1] *Department of Environmental Science and Engineering, SRM University-AP, Amaravati, Andhra Pradesh, India*
[2] *CSIR-National Environmental Engineering Research Institute (CSIR-NEERI), Nagpur, Maharashtra, India*
[3] *Academy of Scientific and Innovative Research (AcSIR), Ghaziabad, Uttar Pradesh, India*
[4] *Civil Engineering Department, MVGR College of Engineering (A), Vizianagaram, Andhra Pradesh, India*

Abstract

In twenty-first century, the world has become dependent and addicted to electronic gadgets like mobile phones, laptops, tablets, etc. The increased use of gadgets has led to a huge generation of E-waste (Electrical and electronic waste). Therefore, management of E-waste has become a major concern at the global level. Life cycle assessment (LCA) evaluates and quantifies the environmental impact of an electronic product from its raw material till its disposal. The techno-economic analysis (TEA) recognizes the technical modifications to overwhelm existing obstacles in commercial E-waste recycling units. The present chapter deals with the LCA and techno-economic analysis of E-waste recycling. This "Lifecycle Evaluation" (combination of LCA and TEA) would help in understanding the environmental and social impacts, which would facilitate the E-waste recycling processes. This approach would prove to be an important tool in developing E-waste management schemes in a sustainable way.

Keywords *environmental impact; economic analysis; recycling; sustainable development*

8.1 Introduction

In the world of rapid advancement in science and technology, there has been a swift revolution of information and communication technologies in the late twentieth century (Awasthi and Li 2017; Shahabuddin et al. 2022). The electronics

Management of Electronic Waste: Resource Recovery, Technology and Regulation, First Edition. Edited by Anshu Priya.

market is one of the determinedly expanding markets of the currently developing nations (Kumar and Dixit 2018). The development is mixed with rapid urbanization and change in technologies, which further adds to the price-dropping trend and constant replacement of electronic gadgets with new ones as for consumer's habits (Dwivedy and Mittal 2012; Kumar and Dixit 2018). As per the literature, Asia is one of the leading producers of E-waste, followed by Europe, America, Oceania, and Africa. India is a growing country that occupies the third position on the global scale and second position in Asia, which has a significantly growing E-waste production (Kumar and Dixit 2018; Andeobu et al. 2021; Shahabuddin et al. 2022). The need for a befitting waste management technique and safe disposal of E-waste has turned out to be a global priority due to the constant discarding of it to landfills, which is the common approach that certainly causes the degradation of the environment (Song and Li 2015; Islam et al. 2016; Kumar and Dixit 2018; Andeobu et al. 2021; Shahabuddin et al. 2022).

The development activities leading to industrialization and urbanization have transformed resource recovery from ore minerals to urban mining. E-waste has been considered as a rich source of base metals, precious metals (PMs), and rare earth metals for its reuse and further utilization (Tanskanen 2013). Studies are present that showed the availability of over 1000 substances in E-waste and many of these are toxic and cause serious environmental pollution due to improper disposal (Puckett et al. 2002; Ikhlayel 2017). Therefore, e-waste requires proper recycling to avoid the loss of valuable and PMs and can further be used to produce new equipment (Kiddee et al. 2013; Baldé et al. 2017).

8.1.1 Life Cycle Assessment

To improve the environmental performance of recycling activities for a more sustainable E-waste management, life cycle assessment (LCA) is considered to be an effective and growing environmental management tool that helps to identify and evaluate environmental impacts that are caused by a product or service, and this tool helps in identifying its improvement potentials (Rebitzer et al. 2004; Xue and Xu 2017; Goh et al. 2022). LCA has four stages specifically: goal and scope definition, life cycle inventory analysis, life cycle impact assessment, and interpretation of life cycle. With these four stages, the product is being defined properly (ISO 2006; Xue and Xu 2017; Muralikrishna and Manickam 2017). The effects of human activities that impose a burden on the environment can be ascertained by the accounting of resources and energy utilized by every phase in the life cycle of the product, which interprets the release of pollutants and waste from electronics.

8.1.2 Techno-Economic Analysis

Techno-economic analysis is a method to analyze the economic performance of an industrial process, product, or service. It utilizes modeling software to estimate

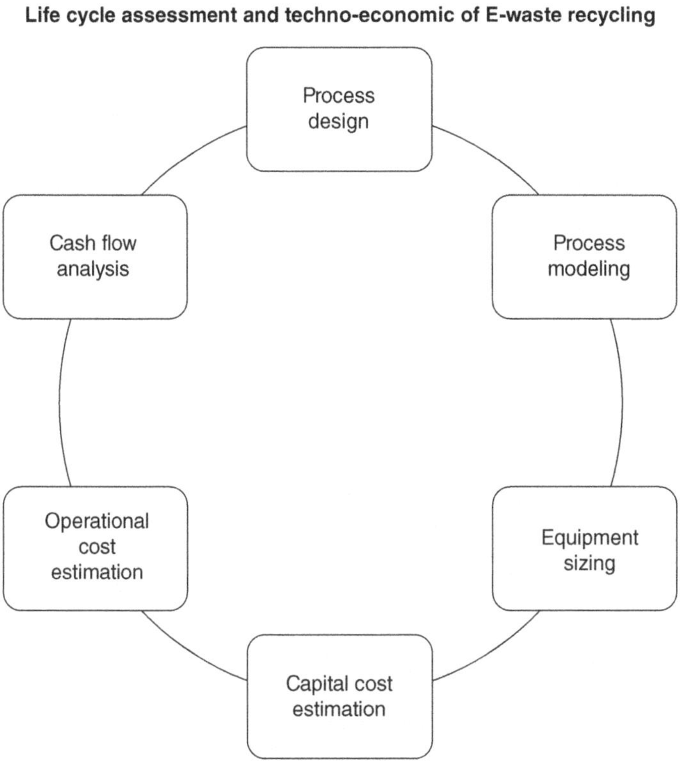

Figure 8.1 Elements of techno-economic model.

investment cost, operational and maintenance cost, and revenue generated based on input financial parameters. The results of such analysis can be visually depicted using visualization tools like sensitivity analysis graphs and tornado diagrams.

Techno-economic analysis can be used to optimize both current and new technologies.

Techno-economic analysis helps industries decide whether methodologies, processes, and techniques that are being adopted are economically feasible to utilize or require more R&D to improve profitability while reducing uncertainty and risk.

Techno-economic model is an integration of process and cost models, which combine various elements, such as given in Figure 8.1.

E-waste is a source of various critical base metals, PMs, and rare earth metals along with nonmetals that have high economic values. Recovery of this high-economic material does not always confirm profitability because of multiple factors like capital cost to set up the recovery units, logistic cost for channel flow,

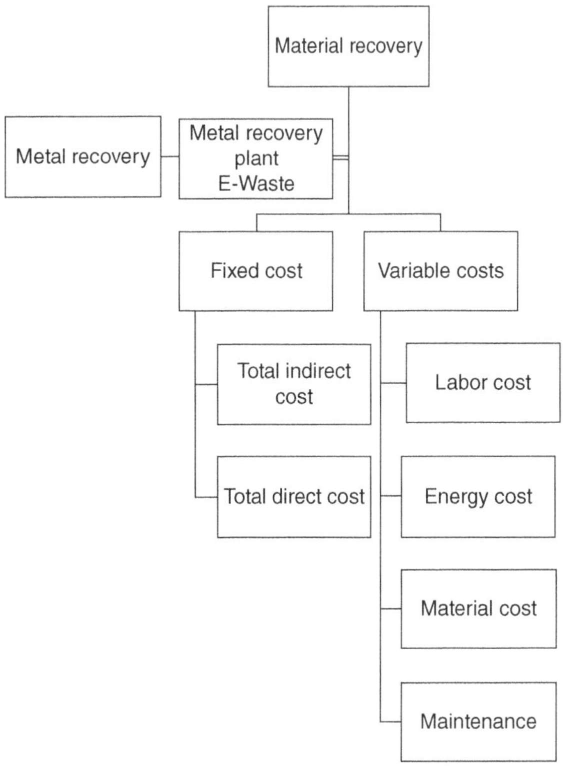

Figure 8.2 Model of the costs and revenues in an MRP (metal recovery process) operation for E-waste. Source: Adapted from Ghodrat et al. (2016).

operational and maintenance costs to run the facility smoothly, uneconomical technologies and recovery methods, and disposal cost of waste and toxic material, etc. Due to all these factors, techno-economic analysis of E-waste is very important so that recyclers can analyze net profit and net expenditure to optimize their pathway accordingly.

As such, techno-economic analysis not only leads toward innovation in technologies but also invites investors to invest in businesses that are profitable, which is why techno-economic analysis of E-waste could play a critical role in dealing with the E-waste problem, which is rising at a rapid pace (Figure 8.2).

Techno-economic analysis sometimes gives multiple routes to proceed with different consequences and benefits take for example, the recovery of base metals from printed circuit boards (PCBs) reduces carbon intensity but limits profitability meanwhile recovery of PMs increases profitability but increases carbon intensity. As we can see, one route gives limited profit but reduces environmental impact, while the other route gives high profit but higher environmental impact as well. From these, we can understand the importance of techno-economic analysis (Lam et al. 2021).

8.1.3 System Application in E-Waste System

As E-waste contains various metals like PMs and rare earth elements (REEs), which makes it one of the best secondary sources. However, it is also a source of toxic substances, if it is treated, disposed of, or handled in an inappropriate manner (Xue and Xu 2017). For the recovery of PMs and REEs, the existing recycling techniques only focused on economic feasibility rather than environmental and social feasibilities. These feasibilities could have occurred in any stage of E-waste management like generation, transportation, logistics, recycling, and disposal. LCA in E-waste is a crucial task in the present era, where we are focusing on resource recovery from E-waste. During the recovery process, the various stages from the generation to the final disposal of E-waste follow the "Cradle to Grave" concept to assess the impacts of pollutants released during the various activities. The LCA in E-waste systems considered all the activities like generation source, collection, transportation, segregation, dismantling, treatment process, refinement of recovered metals, and disposal activities. Each of these activities has both visible and hidden impacts, which can be neglected during the planning process as true potential impacts could not be determined without the application of LCA. LCA plays an important role in estimating the true potential of any E-waste system based on its sustainability, economy, and low impact on the environment (Table 8.1).

Some of the important applications of LCA on E-waste systems are as follows:

- Analysis of the life cycle stages and their contribution to environmental load with a general aim to improve product or process or activity.
- To compare various products or methodologies during internal use.

To further understand the benefits of LCA applications in the E-waste system, some case studies are considered.

- **LCA of monitors (recycling versus landfilling):** Rocchetti and Beolchini (2014) carried out LCA studies of both CRT (Cathode Ray Tube) and LCD (Liquid Crystal Display) monitors and evaluated the total environmental burden generated for both products in each case of recycling and landfilling. As a result, it was found that the recycling process offers more environmental benefits than landfilling.
- **LCA of Lithium-ion (Li-ion) battery (management strategies):** Boyden et al. (2016) carried out LCA to compare the environmental impact of the three management strategies for Li-ion batteries, i.e. pyrometallurgy, hydrometallurgy, and landfilling. Based on the studies, it was found that the pyrometallurgical process has the largest impact caused due to plastic present in batteries, whereas in the case of landfilling, toxicity is the greatest factor. Furthermore, the hydro-metallurgical process shows a greater impact compared to the other two processes (landfilling and pyrometallurgy process)

Table 8.1 Real-time LCA application on various E-waste management strategies from the year 2005 to 2018.

S. no.	LCA application on various E-waste management strategies	Number of studies on LCA application
1	Evaluation of various recycling, recovery, and disposal strategies	5
2	Evaluation of various re-use strategies	3
3	Evaluation of various re-use, recycling, and recovery strategies	2
4	Evaluation of various recycling and recovery strategies: collection and/or recycling rates	6
5	Evaluation of waste electrical and electronic equipment (WEEE) and recycling system – whole system	11
6	Evaluation of WEEE treatment system – partial	13
7	Evaluation of WEEE collection and transportation system – partial	3
8	Evaluation of the hazardous potential of WEEE	3
9	Evaluation of the hazardous potential of WEEE management	2

Source: Adapted from Ismail and Hanafiah (2019).

based on the global warming potential. Overall, the comparison between the three different processes gave the idea that for recycling process for Li-ion batteries should be capable of recovering valuable metals as well as plastics.

- **LCA of PCBs:** A LCA study was performed by Rubin et al. (2014) on recovering copper from PCBs by leaching method using two different leaching agents, i.e. aqua regia and sulfuric acid. The results showed that aqua regia has a much less environmental impact as compared to sulfuric acid.
- **LCA of waste mobile phones (re-use versus recovery):** The LCA studies suggest that the reusability of components of waste mobile phones is more environmentally and cost-effective. At the same time, the social LCA creates conflicts between job creation in reusability and health risk that occurs during the reusability (Lu et al. 2014).

The application of LCA in E-waste management significantly varied from study to study. The first category includes studies that provided comprehensive literature evaluations of LCA studies relating to E-waste. Studies that have offered unique methodological frameworks, models, and methodologies for managing E-waste utilize a life cycle approach. These studies may not have used the LCA tool to generate a conventional LCA of E-waste product(s), but rather a management approach or process (Withanage and Habib 2021).

8.2 Life Cycle Assessment of E-waste Systems

The LCA method is a decision-making technique that examines the impacts of each waste recycling treatment method of E-waste from manufacturing to end-of-life (EoL). LCA studies based on E-waste treatment in formal recycling facilities have favorable environmental benefits as compared to informal recycling facilities. The formal sector uses different methods, procedures, and inputs, which are ignored in informal recycling. Also, the burning and inappropriate disposal of E-waste in informal recycling, are likely to have significant effects on the environment (Arain et al. 2022). The LCA approach was used to assess how the most recent technologies might affect the environment. An LCA research can be conducted by sequential technique, also referred to as the "bottom-up approach," which is focused on a scenario through the EoL stage of a waste management system and uses a simplified linear input and output. Software systems, including models based on Microsoft Excel sheets, commonly employ the sequential approach (Klöpffer and Grahl 2014). Financial life cycle cost (LCC) is a method used to examine the economic consequences of an LCA system and may be seen as an analytical tool parallel to an LCA.

Moreover, LCA is a tool that can measure the environmental sustainability, for example, greenhouse gas (GHG) of E-waste treatment operations. To evaluate the economic feasibility of a product or product system, LCA and LCC must be used in combination (Mangmeechai 2022). Due to the complexity of E-waste in the waste stream, a comprehensive methodology is required for an accurate assessment of the associated environmental impacts, which is best met by the application of LCA (Withanage and Habib 2021). The Society of Environmental Toxicology and Chemistry (SETAC) defines LCA as "a method to evaluate the environmental risks related to a product, system, or activity by analyzing and quantifying material and energy use and wastes discharged into the environment to assess the consequences of all those materials and energy uses and discharges to the environment; and to determining and assessing possibilities to effect environmental improvements". LCA essentially offers a methodological framework for estimating and evaluating the environmental effects of a product or development from a life cycle viewpoint (Table 8.2). Environmental, social, and economic ramifications of the complete supply chain of products spanning the full life cycle from "cradle to grave" must be considered to achieve sustainable production and consumption (Hou et al. 2015).

8.2.1 LCA Methodology

The International Organization for Standardization (ISO) published the first international standard related to LCA in 1998. Three further standards were then published in the three years that followed: ISO 1999, 2000a, and 2000b. These standards break down the LCA process into four steps: goal and scope definition,

Table 8.2 Details of environmental LCA of different technologies.

Technology	Type of LCA	Functional unit	Database /method used	Geographical location	Highest impact categories
Pyrolysis	Comparison with landfill and incineration	E-waste containing 1 kg of plastic	Ecoinvent	United Kingdom	1. Marine aquatic toxicity 2. Freshwater aquatic toxicity 3. Carcinogens 4. Carbon Deposit 5. Climate change 6. Abiotic depletion 7. Eutrophication 8. Radiation
Pyrometallurgical processes	Comparison of secondary copper smelting with and without E-waste.	An input rate of 12,500 kg per hour of feed materials (48 wt% copper scrap/metal oxides, 48 wt% waste PCB, 3.4 wt% slags, and 0.6 wt% coke)	ReCiPe	Australia	1. Human Toxicity 2. Climate Change 3. Marine eutrophication 4. Freshwater eutrophication 5. Freshwater ecotoxicity 6. Water depletion
Hydrometallurgical processes	Standalone	LCA 100 kg of electronic boards of mobile phones	Data obtained from the pilot plant and SigmaPro	EU/Italy	1. Eutrophication 2. Acidification 3. Global warming 4. Abiotic depletion 5. Human toxicity

Source: Debnath et al. (2018)/Springer Nature.

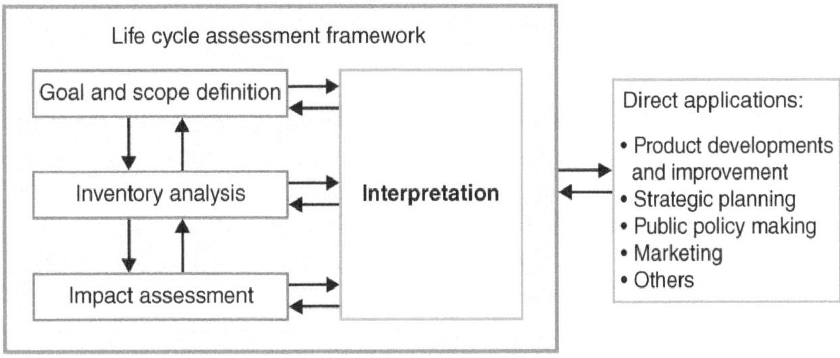

Figure 8.3 Stages of the LCA.

life cycle inventory analysis, life cycle impact assessment, and outcome analysis (Figure 8.3). Since then, LCA development has taken a formal, global approach, establishing the foundation for finding comparisons and implementation across countries. LCA has evolved after over 50 years of development into a useful tool for helping businesses and the government manage the environment (Hou et al. 2015). There are four main stages to an LCA study (Muralikrishna and Manickam 2017):

Stage 1: Goal and scope establish the size of the product life cycle that will be included in the assessment as well as the purpose of the assessment. In this stage, the criteria for system comparison and particular times are described.

Stage 2: Inventory analysis describes the material and energy flows inside the product system at this stage, focusing on how those flows interact with the environment, where raw materials are consumed, and how emissions to the environment are produced.

Stage 3: Impact assessment uses information from the inventory analysis. In this stage, the indicator results for each impact category are broken down, and the significance of each impact category is determined by normalization and ultimately weighting.

Stage 4: Interpreting a life cycle requires critical evaluation, determining the sensitivity of the data, and presenting the outcomes.

8.2.2 Software Used for Modeling

Several types of LCA are performed like product LCA, i.e. LCA, which assesses a product's whole life cycle, from the extraction of raw materials to manufacturing, distribution, transportation, utilization, and EoL. Moreover, LCA can also apply to a particular stage of the lifecycle of a product, such as the EoL stage, this type of LCA is known as waste LCA (Ismail and Hanafiah 2019). E-waste management uses LCA differently from product management. In theory, product LCA may evaluate the complete life cycle to enhance and optimize the development of products, meanwhile waste LCA can investigate multiple solutions to enhance waste management. There are several specialized waste LCA models like OWARE (Sweden) and EASEWASTE (Denmark), and modeling softwares like GaBi and SigmaPro are used to perform LCA, with the combination of different techniques and methods to assess how well E-waste (and other waste) management is being done. One of those techniques was material flow analysis (MFA), which was frequently used to evaluate the management of waste, including E-waste. The LCA has been used in combination with several different methods and tools to assess how well E-waste (and other trash) management is being done. One such approach was MFA, which was frequently used to evaluate the management of trash (including E-waste). The MFA is a methodical accounting approach based on the mass balance concept for the flow and stock of materials within a specified system. By decreasing the number of substances, MFA seeks to promote openness and manageability in analyzing the system under investigation, according to Jeswani et al. (2010). In contrast, the goal of a LCA is comprehensive and includes as many components of the system under investigation (Ismail and Hanafiah 2019; Islam and Huda 2019). The MFA could offer the quantitative data needed by the LCA even though both approaches were in some ways incompatible with one another and inconsistent in terms of direction (such as system boundaries). In the literature, these two techniques were combined to manage E-waste (Haupt et al. 2018; Turner et al. 2016).

8.2.3 Input and Output Modeling Parameters

The background data (life cycle initiative (LCI) data) used in this study reflect prospective emissions from waste treatment and disposal procedures, these technologies can be used for an E-waste management system. In this work, parameter variation, sensitivity analysis, and Monte Carlo analysis were used to assess the applicability of these scenarios and the information in view of significant uncertainties in the evaluation of the data and the scenarios. For the environmental impact analysis of each E-waste management scenario and case, several LCI data were employed. Fundamentally, the method used to design the study boundaries determines where the analysis of a given life cycle starts and finishes, as well as the activities that are a part of the technical system. To facilitate this procedure, a flowchart is frequently employed (Figure 8.4).

8.2.4 Impact Method and Impact Software

The goal of "life cycle impact assessment" (LCIA) is to assess the contribution to impact categories including acidification and global warming. The first step, characterization, involves calculating probable consequences based on the LCI

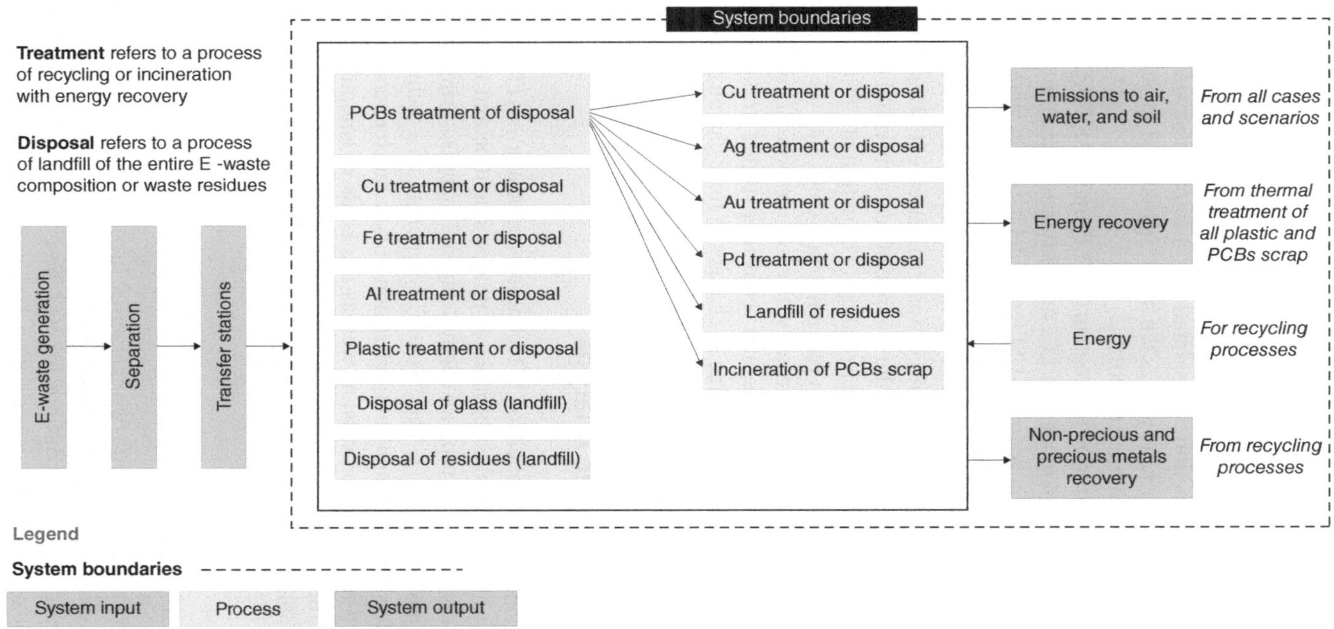

Figure 8.4 A general illustration of system boundaries of all the model scenarios. Source: adapted from Ikhlayel (2017).

results. Normalization and weighing are the next steps as per the ISO standard, but both processes are optional. To compare various environmental effect categories, normalization serves as a base. The term "weighting" refers to the process of giving each impact category a numerical value based on how significant it is. Applying weighting factors has the drawback of distorting the scale of values without improving the overall evaluation. Considerations for impact assessments include thresholds and nonlinearities, as well as temporal, spatial, and monetary dimensions (Muralikrishna and Manickam 2017).

The United Nations Environment Program (UNEP) and the SETAC, guided the development of the USEtox model, an LCI model that is based on the scientific consensus of international academics. USEtox model is used to analyze trends in potential effects on ecological toxicity and human health (cancer and non-cancer sickness). The findings are helpful to eco-design EEEs and offer insightful data to direct the creation of comprehensive E-waste management plans. This scientific model was employed to analyze the possible effects of harmful chemicals in products on ecotoxicology and human toxicology. The output of USEtox modeling includes criteria for human toxicity in relation to outcomes, like the incidence of cancer and non-cancer disorders connected with concentrations of toxic emissions and multimedia exposure evaluations, and ecotoxicity related to the potentially affected fraction (PAF) of species owing to changes in toxic chemical concentrations (Singh et al. 2019).

8.3 Techno-Economic Analysis

In E-waste management, application of techno-economic analysis (TEA) method is to determine how profitable a product or process might be and how to boost profitability. Information on product engineering and manufacturing is the basis of TEA (Mahmud et al. 2021). The "techno" component of a techno-economic assessment refers to a thorough comprehension of how much energy and material are utilized in a given process, as well as the kinds of equipment that are employed. Products may be produced using less energy or material or with the aid of more affordable or effective machinery thanks to technological improvements (Sorrell 2015).

8.3.1 Cost Estimation

Costs are sometimes divided into two categories: (a) on-going or operational costs and (b) capital costs. Operating expenses (OpEx) include the cost of labor and raw materials because firms must continue to cover these as long as they manufacture goods. Costs associated with construction and manufacturing equipment are classified as capital expenses (CapEx) since they are upfront payments made by the business in order to produce goods for a long time. This distinction is important since CapEx are frequently financed with debt, which results in interest payments (Ghodrat et al. 2016). It's crucial to account for interest because it raises the sum that must be repaid in order to fully satisfy a debt commitment, raising the cost of doing business. The costs that buyers would incur if they buy the product could also be a subject of TEA. LCC is a technique used to accomplish this. LCC outlines every expense a user of a product will incur over the course of ownership. An LCC for the transportation of E-waste, for instance, would consider not just the upfront cost of buying the automobile, but also the cost of fuel, maintenance, and other costs (Kobos et al. 2020). With the aid of this kind of study, businesses and consumers may better comprehend how a novel product – while perhaps more expensive up front – might really result in long-term cost savings. Thus, it is still another crucial aspect of comprehending a product's economic performance.

8.3.2 Process Modeling

TEA will evaluate the economic and technical performance of the technological product/system. Prior to the investigation, metrics or indicators are identified to assess the performance of the product/system. The assessment of indicators will be performed by developing an inventory database on material and energy requirements, capital equipment, manpower, and other pertinent elements (Kobos et al. 2020). To assist decision-makers in understanding how to enhance overall performance and accelerate the scaling of technology in its early stages, the indicators are frequently studied in a variety of ways. Framework for techno-economic assessment, encompassing objectives, scope, methods for calculating indicators, interpretation, and reporting. Once the objective and scope are defined, the data collection phase, (the inventory phase), can start. Once the correct inventory data has been located, indicators can be calculated. Throughout the course of the study, interpretation is a process that is iterated upon as the analysis develops. For instance, the interpretation phase requires that the aim be updated in light of the fact that the investigators performing the study were unable to locate the inventory data required to achieve a particular goal. Once the indicators are decided upon and subsequent analysis (including scenario, uncertainty, and sensitivity assessments) is completed, the project can then be concisely summarized in a report for decision-making. As it directs the other elements of the assessment, the TEA's purpose is essential. Clear, transparent objectives can help simplify the outcomes and comparison with other reports to meet the predetermined objectives. Many objectives are, at their most fundamental level, tied to better understanding a specific element of the product or process in order to maximize the possibility that the technology will be accepted and have an impact (Awasthi et al. 2021).

The purpose should outline the intended use and the motivations for the TEA in addition to the time frame, location, production scale, target market, commissioners, and any potential usability restrictions brought on by the assessment's design. All of this data facilitates peer review and research replication procedures while providing helpful context for the remaining analyses (Kobos et al. 2020). Scenario analysis, another essential element of TEA, should be learned about during the goal-setting phase. Scenarios are speculative but physically and economically plausible future alterations that might affect the study's major performance indicators and, as a result, the commercialization of the technology.

Building on the previous aim phase, the scope phase defines how the technology will be assessed and compared to the benchmark technology. This information makes clear the crucial factors in the creation and use of the technology, and it also enables the gathering of inventory data. All of these subjects are covered in the scope phase of the TEA, but if the person or group conducting the TEA differs from the ones who are developing the technology, they typically need to spend a lot of time making sure they fully comprehend the technology, how it functions, and its intended uses. The product application is the first thing that needs to be defined as a part of the assessment's scope. The final user or consumer will use the product in this manner. Because some materials, like concrete, have several potential uses, the application of a product is often described in terms of the volume of the product rather than specific usage. Once the application has been specified, the functional unit and reference flow must be specified. The functional unit is a numerical description of the application. The reference flow is the quantifiable amount of the product required to accomplish the functionality indicated by the functional unit.

Ensure the data is of a high enough caliber to compute the indicators and accomplish the study's goal. Users of the assessment's findings benefit from knowing the assessment's assumptions and sources, as the study's usefulness and conclusions are constrained by low quality or the need to guess at certain elements. The three types of data employed in TEA are generic, typical, and process-specific. Process-specific data for a particular component of the product system is known with a high degree of certainty when measured from a recognized process. Average data, which is often based on data that is unique to a certain process and is gathered as part of another process, reflects general industry assumptions. Generic data is estimated using techniques like stoichiometry or domain knowledge rather than being generated from a measured source. The three primary types of information employed in TEA are process-specific, average, and generic data. Process-specific information is gathered from a known process, making it possible to predict it with reasonable certainty for a given part of the product system. Average data, which is data that is specific to a process and is evaluated as part of another process, is often the foundation for assumptions made about the industry as a whole. Generic data is information that has not been measured; rather, it has been guessed using tools like stoichiometry or professional experience. OpEx, CapEx, and general expenses are the most typical categories of economic data to be gathered. The costs of energy, labor, and raw materials are examples of ongoing expenses spent throughout the normal operation of the plant. Capital costs typically refer to up-front costs for real estate, machinery, or other pieces of equipment that together enable manufacturing. Other costs, such as those associated with marketing, sales, and research, that are typically difficult to separate from production are known as general expenses (Cheela et al. 2021a). Due to their high uncertainty and low importance, general expenses are often not taken into account in early-stage TEAs that are concentrated on identifying economic hotspots and enhancing the viability of new technologies (Thomassen et al. 2019).

Indicators are chosen at the scope phase and modified as needed throughout the study to provide measurements that convey major study outcomes. They can be of the technical or economic variety, or they can be techno-economic if they combine the two in the production process. Economic indicators typically have something to do with the project's running costs, capital needs, or profitability. Operating costs, which include costs for energy, labor, and raw materials, are typical of most industrial processes. While certain operational expenses will be variable, or change as production volume varies, others will be fixed, or not changing as production volume increases. The initial purchases of machinery, land, the factory, contingencies, engineering, start-up costs, and the infrastructure needed to connect the facility to the utilities account for the majority of capital expenditures. Indicators can incorporate the technical, economic, and environmental elements of the technology under consideration. The opportunity to correlate TEA results with relevant LCA data is at this point. A useful metric that could enable comparisons across various technologies is the levelized cost of carbon abatement (LCCA), which divides the cost of implementing the new technology by the difference in emissions to reflect the cost or benefit per metric tonne of CO_2 decreased or removed (Cheela et al. 2021b; Faber et al. 2021).

Following successfully computing the indicators, uncertainty and sensitivity analyses are performed as part of the interpretation process. While the interpretation phase's components are present in each TEA step to evaluate for consistency, data quality, and whether the inventory data can adequately support the study's objectives. Scenario analysis using the scenarios discovered during the objective phase should also be done as part of the interpretation.

8.4 Conclusion

This review provides a broad overview of E-waste relating to the in-depth analysis of research that used LCA as a tool to analyze the environmental impacts involved in E-waste handling and its treatment strategies. Techno-economic analysis is done to assess the economic performance of an industrial process, product, or service. These two studies are helpful to students, researchers, scientists, and people working in industries to get an idea about the two analyses (LCA and TEA). These studies can be extended to simulate the results obtained from the experimental studies to get a range of performance metrics.

References

Andeobu, L., Wibowo, S., and Grandhi, S. (2021). A systematic review of e-waste generation and environmental management of Asia Pacific countries. *International Journal of Environmental Research and Public Health* 18 (17): 9051.

Arain, A.L., Neitzel, R.L., Nambunmee, K. et al. (2022). Material flow, economic and environmental life cycle performances of informal electronic waste recycling in a Thai community. *Resources, Conservation and Recycling* 180: 106129.

Awasthi, A.K. and Li, J. (2017). Management of electrical and electronic waste: a comparative evaluation of China and India. *Renewable and Sustainable Energy Reviews* 76: 434–447.

Awasthi, A.K., Cheela, V.R.S., D'Adamo, I. et al. (2021). Zero waste approach towards a sustainable waste management. *Resources, Environment and Sustainability* 3: 100014.

Baldé, C.P., Forti, V., Gray, V. et al. (2017). *The Global E-waste Monitor – 2017*. Bonn/Geneva/Vienna: United Nations University (UNU), International Telecommunication Union (ITU) & International Solid Waste Association (ISWA). ISBN: 978-92-808-9054-9.

Boyden, A., Soo, V.K., and Doolan, M. (2016). The environmental impacts of recycling portable lithium-ion batteries. *Procedia Cirp* 48: 188–193.

Cheela, V.R.S., Ranjan, V.P., Goel, S. et al. (2021a). Pathways to sustainable waste management in Indian Smart Cities. *Journal of Urban Management* 10 (4): 419–429.

Cheela, V.R.S., John, M., Biswas, W.K., and Dubey, B. (2021b). Environmental impact evaluation of current municipal solid waste treatments in India using life cycle assessment. *Energies* 14 (11): 3133.

Debnath, B., Chowdhury, R., and Ghosh, S.K. (2018). Sustainability of metal recovery from E-waste. *Frontiers of Environmental Science & Engineering* 12 (6): 1–12.

Dwivedy, M. and Mittal, R.K. (2012). An investigation into e-waste flows in India. *Journal of Cleaner Production* 37: 229–242.

Faber, G., Mangin, C., and Sick, V. (2021). Life cycle and techno-economic assessment templates for emerging carbon management technologies. *Frontiers in Sustainability* 2: 764057.

Ghodrat, M., Rhamdhani, M.A., Brooks, G. et al. (2016). Techno economic analysis of electronic waste processing through black copper smelting route. *Journal of Cleaner Production* 126: 178–190.

Goh, P.G., Maghfouri, M., Onn, C.C., and Loo, S.C. (2022). Life cycle assessment on recycled e-waste concrete. *Case Studies in Construction Materials* 17: e01412.

Haupt, M., Kägi, T., and Hellweg, S. (2018). Modular life cycle assessment of municipal solid waste management. *Waste Management* 79: 815–827.

Hou, Q., Mao, G., Zhao, L. et al. (2015). Mapping the scientific research on life cycle assessment: a bibliometric analysis. *The International Journal of Life Cycle Assessment* 20 (4): 541–555.

Ikhlayel, M. (2017). Environmental impacts and benefits of state-of-the-art technologies for E-waste management. *Waste Management* 68: 458–474.

Islam, M.T. and Huda, N. (2019). Material flow analysis (MFA) as a strategic tool in E-waste management: applications, trends and future directions. *Journal of Environmental Management* 244: 344–361.

Islam, M.T., Abdullah, A.B., Shahir, S.A. et al. (2016). A public survey on knowledge, awareness, attitude and willingness to pay for WEEE management: case study in Bangladesh. *Journal of Cleaner Production* 137: 728–740.

Ismail, H. and Hanafiah, M.M. (2019). An overview of LCA application in WEEE management: current practices, progress and challenges. *Journal of Cleaner Production* 232: 79–93.

ISO (International Organization for Standardization) (2006). *Environmental Management: Life Cycle Assessment; Principles and Framework*. ISO.

Jeswani, H.K., Azapagic, A., Schepelmann, P., and Ritthoff, M. (2010). Options for broadening and deepening the LCA approaches. *Journal of Cleaner Production* 18 (2): 120–127.

Kiddee, P., Naidu, R., and Wong, M.H. (2013). Electronic waste management approaches: an overview. *Waste Management* 33 (5): 1237–1250.

Klöpffer, W. and Grahl, B. (2014). *Life Cycle Assessment (LCA): A Guide to Best Practice*. John Wiley & Sons.

Kobos, P.H., Drennen, T.E., Outkin, A.V., Webb, E.K., Paap, S.M., and Wiryadinata, S. (2020). *Techno-Economic Analysis: Best Practices and Assessment Tools (No. SAND-2020-13473)*. Sandia National Lab.(SNL-CA), Livermore, CA (United States); Sandia National Lab.(SNL-NM), Albuquerque, NM (United States); Hobart and William Smith Colleges, Geneva, NY (United States).

Kumar, A. and Dixit, G. (2018). An analysis of barriers affecting the implementation of e-waste management practices in India: a novel ISM-DEMATEL approach. *Sustainable Production and Consumption* 14: 36–52.

Lam, M., O'Gorman, A., Schleehahn, R., and Elkamel, A. (2021). Techno-economic analysis of electrochemical recovery of metals from waste printed circuit boards (PCBs). In: *Proceedings of the 4th European International Conference on Industrial Engineering and Operations Management Rome, Italy*.

Lu, B., Li, B., Wang, L. et al. (2014). Reusability based on life cycle sustainability assessment: case study on WEEE. *Procedia Cirp* 15: 473–478.

Mahmud, R., Moni, S.M., High, K., and Carbajales-Dale, M. (2021). Integration of techno-economic analysis and life cycle assessment for sustainable process design – a review. *Journal of Cleaner Production* 317: 128247.

Mangmeechai, A. (2022). The life-cycle assessment of greenhouse gas emissions and life-cycle costs of e-waste management in Thailand. *Sustainable Environment Research* 32 (1): 1–13.

Muralikrishna, I.V. and Manickam, V. (2017). Life cycle assessment, Chap. 5. In: *Environmental Management* (ed. I.V. Muralikrishna and V. Manickam), 57–75. ISBN: 9780128119891.

Puckett, J., Byster, L., Westervelt, S., Gutierrez, R., Davis, S., Hussain, A., and Dutta, M. (2002). *Exporting Harm: The High-Tech Trashing of Asia*. Seattle.

Rebitzer, G., Ekvall, T., Frischknecht, R. et al. (2004). Life cycle assessment: part 1: framework, goal and scope definition, inventory analysis, and applications. *Environment International* 30 (5): 701–720.

Rocchetti, L. and Beolchini, F. (2014). Environmental burdens in the management of end-of-life cathode ray tubes. *Waste Management* 34 (2): 468–474.

Rubin, R.S., de Castro, M.A.S., Brandão, D. et al. (2014). Utilization of Life Cycle Assessment methodology to compare two strategies for recovery of copper from printed circuit board scrap. *Journal of Cleaner Production* 64: 297–305.

Shahabuddin, M., Uddin, M.N., Chowdhury, J.I. et al. (2022). A review of the recent development, challenges, and opportunities of electronic waste (e-waste). *International Journal of Environmental Science and Technology* 1–8.

Singh, N., Duan, H., Ogunseitan, O.A. et al. (2019). Toxicity trends in E-Waste: a comparative analysis of metals in discarded mobile phones. *Journal of Hazardous Materials* 380: 120898.

Song, Q. and Li, J. (2015). A review on human health consequences of metals exposure to e-waste in China. *Environmental Pollution* 196: 450–461.

Sorrell, S. (2015). Reducing energy demand: a review of issues, challenges and approaches. *Renewable and Sustainable Energy Reviews* 47: 74–82.

Tanskanen, P. (2013). Management and recycling of electronic waste. *Acta Materialia* 61 (3): 1001–1011.

Thomassen, G., Van Dael, M., Van Passel, S., and You, F. (2019). How to assess the potential of emerging green technologies? Towards a prospective environmental and techno-economic assessment framework. *Green Chemistry* 21 (18): 4868–4886.

Turner, D.A., Williams, I.D., and Kemp, S. (2016). Combined material flow analysis and life cycle assessment as a support tool for solid waste management decision making. *Journal of Cleaner Production* 129: 234–248.

Withanage, S.V. and Habib, K. (2021). Life cycle assessment and material flow analysis: two under-utilized tools for informing E-waste management. *Sustainability* 13 (14): 7939.

Xue, M. and Xu, Z. (2017). Application of life cycle assessment on electronic waste management: a review. *Environmental Management* 59 (4): 693–707.

9

E-waste Recycling: Transition from Linear to Circular Economy

Abhinav Ashesh

KPMG India, Gurgaon, Haryana, India

Abstract

Technological advancement has led to an increase in e-waste, with mobile phones being the most common source of it. More phones are entering the market than the number of consumers, necessitating a transition towards an economy based on the principles of reducing, reusing, recycling, and recovering. The circular economy (CE) approach is in line with sustainable environmental development. However, implementing the CE-based business models has yet to be mainstreamed. Business models based on leasing and renting a product, reverse logistics, take-back strategies, and sharing of resources will play an integral role in such a transition. Through CE, companies can create new profit pools along the reverse supply chain and new channels for product remarketers and resellers. It could also address constraints related to shorter product life cycles, supply risks, and intensified cost-price squeezes. On the other hand, customers will benefit from lower costs and greater convenience.

Keywords *circular economy; e-waste; closed loop; 4R framework; business strategy; eco-innovation; reverse supply chains*

9.1 Introduction

Societal needs and our economy have always been dependent on the use of natural resources; however, with the ever-increasing demand, there has been an increased pressure on the limited resources. While this demand is posed to increase in the coming years, the economies have been focusing on the development of alternate economic models based on the principles of sustainability to manage existing resources. The economy till date has been mostly operating on a linear, extraction–production–consumption model. However, the sustainability of such a

Management of Electronic Waste: Resource Recovery, Technology and Regulation, First Edition. Edited by Anshu Priya.

Table 9.1 Types of industrial economies.

Type	Description
Linear economy	The linear economy is based on the "take, make, and dispose" model, where the raw material is extracted and turned into a finished product through a series of value-adding steps. Once the product is sold, the entire ownership and responsibility are transferred to the buyer. The user then decides the end-of-life treatment of the product, which is either dumping, reusing, or recycling.
Circular economy	The circular economy forms a closed loop through the reprocessing of goods. This generates jobs and saves energy while reducing resource consumption and waste.
Performance economy	In a performance economy, the ownership of the goods is not transferred: rather, it is leased out or rented to the consumer. The performance economy model focuses on reusing the product to generate profitability and reduce waste.

model is challenging and has necessitated a shift toward other economic models, mainly, circular and performance economies. Table 9.1 describes the three types of industrial economies: linear, circular, and performance.

The OECD's Global Material Resources Outlook to 2060 expects that the consumption of natural resources such as metals and minerals, fossil fuels, and biomass will double in the next 40 years. Further, the annual waste generation is expected to increase by 70% by 2050. To accelerate the transition toward a regenerative growth model, the European Commission published in March 2022, the "Circular Economy Action Plan." It estimates that the circular economy (CE) has the potential to increase the GDP of the European Union by an additional 0.5% by 2030 and create 700,000 new jobs. Globally, CE is expected to generate US$ 4.5 trillion in economic growth by 2030.

While many countries, such as China and Netherland have adopted a law for CE, the implementation of CE-based business models is yet to be mainstreamed (CIRAIG 2015). Adoption of CE becomes more important considering that the existing linear economic model is unsustainable and would lead to increased waste. The CE approach is in line with sustainable environmental and economic development (EMAF 2015; EMAF (Ellen MacArthur Foundation) 2013; EMAF 2012; CIRAIG 2015; COM 2014, 2015) as it emphasizes reuse, refurbishment, and remanufacturing of products.

9.2 Linear Economy and its Limitations

The existing linear economy, based on extraction–production–consumption model, has various limitations, as outlined below.

- As there is a transfer of ownership and associated risks of the product to the owner, the reuse or recycling decision of the product becomes dependent on the end user.
- With the increasing consumption and the scarcity of natural resources, the price of raw materials is expected to rise, thus increasing the cost of products. Further, it has an environmental cost associated with it.
- Unlike CE, which is more dependent on unlimited resources, like labor, the linear economy is dependent on natural resources, which have a limited supply.
- In a linear economy, a significant volume of waste is generated across the value chain, that is from the extraction of raw materials to the manufacturing of finished products. Further, a significant proportion of the raw material is lost during the process of conversion. The Sustainable Europe Research Institute (SERI) estimates that more than 21 billion tons of materials are consumed each year across the OECD countries during the process of manufacturing of products. Interestingly, these materials do not themselves are part of the final product.
- According to a UNEP report, the recycling rate of only one-third of the 60 metals it studied, was more than 25%. Further, for the metals having higher recycling rates, which include copper, gold, aluminum, and silver, it is estimated that there is an annual loss of about US$ 52 billion, US$ 34 billion, US$ 15 billion, and US$ 7 billion, respectively.
- In a linear model, once the product is dumped, its residual energy is lost, whereas reuse or recycling of the product would have saved significant energy. The processes in the upstream of the supply chain of a product are highly energy intensive; for example, the conversion of bauxite into semi-finished aluminum accounts for 80% of the energy consumed. By reusing or recycling the product, much of this energy can be saved.

9.3 Circular Economy – Need of the Hour

CE is viewed as an approach toward implementation of the concept of sustainable development and has gathered interest from both scholars and practitioners (Ghisellini et al. 2016; Murray et al. 2017). A product designed under a CE approach takes into consideration its end-of-life use, which includes ease of reuse, refurbishment, or recycling. The CE focuses on reclaiming the materials used in the manufacturing of the product instead of extracting resources. Thus, CE presents itself as an economic strategy that promotes innovation in product development to move from a consumption-based linear economy to a more sustainable economy through material savings.

The strategies to implement CE are mainly based on the 4R framework. These are:

1. **Reduce:** Preventing or minimizing the use of resources and/or increasing the lifespan of products through redesigning,
2. **Reuse:** Refurbishing a used product or reusing the useful resources present in the product,
3. **Recycle:** Reusing the waste or remanufacturing,
4. **Recover:** Incineration of materials with energy recovery.

In the absence of a shift toward a dominant business model where a product can be used as a service, practitioners view reduction as restraining consumption and hence economic growth (George et al. 2015; Lacy et al. 2014). Lieder and Rashid (2016) also note that CE supply chains need to consider consumption and not only the various production and distribution processes. Yuan et al. (2006, p. 5) also highlight that CE would require novel production, processes, and consumption activities. Further, it would be critical to promote consumer responsibility in a CE (Ghisellini et al. 2016, p. 19) (Figure 9.1).

CE creates a win–win situation for the environment, society, and the economy. For economies, there are net material cost savings, reduced supply risks, and price volatility, as well as possible employment benefits. It is estimated that higher remanufacturing rates and reuse of mobile phones could lead to the eradication of at least 1.3 million tons of CO_2 annually at 2010 production levels across the European Union. The environment will benefit from the exclusion of water- or energy-intensive production and the use of less toxic materials, thus experiencing reduced pressure on greenhouse gas emissions, usage of water, and biodiversity. Some of the major benefits are highlighted in Table 9.2.

Through CE, companies will be able to create new profit pools along the reverse supply chain. There will be new channels for product remarketers and sellers, such as eBay and Craigslist, that focus on the trading of used or refurbished goods

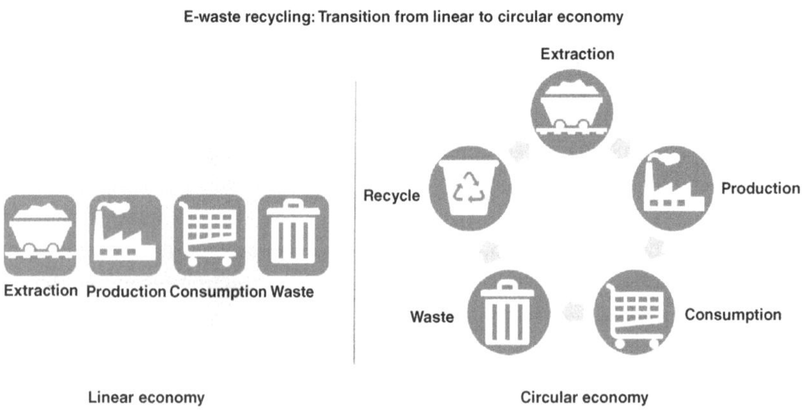

Figure 9.1 Linear and circular economies.

Table 9.2 Environmental, social, and economic benefits of CE.

Environmental benefit	Social benefit	Economic benefit
• Waste reduction • Reduced emission of harmful chemicals • Enhanced resource utilization	• New employment opportunities • Development of innovative business models based on "shared economy" • Increased awareness among consumers regarding e-waste management	• Reduced raw material and energy costs • Reduced waste management costs • Reduced costs in complying with environmental regulations and taxes

online. CE could address constraints related to shorter product life cycles, supply risks, intensified cost-price squeeze, and decreasing customer loyalty. On the other hand, customers will benefit from lower costs and greater convenience.

9.4 The Transition from Linear to Circular Economy

A transition to CE necessitates the promotion of "functional service" model, where the ownership of a product is retained with the manufacturer or retailer and they act as a service provider, that is, as a seller of the "use" of the product instead of selling the product itself. Welford and Tukker (see e.g. Welford 1998; Tukker 2015) note that new consumption systems would have groups and communities who share the value of physical products, their functions, and services as against the individuals who hold ownership and consume the physical products.

New business models based on leasing and renting of a product, reverse logistics, take-back strategies, and sharing of resources will play an integral role in transitioning toward CE. Some of the early-stage CE-compatible business models include those focused on recycling and remanufacturing (Geyer and Jackson 2004; Ongondo et al. 2013; Zhang et al. 2011), establishing new businesses through a circular canvas method (Lewandowski 2016), decreasing ownership and increasing the rental of services (Bakker et al. 2014), and practicing dematerialization (Ehrenfeld 2000; Yang et al. 2014), among others.

Further, technological innovation can be developed at (i) the micro level, that is, inside local businesses, (ii) at the meso level, that is where interconnected industries operate, and (iii) the macro level consisting of institutions and the region as a whole. Companies need to tighten their supply chain by convincing the suppliers to ensure high purity and non-toxicity of materials, thereby, enhancing the recycling rates. The last-mile connectivity needs to be activated, as consumption and disposal are typically local activities. Firms with CE approach need to manage the rapid changes in customer preferences and market trends (Van der

Wiel et al. 2012) through strategies such as reverse logistics for waste management and the design of sustainable products. These products shall be designed to ensure their recovery through biological or technical cycles (McDonough et al. 2003). However, Kalmykova et al. (2015) note that CE will not be feasible if the recycling rates are lower compared to the increase in consumption. Further, reverse logistics can be initiated by setting up an outlet for recovered components and materials. Over the last few years, companies and governments have started looking at CE as a hedge against scarce resources.

The success of CE strategies in companies is dependent upon their technological capabilities, market structure, innovations within the organization, and cost savings as a result of efficient production. It is also linked with the capacity of these companies to make their business models sustainable and competitive (Yang et al. 2014) and to the profitability of circular resource use (Andersen 2007). Various leading companies, including Dell Technologies, Energizer, Lockheed Martin Corporation, General Motors, and Renault, have already incorporated the CE approach into their business models.

At the consumer's end, transitioning toward CE will mainly depend on environmental awareness and preferences toward sustainable products. Cohen-Rosenthal 2000, notes that over the long term, implementation of CE will depend on the consumer's perception of value addition and social perception of sustainable products. Table 9.3 summarizes the eight kinds of eco-innovation needed to implement CE.

9.5 Understanding E-Waste Through Smartphones

Technological advancement has led to an increase in e-waste over the last decade. Mobile phones are among the most common sources of e-waste, with more phones entering the market than the number of consumers. On average, a consumer owns 1.1 mobile phones and uses a phone for less than 2.5 years. A mobile phone typically weighs less than 150 g but contains valuable metals such as gold, silver, and some rare earth metals. Though the quantity of precious metals in a single device is relatively small, billions of such devices can be a significant source of these metals (Silveira and Chang 2010). In a feature phone, palladium and cobalt are the main minerals, whereas a smartphone, generally, also contains beryllium, antimony, platinum, praseodymium, and neodymium (He et al. 2018). In a linear economic setup, almost the entire material within the phone is lost owing to the low collection and recycling rates.

China, the world's largest manufacturer and consumer of mobile phones (Chen et al. 2018; He et al. 2018), had a mobile-cellular subscription of 1.32 billion units (MIIT 2017), whereas the global numbers stood at 7.38 billion units in 2017 (ITU 2017a, 2017b). In China, the number of waste mobile phones grew to

Table 9.3 Eco-innovations for implementation of CE.

Business model innovations	Innovation in the way a company operates and creates value
Network innovations	Innovations within a network of related companies
Organizational structure innovations	Development of new organizational and management practices to support environmental strategy
Process innovations	The way company creates a product or offers its services
Product innovations	Focus is on the quality and functionality of the product
Service innovations	Shift from the "ownership" model toward "shared" services model
Market innovations	Marketing and positioning of the product
Customer engagement innovations	Focus is on customer needs and preferences

739.98 million in 2012 from 47.92 million in 2002 (Li et al. 2015). Considering the increasing number of waste phones, it becomes imperative to collect them and ensure their end-of-life treatment.

In Europe, the collection rates hovered around 15%, with hardly any component being reused or remanufactured. Further, the secondary mobile phone market, though fast-growing, is almost negligible at around 6% of the primary market.

The depleting resources, including some rare earth metals and precious metals, have accelerated the recycling of mobile phones over the past year. This would further need to be stepped up through the creation of infrastructure that allows manufacturers, remanufacturers, and sellers to gain scale. Once the scale is achieved, it will attract investments in larger and more streamlined facilities, subsequently leading to increased efficiency.

To encourage collection, lease/buyback models need to be promoted with an improved customer dialogue. Further, this may be complemented with more automated treatments and extraction systems. In an advanced circularity stage, the industry players comprising original equipment manufacturers (OEMs), manufacturers, retailers, and reverse logistics companies would need to form joint collection systems to achieve greater efficiency.

9.5.1 Increasing Circularity in the Smartphone Market

Collaboration among various stakeholders within and across sectors is imperative for scaling up the circularity of e-waste. Further, the collaboration needs to be accompanied by transparency in the value chain, setting up industry benchmarks, and appropriate distribution of profit margins among the value chain actors. Some of the key determinants of circularity are discussed below.

- **Reuse – developing the secondary phone market:** This will ensure a longer life for phones through minimal cleaning and repackaging. However, the first-hand user of the phone needs to be confident enough that the personal data are completely erased by the manufacturer or the reseller through reliable software or through an automated process.
- **Remanufacturing:** Phones may be designed in a way so as to make it easy to strip out usable components from it and re-manufacture. The new design shall promote easy and cost-effective disassembly of usable components. These components may include batteries, chargers, displays, and cameras. These are among the most valuable components of the phone and could be used in the production of new devices or for sale in the secondary markets. However, the success of such a model would depend on a consensus among the manufacturers to standardize these components across various models and brands.
- **Development of short reverse supply chains:** Phones, especially smartphones, contain high-value, sophisticated embedded components, and their value may degrade while passing through multiple channels in the reverse supply chains. Further, short reverse supply chains will ensure that the cost of retrieving the embedded components is economically attractive.
- **Accessibility of the used components in the business-to-business (B2B) context:** As the phone market is controlled by a limited number of players, access to usable components at market prices may be challenging in the business-to-business (B2B) context. Cost-effectiveness and easy accessibility of used components for new and smaller players are needed to achieve scale in the refurbishment of smartphones.
- **Incentives to boost the collection rate:** Incentives in the form of buy-back systems or through attractive exchange offers for customers may boost the collection rate of phones. However, as discussed in the earlier section, as a prerequisite, the first-hand customers would need assurance that their personal data cannot be retrieved or abused.

9.6 Conclusion

The economies across the globe have been relying on a system where the entire production economy, regulations, and consumer mindset favor the linear model.

Reusing of products or their increased usable life will mean lesser demand and consequently less revenue and profits for the companies. In such a scenario, the shift toward CE will be dependent on the development of a well-established secondary market offering attractive margins for companies involved in selling refurbished products and components.

There also needs to be a change in consumer's perception of refurbished products and the reverse supply chain. Stepping up education and conducting consumer awareness campaigns could be helpful in creating a behavioral change among them. Further, technological interventions related to data privacy and tracking the health of the refurbished products will boost their confidence in the CE model. The shift toward CE is expected to ignite innovation and entrepreneurship, thus leading to a bottom-up revolution in response to scarce resources and tightening regulations.

References

Andersen, M.S. (2007). An introductory note on the environmental economics of the circular economy. *Sustainability Science* 2 (1): 133–140.

Bakker, C., Wang, F., Huisman, J., and den Hollander, M. (2014). Products that go round: exploring product life extension through design. *Journal of Cleaner Production* 69: 10–16.

Chen, Y., Chen, M., Li, Y. et al. (2018). Impact of technological innovation and regulation development on e-waste toxicity: a case study of waste mobile phones. *Scientific Reports* 8 (1): 7100.

CIRAIG (2015). *Circular Economy: A Critical Literature Review of Concepts*. Montréal, Québec, Canada: Bibliothèque et Archives nationales du Québec (BAnQ) http://www.ciraig.org/pdf/CIRAIG_Circular_Economy_Literature_Review_Oct2015.pdf.

Cohen-Rosenthal, E. (2000). A walk on the human side of industrial ecology. *The American Behavioral Scientist* 44: 245–264.

COM. (2014). COM (European Commission). Towards a Circular Economy: a Zero Wast Programme for Europe, Brussels. https://www.oecd.org/env/outreach/EC-Circular-econonomy.pdf

COM (2015). *COM (European Commission)*. Brussels: Closing the Loop - an EU Action Plan for the Circular Economy https://www.eea.europa.eu/policy-documents/com-2015-0614-final.

Ehrenfeld, J.R. (2000). Industrial ecology: paradigm shift or normal science? *American Behavioral Scientist* 44 (2): 229–244.

EMAF. (2012). Towards the Circular Economy, London. https://www.werktrends.nl/app/uploads/2015/06/Rapport_McKinsey-Towards_A_Circular_Economy.pdf

EMAF. (2015). Towards a circular economy: Business rationale for an accelerated. https://www.ellenmacarthurfoundation.org/assets/downloads/TCE_Ellen-MacArthur-Foundation_9-Dec2015.pdf

EMAF (Ellen MacArthur Foundation), 2013, *Towards the Circular Economy*, London.

George, D.A., Lin, B.C.A., and Chen, Y. (2015). A circular economy model of economic growth. *Environmental Modelling and Software* 73: 60–63.

Geyer, R. and Jackson, T. (2004). Supply loops and their constraints: the industrial ecology of recycling and reuse. *California Management Review* 46: 55.

Ghisellini, P., Cialani, C., and Ulgiati, S. (2016). A review on circular economy: the expected transition to a balanced interplay of environmental and economic systems. *Journal of Cleaner Production* 114: 11–32.

He, P. et al. (2018). The present and future availability of high-tech minerals in waste mobile phones: evidence from China. *Journal of Cleaner Production*.

ITU. (2017a). Global ICT developments http://www.itu.int/en/ITU-D/Statistics/Pages/stat/default.aspx (2017)

ITU. (2017b). Key 2005-2016 ICT data for the world. http://www.itu.int/en/ITU-D/Statistics/Documents/statistics/2016/ITU_Key_2005-2016_ICT_data.xls (2017)

Kalmykova, Y., Patrício, J., Rosado, L., and Berg, P.E.O. (2015). Out with the old, out with the new—The effect of transitions in TVs and monitors technology on consumption and WEEE generation in Sweden 1996–2014. *Waste Management* 46: 511–522.

Lacy, P., Keeble, J., McNamara, R.J.T.M.A.C.K.P.A.T. et al. (2014). *Circular Advantage: Innovative Business Models and Technologies to Create Value in a World Without Limits to Growth*, 24. Accenture: Chicago, IL, USA.

Lewandowski, M. (2016). Designing the business models for circular economy—Towards the conceptual framework. *Sustainability* 8 (1): 43.

Li, B., Yang, J., Lu, B., and Song, X. (2015). Estimation of retired mobile phones generation in China: a comparative study on methodology. *Waste Management* 35: 247–254.

Lieder, M. and Rashid, A. (2016). Towards circular economy implementation: a comprehensive review in context of manufacturing industry. *Journal of Cleaner Production* 115: 36–51.

McDonough, W., Braungart, M., Anastas, P.T., and Zimmerman, J.B. (2003). Peer reviewed: applying the principles of green engineering to cradle-to-cradle design. *Environmental Science and Technology* 37 (23): 434A–441A.

MIIT (2017). *MIIT Communication Operation Industry Statistical Bulletin*. Beijing, China: Ministry of Industry Information Technology.

Murray, A., Skene, K., and Haynes, K. (2017). The circular economy: an interdisciplinary exploration of the concept and application in a global context. *Journal of Business Ethics* 140 (3): 369–380.

Ongondo, F.O., Williams, I.D., Dietrich, J., and Carroll, C. (2013). ICT reuse in socioeconomic enterprises. *Waste Management* 33: 2600–2606.

Silveira, G.T. and Chang, S.Y. (2010). Cell phone recycling experiences in the United States and potential recycling options in Brazil. *Waste Management* 30 (11): 2278–2291.

Tukker, A. (2015). Product services for a resource-efficient and circular economy – a review. *Journal of Cleaner Production* 97: 76–91.

Van der Wiel, A., Bossink, B., and Masurel, E. (2012). Reverse logistics for waste reduction in cradle-to-cradle-oriented firms: waste management strategies in the Dutch metal industry. *International Journal of Technology Management* 60: 96–113.

Welford, R. (1998). Corporate environmental management, technology and sustainable development: postmodern perspectives and the need for a critical research agenda. *Business Strategy and the Environment* 7 (1): 1–12.

Yang, M., Vladimirova, D., Rana, P., and Evans, S. (2014). Sustainable value analysis tool for value creation. *Asian Journal of Management Science and Applications* 1 (4): 312–332.

Yuan, Z., Bi, J., and Moriguichi, Y. (2006). The circular economy: a new development strategy in China. *Journal of Industrial Ecology* 10 (1-2): 4–8.

Zhang, T., Chu, J., Wang, X. et al. (2011). Development pattern and enhancing system of automotive components remanufacturing industry in China. *Resources, Conservation and Recycling* 55: 613–622.

10

E-Waste Valorization and Resource Recovery

Anusha Vishwakarma and Subrata Hait

Indian Institute of Technology Patna, Department of Civil and Environmental Engineering, Patna, Bihar, India

Abstract

Excessive demand for electrical and electronic equipment (EEE) has increased global e-waste generation, posing a serious environmental threat. However, in the global market, e-waste represents a substantial economic opportunity due to its diversified composition. Regardless of its intrinsic value, only a small fraction of e-waste is collected and recycled, necessitating sustainable recycling practices for efficient resource utilization. As a result, urban mining of e-waste has gained attention to reduce the pressure on virgin mines as well as avoid their unsafe disposal. Therefore, it is crucial to sustainably manage e-waste and transform it into value-added products to achieve a global circular economy. In this context, the chapter aims to comparatively assess the widely adopted resource recovery techniques, focusing on the valorization of recycling materials from end-of-life (EoL) EEE with selected case studies. Furthermore, the opportunities and challenges of different recycling techniques have been elucidated.

Keywords *electronic waste; valorization; resource recovery; urban mining; circular economy; recycling; value-added products*

10.1 Introduction

In recent decades, the technological revolution has benefited society in various ways at numerous levels, contributing to a more interconnected world. The Information and Communication Technology (ICT) revolution has significantly increased the use of electrical and electronic equipment (EEE). However, the highly unsustainable pattern of production and consumption of EEE increased the obsolete electronic items, i.e. electronic waste (e-waste). The term e-waste

Management of Electronic Waste: Resource Recovery, Technology and Regulation, First Edition. Edited by Anshu Priya.

refers to all types of waste generated from EEE and its components after they have reached the end of their useful lives. The technological advancement, coupled with short innovation cycles and company policies, is the primary driving force, which shortens the lifespan of the EEE. In addition, other essential factors significantly accelerated the e-waste generation including rapid economic growth, consumers' higher dependency on EEE, and improved living standards. In consequence, e-waste volume is increasing exponentially every year. For instance, since 2014, e-waste production has surged by 9.2 million metric tons (Mt) and is forecast to reach 74.7 Mt by 2030 (Forti et al. 2020). In 2019, the world generated 53.6 Mt of e-waste, of which only 17.4% was formally documented and recycled (Forti et al. 2020). Asia-Pacific nations (such as China, India, Japan, and Indonesia) have collectively contributed 47% of the globally generated e-waste. Next, the Americas, Europe, Africa, and Oceania come in with relative production rates of 25%, 22%, 5%, and 1% of the world's e-waste (Forti et al. 2020). However, in many developed and developing economies, the WEEE Directive (2012/19/EU), which is a global e-waste regulatory framework formulated by the European Union (EU), acts as a blueprint for managing e-waste. The aforesaid regulatory guidelines have been introduced to promote circular economy, extend producer responsibility (EPR), producer responsibility organization (PRO), and restriction of hazardous substances (RoHS).

E-waste compositional characteristics are diverse and may vary from device to device (Ottoni et al. 2020). It contains numerous hazardous materials, including heavy metals and toxic organic compounds (Kidee et al. 2013). Many EEEs have significant levels of harmful elements, such as Hg in fluorescent lamps, Pb in cathode ray tubes (CRTs), Cd in photocells, Ni, Be, and Zn in printed circuit boards (PCBs) and monitors, and brominated compounds as flame retardants (FRs) in a number of devices (Needhidasan et al. 2014; Williams 2010). However, e-waste is considered a secondary resource because of its hazardous nature due to its rich metallic and nonmetallic fractions. Therefore, using unconventional methods, the informal handling of such devices at scrap yards is a severe concern for both developed and developing nations (Ikhlayel 2018; Kyere et al. 2018). The variety of e-waste and its low compatibility with the environment make processing difficult and necessitate special handling. Thus, the complexities of waste electrical and electronic equipment (WEEE) require a multidisciplinary recycling strategy to increase the value of the fiber, metal, and plastic fractions and reduce environmental impact. The metal content in e-waste, particularly in waste printed circuit boards (WPCBs), is comparatively higher than in primary ores (Priya and Hait 2017a). Therefore, urban mining of metals from this valuable waste will minimize the pressure on virgin ore.

Currently, handling of e-waste involves incineration, dumping in landfills, or shipping it to underdeveloped nations, which are not eco-friendly and are against international regulations. However, researchers have introduced different scientific techniques to recover metallic and nonmetallic fractions. For instance, in recycling materials from e-waste, many traditional thermal, physiochemical, and mechanical processes are available. Advanced metallurgical techniques, such as pyrometallurgy, hydrometallurgy, and electrometallurgical methods, are also in practice. In the pyrometallurgical process, high temperatures (more than 2000 °C) are used to melt metals and accelerate chemical reactions that turn compounds into metal alloys (Kaya 2019). Moreover, the hydrometallurgical process uses either acidic or basic aqueous solutions to leach the metals in a leaching solvent. Nevertheless, these methods are costly, energy-intensive, and contribute to secondary pollutants. It is indeed a significant challenge to recycle e-waste using cost-effective and eco-friendly techniques. To overcome these constraints, researchers introduced an eco-friendly approach called biohydrometallurgy to recycle WPCBs, which leaches metal using bacteria and enzymes. However, the slow kinetics of the bioleaching process create a barrier to its commercial application.

The valorization of e-waste is a promising strategy to reduce manufacturing costs and pollution load, optimize the cycles in the production chain, and contribute significantly to the supply chain network. Furthermore, to aid the global circular economy, it is important to manage e-waste and convert it into value-added products through sustainable practices. Therefore, the present chapter provides a state-of-the-art review of various recycling approaches to recycling metallic and nonmetallic substances from valuable components of WEEE. Additionally, it outlines the advantages and disadvantages of the operational steps involved in processing WEEE by physical, chemical, and biological means. This chapter also intends to provide the recovery and utilization of valuable metals and nonmetals from different end-of-life (EoL) EEE, viz., computers, televisions, and mobile phones. Furthermore, the opportunities and challenges of different recycling and resource recovery techniques have been elucidated.

10.2 E-Waste Composition

EEE is classified into white, brown, and gray goods based on their characteristics and usefulness (Nnorom and Osibanjo 2008). White goods are primarily large consumer electronics and household appliances, usually coated with glossy white materials. However, brown goods are often light-weighted consumer equipment, including telecommunication devices, telephones, mobiles, etc. (Saini and Taneja 2012). In addition, gray goods are branded electronic products sold by unauthorized sellers outside the authorized market at a discount from the manufacturer's price (Priya and Hait 2018a). EoL photovoltaic solar panels are a new category of e-waste. The global volume of waste solar panels is expected to reach 78 million tonnes by 2050 (Chowdhury et al. 2020).

E-waste composition primarily depends on the age, type, and category of EEEs (Tanskanen 2013). The complex composition of e-waste comprises more than 1000 different materials, including some precious metals such as platinum (Pt), palladium (Pd), indium (In), gold (Au), silver (Ag), copper (Cu), aluminum (Al), and iron (Fe) (Zeng et al. 2017; Widmer et al. 2005). Moreover, the composition and characteristics of e-waste also depend on the sector to which it belongs. For instance, e-waste from the telecommunications and IT sectors comprises more valuable metals than ordinary electronic scraps from daily life. A mobile phone consists of more than 40 different metals and metalloids, including precious metals and base metals (Ghimire and Ariya 2020). However, the presence of various hazardous components, generally including batteries, bare boards, CRT monitors, LCD (liquid crystal display) screens, and LED (light-emitting diode) displays, makes them toxic as well. These components contain multiple poisonous and heavy metals such as arsenic (As), antimony (Sb), beryllium (Be), chromium (Cr), cadmium (Cd), cobalt (Co), lead (Pb), mercury (Hg), nickel (Ni), and thallium (Tl). Besides, several other toxic compositions are also present, such as plastics with brominated flame retardants (BFRs), polychlorobiphenyl, and polyvinyl chloride (Othman et al. 2008; Julander et al. 2014). The distinctive components of e-waste and the average metal and material fractions in WPCBs are illustrated in Figure 10.1.

WPCB accounts for approximately 3–6% of total e-waste generated globally (Priya and Hait 2017a; Faraji et al. 2018). Due to the ongoing technological development, a dramatic increase in the volume of WPCB is inevitable (Szałatkiewicz 2014). Based on environmental restrictions and economic value, WPCB has been considered as a substantial component of discarded electronic scrap (Das et al. 2009; Kaya 2017). Furthermore, this represents as most valuable component of e-waste, accounting for over 40% of the total metal value, as depicted in Figure 10.1 (Kumar et al. 2017). A large proportion of the precious metals in PCBs make it a rich source of secondary raw metals called urban mines for metal recovery and recycling (Dave et al. 2018). WPCB has a complex elemental composition, usually consisting of 60 different types of elements, which generally vary depending on the type of PCB (Kaya 2019; Szałatkiewicz 2014). The composition of WPCB varies based on the categories of PCB, kind of electronic item, manufacturing year, and manufacturing company (Luda 2011; Szałatkiewicz 2014). Table 10.1 presents the typical metallic and nonmetallic compositions of various electronic devices.

The structure of PCB is composed of three different types of materials, viz., metals, nonmetals, and organic substances (Szałatkiewicz 2014; Dave et al. 2018). The metal portion comprises 28% of the total weight of the PCB, including rare earth elements (REEs), base metals, and precious metals (Kavitha et al. 2017; Priya and Hait 2018a). At the same time, 72% is covered by a nonmetallic fraction, including ceramics, glass fibers, various types of FRs (such as Sb_2O_3, phosphates,

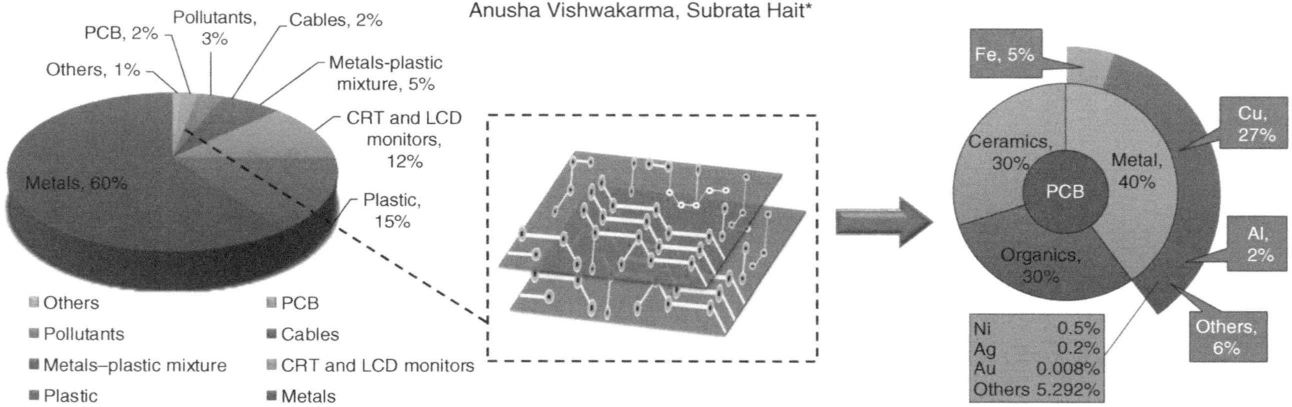

Figure 10.1 Typical composition of e-waste and WPCBs.

Table 10.1 Material composition of different types of WEEE.

S. no.	Electronic device	Average relative weight of the material fractions (unit: % w/w)						References
		Ferrous metals	Nonferrous metals	Plastic	PCBs	Cables	Others (e.g. textile and wood,)	
1	Mobile phone	1.4	3.9	25.6	22.1	0	6.2	Ernst et al. (2003), Chancerel et al. (2009)
2	Coffee machine	7.5	10.2	64.2	0.3	11.6	0.2	Chancerel and Rotter (2009)
3	Personal computer	58	0.4	3.5	8.9	2.6	NA	Keller (2006), Hagelüken (2006), Chancerel et al. (2009)
4	Iron	3.9	44.8	37.1	NA	14.3	0	Chancerel and Rotter (2009)
5	LCD TV	0.37	0.01	0.32	0.08	NA	0.22	Babbitt et al. (2017)
6	LED TV	0.43	0	0.31	0.02	NA	0.24	Babbitt et al. (2017)
7	Laptop	6.8	2.7	19.5	17.1	3.2	NA	Chancerel and Rotter (2009)
8	Mixer	2.7	0	20.4	0.7	6.8	0.1	Chancerel and Rotter (2009)
9	Printer	9.9	0.3	34.4	6.6	0.8	0.6	Huisman et al. (2007)
10	Calculator	5.4	10.2	58.9	14.1	0	10.1	Chancerel and Rotter (2009)
11	Hair dryer	3.5	0.2	35	0.1	24.2	0.3	Chancerel and Rotter (2009)
12	Radio set	3.5	0.7	42	20.5	3.7	0	Chancerel and Rotter (2009)
13	Desktop	0.58	0.15	0.12	0.14	NA	0.02	Babbitt et al. (2017)
14	DVD player	0.59	0.05	0.17	0.19	NA	0.01	Huisman et al. (2007), Babbitt et al. (2017)
15	Water heater	12.1	8.8	55.9	NA	19.9	NA	Chancerel and Rotter (2009), Chancerel et al. (2009)
16	Gaming console	0.36	0.14	0.34	0.16	NA	0.01	Babbitt et al. (2017)
17	Washing machine	37	11	24	1.385	NA	27	Babbitt et al. (2017), Priya, and Hait (2018a)

and BFRs), and polymers (Zhou and Qiu 2010; Priya and Hait 2017a; Zhang et al. 2021). Among the metallic constituents, an ample amount is covered by valuable critical and base metals such as Au, Ag, Cu, Ni, Pd, Pt, Pb, Sn, and Zn (Priya and Hait 2017a; Kaya 2017). About 90% of the inherent value of WPCB is derived from the presence of Ag and Pd (Luda 2011; Kaya 2017). Precious metals obtained from recycling WPCBs are purer than the raw materials obtained from natural resources. Hence, it is of utmost importance to recycle the waste PCBs to recover the metallic and nonmetallic fractions and contribute to value-added products.

10.3 Resource Recovery Techniques

Effective recycling of WEEE keeps them out of landfills to make a safer and contaminant-free environment. However, dealing with e-waste as a single stream, comprising a wide range of electronic items, makes e-waste management highly problematic. Furthermore, separating materials such as plastic, metal, and glass from e-waste is challenging, costly, and time-consuming. According to the literature, four recycling techniques have been employed to recover valuable material fractions, including mechanical, pyrometallurgical, hydrometallurgical, and biohydrometallurgical processes. E-waste recycling typically begins with mechanical processing, which includes the following steps: disassembly, shredding, and upgrading, followed by refining of metals as the final step. Moreover, metallurgical processes, such as pyrometallurgy, hydrometallurgy, electrometallurgy, and biometallurgy, and their combinations are used to refine the metals (Cui and Zhang 2008; Ilyas et al. 2010; Priya and Hait 2017a). However, the release of toxic pollutants and high volume of lixiviants are the drawbacks associated with metallurgical techniques, prompting the need for efficient treatment methods with minimum harmful impacts on the environment (Abdelbasir et al. 2018).

10.3.1 Mechanical Methods

In WEEE's pretreatment, the mechanical method is used to separate the different materials, with multiple separation steps based on the e-waste components (Veit et al. 2002; Zhang and Forssberg 1997). Several studies have been conducted to mechanically separate materials from e-waste based on density differences, particle sizes, magnetic and electrical characteristics, and other factors (Hayes 1993; Wills 1988). Mechanical processing includes several steps, from comminution of waste to size distribution, gravity separation, magnetic separation, water separation, and electrostatic separation (Veit 2015). Magnetic and water separation are two critical stages in mechanical processing. In magnetic separation, shredded e-waste passes through a giant magnet to filter ferrous metals from nonferrous material. Further, nonferrous metals are then separated using an eddy current

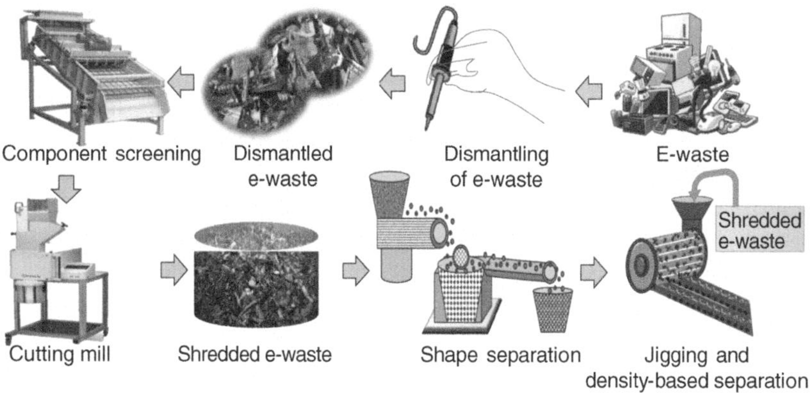

Figure 10.2 Individual sub-processes involved in the pre-treatment of e-waste.

separator. Alternatively, water separation uses water to separate the nonmetallic portion of e-waste, which primarily consists of plastic and glass. After mechanical separation, the separated materials are then sent to dedicated recycling plants for smelting and further purification. Likewise, the metal fraction obtained after mechanical processing is further processed by hydro, electro, and pyrometallurgical methods.

Mechanical processing offers an alternative means of retrieving valuable components based on the properties of WEEE. A complete material recovery, including plastics, is possible using a mechanical technique to improve WEEE recycling. Mechanical recycling methods can effectively separate e-waste materials based on variations in their physical properties. In addition, it is also important to know the physical, magnetic, and electrical properties of the components of the EEE for mechanical separation. Hence, the characterization of specific items in the e-waste stream is highly warranted. Further, the literature also suggested that particles' shape, size, and liberation degree play an important role in mechanical recycling. In contrast, differences in particle shape have also been utilized in shape-sorting techniques (Cui and Forssberg 2003). The individual sub-processes, such as dismantling, screening, and separation, including electrical conductivity-based separation, density-based separation, and jigging, are widely involved in the mechanical recycling process, as summarized in Figure 10.2.

10.3.2 Pyrometallurgy

Currently, pyrometallurgical recycling methods predominate; rather than mechanical processing, smelters handle more than 70% of WEEE (Cui and Zhang 2008; Khaliq et al. 2014). Pyrometallurgy, notably smelting, is a traditional method of recovering nonferrous and precious metals from e-waste. In this method, metals are sorted based on their chemical and metallurgical properties. The pyrometallurgical processing thermally treats e-waste at high temperatures (>1000 °C) without aqueous media to concentrate the metals in a metallic phase and reject undesirable elements as slag and gas (Hsu et al. 2019). In contrast, the plastic parts are more challenging to recover due to more complex polymer composites. Mechanically crushed electronic scrap is transferred for smelting in the furnace or molten bath to recover metals and remove polymeric substances. The high stability behavior of alloys during melting requires more energy, making it challenging to recover pure metal (Nakajima et al. 2010). Pyrometallurgical techniques are economically efficient, with other advantages including efficient recovery of precious metals, no need for pre-treatment, being applicable to a wide range of e-waste, and requiring fewer processing steps, making them easier to operate. However, the high energy demand, high investment, and other problems associated with this technique, including toxic gas emissions such as dioxins and furans, loss of precious and base metals, and slag formation, are the major drawbacks of this method. In addition, the loss of resources like ceramics, polymers, and glass, and low recovery rates for some particular metals like Sn, Pb, Al, and Zn, are considered as some other disadvantages of this technique (Veit 2015; Murugappan and Karthikeyan 2021).

Typical e-waste contains a significant amount of Al, Cu, Fe, and Pb. Therefore, in order to separate them, WEEE undergoes Pb and Cu smelting processes. According to the literature, Cu smelting can be done by "bath" or "flash" smelting, a molten liquid bath offered to smelt concentrates exposed to liquid slag. Further, the liquid slag is then converted to solid metals by injecting air over the bath. However, in flash smelting, the liquid slag is transformed into an air stream (Ramanayaka et al. 2020). Earlier literature reported on the techno-economic study of pyrometallurgy revealed the potential value of integrated e-waste recycling process of Cu smelting, which is economically viable and can process 30 kt of e-waste/annum (Ramanayaka et al. 2020).

10.3.3 Hydrometallurgy

Hydrometallurgy has been seen as a widely adopted metal extraction method in conventional metallurgical processes. The high energy requirement in the pyrometallurgical process has made hydrometallurgy a promising alternative. Hydrometallurgy is considered the most efficient technique in order to treat various impurities and metals at different processing stages and is also economically viable for refining chemicals (Sadegh-Safarzadeh et al. 2007). In this process, strong acids are generally used to leach the targeted metals into the desired concentration solution. This method involves the metal dissolution of comminuted e-waste into suitable leaching reagents. For instance, chemical lixiviants, generally acidic or alkaline, are often used to solubilize the metal from the solid matrix in the solution phase (Cui and Zhang 2008; Silvas et al. 2015). According to the literature, the popular hydrometallurgical lixiviants used to extract metals from crushed e-waste are nitric acid (HNO_3), sulfuric acid (H_2SO_4), aqua regia (HNO_3/H_2SO_4), thiourea, thiosulfate, cyanide, and halides (Huang et al. 2009; Pant et al. 2012). Further, the chemically dissolved metals with few impurities are extracted from the pregnant leach solution using different extraction techniques, such as solvent extraction, ion exchange, and absorption (Cui and Zhang 2008; Khaliq et al. 2014; Abdelbasir et al. 2018). Finally, electrorefining or electrochemical reduction processes are performed to fully recover the purified solid metals from e-waste (Ashiq et al. 2019).

Over the years, cyanide, as a lixiviant, has extensively been employed to retrieve precious metals with a high recovery rate. As of 2000, >90% of the gold and silver obtained from 875 gold and silver mines was processed by cyanide liquefaction (Akcil et al. 2010). However, the risk associated with the technique for humans and the environment restricts its application (Tuncuk et al. 2012). Nevertheless, non-cyanidation lixiviants such as thiosulfate and thiourea have also shown a fast leaching rate, indicating a promising alternative to cyanide leaching. Thiosulfate complexes, on the other hand, have low chemical stability and easily decompose into metallic sulfides (Brandl et al. 2008). Metal leaching has also been investigated using diethylenetriaminepenta acetate (DTPA), ethylene diaminetetramaide (EDTA), and other chemical lixiviants such as citric acid and oxalic acid (Pant et al. 2012; Priya and Hait 2017a). The hydrometallurgical technique is relatively easy to perform while extracting selective metals with the aid of the correct amount and concentration of acid (Ghosh et al. 2015; Ashiq et al. 2019). Apart from its benefits of high metal selectivity and low operating cost, some disadvantages associated with this process include difficulty in processing more complex composites and the generation of large volumes of toxic wastewater and solid waste (Veit 2015).

10.3.4 Biohydrometallurgy

Biotechnology integration into the hydrometallurgical process, collectively referred to as biohydrometallurgy or bioleaching, has emerged as an efficient metallurgical process for recovering metals from valuable secondary resources, viz. e-waste (Liu et al. 2016). Several studies in recent years have shown that using biocatalysts in place of harmful chemicals facilitates metal leaching, which has gained wide attention. Biocatalysts, such as microorganisms or enzymes, are a cornerstone of the biohydrometallurgical process (Habibi et al. 2020; Wang et al. 2009). This process utilizes a diverse range of microorganisms, such as autotrophic and heterotrophic bacteria, chemolithoautotrophs, heterotrophic fungi, and archaea (Liu et al. 2016). In the bioleaching process, microorganisms interact with comminuted metallic waste to oxidize and dissolve metals from complex metallic compounds into a soluble form (Priya and Hait 2017b). The dissolved metal is further extracted and purified by employing different well-established processes such as precipitation, adsorption, ion exchange, or electrowinning.

Chemolithoautotrophs are the most prevalent group of bacteria, using inorganic substances such as reduced S and Fe^{2+} as their energy source and CO_2 as a carbon source to solubilize metals (Brandl et al. 2001; Le et al. 2006). Biohydrometallurgy is an eco-friendly technique with key technical advantages over alternative treatment methods, including low operating costs, low energy requirement, less demand for skilled personnel, and the generation of a small amount of secondary waste (Karwowska et al. 2014; Habibi et al. 2020). In bioleaching, the bio-oxidation of metals occurs through two different mechanisms: direct and indirect mechanisms. In direct mechanism, bacterial cell adheres to the surface of a solid substrate that oxidizes the metals enzymatically. However, in the indirect mechanism, bacterial lixiviants (such as Fe^{3+} and H_2SO_4) oxidize the metal, causing it to dissolve in solution (Priya and Hait 2017b; Habibi et al. 2020). According to the literature, iron- and sulfur-oxidizing chemolithotrophic acidophilic autotrophs such as *Acidithiobacillus thiooxidans*, *Acidithiobacillus ferrooxidans*, and *Leptospirillum ferrooxidans* are the most widely used bacteria. Besides these, heterotrophic microorganisms such as *Aspergillus* species (fungi) and *Penicillium* species (fungi), and bacteria like *Chromobacterium violaceum* and *Pseudomonas* species have also been widely used (Xin et al. 2009; Priya and Hait 2018b). From an economic and environmental perspective, the bioleaching process offers the most efficient approach to recover precious metals from e-waste. However, its slow kinetics and inefficacy in treating high pulp density are significant drawbacks of this method (Işıldar et al. 2017; Habibi et al. 2020).

10.4 Valorization of E-Waste for Circular Economy

E-waste valorization refers to any industrial processing activities to reuse and recycle waste into valuable products. Valorization of e-waste typically involves urban mining of metals, turning waste into raw materials, reusing discarded items, and adding waste into value-added products. It is quite challenging to produce high-value products from e-waste due to its complexity and diversity, using efficient and sustainable recycling techniques. Furthermore, a circular economy model that is a closed-loop approach works on the principle of circulating products from manufacturing to be reintroduced to the supply chain at the end of their life (Kirchherr et al. 2017). This model is best suited for resource conservation and remanufacturing products with minimal natural resource exhaustion. With respect to the circular economy concept, recycling and recovering materials from e-waste obviate the need to obtain raw materials from

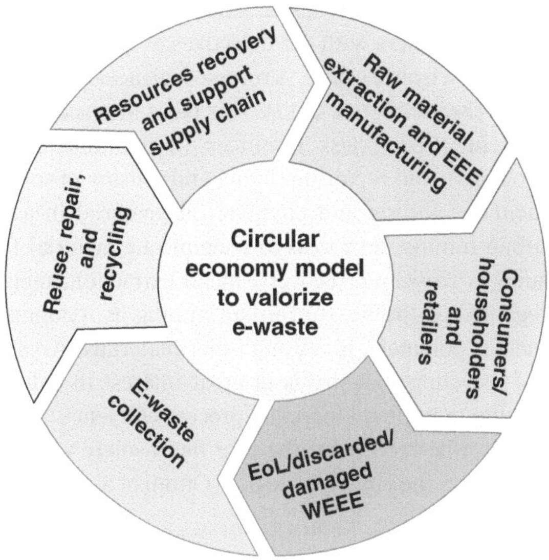

Figure 10.3 A circular economy approach toward e-waste management and valorization.

mining with significant CO_2 emissions. As long as the e-waste comprises valuable materials, it will be a potential input to the subsequent supply chain. The circular economy model prioritizes quality over quantity and switches the industry's emphasis from "product offering" to "service offering" (Naor et al. 2015; Ottoni et al. 2020). Some alternative options, such as preserving resources close to their original state or recycling or reusing EoL electronic products with minimal loss, must be considered while valorizing e-waste in the circular economy model (Reike et al. 2018). It is noteworthy that e-waste carries an intrinsic value. Therefore, applying a circular economy approach to manage e-waste will retain its value and give an impetus to the evolution of the recycling industry (Figure 10.3) (Ottoni et al. 2020; MeitY 2021).

10.4.1 Benefits of Valorization

Economically emerging nations like India, China, Brazil, and others are inefficient in managing the vast volume of e-waste generated from discarded EEE. Correspondingly, valorizing secondary raw materials (SRMs) recovered from e-waste is an exciting prospect for creating economically valuable materials (Ottoni et al. 2020). However, the introduction of the circular economy concept in the early 2000s has focused industrial practices on using secondary resources as value-added materials (Nzihou 2010). The circular economy model emphasizes the modular design of electronic devices, which allows simple disassembly and easy replacement of damaged parts.

E-waste valorization is inextricably linked to management strategies adopted by decision-makers with the objectives of system optimization and maximum value extraction from e-waste. A reverse logistical approach is used to recover SRM from waste (Serranti et al. 2012). This takes into account the 4-R strategies – refusing, reducing (using less resources in manufacturing), reusing (using outdated products), and repairing (fixing and reusing obsolete items) – to reduce environmental pollution and circulate the material in a circular pattern. In addition, urban mining, a process to reclaim raw material from waste, covers operations such as recycling, recovery, metal extraction, metal processing, etc. E-waste is regarded as the heart of urban mining as it contains a considerable amount of metals, polymers, glass, and other materials (Arya and Kumar 2020). Therefore, the recycling industry is of great interest in valorizing e-waste through urban mining to improve recycling process efficiency. The valorization of obsolete EEE can also be retained if the user deals solely with the defective part rather than discarding the complete device (Ottoni et al. 2020).

10.4.2 Comparison of Resource Recovery Technique

As aforementioned, WEEE has significant economic interest due to its high concentration of valuable metals (copper, iron, gold, silver, nickel, palladium, platinum, etc.). Moreover, the presence of other materials like plastic, glass, ceramics, and other constituents may require further processing to recover and support the economy. Thus, it becomes more challenging to compare different resource recovery techniques. Current e-waste disposal methods include landfilling and incineration. E-waste is mechanically piled up or dumped into landfill sites with other solid waste. Such practices release toxic substances into the environment. Further, incineration of e-waste is technically advantageous since it produces heat and reduces volume quickly. However, incomplete combustion emits poisonous gases such as polycyclic aromatic hydrocarbons, carbon monoxide, and dioxins, while complete oxidation also releases CO_2, a greenhouse gas, in the effluent (Jung et al. 2022). As discussed in the above section, various metallurgical options for e-waste metal extraction are available, including pyrometallurgical, hydrometallurgical, and biohydrometallurgical processes. Figure 10.4 shows a typical schematic representation of metallurgical processes for the recovery of metals. The feasibility of these technologies for practical implications depends on their relative advantages and challenges.

Pyrometallurgy is a traditional and widely used method for recovering nonferrous metals from e-waste. The ability to burn any kind of WEEE is the key benefit of the incineration process, as the organic matter from e-waste is already burned as fuel. However, the formation of toxic gases in the burning process may lead to environmental harm, necessitating off-gas treatment (Tuncuk et al. 2012). Additionally, high energy consumption to melt the metals and the production of

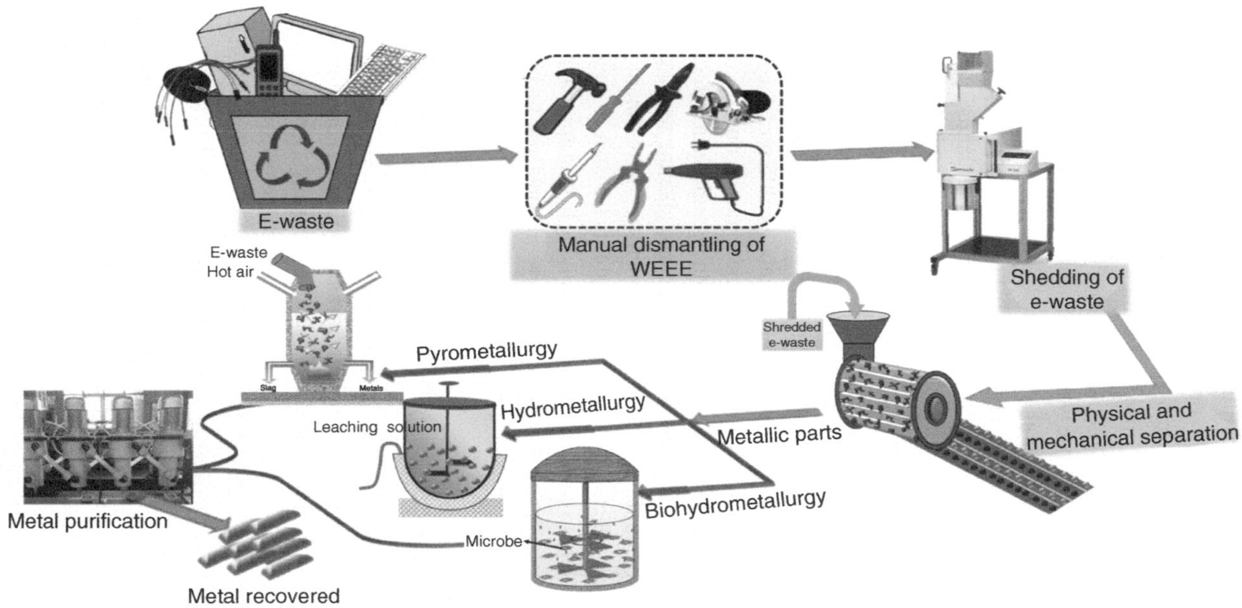

Figure 10.4 Schematic representation of steps followed to recover metallic fraction employing different metallurgical approaches.

large quantities of slag are the inherent drawbacks of the incineration process (Ramanayaka et al. 2020). In contrast, hydrometallurgy is often considered more eco-friendly than pyrometallurgical processes because hydrometallurgical processes demand less energy to operate. Further, the hydrometallurgical technique is simpler and releases less toxic gases. However, the process becomes challenging as the feedstock consists of complex mixed materials. In addition, some limitations of this method include the multiple processing steps, increased chemical use, production of toxic wastewater, and solid waste (leaching residues), which further require additional treatment (Coman et al. 2013). In contrast, biohydrometallurgy has several advantages over traditional metallurgical methods, including lower operating costs, lower energy demands, no need for a skilled workforce, ease of handling, and less secondary pollution (Wang et al. 2013). Despite significant advances, the commercial application of metallurgical techniques for e-waste recycling, including bioleaching, remains constrained by technological, economic, and environmental concerns (Priya and Hait 2018b). Besides technical and environmental viability, a comprehensive comparison based on energy–exergy balance, economic feasibility, different waste emissions, and specific regulatory requirements must be taken into account.

In general, hydrometallurgical and bioleaching operations do not emit harmful gases but often produce large volumes of acidic and alkaline waste that must be purified, neutralized, and recycled later. The downstream treatment required for these processes may affect their scale-up applicability for e-waste recycling. Thus, it seems more efficient to manage waste by employing hybrid approaches, combining two or more techniques to minimize negative impacts and increase material recovery values, rather than focusing on individual processes.

10.4.3 Case Studies

E-waste includes high-value metals and other valuable substances, stimulating market interest in strategically tackling urban mining. Therefore, recovering high-value metals is the most attractive option for e-waste valorization. Interestingly, electronic residues are a good source of carbon and silicon that can be used to prepare nanomaterials. In this regard, several approaches have been explored, combining pyrolytic and catalytic processes to convert various components of e-waste into added-value materials (Table 10.2). According to EPA, 1 Mt of WPCB contains gold 800 times more than 1 Mt of natural ore (Duan and Zhu 2022). Furthermore, according to Forti et al. (2020), the raw material value of the total volume of e-waste in 2019 was US $57 billion, a major part of this value was generated due to the presence of primarily Cu, Fe, and Au. Besides, the raw material value of formally collected and recycled e-waste is US $10 billion, accounting for 17.4% (4 million) of the total volume (Forti et al. 2020). Urban mining of recyclable materials is extensively exploited in many economically rich countries

Table 10.2 The summary of value-added products prepared using different parts of WEEE.

WEEE	Value-added product	Method	Application	References
WPCBs	Cu NPs	Microemulsion under controlled conditions, i.e. ambient temperature and pressure	Nanofluid consisting of Cu NPs for thermal conductivity applications	Mdlovu et al. (2018)
		Hydrometallurgical method: selective leaching and precipitation of Cu at room temperature	Potential applications: electronics inking and electrochemical sensors	El-Nasr et al. (2020)
	Cu–Sn-based alloy	Two-step thermal transformation at high (1200 °C) and low (400 °C) temperatures	In local microfactories as value-added metal alloys	Shokri et al. (2017a)
	Cu–Sn NPs	Thermal transformation at high temperature (900 °C)	Pb separation from WPCB	Shokri et al. (2017b)
	SnO_2 NPs and PbO_2 microparticles	Cost-effective method with low preparation cost	Catalytic, electrochemical, and biomedical applications	Tatariants et al. (2018)
	Cu-based catalyst	Pyrolysis at 350 °C	Carbon capture, storage, and utilization	Wang et al. (2019)
	Cu oxide NPs	Biohydrometallurgical leaching	Adsorption and removal of methylene blue from aqueous solution	Nithya et al. (2018)
	WPCB photocatalysts	Hydrochemical processing and doping with Ag^+, Cu^{2+}, and Sn^{2+}	Visible light photocatalysts for degradation of methyl violet and methylene blue	Ramaswamy et al. (2017)

	Reinforcing filler in composites	Mixing and curing at room temperature	Reinforcing filler in unsaturated polyester resin	Hu et al. (2018)
	Nonmetallic particles	Hydrothermal modification, pouring, and hot pressing	Raw materials for epoxy resin and glass fiber resin composites	Gao et al. (2018a)
	Oil and gas	Microwave-induced pyrolysis along with mechanical processing	Fuel gas (with high calorific value) and oil	Sun et al. (2011)
	Oil and gas	Pyrolysis in different conditions at 600 °C	Energy recovery from the organic fraction	Evangelopoulos et al. (2017)
	Oil-base resin	Crushing, debromination pyrolysis at 600 °C under Ar atmosphere	Pyrolysis oil usage for resin production	Gao et al. (2018b)
	Magnetic zeolite coated with carbon fiber	Pyrolysis at 750 °C	Applied as demulsifier for the removal of emulsified oil	Gao et al. (2019)
	Reinforcing filler in composites	Mixing and curing at room temperature	Reinforcing filler in unsaturated polyester resin	Hu et al. (2018)
Cathode-ray tubes	Filler material in concrete	E-waste above 15% the porosity of the concrete increases	Partial replacement for coarse aggregate	Hamsavathi et al. (2020)
Plastic from waste printers	Transformation into sustainable 3D printing filaments	Extrusion of size-reduced plastics into filaments, then subjection to 3 extrusion cycles; effect of repeated extrusion cycles on the polymer properties	3D printing	Gaikwad et al. (2018)

and is highly praised for offering potential alternatives. For example, China is extensively exploiting metals from e-waste to meet the country's resource needs for the production of EEE. In 2018, 81% of the world's total REE were individually produced by China (Leader et al. 2019). According to UNU (2018), Europe has the potential to mine 2 Mt of spent batteries per year. This equates to an ample supply of critical metals such as 0.078 Mt of lithium, 0.021 Mt of cobalt, and 0.114 Mt of manganese. Recently, in Tokyo Olympic and Paralympic Games, all the medals awarded to winners were made from e-waste's recycled metals (Ailbhe 2021). This was accomplished by pyrometallurgical processing of gold, silver, and bronze containing components of WEEE. It thus entails melting the substance to extract the base metal (TOI 2021). By recycling approximately 80 tons of WEEE, the campaign successfully collected 32 kg of gold, 3493 kg of silver, and 2200 kg of bronze (TOI 2021). Similarly, in Rio Olympics, 2016, silver and bronze medals were produced using 30% recycled materials from e-waste (Ailbhe 2021). According to the Business World Online Bureau (2018), effective urban mining of e-waste has the potential to save around US $21 billion. A report by UNU (2018) states that a smartphone contains 40 important metals, of which the extracted gold has a purity of 25–30 times higher than that of primary ore (Singh et al. 2018; Arya and Kumar 2020).

In the global e-waste management market, some leading companies are working on resource recovery by using different treatment techniques. Some of the top-leading e-waste recycling companies are Aurubis AG (Germany), Boliden AB (Sweden), Enviro-Hub Holdings Limited (Singapore), Electronic Recyclers International (ERI) (United States), Global Electric Electronic Processing (GEEP) (Canada), Michael Biddle and Association (MBA) Polymers (United States), Sims Metal Management Limited (Australia), Stena Technoworld AB (Sweden), and Umicore S.A. (Belgium) (Yken et al. 2021). For instance, Boliden (Sweden), Aurubis (Germany), Umicore (Belgium), Dowa (Japan), and Noranda (Quebec) are top-leading companies that use pyrometallurgical techniques for managing e-waste and resource recovery (Arya and Kumar 2020; Yken et al. 2021). A detailed description of these companies in terms of their treatment approach and capacity is given in Table 10.3. MBA Polymers, a US-based company, works in recycling plastics from waste materials such as e-waste. In 2017, MBA Polymers founded a plant in Pune, India, with a recycling capacity of 100 Mt to produce post-consumer plastic from waste electronic and automotive plastic products (Straits Research 2022). Despite this, a substantial portion of the e-waste produced globally in 2019, which accounts for 82.6% of total e-waste, was not formally collected, treated, and disposed of in an ecologically sound manner (Forti et al. 2020). This indicates the nonavailability of proper collection channels and the slow pace of recycling activities worldwide in managing the huge volume of e-waste, which need to be addressed.

Table 10.3 Top leading companies working in the global e-waste management and resource recovery, with their treatment approach and capacity.

Company	Location	Metals and materials recovered	Process overview	Process capacity for e-waste	Revenue (USD)	Reference
Aurubis AG	Germany	Cu and precious metals, such as Ag, Au, Cu, Pb, Sn, and Zn	Top-submerged lance (TSL) bath, Cu smelting, electrorefining, PMs refining	300 kt/annum	1.12 billion (2018)	Straits Research (2020), Yken et al. (2021)
Boliden AB	Sweden	Recycling and the production of Zn, Cu, and precious metals	Dismantling, removal of plastic, iron, and Al, shredding and then sent to Kaldo furnace, smelting in Rönnskär smelter	120 kt/annum	5.1 billion (2018)	Straits Research (2020)
Enviro-Hub Holdings Limited	Singapore	Platinum group metals (PGMs)	Refining of PGM, construction works, conversion of plastics to fuel, and rental and servicing of machinery and equipment	3.5 kg of PMs/tonnes of Cu scrap	0.026 billion (2018)	Straits Research (2020), Tan (2007)
Electronic Recyclers International (ERI)	United States	Metals, plastic, and glass	Metal sorting by AI-driven Super Automated Machine (SAM)	124.737 kt/annum	—	Straits Research (2020), NERC (2022), Toto (2021)

(Continued)

Table 10.3 (Continued)

Company	Location	Metals and materials recovered	Process overview	Process capacity for e-waste	Revenue (USD)	Reference
Global Electric Electronic Processing (GEEP)	Canada	Cu, Al, and other metals, steel, and glass	Material shredding, sorting, hammer mill processing, size reduction, mechanical separation of materials into ferrous, nonferrous, and plastic materials	5.896 kt/annum	—	Krikke (2008)
Michael Biddle and Association (MBA) Polymers	United State	Consumer plastics, viz. ABS, HIPS, PP, and PC/ABS	Separating, cleaning, and finishing	20 kt/annum	—	Straits Research (2020)
SIMS Metal Management Limited	Australia	Cu, Al, Zn, Ni, Pb, steel, Fe	Manual sorting, dismantling, and shredding, fed into TSL lead and Cu smelting furnaces, metal refining	10,000 kt/annum (in 2017)	6.64 billion (2018)	Straits Research (2020)
STENA Techno World AB	Sweden	PS/ABS plastics, shredded steel, Al, PMs	Recycling and refining of e-waste, producing secondary raw materials, steel products, and marine fuels	2.5 kt/annum (Victoria-based plant)	3.8 billion (2018)	Straits Research (2020)

Umicore S.A.	Belgium	Au, Pb, Ag, Cu, platinum group metals (Pd, Pt, Ru, Rh, Ir), paramagnetic (Ni, Co), special metals (Se, Te, In), secondary metals (Sn, Sb, As, Bi)	Combined pyrometallurgical, hydrometallurgical, and electrochemical processes, ISA SMELT™ smelting, leaching, electrowinning, PMs refining	250 kt/annum	PGMs:50 t/yr., Au:100 t/yr., and Ag:2400 t/yr.	Islam et al. (2020), Khaliq et al. (2014), Yken et al. (2021)
Noranda	Canada	Ag, Au, Cu, Ni, Pd, Pt, Se, Te	Smelting, electrorefining, PMs refining	100 kt/annum	—	Khaliq et al. (2014), Yken et al. (2021)
Trojan Electronics Limited	Europe	Manufacturing of electronics and IT asset disposition (ITAD services)	Offers e-waste management and ITAD services	—	—	Straits Research (2020)
Rönnskär	Sweden	Ag, Au, As, Bi, Cu, In, Ir, Ni, Pb, Pd, Pt, Rh, Ru, Sb, Se, Sn	Smelting in Kaldo reactor, PMs refining, purification	120 kt/annum	—	Khaliq et al. (2014), Yken et al. (2021)
DOWA smelter	Japan	Ag, Au, Bi, Cu, Ni, Pb, Sb, Sn, Te	TSL smelting, Cu refining, PMs refining	150 kt/annum	62.06 billion	Yken et al. (2021)

10.5 Opportunities and Challenges of Valorization of E-Waste

WEEE is an "urban mine" of critical metals (like gold, silver, copper, aluminum, platinum, etc.) and nonmetals (plastic and glass), which can be used as secondary materials. Most electronic devices are completely recyclable. Reclaiming valuable materials from e-waste, rather than dumping them in landfills, would preserve natural resources. In the present state of affairs, e-waste recycling rates are alarmingly low compared to the amount of e-waste generated. According to a report provided by the UN, a minimum of US $10 billion worth of precious metals (including gold, platinum, and other metals) are dumped annually into e-waste piles (Carrington 2020), leading to the loss of vast amounts of valuable resources. At the same time, recycling BFR from e-waste represents a major challenge, as its separation from other polymers incurs additional costs (Forti et al. 2020). Additionally, relative legislation in any nation significantly impacts countries' waste management. In the Asian continent, particularly in South Asia, e-waste management is dominated by informal sector activities for collection, dismantling, and recycling of e-waste. However, formal recycling capacity remains underutilized, as the informal sector still handles most of the e-waste. Lack of collection services and consumers' limited awareness are the main reasons for channeling e-waste to informal recycling sectors in underdeveloped nations (Forti et al. 2020). Meanwhile, enforcing stringent guidelines, norms, and standards remains challenging for authorized recyclers to conduct the proper recycling process smoothly. Lack of reliable information, informality in the segment, and the absence of an adequate system for e-waste reverse logistics are realities in most metropolitan areas (Souza et al. 2016). Thus, to effectively meet all challenges, systematic e-waste collection solutions are urgently required to avail the unavailability of logistics infrastructure (Akcil et al. 2019).

10.6 Conclusion

Technological innovation has made our lives easier, simpler, and happier. Accordingly, per capita demand for electronic gadgets has increased dramatically. As a result, the consumption of critical metals such as REE, precious metals, and platinum group metals (PGMs) is in demand. Also, the rapid development in electronic industries has caused a massive surge in global e-waste production, becoming a significant concern. Since e-waste is both a secondary resource and an environmental hazard, its sustainable management, valorization, and resource utilization are essential for EEE manufacturing industries to meet the shortage of primary resources. Almost all recycling processes have some merits and demerits. Although hydrometallurgical methods have shown more promising results for resource recovery, generating secondary waste inhibits their application. In this regard, the bioleaching of metals from e-waste has received considerable interest due to its low operating cost, minimal use of chemicals, and easy handling of detoxifying sludge. Recently, adding extra value to final recovered products from WPCB has been trending by turning them into catalysts, composites, nanomaterials, and other valuable products. Therefore, comprehensive recycling and valorization of e-waste, notably of WPCB, concurrently addresses issues including natural resource depletion, supply chain crises for critical metals, and insufficient disposal facilities. Additionally, by recovering metals from e-waste, industries have many opportunities to ease the strain on raw metal mines. Therefore, more sustainable resource utilization techniques are needed to explore fast and efficient recycling of WEEE's metallic and nonmetallic components.

References

Abdelbasir, S.M., Hassan, S.S., Kamel, A.H., and El-Nasr, R.S. (2018). Status of electronic waste recycling techniques: a review. *Environmental Science and Pollution Research* 25 (17): 16533–16547. https://doi.org/10.1007/s11356-018-2136-6.

Ailbhe, G. (2021). Urban Mining: The Hidden Value of e-Waste. https://theintelligentminer.com/2021/08/25/urban-mining-the-hidden-value-of-e-waste (accessed 7 July 2022).

Akcil, A., Deveci, H., Jain, S., and Khan, A. (2010). Mineral biotechnology of sulphides. In: *Geomicrobiology* (ed. M.K. Rai), 101–137. Enfield: Science Publishers. ISBN: 9780429192197.

Akcil, A., Agcasulu, I., and Swain, B. (2019). Valorization of waste LCD and recovery of critical raw material for circular economy: a review. *Resources, Conservation and Recycling* 149: 622–637. https://doi.org/10.1016/j.resconrec.2019.06.031.

Arya, S. and Kumar, S. (2020). Bioleaching: urban mining option to curb the menace of E-waste challenge. *Bioengineered* 11 (1): 640–660. https://doi.org/10.1080/21655979.2020.1775988.

Ashiq, A., Kulkarni, J., and Vithanage, M. (2019). Hydrometallurgical recovery of metals from E-waste. In: *Electronic Waste Management and Treatment Technology*, 225–246. Butterworth-Heinemann https://doi.org/10.1016/B978-0-12-816190-6.00010-8.

Babbitt, C., Althaf, S., and Chen, R. (2017). Sustainable Materials Management for the Evolving Consumer Technology Ecosystem. Summary report of phase, 1.

Brandl, H., Bosshard, R., and Wegmann, M. (2001). Computer-munching microbes: metal leaching from electronic scrap by bacteria and fungi. *Hydrometallurgy* 59: 319–326. https://doi.org/10.1016/S0304-386X(00)00188-2.

Brandl, H., Lehmann, S., Faramarzi, M.A., and Martinelli, D. (2008). Biomobilization of silver, gold, and platinum from solid waste materials by HCN-forming microorganisms. *Hydrometallurgy* 94: 14–17. https://doi.org/10.1016/j.hydromet.2008.05.016.

Business World Online Bureau (2018). *Urban Mining Has the Potential to Unlock India's E-waste Economy*. Business World https://www.businessworld.in/article/Urban-Mining-Has-The-Potential-To-Unlock-India-s-E-Waste-Economy/15-12-2018-165172 (accessed 18 July 2022).

Carrington, D. (2020). *$10bn of Precious Metals Dumped Each Year in Electronic Waste, Says UN*. The Guardian https://www.theguardian.com/environment/2020/jul/02/10bn-precious-metals-dumped-each-year-electronic-waste-un-toxic-e-waste-polluting. (accessed 17 July 2022).

Chancerel, P. and Rotter, S. (2009). Recycling-oriented characterization of small waste electrical and electronic equipment. *Waste Management* 29 (8): 2336–2352. https://doi.org/10.1016/j.wasman.2009.04.003.

Chancerel, P., Meskers, C.E., Hagelüken, C., and Rotter, V.S. (2009). Assessment of precious metal flows during preprocessing of waste electrical and electronic equipment. *Journal of Industrial Ecology* 13 (5): 791–810. https://doi.org/10.1111/j.1530-9290.2009.00171.x.

Chowdhury, M.S., Rahman, K.S., Chowdhury, T. et al. (2020). An overview of solar photovoltaic panels' end-of-life material recycling. *Energy Strategy Reviews* 27: 100431. https://doi.org/10.1016/j.esr.2019.100431.

Coman, V., Robotin, B., and Ilea, P. (2013). Nickel recovery/removal from industrial wastes: a review. *Resources, Conservation and Recycling* 73: 229–238. https://doi.org/10.1016/j.resconrec.2013.01.019.

Cui, J. and Forssberg, E. (2003). Mechanical recycling of waste electric and electronic equipment: a review. *Journal of hazardous materials* 99 (3): 243–263. https://doi.org/10.1016/S0304-3894(03)00061-X.

Cui, J. and Zhang, L. (2008). Metallurgical recovery of metals from electronic waste: a review. *Journal of hazardous materials* 158 (2–3): 228–256. https://doi.org/10.1016/j.jhazmat.2008.02.001.

Das, A., Vidyadhar, A., and Mehrotra, S.P. (2009). A novel flowsheet for the recovery of metal values from waste printed circuit boards. *Resource Conservation and Recycling* 53 (8): 464–469. https://doi.org/10.1016/j.resconrec.2009.03.008.

Dave, S.R., Sodha, A.B., and Tipre, D.R. (2018). Microbial technology for metal recovery from e-waste printed circuit boards. *Journal of Bacteriology and Mycology* 6 (4): 241–247.

Duan, H. and Zhu, X.N. (2022). Recent advances in recovering technology for recycling gold from waste printed circuit boards: a review. *Energy Sources, Part A: Recovery, Utilization, and Environmental Effects* 44 (1): 1640–1659. https://doi.org/10.1080/15567036.2022.2056271.

El-Nasr, R.S., Abdelbasir, S.M., Kamel, A.H., and Hassan, S.S. (2020). Environmentally friendly synthesis of copper nanoparticles from waste printed circuit boards. *Separation and Purification Technology* 230: 115860. https://doi.org/10.1016/j.seppur.2019.115860.

Ernst, T., Popp, R., Wolf, M., and van Eldik, R. (2003). Analysis of eco-relevant elements and noble metals in printed wiring boards using AAS, ICP–AES and EDXRF. *Analytical and Bioanalytical Chemistry* 375 (6): 805–814. https://doi.org/10.1007/s00216-003-1802-8.

Evangelopoulos, P., Kantarelis, E., and Yang, W. (2017). Experimental investigation of pyrolysis of printed circuit boards for energy and materials recovery under nitrogen and steam atmosphere. *Energy Procedia* 105: 986–991. https://doi.org/10.1016/j.egypro.2017.03.435Evangelopoulos.

Faraji, F., Golmohammadzadeh, R., Rashchi, F., and Alimardani, N. (2018). Fungal bioleaching of WPCBs using *Aspergillus niger*: observation, optimization and kinetics. *Journal of Environmental Management* 217: 775–787. https://doi.org/10.1016/j.jenvman.2018.04.043.

Forti, V., Balde, C.P., Kuehr, R., and Bel, G. (2020). The Global E-waste Monitor 2020: Quantities, flows and the circular economy potential. (Bonn, Geneva and Rotterdam: United Nations University/United Nations Institute for Training and Research, International Telecommunication Union, and International Solid Waste Association, 2020). ISBN: 9789280891140.

Gaikwad, V., Ghose, A., Cholake, S. et al. (2018). Transformation of E-waste plastics into sustainable filaments for 3D printing. *ACS Sustainable Chemistry and Engineering* 6 (11): 14432–14440. https://doi.org/10.1021/acssuschemeng.8b03105.

Gao, X., Li, Q., and Qiu, J. (2018a). Hydrothermal modification and recycling of nonmetallic particles from waste print circuit boards. *Waste Management* 74: 427–434. https://doi.org/10.1016/j.wasman.2018.01.004.

Gao, R., Liu, Y., and Xu, Z. (2018b). Synthesis of oil-based resin using pyrolysis oil produced by debromination pyrolysis of waste printed circuit boards. *Journal of cleaner production* 203: 645–654. https://doi.org/10.1016/j.jclepro.2018.08.228.

Gao, R., Liu, B., and Xu, Z. (2019). Fabrication of magnetic zeolite coated with carbon fiber using pyrolysis products from waste printed circuit boards. *Journal of cleaner production* 231: 1149–1157. https://doi.org/10.1016/j.jclepro.2019.05.220.

Ghimire, H. and Ariya, P.A. (2020). E-wastes: bridging the knowledge gaps in global production budgets, composition, recycling and sustainability implications. *Sustainable Chemistry* 1 (2): 154–182. https://doi.org/10.3390/suschem1020012.

Ghosh, B., Ghosh, M.K., Parhi, P. et al. (2015). Waste printed circuit boards recycling: an extensive assessment of current status. *Journal of cleaner production* 94: 5–19. https://doi.org/10.1016/j.jclepro.2015.02.024.

Habibi, A., Shamshiri Kourdestani, S., and Hadadi, M. (2020). Biohydrometallurgy as an environmentally friendly approach in metals recovery from electrical waste: a review. *Waste Management and Research* 38 (3): 232–244. https://doi.org/10.1177/0734242X1989532.

Hagelüken, C. (2006). Recycling of electronic scrap at Umicore's integrated metals smelter and refinery. *Erzmetall* 59 (3): 152–161.

Hamsavathi, K., Prakash, K.S., and Kavimani, V. (2020). Green high strength concrete containing recycled Cathode Ray Tube Panel Plastics (E-waste) as coarse aggregate in concrete beams for structural applications. *Journal of Building Engineering* 30: 101192. https://doi.org/10.1016/j.jobe.2020.101192.

Hayes, P.C. (1993). *Process Principles in Minerals and Materials Production*, 29. Brisbane: Hayes Publishing Co.

Hsu, E., Barmak, K., West, A.C., and Park, A.H.A. (2019). Advancements in the treatment and processing of electronic waste with sustainability: a review of metal extraction and recovery technologies. *Green chemistry* 21 (5): 919–936. https://doi.org/10.1039/C8GC03688H.

Hu, D., Jia, Z., Li, J. et al. (2018). Characterization of waste printed circuit boards nonmetals and its reutilization as reinforcing filler in unsaturated polyester resin. *Journal of Polymers and the Environment* 26: 1311–1319. https://doi.org/10.1007/s1092.

Huang, K., Guo, J., and Xu, Z. (2009). Recycling of waste printed circuit boards: a review of current technologies and treatment status in China. *Journal of Hazardous Material* 164: 399–406. https://doi.org/10.1016/j.jhazmat.2008.08.051.

Huisman, J., Magalini, F., Kuehr, R., Maurer, C., Ogilvie, S., Poll, J., Delgado, C., Artim, E., Szlezak, J., and Stevels, A. (2007). *2008 Review of Directive 2002/96 on Waste Electrical and Electronic Equipment (WEEE).*

Ikhlayel, M. (2018). An integrated approach to establish e-waste management systems for developing countries. *Journal of Cleaner Production* 170: 119–130. https://doi.org/10.1016/j.jclepro.2017.09.137.

Ilyas, S., Ruan, C., Bhatti, H.N. et al. (2010). Column bioleaching of metals from electronic scrap. *Hydrometallurgy* 101 (3–4): 135–140. https://doi.org/10.1016/j.hydromet.2009.12.007.

Işıldar, A., Vossenberg, J.V.D., Rene, E.R. et al. (2017). Biorecovery of metals from electronic waste. In: *Sustainable Heavy Metal Remediation*, 241–278. Cham: Springer. https://doi.org/10.1007/978-3-319-61146-4_8.

Islam, A., Ahmed, T., Awual, M.R. et al. (2020). Advances in sustainable approaches to recover metals from e-waste – a review. *Journal of Cleaner Production* 244: 118815. https://doi.org/10.1016/j.jclepro.2019.118815.

Julander, A., Lundgren, L., Skare, L. et al. (2014). Formal recycling of e-waste leads to increased exposure to toxic metals: an occupational exposure study from Sweden. *Environment International* 73: 243–251. https://doi.org/10.1016/j.envint.2014.07.006.

Jung, S., Lee, S., Song, H. et al. (2022). Sustainable valorization of E-waste plastic through catalytic pyrolysis using CO_2. *ACS Sustainable Chemistry & Engineering* 10 (26): 8443–8451. https://doi.org/10.1021/acssuschemeng.2c01469.

Karwowska, E., Andrzejewska-Morzuch, D., Lebkowska, M. et al. (2014). Bioleaching of metals from printed circuit boards supported with surfactant producing bacteria. *Journal of Hazardous Material* 264: 203–210. https://doi.org/10.1016/j.jhazmat.2013.11.018.

Kavitha, K.K., Senthil, K.P., and Pradeepa, R. (2017). Review on microbial remediation of heavy metals from E-waste. *International Journal of Agricultural and Life Sciences* 3: 123–130. ISSN: 2454-6127.

Kaya, M. (2017). Recovery of metals and non-metals from waste printed circuit boards (PCBs) by physical recycling techniques. *Energy Technology* 433–451. https://doi.org/10.1007/978-3-319-52192-3_43.

Kaya, M. (2019). *Electronic Waste and Printed Circuit Board Recycling Technologies*. Berlin: Springer. ISBN: 978-3-030-26593-9.

Keller, M. (2006). *Assessment of Gold Recovery Processes in Bangalore, India and Evaluation of an Alternative Recycling Path for Printed Wiring Boards*. Institute for Spatial and Landscape Planning, Regional Resource Management; Empa, Technology and Society Lab.

Khaliq, A., Rhamdhani, M.A., Brooks, G., and Masood, S. (2014). Metal extraction processes for electronic waste and existing industrial routes: a review and Australian perspective. *Resources* 3 (1): 152–179. https://doi.org/10.3390/resources3010152.

Kidee, P., Naidu, R., and Wong, M.H. (2013). Electronic waste management approaches: an overview. *Waste Management* 33: 1237–1250. https://doi.org/10.1016/j.wasman.2013.01.006.

Kirchherr, J., Reike, D., and Hekkert, M. (2017). Conceptualizing the circular economy: an analysis of 114 definitions. *Resources, Conservation and Recycling* 127: 221–232. https://doi.org/10.1016/j.resconrec.2017.09.005.

Krikke, J. (2008). Recycling e-waste: the sky is the limit. *IT Professional* 10 (1): 50–55. https://doi.org/10.1109/MITP.2008.16.

Kumar, A., Holuszko, M., and Espinosa, D.C.R. (2017). E-waste: an overview on generation, collection, legislation and recycling practices. *Resources, Conservation and Recycling* 122: 32–42. https://doi.org/10.1016/j.resconrec.2017.01.018.

Kyere, V.N., Greve, K., Atiemo, S.M. et al. (2018). Contamination and health risk assessment of exposure to heavy metals in soils from informal E-waste recycling site in Ghana. *Emerging Science Journal* 2 (6): 428–436. https://doi.org/10.28991/esj-2018-01162.

Le, L., Tang, J., Ryan, D., and Valix, M. (2006). Bioleaching nickel laterite ores using multi metal tolerant *Aspergillus foetidus* organism. *Minerals Engineering* 19: 1259–1265. https://doi.org/10.1016/j.mineng.2006.02.006.

Leader, A., Gaustad, G., and Babbitt, C. (2019). The effect of critical material prices on the competitiveness of clean energy technologies. *Materials for Renewable and Sustainable Energy* 8 (2): 1–17. https://doi.org/10.1007/s40243-019-0146-z.

Liu, R., Li, J., and Ge, Z. (2016). Review on *Chromobacterium violaceum* for gold bioleaching from e-waste. *Procedia Environmental Sciences* 31: 947–953. https://doi.org/10.1016/j.proenv.2016.02.119.

Luda, M.P. (2011). Recycling of printed circuit boards. In: *Integrated Waste Management*, vol. II. IntechOpen. ISBN: 978-953-307-447-4.

Mdlovu, N.V., Chiang, C.L., Lin, K.S., and Jeng, R.C. (2018). Recycling copper nanoparticles from printed circuit board waste etchants via a microemulsion process. *Journal of Cleaner Production* 185: 781–796. https://doi.org/10.1016/j.jclepro.2018.03.087.

MeitY (2021). Circular economy. In: *Electronics and Electrical Sector*. Ministry of Electronic and Information Technology www.meity.gov.in/writereaddata/files/Circular_Economy_EEE-MeitY-May2021-ver7.pdf (accessed 21 June 2022).

Murugappan, R.M. and Karthikeyan, M. (2021). Microbe-assisted management and recovery of heavy metals from electronic wastes. *Environmental Management of Waste Electrical and Electronic Equipment* 65–88. https://doi.org/10.1016/B978-0-12-822474-8.00004-0.

Nakajima, K., Takeda, O., Miki, T. et al. (2010). Thermodynamic analysis of contamination by alloying elements in aluminium recycling. *Environmental Science and Technology* 44: 5594–5600. https://doi.org/10.1021/es3043559.

Naor, M., Bernardes, E.S., Druehl, C., and Shiftan, Y. (2015). Overcoming barriers to adoption of environmentally-friendly innovations through design and strategy: learning from the failure of an electric vehicle infrastructure firm. *International Journal of Operations and Production Management* 35 (1): 26–59. https://doi.org/10.1108/IJOPM-06-2012-0220.

Needhidasan, S., Samuel, M., and Chidambaram, R. (2014). Electronic waste – an emerging threat to the environment of urban India. *Journal of Environmental Health Science and Engineering* 12 (1): 1–9. https://doi.org/10.1186/2052-336X-12-36.

NERC (2022). *Electronic Recyclers International*. Northeast Recycling Council https://nerc.org/advisory-members/member-spotlight/2015/03/electronic-recycling-international-(eri). (accessed 22 August 2022).

Nithya, R., Sivasankari, C., Thirunavukkarasu, A., and Selvasembian, R. (2018). Novel adsorbent prepared from bio-hydrometallurgical leachate from waste printed circuit board used for the removal of methylene blue from aqueous solution. *Microchemical Journal* 142: 321–328. https://doi.org/10.1016/j.microc.2018.07.009.

Nnorom, I.C. and Osibanjo, O. (2008). Overview of electronic waste (e-waste) management practices and legislations, and their poor applications in the developing countries. *Resources, Conservation and Recycling* 52 (6): 843–858. https://doi.org/10.1016/j.resconrec.2008.01.004.

Nzihou, A. (2010). Toward the valorization of waste and biomass. *Waste and Biomass Valorization* 1 (1): 3–7. https://doi.org/10.1007/s12649-010-9014-x.

Othman, N., Basri, N.E.A., Yunus, M.N.M., and Sidek, L.M. (2008). Determination of physical and chemical characteristics of electronic plastic waste (Ep-Waste) resin using proximate and ultimate analysis method. *International Conference on Construction and Building Technology* 16: 169–180.

Ottoni, M., Dias, P., and Xavier, L.H. (2020). A circular approach to the e-waste valorization through urban mining in Rio de Janeiro, Brazil. *Journal of Cleaner Production* 261: 120990. https://doi.org/10.1016/j.jclepro.2020.120990.

Pant, D., Joshi, D., Upreti, M.K., and Kotnala, R.K. (2012). Chemical and biological extraction of metals present in e-waste: a hybrid technology. *Waste Management* 32: 979–990. https://doi.org/10.1016/j.wasman.2011.12.002.

Priya, A. and Hait, S. (2017a). Comparative assessment of metallurgical recovery of metals from electronic waste with special emphasis on bioleaching. *Environmental Science and Pollution Research* 24 (8): 6989–7008. https://doi.org/10.1007/s11356-016-8313-6.

Priya, A. and Hait, S. (2017b). Qualitative and quantitative metals liberation assessment for characterization of various waste printed circuit boards for recycling. *Environmental Science and Pollution Research* 24 (35): 27445–27456. https://doi.org/10.1007/s11356-017-0351-1.

Priya, A. and Hait, S. (2018a). Comprehensive characterization of printed circuit boards of various end-of-life electrical and electronic equipment for beneficiation investigation. *Waste Management* 75: 103–123. https://doi.org/10.1016/j.wasman.2018.02.014.

Priya, A. and Hait, S. (2018b). Extraction of metals from high grade waste printed circuit board by conventional and hybrid bioleaching using *Acidithiobacillus ferrooxidans*. *Hydrometallurgy* 177: 132–139. https://doi.org/10.1016/j.hydromet.2018.03.005.

Ramanayaka, S., Keerthanan, S., and Vithanage, M. (2020). Urban mining of E-waste: treasure hunting for precious nanometals. In: *Handbook of Electronic Waste Management*, 19–54. Butterworth-Heinemann. https://doi.org/10.1016/B978-0-12-817030-4.00023-1.

Ramaswamy, K., Radha, V., Malathi, M. et al. (2017). Degradation of organic pollutants by Ag, Cu and Sn doped waste non-metallic printed circuit boards. *Waste Management* 60: 629–635. https://doi.org/10.1016/j.wasman.2016.09.026.

Reike, D., Vermeulen, W.J., and Witjes, S. (2018). The circular economy: new or refurbished as CE 3.0? - exploring controversies in the conceptualization of the circular economy through a focus on history and resource value retention options. *Resources, Conservation and Recycling* 135: 246–264. https://doi.org/10.1016/j.resconrec.2017.08.027.

Sadegh-Safarzadeh, M., Bafghi, M.S., Moradkhani, D., and Ojaghi, I.M. (2007). A review on hydrometallurgical extraction and recovery of cadmium from various resources. *Minerals Engineering* 20 (3): 211–220. https://doi.org/10.1016/j.mineng.2006.07.001.

Saini, A.K. and Taneja, A. (2012). Managing E-waste in India – a review. *International Journal of Applied Engineering Research* 7 (11): 1–6.

Serranti, S., Di Maio, F., Rem, P., and Bonifazi, G. (2012). Innovative technologies and processing architectures in Urban Mining: two key issues to ensure secondary raw materials supply. In: *Urban Mining: A Global Cycle Approach to Resource Recovery from Solid Waste*, 23–47. Padova: CISA Publisher.

Shokri, A., Pahlevani, F., Cole, I., and Sahajwalla, V. (2017a). Selective thermal transformation of old computer printed circuit boards to Cu-Sn based alloy. *Journal of Environmental Management* 199: 7–12. https://doi.org/10.1016/j.jenvman.2017.05.028.

Shokri, A., Pahlevani, F., Levick, K. et al. (2017b). Synthesis of copper-tin nanoparticles from old computer printed circuit boards. *Journal of Cleaner Production* 142: 2586–2592. https://doi.org/10.1016/j.jclepro.2016.11.017.

Silvas, F.P.C., Correa, M.M.J., Caldas, M.P.K. et al. (2015). Printed circuit board recycling: physical processing and copper extraction by selective leaching. *Waste Management* 46: 503–510. https://doi.org/10.1016/j.wasman.2015.08.030.

Singh, N., Duan, H., Yin, F. et al. (2018). Characterizing the materials composition and recovery potential from waste mobile phones: a comparative evaluation of cellular and smart phones. *ACS Sustainable Chemistry and Engineering* 6 (10): 13016–13024. https://doi.org/10.1021/acssuschemeng.8b02516.

Souza, R.G., Clímaco, J.C.N., Sant'Anna, A.P. et al. (2016). Sustainability assessment and prioritization of e-waste management options in Brazil. *Waste Management* 57: 46–56. https://doi.org/10.1016/j.wasman.2016.01.034.

Straits Research (2020). Top 10 E-waste Management Companies in the World. https://straitsresearch.com/blog/top-10-e-waste-management-companies-in-the-world. (accessed 20 August 2022).

Straits Research (2022). E-waste Management Market: Information by Types (Trashed and Recycled), Source Type (Household Appliances, Entertainment, IT), Application (Disposal) and Regional Outlook—Forecast till 2026. https://straitsresearch.com/report/e-waste-management-market. (accessed 21 July 2022).

Sun, J., Wang, W., Liu, Z., and Ma, C. (2011). Recycling of waste printed circuit boards by microwave-induced pyrolysis and featured mechanical processing. *Industrial and Engineering Chemistry Research* 50: 11763–11769. https://doi.org/10.1021/ie2013407.

Szałatkiewicz, J. (2014). Metals content in printed circuit board waste. *Polish Journal of Environmental Studies* 23 (6): 2365–2369.

Tan, H. T. S. (2007). *Recycling of Used Electronics in Singapore.*

Tanskanen, P. (2013). Management and recycling of electronic waste. *Acta Materialia* 61 (3): 1001–1011. https://doi.org/10.1016/j.actamat.2012.11.005.

Tatariants, M., Yousef, S., Skapas, M. et al. (2018). Industrial technology for mass production of SnO_2 nanoparticles and PbO_2 microcube/microcross structures from electronic waste. *Journal of Cleaner Production* 203: 498–510. https://doi.org/10.1016/j.jclepro.2018.08.283.

TOI (2021). *Tokyo 2020: Olympic Medals Made of Old Electronic Devices*. Times of India https://timesofindia.indiatimes.com/life-style/spotlight/tokyo-2020-olympic-medals-made-of-old-electronic-devices/articleshow/84884339.cms, (accessed 7 July 2022).

Toto, D. (2021). ERI improves recovery with robotics. In Recycling Today. (accessed 20 July 2022). https://www.recyclingtoday.com/article/electronics-recycler-improes-recovery-robotics-ai

Tuncuk, A., Akcil, A., Yazici, E.Y., and Devici, H. (2012). Aqueous metal recovery techniques from e scrape: hydrometallurgy in recycling. *Minerals Engineering* 25: 28–37. https://doi.org/10.1016/j.mineng.2011.09.019.

UNU, (2018). *New EU "Urban Mining" Tools Map Valuable Resources in E-waste, Scrap Vehicles, Mine Waste*. United Nations University. https://unu.edu/media-relations/releases/new-euurban-mining-tools-prosum.html

Veit, H.M. (2015). Processing techniques. In: *Electronic Waste*, 13–17. Cham: Springer https://doi.org/10.1007/978-3-319-15714-6_3.

Veit, H.M., Pereira, C.C., and Bernardes, A.M. (2002). Using mechanical processing in recycling printed wiring boards. *Journal of Management* 54 (6): 45–47. https://doi.org/10.1007/BF02701850.

Wang, J., Bai, J., Xu, J., and Liang, B. (2009). Bioleaching of metals from printed wire boards by *Acidithiobacillus ferrooxidans* and *Acidithiobacillus thiooxidans* and their mixture. *Journal of Hazardous Material* 172: 1100–1105. https://doi.org/10.1016/j.jhazmat.2009.07.102.

Wang, X., Lu, X., and Zhang, S. (2013). Study on the-waste liquid crystal display treatment: focus on the resource recovery. *Journal of Hazardous Material* 244–245: 342–347. https://doi.org/10.1016/j.jhazmat.2012.11.059.

Wang, C., Cao, Y., and Wang, H. (2019). Copper-based catalyst from waste printed circuit boards for effective Fenton-like discoloration of Rhodamine B at neutral pH. *Chemosphere* 230 (2019): 278–285. https://doi.org/10.1016/j.chemosphere.2019.05.068.

Widmer, R., Oswald-Krapf, H., Sinha-Khetriwal, D. et al. (2005). Global perspectives on e-waste. *Environmental Impact Assessment Review* 25 (5): 436–458. https://doi.org/10.1016/j.eiar.2005.04.001.

Williams, P.T. (2010). Valorization of printed circuit boards from waste electrical and electronic equipment by pyrolysis. *Waste and Biomass Valorization* 1 (1): 107–120. https://doi.org/10.1007/s12649-009-9003-0.

Wills, B.A. (1988). *Mineral Processing Technology*, 4ee. Great Britain: Pergamon Press.

Xin, B., Zhang, X., Xia, Y. et al. (2009). Bioleaching mechanism of Co and Li from spent lithium-ion battery by the mixed culture of acidophilic sulfur-oxidizing and iron-oxidising bacteria. *Bioresource Technology* 100: 6163–6169. https://doi.org/10.1016/j.biortech.2009.06.086.

Yken, V.J., Boxall, N.J., Cheng, K.Y. et al. (2021). E-waste recycling and resource recovery: a review on technologies, barriers and enablers with a focus on Oceania. *Metals* 11 (8): 1313. https://doi.org/10.3390/met11081313.

Zeng, X., Wang, F., Li, J., and Gong, R. (2017). A simplified method to evaluate the recycling potential of e-waste. *Journal of Cleaner Production* 168: 1518–1524. https://doi.org/10.1016/j.jclepro.2017.06.232.

Zhang, S. and Forssberg, E. (1997). Mechanical separation-oriented characterization of electronic scrap. *Resource Conservation Recycling* 21: 247–269. https://doi.org/10.1016/S0921-3449(97)00039-6.

Zhang, J., Wu, X., Guo, H. et al. (2021). Pollution of plastic debris and halogenated flame retardants (HFRs) in soil from an abandoned e-waste recycling site: do plastics contribute to (HFRs) in soil? *Journal of Hazardous Material* 410: 124649. https://doi.org/10.1016/j.jhazmat.2020.124649.

Zhou, Y. and Qiu, K. (2010). A new technology for recycling materials from waste printed circuit boards. *Journal of Hazardous Material* 175: 823–828. https://doi.org/10.1016/j.jhazmat.2009.10.083.

11

Hydrometallurgical Processing of E-waste and Metal Recovery

Amilton Barbosa Botelho Junior[1], *Ummul Khair Sultana*[2], *and James Vaughan*[2]

[1] *Department of Chemical Engineering, Polytechnic School, University of Sao Paulo, Sao Paulo, Brazil*
[2] *School of Chemical Engineering, Faculty of Engineering, Architecture and Information Technology, University of Queensland, Brisbane, QLD, Australia*

Abstract

Demand for electronic devices has been rapidly increasing, and the types of devices purchased and disposed of are continually evolving. It results in increased need for primary extractive processes to supply the raw materials required. Increased e-waste recycling rates and efficiency of recycling processes are required to move toward a circular economy. Hydrometallurgical processing is especially effective for selective metal separations, and the process can more readily be deployed on a small scale. Hydrometallurgical extraction of the metals can be less greenhouse gas intensive than pyrometallurgical processes where high heat energy inputs or carbon-based reductants are used. Hydrometallurgy is also required to obtain high-purity products. This book chapter presents an overview of select hydrometallurgical operations for recycling of e-waste such as batteries, printed circuit boards (PCBs), screens, and lamps. Novel technologies are being explored as use of ionic liquids (ILs), deep eutectic solvents (DESs), supercritical fluids, and nanohydrometallurgy is discussed.

Keywords *hydrometallurgy; leaching; battery; PCBs; LCD; LED; nanotechnology*

11.1 Introduction

The increase in demand for sustainable technologies to supply different industrial sectors caused the search of different metals from primary sources. The main technologies are: Li-ion batteries, fuel cells, wind energy, photovoltaic panels, robotics, 3D printers, and digital technologies. Due to the growing demand for these technologies, the key materials used for manufacturing were classified as

Management of Electronic Waste: Resource Recovery, Technology and Regulation, First Edition. Edited by Anshu Priya.

critical (Bobba et al. 2020). For instance, the European Union listed, in 2020, 30 materials as critical due to the risk of supply interruption in a medium and long term (Martins et al. 2021b). The United States of America (U.S. Department of Energy) has a list including 35 materials (U.S. Department of Energy 2010), and the list of Brazil includes 25 materials ("strategic materials") (MCTIC 2018). In all these lists are included the rare earth elements (REEs) Mg, Nb, Ge, Sc, Sr, Co, PGMs (platinum k-group metals), In, V, Li, W, Ti, Ga, Hf, Si, Mn, Cr, Zr, Te, Ni, and Cu.

In addition, these critical materials/metals are crucial for printed circuit boards (PCBs) (Rao et al. 2022), permanent magnetics (Fe-Nd-B magnets), fluorescent (Botelho Junior et al. 2021a), and LEDs/OLED lamps, and LCD/LED/OLEDs screens (Cenci et al. 2021; Oliveira et al. 2020). The literature reports different recycling routes to obtain these critical metals. Among those, there are three main processes used on an industrial scale and evaluated in academia: pyrometallurgy, hydrometallurgy, and biohydrometallurgy (Kumar et al. 2017). Pyrometallurgical processing involves the use of high temperatures for organic compound removal and to form alloy and slag phases (Assefi et al. 2020; Ma 2019). For instance, the recycling of Li-ion batteries starts at 150–500 °C for removal of electrolytes and organic solvents, and the alloy phase is obtained at 1400–1700 °C. Li is concentrated in slag phase. Both alloy and slag are composed of different metals, and the production of high-purity products is only possible by hydrometallurgical processing (Assefi et al. 2020).

Hydrometallurgy is the process of extracting metals from metal-bearing material into aqueous solution with subsequent solution purification, metal separation, metal concentration upgrading, and metal product recovery. Figure 11.1 shows a generic hydrometallurgical process flowsheet with a range of metal-bearing feed types entering the leaching or anodic dissolution steps where metals are dissolved into solution.

The first step is leaching, which aims at the solubilization of the material; the goal is the leaching of target metals or contaminants present in raw material (mineral or e-waste, for instance). Anodic dissolution is a part of metal electrorefining where impure metals are upgraded, for example, copper anodes are transformed into copper cathodes and impurities or other valuable metals are collected in the anode slimes. More commonly, the materials are leached into solutions in reactors such as stirred tanks, autoclaves, or heaps. The leaching chemistry can vary widely depending on the feed composition and target metal(s). The most common leaching solution is acid, but alkaline solutions can be employed, for example, in the Bayer process for refining bauxite to alumina (Havlík 2008). Once the metals are in solution, separation and purification methods are further applied to separate valuable metals from impurities and then obtain high-purity products, which can be, for instance, metallic, salts, or oxides. Solution purification can be achieved by precipitation (impurities or intermediate-value metal concentrates),

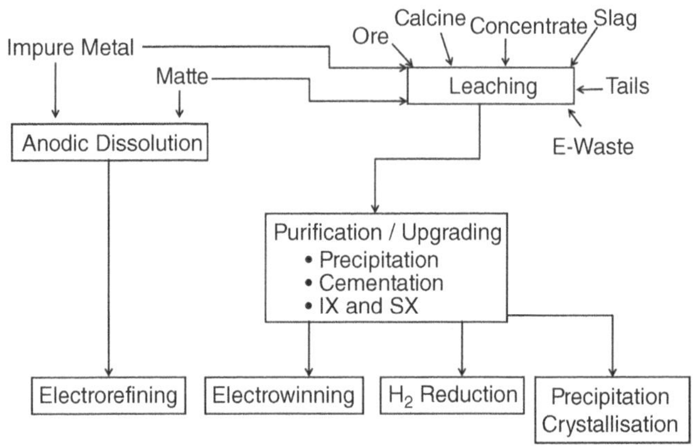

Figure 11.1 Hydrometallurgical process flowsheet.

cementation (an electrochemical process where a reactive metal is oxidized and dissolves while noble metals in solution are reduced to the solid metallic form), liquid–liquid solvent extraction (SX), or ion exchange resin (IX). Finally, once a sufficiently concentrated a pure solution is obtained, metal products are typically recovered by electrowinning, hydrogen reduction, precipitation, or crystallization (Gupta and Mukherjee 1990; Vignes 2013).

Hydrometallurgical route is used for extraction of nonferrous metals from primary sources, such as Ni, Co, platinum-group metals (PGMs) (Crundwell et al. 2011), and rare earth elements (REEs) (Qi 2018). There are several advantages of hydrometallurgical processing in comparison to pyrometallurgy: less energy intensive, lower material losses, obtaining high-purity products, and less carbon footprint. On the other hand, the process requires vast amounts of reagents and strictly controlled process/reaction (Velázquez-Martínez et al. 2019).

In the case of e-waste, the growing demand requires proper treatment than landfill, since these residues are composed of valuable and critical metals (Lai et al. 2021). The e-waste generation increased from 55.5 Mt in 2020 to 57.4 Mt in 2021 (Forti et al. 2020). Its generation rate is threefold higher than all other types of residues (Kumar et al. 2017). In comparison, the waste from construction sector is the most voluminously generated in the world (Zhang et al. 2020). Indeed, the COVID-19 pandemic affected wastes generation worldwide (Ibn-Mohammed et al. 2021; Naughton 2020).

There are several hydrometallurgical techniques to recover metals from e-waste, which mostly comprise leaching, separation, and purification techniques (Havlík 2008). The choice of reagents and techniques depends on the type of e-waste, its composition, and the target elements. For this reason, the present chapter aims to provide the literature review of recycling technologies of e-waste.

The characterization (Chapter 2) is the first step to evaluate the potential metals and hydrometallurgical approach. Different leaching strategies (Chapter 3) may be performed depending on target metals, which include acid, inorganic (3.1.1) or organic (3.1.2), alkaline (3.2), cyanide (3.3), thiosulfate, and thiourea (3.4) leaching. In this chapter, technologies for hydrometallurgical processing without thermal treatment will be discussed.

After dissolution, target metals and impurities are present in the aqueous media. For this reason, separation and purification methods are used, such as precipitation (4.1), solvent extraction (4.2), and ion exchange resins (4.3), where the target metal is purified and concentrated to be obtained in electrodeposition step (4.4), for instance.

In addition, there are several emerging techniques to improve extraction and separation efficiencies, decrease the consumption of reagents and energy, and use of less harmful and eco-friendly compounds. Among those, ionic liquids (ILs) (5.1), deep eutectic solvents (DESs) (5.2), supercritical fluids (5.3), and nanohydrometallurgy (5.4) have been explored for e-waste recycling. Further chapters will discuss each topic of hydrometallurgical technique for e-waste. At the process research and development stage, there are several emerging techniques being explored with the aim to improve some aspects of extraction and separation efficiencies, decrease the consumption of reagents and energy, and to improve process safety and environmental outcomes. Among those, ILs (5.1), DESs (5.2), supercritical fluids (5.3), and nanohydrometallurgy (5.4) have been explored for e-waste recycling. Further chapters will discuss each topic of hydrometallurgical technique for e-waste.

11.2 Characterization

The first step to designing a recycling process is to undertake a detailed characterization of the proposed e-waste feed. Each feed sample has singular composition; also, even similar types of feed may contain differences. Table 11.1 depicts the elementary composition of different e-waste found in the literature. Even PCBs have different compositions, for example Dutta et al. (2018) collected and characterized PCBs from personal computers before acid leaching, and determined the material to be 1.0% Pb, 19% of Cu, 6.9% Fe, 2.2% Sn, and 0.26% Ni (Dutta et al. 2018). However, in 2002, a European directive (2002/95/CE RoHS) restricted the use of Pb, Cd, and Cr, for instance, in electronic devices. For this reason, the PCBs studied by Martins et al. (2021a) did not contain Pb and were classified as *lead-free*. The Pb was replaced by Ag or Sn in the soldering, which explains why PCBs from Dutta et al. (2018) analysis may not contain Ag. According to data presented by Martins et al. (2021a), the Ag content in the PCB memory chips and mother board is up to 0.04%. Despite that, Pb could continue to be expected in recycled material feed

Table 11.1 Literature review of the chemical composition of the main elements in e-waste (in mass percentage).

%wt	PCBs (Dutta et al. 2018)	PCB-mother[a] (Martins et al. 2021a)	PCB-memory[a] (Martins et al. 2021a)	LEDs[b] (Cenci et al. 2020b)	LEDs modules[b] (Cenci et al. 2020b)	Fluorescent lamps (Botelho Junior et al. 2021a; Innocenzi et al. 2018)	Li-ion battery cell phone – anode (Takahashi et al. 2020)	Li-ion battery cell phone – cathode (Takahashi et al. 2020)	Li-ion battery – NMC type[c] (Xuan et al. 2021)	Li-ion battery – LFP type (Kumar et al. 2020)	Fe-Nd-B magnets (München et al. 2018)	Photovoltaic panels[d] (Dias et al. 2016)	LCD screens[e] (Gabriel et al. 2020)
Ag	–	0.045	0.053	0.482–0.718	0.002–0.004	–	–	–	–	–	–	1.4	–
Al	–	2.66	1.403	0.194–0.252	20.775–54.728	0–7.01	0.08	0.65	–	0.9	0.47	–	11.8
As	–	–	–	–	–	–	–	–	–	–	–	–	0.98
Au	–	0.002	0.069	0.054–0.054	0.028–0.030	–	–	–	–	–	–	–	–
Ba	–	–	–	–	–	0–0.41	–	–	–	–	0.9	–	–
C	–	–	–	–	–	–	96.8	12.5	–	–	–	–	–
Ca	–	–	–	–	–	8.13–29.49	–	–	–	–	–	–	6.76
Ce	–	–	–	0.009	–	0.23–0.49	–	–	–	–	–	–	–
Co	–	–	–	–	–	0–0.05	0.15	54.6	5.65–19.2	–	0.87	–	–
Cr	–	–	–	–	–	–	–	–	–	–	–	0.14	–
Cu	19.34	19.65	13.83	54.65–59.03	2.963–5.006	–	0.05	0	–	0.7	0.26	0.04	–
Dy	–	–	–	–	–	–	–	–	–	–	0.98	–	–
Eu	–	–	–	–	–	0–0.44	–	–	–	–	–	–	–
Fe	6.89	8.34	1.47	0.480–0.741	0.13–0.38	0–0.58	0.08	0.16	–	30.1	65.31	–	–
Ga	–	–	–	0.084–0.381	–	–	–	–	–	–	–	–	–
Gd	–	–	-	–	–	0–0.25	–	–	–	–	–	–	–
In	–	–	–	–	–	–	–	–	–	–	–	–	0.036
K	–	–	–	–	–	0–0.08	–	–	–	–	–	–	0.68

(Continued)

Table 11.1 (Continued)

%wt	PCBs (Dutta et al. 2018)	PCB-mother[a] (Martins et al. 2021a)	PCB-memory[a] (Martins et al. 2021a)	LEDs[b] (Cenci et al. 2020b)	LEDs modules[b] (Cenci et al. 2020b)	Fluorescent lamps (Botelho Junior et al. 2021a; Innocenzi et al. 2018)	Li-ion battery cell phone – anode (Takahashi et al. 2020)	Li-ion battery cell phone – cathode (Takahashi et al. 2020)	Li-ion battery – NMC type[c] (Xuan et al. 2021)	Li-ion battery – LFP type (Kumar et al. 2020)	Fe-Nd-B magnets (München et al. 2018)	Photovoltaic panels[d] (Dias et al. 2016)	LCD screens[e] (Gabriel et al. 2020)
La	–	–	–	–	–	0.07–0.38	–	–	–	–	–	–	–
Li	–	–	–	–	–	–	2.2	7.4	7.19–7.33	3.7	–	–	–
Mg	–	–	–	–	–	0–17.58	–	–	–	–	–	–	–
Mn	–	–	–	–	–	0–0.45	0.01	0.01	5.56–18.3	–	–	–	–
Nd	–	–	–	–	–	–	–	–	–	–	23.95	–	–
Ni	0.26	0.27	0.32	0.082–0.183	0.001– 0.038	–	–	–	19.7–48.3	–	2.86	–	–
P	–	–	–	–	–	–	–	–	–	16.8	–	–	–
Pb	1.01	–	–	0–1.025	0.001–0.011	–	–	–	–	–	–	0.08	–
Pr	–	–	–	–	–	–	–	–	–	–	4.2	–	–
Si	–	–	–	–	–	12.91–24.65	–	–	–	–	–	98.2	67.48
Sn	2.16	1.99	1.7	6.81–10.96	1.25–5.83	–	–	–	–	–	–	–	0.013
Tb	–	–	–	–	–	0–0.27	–	–	–	–	0.19	–	–
Ti	–	–	–	0.019–0.058	0.025–0.079	–	–	–	–	–	–	–	–
Y	–	–	–	0.290–0.683	–	2.14–11.21	–	–	–	–	–	–	–
Zn	0.11	2.55	0.043	–	–	–	–	–	–	–	–	–	–

a) Lead-free.
b) Tubular and bulb lamps.
c) NMC 811, 622, 532 and 111 activated materials.
d) Semiconductor extracted from the modules.
e) Percentage for oxides.

due to older PCBs or those continuing to be fabricated outside the jurisdiction of this directive.

The Ni content in PCBs studied by Dutta et al. (2018) and Martins et al. (2021a) was similar (0.2–0.3%). However, different types of PCBs might also contain contrasting metal contents. For instance, Cu content of PCBs analyzed by Dutta et al. (2018) and Martins et al. (2021a) was similar in the case of mother boards (up to 19%) but lower for the memory board (14%). The composition was similar for Fe and Sn. As observed, despite the content differences, PCBs *lead-free* may be processed together in a unique industrial route.

Commonly, residues from lighting systems are also classified as e-waste, including LED and fluorescent lamps. Cenci et al. (2020a) characterized three parts of LED lamps: the PCBs (used for voltage conversion and stabilization/regulation of the electric current), LED module (where the LEDs are fixed), and LEDs. The configuration is the same in tubular and bulb lamps. LEDs concentrate the most valuable metals, such as Ag, Au, Cu, Ni, Ti, Ga, and REEs (mainly Y). Moreover, there is the presence of Pb from soldering (Cenci et al. 2020a). Some manufacturers have replaced Pb with Sn, Bi, Cu–Sn, or Sn–Ag–Cu (Cheng et al. 2017). Fluorescent lamps also contain REEs, such as Y, La, Ce, Eu, and Gd (Botelho Junior et al. 2021a; Innocenzi et al. 2018). Similar compositions and physical characteristics among LED and fluorescent lamps may justify a recycling route for both materials.

There are several types of Li-ion batteries on the market for both electronic devices and electric or hybrid vehicles. These batteries are structured in case (external structure), cathode, anode, electrolyte (responsible for Li-ion transport), and polymeric separators (avoiding short circuit). The main difference among these batteries is the cathode composition, which may be mixed oxides ($LiCoO_2$ – LCO, $LiMnO_2$ or $LiMn_2O_4$ – LMO, $LiNi_xCo_yMn_zO_2$ – NMC, $LiNi_{0.8}Co_{0.15}Al_{0.05}O_2$ – NCA), or iron phosphate ($LiFePO_4$ – LFP) (Velázquez-Martínez et al. 2019; Botelho Junior et al. 2021c). Anodes are composed of carbon (graphite) in most Li-ion batteries found in the market (Velázquez-Martínez et al. 2019; Yi et al. 2021). In small devices, such as notebooks and cellphones, cathode materials commonly used are $LiCoO_2$, while Al and Cu metallic foils are bonded to cathode and anode as electron collectors (Takahashi et al. 2020). NMC batteries have different configurations varying the content of Ni, Mn, and Co (811, 622, 532, and 111, for instance), which explain the large range of metal content (Table 11.1) (Xuan et al. 2021).

Nd-Fe-B magnets are widely used in electronic devices and contain many REEs, such as Dy (0.98%), Nd (23.95%), Pr (4.2%), and Tb (0.19%), as characterized by München et al. (2018) from post-consumer HDDs, desktops, and notebooks. Common elements such as Al, Ni, and Fe are also found in these magnets (München et al. 2018).

An increasingly voluminous e-waste is the photovoltaic panel for energy generation, which contains up to 98% of crystalline Si, 1.4% Ag, and 0.08% Pb

(Dias et al. 2016). Finally, LCD screens uses has been declined due to the replacement by LED, and the presence of In (0.036%) makes the recycling crucial for global supply (Gabriel et al. 2020), since the element is obtained almost exclusively as a by-product from select primary Zn production facilities (Werner et al. 2015; Jowitt et al. 2018). For this reason, the recycling is increasingly important to satisfy new supply of In used in solar cells and indium-tin oxide (ITO) coatings (Werner et al. 2015).

11.3 Leaching Techniques

Because of the wide range of post-consumer electronic waste, the approach to leaching the metals into solution is similarly varied. Before leaching, the feed material is usually reduced in size and processed through physical separations where possible. Equipment such as hammer, knife, ceramic balls mills, jaw crusher, and shredder are commonly used for e-waste processing (Cenci et al. 2021; Dias et al. 2021). Physical separations can then be achieved based on material properties such as size, density, magnetic susceptibility, or surface properties. Generally, it is less costly to effect physical separation of materials compared with chemical processing. The resulting metal-bearing material concentrates are then sent to the hydrometallurgical (chemical) processing. The leaching is the first step of the hydrometallurgical process; acknowledging that many processes have pyrometallurgical (high-temperature chemical processing) step initially. In the present chapter, the direct leaching of e-waste will be emphasized (Havlík 2008; Kaya 2019). The step aimed at the solubilization of target metals present in e-waste or impurities to concentrate the valuable metals in solid phase (Crundwell et al. 2011; Addison and Roe 2018).

There are different types of leaching agents, which depend on target metals, impurities, physical characteristics of e-wastes, and further separation of metals (Chapter 4). Most common leaching agents are inorganic acids, such as H_2SO_4, HNO_3, HCl, and H_3PO_4, since most metals are leached in acid conditions (Pourbaix 1974). These reagents have widely been used in mining industry for leaching of minerals, as occurs for Ni and Cu processing (Crundwell et al. 2011). Mixtures of acids have been explored, but they impact directly separation and purification techniques due to the presence of different acids and corrosion issues (Espinosa and Mansur 2019). However, several issues are intrinsic to their use, such as corrosion and emission of SO_2, NO_x, and Cl_2 (Tunsu et al. 2016). For this reason, organic acids have been studied to replace inorganics, such as citric, DL-malic, ascorbic, lactic, butyric, acetic, propionic, and tartaric acids, and glycine (amino acid) (Botelho Junior et al. 2021c).

Alkaline leaching has been studied as well, mainly for Al leaching in recycling of batteries. In the cathode, metallic Al foils are used as electron collectors (as discussed in Chapter 2), and they may be leached in alkaline solutions (pH above 10–12) leaving most of other metals in the solid phase (Velázquez-Martínez et al. 2019). As most e-waste, including PCBs, contains Au in its composition, cyanide is still used for leaching despite the environmental and human hazards (Zhang et al. 2022b). Thiosulfate and thiourea leaching have been explored replacing cyanide for Ag and Au leaching from e-waste (Feng and Van Deventer 2011; Jing-ying et al. 2012).

Leaching step may be carried out in different configurations, with stirring and autoclave types being the most common for e-waste, and the variables solid–liquid ratio (S/L ratio), time, temperature, leaching agent concentration, pressure, and particle size morphology and size are crucial for the process (Kaya 2019). The next chapters will discuss in depth each leaching technique for e-waste.

11.3.1 Acid Leaching

Acid leaching is the most common approach to solubilize metals from e-waste, which is widely carried out by inorganics (H_2SO_4, HNO_3, HCl, and H_3PO_4). In addition, the literature reports the use of organic acids as an alternative. Further chapters will present examples from the literature for different types of e-waste.

11.3.1.1 Inorganic Acids

Sulfuric acid is one of the least expensive acids that can be produced from elemental sulfur, a by-product of oil and gas refining. Sulfuric acid is also a by-product of primary metal production from sulfide ore, where in the high-temperature smelting and converting steps, sulfur dioxide must be sequestered due to environmental considerations. Sulfate solutions resulting from sulfuric acid leaching processes can largely be treated with inexpensive limestone or lime, resulting in a relatively benign gypsum-containing waste solid. For these reasons, sulfuric acid is generally the first choice for acid leaching processes.

In recycling e-waste, sulfuric acid has some shortcomings. In particular, many precious metals are not readily soluble in sulfuric acid, for example gold, platinum, and palladium. In order to solubilize the precious metals, hydrochloric acid can be used, or chlorides can be included in the leach solution to enable the formation of soluble metal chloro-complexes. Hydrochloric acid tends to be more expensive than sulfuric acid. Additional considerations when using chlorides are how to manage the chloride-containing waste, and acid chloride solutions may also require additional considerations regarding equipment materials and construction due to enhanced corrosion.

Nitric acid is generally even more expensive than sulfuric acid and hydrochloric acid, although it can potentially be more easily recycled through generation of NOx gases and subsequent oxy-hydrolysis. Nitric acid can play a role in refining precious metals where selective dissolution of silver is required.

11.3.1.2 Organic Acids

Organic acid leaching can be an eco-friendly and cost-effective method over other pyro and hydrometallurgical processes for recovering PMs from the electronic waste. Several organic acids such as ascorbic acid (Lie and Liu 2021; Li et al. 2012), citric acids (Li et al. 2010b), malic acid (Li et al. 2010a), aspartic acid (Li et al. 2013), and acetic acid (Golmohammadzadeh et al. 2017) have been investigated for PMs recovery from PCBs, spent batteries, and mobile devices. The benefit of using these organic acids is that they are easily degradable, and no toxic gas is evolved during the leaching process. In general, leaching operation is carried out after mechanical separation of metals and nonmetals and pulverization. However, the major disadvantage of physical separation is the significant amount of valuable metal loss as metal liberation from the plastics might be insufficient as well as inefficient recovery of metals might happen due to the generation of fine particles (Tuncuk et al. 2012).

Sodium hydroxide pre-treatment can be an alternative route for avoiding the extensive mechanical-physical pre-treatment and enabling the lixiviant to react with the metals present in the PCBs (Jadhav and Hocheng 2015). Citric acid leaching, where hydrogen peroxide was a reducing agent, was found to be a successful lixiviant for the complete recovery of Li along with 90% recovery of cobalt (Li et al. 2010b). However, the use of organic acid leaching may be limited by the strength of acids, temperature, leaching period, and solid–liquid ratios.

11.3.2 Alkaline Leaching

Most of metals present in e-waste are leached by acid solutions. However, a few of them may be leached in alkaline conditions. For instance, Pb and Al may be leached at pH above 10, as shown in the Pourbaix Diagram of both elements (Pourbaix 1974). Senćanski et al. (2017) removed Al foils from Li-ion batteries by boiling 5 mol/L of NaOH solution for 30 min. Similar experimental procedure was carried out by Zhang et al. (2016) for Pb leaching from spent leaded glass of CRT at 80 °C. In addition, Yao et al. (2016) combined alkaline media with S source (NaS) for leaching of Si and Pb with 6 mol/L of NaOH, S/Pb molar ratio 1:1, at 180 °C within 8 h.

Leaching by glycine may occur in alkaline conditions. Li et al. (2020) evaluated the leaching of PCBs at different pH values. According to the authors, the extraction of Cu is related to the glycinate anion due to bidentate chelation complexes, which achieve the maximum value at pH 12. It is more stable than complexes formed with Cu and cationic glycine ($^{+}H_3NCH_2COOH$) or zwitterion through monodentate coordination (by the carboxylate group). Under O_2 atmosphere, metallic Cu from PCBs forms CuO, which reacts with glycinate anion forming a chelate (11.1). The authors claimed that Cu leaching by glycine in alkaline media achieved its maximum efficiency at pH 10.0, above which CuO is not soluble (Li et al. 2020).

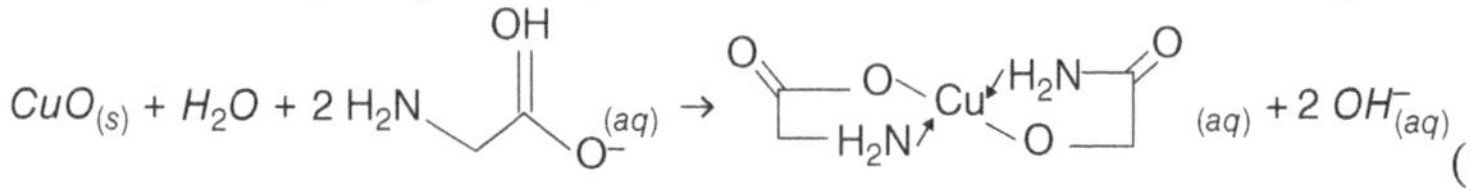

(11.1)

Chen et al. (2021) studied the leaching of Li-ion batteries (LCO cathode) by glycine (11.2). Glycine-Co complexes are stable at pH 7.0–9.5, and as the concentration of glycine increases, the leaching efficiency also increases as more reagent is available to form complexes (Chen et al. 2021). Results have demonstrated that within 7 h of reaction, the leaching of Li and Co achieved similar results with H_2SO_4 in 1.5 h (Chen and Ho 2018) and HCl in 2 h (Xuan et al. 2021). However, the use of green reagent and mild reaction would make the recycling process even more sustainable (Li et al. 2020).

$$3\ NH_2CH_2COO^- + LiCoO_2 \rightarrow NH_2CH_2COOLi + (NH_2CH_2COO)_2Co \quad (11.2)$$

Other inorganic reagents have been explored, such as NH_3-$(NH_4)_2SO_4$ or CO_3, and $C_5H_8NO_4Na$ (Chen et al. 2021). The leaching of base metals such as Cu and Ni is favorable in alkaline media when ammonia is present and metal–ammine complexes are formed (11.3 and 11.4) (Table 11.2). These species are stable at pH 9–11 (Jadhao et al. 2021), and leaching efficiency is similar to found for inorganic acid (Martins et al. 2021a). The increase of $(NH_4)_2SO_4$ concentration is beneficial to the process acting as reducing agent of metallic Cu and Ni (Jadhao et al. 2021). Perea et al. (2021) studied the leaching of PCBs by monosodium glutamate ($C_5H_8NO_4Na$) at pH 9–11, where Cu ions form complexes as $Cu(C_5H_8NO_4Na)_2$. The leaching agent is also a low-cost and low-toxicity alternative for hydrometallurgical processes (Perea et al. 2021; Jadhao et al. 2021).

$$Cu + \frac{1}{2}O_2 + 2NH_3 + 2NH_4^+ \rightarrow Cu(NH_3)_4^{2+} + H_2O \quad (11.3)$$

$$Ni + \frac{1}{2}O_2 + 4NH_3 + 2NH_4^+ \rightarrow Ni(NH_3)_6^{2+} + H_2O \quad (11.4)$$

11.3.3 Cyanide Leaching

Cyanide (HCN) leaching is still studied and applied for leaching of Au and Ag, where cyanide reacts and forms soluble complexes in the alkaline pH range (Figure 11.2). Cyanide leaching has been used since 1887, when John Stewart MacArthur (1856–1920) patented the cyanidation process (Habashi 2005). Cyanide is the conventional lixiviant for gold due to the high stability of gold cyanide complexes and their great selectivity over many impurities. However, the potential health hazards related to HCN poisoning and environmental damage associated with cyanide discharge have motivated the search for alternatives to cyanide for both primary production of gold and in e-waste recycling applications.

Table 11.2 Literature review of the alkaline leaching of e-waste.

References	E-waste	Leaching agents	Conditions	Leaching efficiency
Jadhao et al. (2021)	PCBs	$[NH_3\text{-}(NH_4)_2SO_4]$	90g/L NH_3, 180g/L $(NH_3)_2SO_4$, 0.4 mol/L H_2O_2, S/L ratio 20 mL/g, 700rpm, 4 h, 80 °C	*Cu – 100%* *Ni – 90%*
Li et al. (2020)	*PCBs*	*Glycine (NH_2CH_2COOH)*	0.5 mol/L glycine, S/L ratio 1/50, <2 mm particle size, pH 10, ambient O_2, 25 °C, 72 h	*Cu – 96.5%*
Chen et al. (2021)	*Li-ion batteries*	*Glycine (NH_2CH_2COOH)*	300g/L glycine, 10% H_2O_2, S/L ratio 1:100, pH range 7.0–9.5, 80 °C, 7 h	*Co – 97.0%* *Li – 91.1%.*
Senćanski et al. (2017)	*Li-ion batteries*	*NaOH*	5 mol/L NaOH, 30 min	*Al – 100%*
Zhang et al. (2016)	*Waste glass from CRT*	*NaOH*	5 mol/L NaOH, 80 °C, 2 h	*Pb – >90%*
Yao et al. (2016)	*Waste glass from CRT*	*NaOH + NaS/S*	NaOH 6 mol/L, S/Pb molar ratio 1.0:1, 180 °C, 8 h	*Pb – 80%*
Perea et al. (2021)	*Computer pins (electrical contacts composed mainly of a copper-tin alloy with a gold coating or bath)*	*monosodium glutamate – $C_5H_8NO_4Na$*	0.5 mol/L glutamate, 0.03 mol/L H_2O_2, pH 9.4–11, 2 h	*Cu – 92%*

Cyanide is still used in artisanal mining combined with Hg amalgamation techniques to obtain gold, often practiced with low standards in terms of health and safety and for the environment (Smith 2019; García et al. 2015). Liquid Hg is used for amalgamation and then heated for its removal to obtain gold with

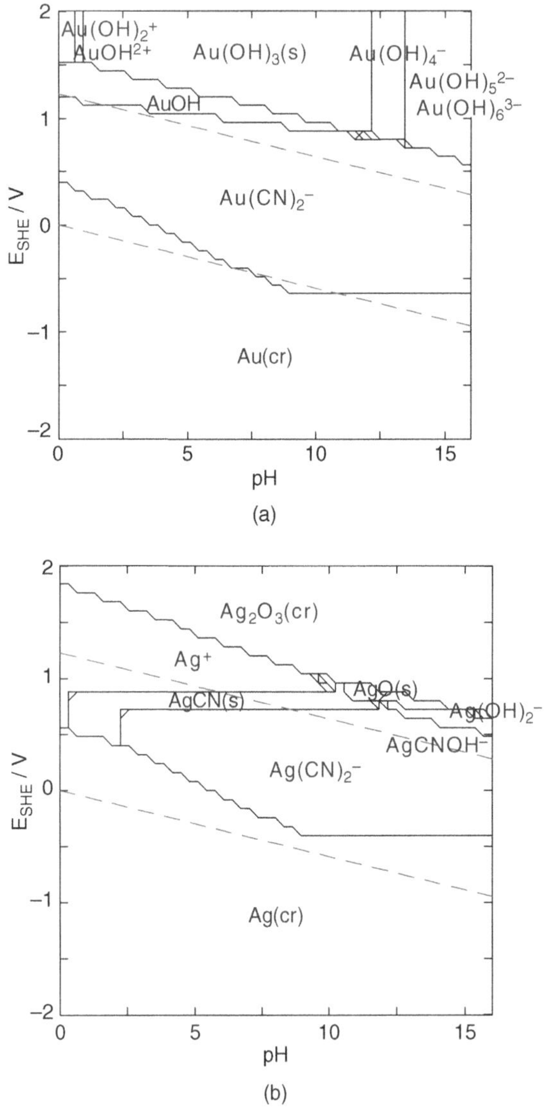

Figure 11.2 Pourbaix equilibrium potential–pH Diagram of (A) Au and (B) Ag plotted using Hydra-Medusa software, displaying the conditions required for cyanide leaching at 25 °C.

an extraction percentage of 30%. The remaining gold is leached with cyanide from Hg-contaminated tailings. Extraction efficiency is up to 90% (UNEP 2013; Marsden and House 2006). The reaction occurs at pH above 10, where cyanide is present in the majority as CN^- ions. As the pH decreases, the concentration of free cyanide ions declines. For this reason, the pH control is crucial to the process using NaOH, for instance (Drace et al. 2016; Aliprandini et al. 2020).

When aiming to recover Au and Ag, the cyanide leaching is applied to e-waste recycling. Eqs. (11.5) and (11.6) depicts the cyanide leaching for Au and Ag extraction, respectively, in alkaline media, where cyanide complexes are formed ($Ag(CN)_2^-$ and $Au(CN)_2^-$) (Gönen et al. 2004). According to Zhang et al. (2012b), the reaction is effective for Au and Ag leaching at the surface of of e-waste, mainly PCBs. Despite its high environmental impacts, the process is still studied due to its good economic feasibility.

$$4\,Au + 8\,CN^- + O_2 + 2\,H_2O \rightarrow 4\,Au(CN)_2^- + 4\,OH^- \tag{11.5}$$

$$4\,Ag + 8\,CN^- + O_2 + 2\,H_2O \rightarrow 4\,Ag(CN)_2^- + 4\,OH^- \tag{11.6}$$

Atmospheric oxygen is applied as the oxidant, and the degree of solution aeration is sometimes the leaching process rate-limiting reagent (Gönen et al. 2004; Hernandez et al. 2014). Cyanide concentration, temperature, pH (as aforementioned), and surface area affect the leaching efficiency (Bisceglie et al. 2017). Sodium or potassium cyanide are typically used as the cyanide source or to increase the CN^--ion concentration and improve the leaching efficiency (11.5 and 11.6) (Bisceglie et al. 2017).

In order to avoid the use of cyanide or salts for leaching, bioleaching has been considered to extract metals from e-waste. The cyanogenic bacteria produce cyanide and then leach Ag and Au from waste PCBs. Merli et al. (2022) evaluated the *Pseudomonas aeruginosa* for cyanide leaching, achieving 90% and 20% of Ag and Au extraction, respectively. Cyanide produced by bacteria achieved the maximum concentration in 20 hours, when the pH kept between 8.4 and 9.2. The leaching process occurred at pH 9.0, and the reaction achieved maximum efficiency in 168 hours (Merli et al. 2022).

According to Yuan et al. (2018), the bacteria strain *Pseudomonas fluorescens* may be used as a cyanide source for leaching of PCBs. Wang et al. (2021) report the use of *C. violaceum*, *B. megaterium*, *P. putida*, *P. chlororaphis*, and *P. plecoglossicida*, which produce cyanide and may be used for Ag and Au leaching. The literature also reports the use of eukaryotic (fungi, algae, and plants) and prokaryotic (Archaea, Eubacteria) organisms for cyanide production (Wang et al. 2021).

The incorrect use of cyanide is potentially lethal; however, with appropriate safety protocols, the leaching reagent can be used safely, primarily by avoiding chemical conditions where hydrogen cyanide can be formed. Environmental issues associated with cyanide can be avoided by using cyanide destruction technologies, which are standard practice in primary gold production operations (Akcil et al. 2015). Social, environmental, and economical issues remain challenges to cyanide-based leaching technologies. As a result, other reagents are considered as alternatives to cyanide, such as chloride, thiourea, thiosulfate, bromide, or iodide leaching (Zhang et al. 2022b, 2012b).

11.3.4 Thiosulfate and Thiourea Leaching

There has been a significant desire for a suitable alternative leaching agent for extraction and recovery of gold and silver, which can be found in elevated concentrations in computers and mobile phones, due to the major health and environmental concerns from using cyanide (Abbruzzese et al. 1995). Recycling of mobile phones requires special environmental protection and resource preservation as they contain other toxic elements such as lead, mercury, nickel, chromium, antimony, beryllium, and arsenic, along with valuable metals such as gold, silver, palladium, and platinum (Osibanjo and Nnorom 2008). There have been early reports on the dissolution of gold and silver in thiosulfate solutions containing oxidizing agents, and thiosulfate leaching technology has been implemented for primary production of gold at Barrick's Goldstrike operation in Nevada, USA. This technology was first applied in the silver mining industry, which is called the Patera process, for recovering silver from high-grade ore. Later, ammonia was added to the leaching solution to improve the dissolution of gold as ammine complexes as well as reduce the consumption of thiosulfate associated with oxidation (Muir et al. 1993; Zipperian et al. 1988). Recently, in response to the environmental regulations and public concerns over the conventional cyanide leaching of gold, thiosulfate leaching has been a subject of great interest in hydrometallurgy; however, its use in primary production is limited to certain niche applications where cyanide performance is compromised, for example when there is pregnant leach liquor robbing of dissolved gold complexes due to the presence of carbonaceous material (Figure 11.3). Thiosulfate is a divalent and type 'A' soft ligand that tends to generate stable complexes with low-spin metal ions (i.e. Cu^{I}, Ag^{I}, Au^{I}, Au^{III}, Hg^{II}) (Wilkinson and Gillard 1987).

During thiosulfate leaching of gold and silver, catalytic oxidation of the metal is required, which may occur via soluble metal complexes, and conventionally, the copper(II)-tetra-amine complexes are used as oxidants. One of the major advantages of using thiosulfate leaching is that this approach limits the interference of foreign cations during leaching as well as allows favors the formation of recoverable complexes, which are stable over a wide range of pH–E_h values (Abbruzzese et al. 1995). The target pH presence of ammonia not only impedes the dissolution

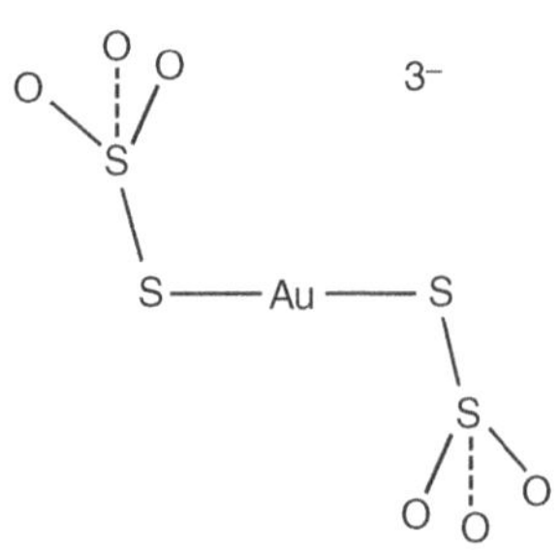

Figure 11.3 Structure of Au-thiosulfate complex, adapted from (Zhao et al. 1998).

of other species such as iron oxide, silica, silicates, and carbonates but can also facilitate the selective oxidation of the target metal as opposed to oxidation of the thiosulfate complexes (Abbruzzese et al. 1995; Ubaldini et al. 2000).

Usually, sodium thiosulfate and ammonium thiosulfate are considered for the recovery of gold and silver from e-waste. Though these two lixiviants are non-toxic and non-corrosive, the consumption rates are higher than what would be the case if cyanide were used. There have been several satisfactory reports on the gold leaching using thiosulfate in the presence of ammonia, where cupric ions acted as oxidants and improved the leaching kinetics, as reported by Aylmore (2001), Grosse et al. (2003), Breuer and Jeffrey (2002), Langhans et al. (1992), Jeffrey et al. (2001), Abbruzzese et al. (1995), and Liu et al. (2020).

Copper (II) is more efficient than other oxidants in the thiosulfate system as it not only acts as an oxidant, but also contributes to enhancing the gold oxidation reaction. However, due to the changing and unreliable reactions between copper(II) and thiosulfate, the solution chemistry tends to follow complex path in the thiosulfate system (Breuer and Jeffrey 2002; Byerley et al. 1973a).

There have already been several reports on the decreasing rate of gold leaching due to the reaction between Cu(II) and thiosulfate, as reported by Byerley et al. (1973b), Byerley et al. (1975), and Breuer and Jeffrey (2002). To stabilize the copper(II) as copper(II)–ammine complex as well as avoid thiosulfate decomposition, ammonia is added to the solution. However, copper(I) and tetrathionate may accumulate due to the oxidation of thiosulfate, which usually occurs due to the absence of oxygen in the system (Byerley et al. 1973b). While considering the leaching of gold, the role of Cu(II) ions can be presented in the form of following reaction where the metallic gold is oxidized to aurous Au^+ ion (Martell and Smith 1989):

$$Au^0 + Cu(NH_3)_4^{2+} + 5S_2O_3^{2-} + 8e^- \rightarrow Au(S_2O_3)_2^{3-} + Cu(S_2O_3)_3^{5-} + 4NH_3 \tag{11.7}$$

However, the degradation of thiosulfate during the leaching of gold can be demonstrated by the following reaction where thiosulfate is oxidized by copper(II) ions to generate tetrathionate (Ha et al. 2010):

$$2Cu(NH_3)_4^{2+} + 8S_2O_3^{2-} + 7e^- \rightarrow 2Cu(S_2O_3)_3^{5-} + 8NH_3 + S_4O_6^{2-} \tag{11.8}$$

This auro-thiosulfate complex is the most stable and soluble gold-bearing species over a wide range of pH (i.e. pH 8.5–10.5) and potential values (i.e. 150–200 mV vs. SHE). However, if the pH becomes more alkaline (i.e. pH > 10.5) gold diamine complex is also formed in the solution:

$$Au(S_2O_3)_2^{3-} + 2NH_3 = Au(NH_3)_2^{+} + 2S_2O_3^{2-} \tag{11.9}$$

Though in thiosulfate leaching of gold, fewer copper ions are required as oxidizing agents, a high consumption of thiosulfate via its easy oxidation by $Cu(NH_3)_4^{2+}$ is considered as a drawback for the $Cu - NH_3 - S_2O_3^{2-} - H_2O$ system

(Zhang et al. 2022c). In order to increase the stability of oxidizing agents or to replace the NH_3-forming complexes, other ligands such as ethylenediaminetetraacetic acid (EDTA), monoethanolamine acid (MEA) and ethylenediamine (en) citrate (cit3–) have also been used by researchers Feng and Van (2010), Puente-Siller et al. (2013), Wang et al. (2019), Wang et al. (2020), Yu et al. (2014). In fact, the non-ammonia copper(II)-complexes are more stable and show less reactivity toward thiosulfate, which in turn reduces the consumption of thiosulfate. However, this benefit of reduced consumption of thiosulfate comes at the cost of slower leaching kinetics (Zhang et al. 2022c). Other transitional metals such as Ni, Co and Fe have also been reported to be effective oxidizing catalysts during thiosulfate leaching of gold, as stated by Arima et al. (2003, 2004), Xu et al. (2016, 2019), Zhang et al. (2021), and Gui et al. (2020).

Thiourea is a sulfur-based organic complexing agent that is able to form stable cationic complexes with gold under acidic conditions and usually in the presence of an oxidant (Zhang et al. 2012b). The extraction through this technique is quite rapid, and recovery could reach 99%. Typically, Fe^{3+} (i.e. $Fe_2(SO_4)_3$) is used as the oxidant. However, it is necessary to maintain an acidic environment for this leaching system, as thiourea decomposes easily in alkaline solutions. During leaching, the gold and silver dissolve selectively by forming stable metal- $(CS(NH_2)_2)_2^+$ aqueous complexes. In the case of the gold leaching, auric ion (i.e. Au^{III}) is not stable in the solution and is reduced to the aurous ion (i.e. Au^{I}) while oxidizing thiourea (Groenewald and Jones 1971).

$$Au^+ + 2SC(NH_2)_2 \rightarrow Au(SC(NH_2)_2)_2^+ \tag{11.10}$$

However, with the help of oxidants such as Fe^{3+}, thiourea is can dissolve gold according to the half-cell reaction presented in Eq. (11.11).

$$Au + 2SC(NH_2)_2 \rightarrow Au(SC(NH_2)_2)_2^+ + e^- \tag{11.11}$$

As it has already been mentioned earlier that typically ferric sulfate is used as an oxidant in the thiourea–gold leaching technique, it is worth giving some attention to the reaction mechanism of ferric ions during leaching. Usually, the redox reaction between thiourea and ferric ions is quite sluggish, as they are susceptible to generate stable complexes in the acid sulfate solutions, which in turn favors gold leaching in thiourea solution (Groenewald and Jones 1971). This phenomenon also helps in slowing the rate of thiourea decomposition, which results in lower consumption of thiourea. Hence, Fe^{3+} has always been an attractive oxidant for the acidic thiourea–gold leaching system.

$$Fe^{3+} + SO_4^{2-} + SC(NH_2)_2 = [FeSO_4SC(NH_2)_2]^+ \tag{11.12}$$

$$Fe^{3+} + SC(NH_2)_2 = Fe(SC(NH_2)_2]^{3+} \tag{11.13}$$

$$Fe^{3+} + 2SC(NH_2)_2 = [Fe(SC(NH_2)_2)_2]^{3+} \tag{11.14}$$

Despite having exhibited high extraction rates and high recovery, the industrial use of this leaching system is limited due to its cost. The quantity of the reactants required for thiourea leaching is higher than that required for conventional cyanide leaching. Additionally, thiourea leaching has another problem of gold surface passivation after some time, which may result in lower leaching kinetics, followed by low extraction. The presence of sulfide minerals, such as pyrite or chalcopyrite and copper also has adverse effects on gold leaching kinetics as they oxidize thiourea as well as catalyze thiourea oxidation by ferric ions (Krzewska and Podsiadły 1980).

11.4 Separation and Recovery

11.4.1 Precipitation

Cementation or the Merrill-Crowe process is considered as the primary process for the recovery of valuable metals (i.e. Au) via a double displacement reaction where a relatively reactive metal is oxidized and dissolves into solution while the noble metal in solution is reduced to its metallic state. During this precipitation process, pulverized metals are added to the system to extract the gold from the pregnant leach liquor. The cementation process involves electron transfer between the ion of the metal that needs to be precipitated and the cementing metals (Havlík 2008). The displacing metal needs to have more negative reduction potential (more reactive) than that of the displaced metal. The precipitation of metals changes the concentration of their corresponding ions as well their potential. For gold precipitation, the reaction can be described as follows:

$$2Au^{+} + M^{0}_{(solid)} \rightarrow 2Au^{0}_{(ppt)} + M^{2+} \qquad (11.15)$$

Hence, caution is required when choosing the precipitant, especially regarding the redox potential, which is the driving force for displacing the precious metals in the solution stoichiometrically. Commonly, zinc, copper, and iron are used as precipitants (Marchbank et al. 1998). However, cementation is not considered as a selective process as other metallic elements present in the solution can be cemented or co-precipitated on the precipitants, which requires further purification. For example, while zinc is used for extracting gold or silver from the cyanide leach or bromide liquor, other elements can be precipitated by zinc, along with gold and silver, such as copper. Besides, in the case of thiosulfate leaching, thiosulfate ions may be reduced by the metallic precipitants, which will produce undesirable cations in solution and complicates the recycling of lixiviants. Using copper as a precipitant can be a solution here, as thiosulfate solutions containing copper ions can be recycled further as leach liquor (Grosse et al. 2003). Other precipitation methods, such as addition of sulfide salts or chemical reduction via hydrogen, sodium borohydride, or sulfur dioxide, weren't successful as these techniques have been reported to be less selective (i.e. precipitate most of the metals from the leach liquor) and the leach liquor cannot be recycled for further use (Deschenes and Ritcey 1990; Awadalla and Ritcey 1991).

11.4.2 Solvent Extraction

Solvent extraction is a liquid–liquid separation technique involving an aqueous phase, where in the hydrometallurgical route it comes from leaching or purification steps, and an organic phase, which contains an extractant compound diluted in an organic solvent (in most cases, kerosene) to decrease the viscosity and enable organic phase coalescence and phase separation. Both phases are mixed and then separated by density difference, where the metallic ions migrate from the aqueous phase to the organic phase during the organic loading step. In the hydrometallurgical route, the organic phase rich in target metal (or metals) can be treated to separate impurities (scrubbing), and then the target metal is recovered (stripped or eluted) from the organic phase using aqueous solution. The organic phase is then reused in a closed-loop process (Kislik 2014; Moldoveanu et al. 2015).

The solvent extraction technique is widely applied in different extractive processes from primary sources, such as nickel, copper, uranium, REEs and PGMs (Crundwell et al. 2011; Kołodyńska et al. 2019). In addition, as most leach solutions from recycling processes are similar in composition to those obtained from mining industries, solvent extraction has been evaluated and applied for recycling of e-waste (Botelho Junior et al. 2021c; Sole et al. 2018). There are three main types of organic extractants: cationic, anionic, solvating, and chelating.

Cationic extractants, known as acid extractants, react by ion exchange between the cation from organic phase and cation from aqueous phase, as depicted in Eq. (11.16). As an example, Cyanex 272 and D2EHPA (phosphonic acid) are a cationic extractants, as shown in Figure 11.4A. Anionic extractants, on the other hand, react by ion exchange involving anions (11.17) as occurs with amine-based extractants such as Alamine 336 (Figure 11.4B). Solvation extractants have as a mechanism the formation of complexes that involve the metal of interest by solvation so that it becomes soluble in the organic phase (11.18), such as the tributyl phosphate (TBP) depicted in Figure 11.4C. Chelating extractants belong to the category of acid extractants that act by involving the metal ion in structures containing rings that are soluble in the organic phase. The spherical formation between the ion and the extractant molecules limits the extraction to ions with specific sizes. This makes these extractants selective for specific metal ions, for instance, Cu from PCBs (11.19). For instance, Cyanex 923 (mixture of four trialkylphosphine acids) is an organic extractant that reacts by solvation (Figure 11.4D) (Kislik 2014; Schlesinger et al. 2011).

$$nLH_{(org)} + M^{n+}{}_{(aq)} \leftrightarrow [M(L)_n]_{(org)} + nH^+{}_{(aq)} \tag{11.16}$$

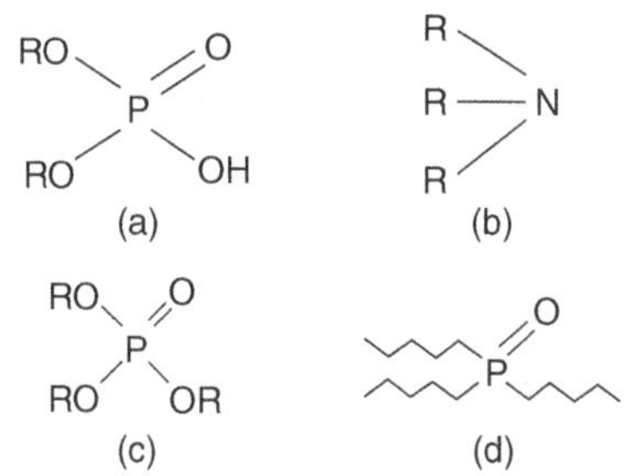

Figure 11.4 Example of chemical structures of common organic extractants used in recycling process of e-waste: (a) phosphonic acid, (b) amine-based, (c) solvation (where R is an alkyl group), and (d) chelating extractants.

$$nLN^{+}A^{-}_{(org)} + B^{-}_{(aq)} \leftrightarrow [(L)_{n}N^{+}A^{-}]_{(org)} + A^{-}_{(aq)} \tag{11.17}$$

$$UO_{2}^{2+}{}_{(aq)} + 2NO_{3}^{-}{}_{(aq)} + TBP_{(aq)} \leftrightarrow [UO_{2}(NO_{3})_{2}(TBP)_{2}]_{(org)} \tag{11.18}$$

$$2LH_{(org)} + Cu^{2+}{}_{(aq)} \leftrightarrow DL_{2(org)} + 2H^{+}{}_{(aq)} \tag{11.19}$$

The use of the solvent extraction technique has been explored for recycling LIBs. For instance, Takahashi et al. (2020) separated Co from Li using Cyanex 272 in kerosene. Suzuki et al. (2012) designed a recycling process using solvent extraction to separate the metals from leach solutions containing Al, Co, Cu, and Li. According to the authors, precipitation was not suitable due to the Co losses with Al or Cu hydroxides. First, Cu is separated using Acorga M5640 (5-nonyl-2-hydroxy-benzaldoxime) extractant at pH 1.5–2.0. Then, Al is separated using PC-88A (2-ethylhexyl 2-ethylhexylphosphonic acid) at pH 2.5–3.5. The remaining solution, which contained Co and Li, was separated using PC-88A mixed with TOA (tri-n-octylamine) (Suzuki et al. 2012). Mixtures of organic extractants are often explored to improve the separation by interaction of the extracts. Mixed extraction organic phases are intended to provide some synergy that improves the overall desired separation. For example, Santanilla et al. (2021) reported an improvement in extraction of Ni by the mixture of Cyanex 272 and Versatic 10 (neodecanoic acid), but a decline was noted for Co. Care needs to be taken when considering synergistic solvent extraction systems so that they are well understood and can be practically maintained in an operating environment.

Vieceli et al. (2020) evaluated the organic extractant D2EHPA (bis(2-ethylhexyl) phosphoric acid) diluted in Isopar L for Mn recovery from LIB recycling. In this study, the authors leached NMC-type batteries and used solvent extraction for selective separation of Mn, which reached 70% of efficiency with losses of Co (5%) (Vieceli et al. 2020). Similar results were found from mobile LIBs (Nadimi and Karazmoudeh 2021). Łukomska et al. (2021) evaluated Cyanex 272 and D2EHPA for alkaline battery recycling, where all Mn remained in solution while all Zn was extracted by Cyanex 272.

The use of solvent extraction in recycling of PCBs has been reported due to the importance of Cu recovery, as the technique is widely used for obtaining Cu from primary sources using oxime extractants (Havlík 2008; Han 2002). Correa et al. (2018) separated Cu from leach solution of PCBs from computers using D2EHPA in kerosene. According to the authors, all Zn, Fe, and Al were separated at pH 3.5, 2:1 A/O, 10% (v/v) D2EHPA from the leach solution. Then, the Cu that remained in aqueous phase was separated at pH 3.5, 1:1 A/O and 20% (v/v) D2EHPA (Correa et al. 2018).

Akbari and Ahmadi (2019) separated Cu from leach solution obtained from bioleaching process using LIX 984N (ketoximes and aldoximes chelating extractant) in kerosene. The extraction achieved 96% in two stages of organic-aqueous contact (Akbari and Ahmadi 2019). Rao et al. (2021) extracted Cu and Au from leach solution of PCBs from mobile phones using Acorga M5640 and amine extractants in kerosene, respectively.

The separation of REEs from e-waste may be carried out using solvent extraction. For instance, Patil et al. (2021) evaluated the Alamine 336 (tricapryl methyl ammonium chloride) and HDEHP (bis (2-ethylhexyl) phosphate) to separate REEs from fluorescent lamp powders. The authors obtained over 99% purity for Y, Eu, and Tb after 1, 25, and 55 extraction stages, respectively, from leach solutions of HCl and HNO_3 (Patil et al. 2021). Pavón et al. (2021) evaluated the Cyanex 923 and Cyanex 572 (mixtures of phosphonic and phosphinic acids) for separation of Y and Eu from leaching by NH_4Cl decomposition and releasing HCl. Such results are important to design recycling process for LED, which technology replaced LCD and fluorescent from TVs and lamps, respectively (de Oliveira et al. 2021).

11.4.3 Ion Exchange Resins

The ion exchange separation by resins is a reversible solid–liquid separation between polymeric resins (solid phase), which exchanges ions with the aqueous phase, whereas in hydrometallurgical route the leach solution is used (Ahamed and Asiri 2019; Zagorodni 2012). Such resins used in hydrometallurgical processes are long chain functionalized polymers which can be tailored to particular applications, as with solvent extraction. The main types of resins are cationic, anionic, and chelating (Zagorodni 2012). Ion exchange resin technology tends to be more applicable for solutions where the target metal is at relatively low concentration; with higher metal concentrations, liquid–liquid solvent extraction tends to be preferred. Ion exchange resin is simpler than solvent extraction from a process equipment and operational point of view and also avoids any hazards associated with the risk of fire with the flammable liquid organics.

Cationic resins release cations from the functional group into the solution. Eq. (11.20) depicts the reaction between a cation in solution (A^+) and a cationic resin, where there is an exchange of the cation from the functional group (H^+), which is released to the solution. Anionic resins react by the exchange between anions from functional group and from solution (11.21). Unlike cationic and

Table 11.3 Examples of chemical groups and formulas of cationic and anionic resins.

Type of exchangers	Chemical groups	Chemical formulas
Strongly acidic	Sulfonate	$-SO_3^-$
Strongly acidic	Phosphorate acid	$-PO_3^-$
Strongly basic	Tetraammonium	$-NR_3^+$
Weakly acidic	Phenolic	$-OH$
Weakly basic exchangers	Primary amine	$-NH_2$
Weakly basic exchangers	Secondary amine	$-NRH$
Medium position between strong and weak acidic	Carboxyl groups	$-COO^-$
Medium position between weak acidic and basic	Tertiary amine	$-NR_2$

Source: Adapted from Inamuddin (2012).

anionic resins, the chelating resins works by the exchange of ions and by chelating reactions, which are interactions by coordinate bonds and electrostatic interactions. These resins have an order of selectivity depending on the functional groups important to separate two similar ions (charge and ionic radii). The literature reports the chelating resins as a subgroup of ion exchange resins. There is also other types of ion exchange materials, such as zeolites, bentonites, membranes, hydrogels, and fibers (functionalized or not) (Inamuddin 2012).

$$R^-H^+_{(resin)} + A^+_{(aq)} \leftrightarrow R^-A^+_{(resin)} + H^+_{(aq)} \tag{11.20}$$

$$R^+OH^-_{(resin)} + B^-_{(aq)} \leftrightarrow R^+B^-_{(resin)} + OH^-_{(aq)} \tag{11.21}$$

Table 11.3 shows examples of chemical groups of cationic and anionic resins. They are also divided into strong and weak exchangers depending on the type of the functional group. As most of leaching reactions in hydrometallurgical processes occur in acidic media (Chapter 3), the cationic resins are the most applied. Strong acid exchangers are active at all pH values, while weak acid exchangers are not active at pH ranges 4–6, both of which are available in Na^+ or H^+ form. Differences between manufacturers for the type of resin occur due to different degrees of cross-linking (Inamuddin 2012).

Table 11.4 presents some common chelating resins applied in hydrometallurgical processes. As most of the leach solution contains different metallic ions, this type of resin is preferable due to the order of selectivity. For instance, iminodiacetate chelating resins such as Lewatit TP 207 and Purolite S930 are more selective for Cu than Co ions, which means that copper will be preferentially loaded, and cobalt will be preferentially stripped. For instance, losses of 17% Ni and 7% Co were observed in separation of Cu by Lewatit TP 207 (Botelho Junior et al. 2019).

Table 11.4 Examples of chemical groups and formulas of chelating resins (Inamuddin 2012; Botelho Junior et al. 2019).

Chemical groups	Formulas	Order of selectivity
Iminodiacetate	$-CH_2N(CH_2COOH)_2$	$Hg^+ > Cu^{2+} > UO_2^{2+} > Pb^{2+} > Fe^{3+} > Al^{3+} > Cr^{3+} > Ni^{2+} > Zn^{2+} > Ag^+ > Co^{2+} > Cd^{2+} > Fe^{2+} > Mn^{2+} > Ba^{2+} > Ca^{2+} > Sr^{3+} > Mg^{2+} > Na^+$
Aminophosphonic acid	$-CH_2NHCH_2CH_2PO_3H$	$Th^{4+} > U^{4+} > UO_2^{2+} > Fe^{3+} > Be^{2+} >$ Rare earth elements $> H^+ > Ag^+ > Cd^{2+} > Zn^{2+} > Cu^{2+} > Ni^{2+} > Co^{2+} > Mn^{2+} > Ca^{2+} > Na^+$
Bis-picolylamine	$-CH_2N-(CHOH)_4CH_2OH \mid CH_3$	$Cu^{+2} > Ni^{+2} > Fe^{+3} > Co^{+2} > K^+ > Ca^{+2} > Na^+ > Mg^{+2} > Al^{+2}$
Thiol	$-SH$	$Ag^+ > Cu^+ \gg Pb^{2+} > Cd^{2+} \gg Zn^{2+} > Ni^{2+} > Fe^{3+} > Ca^{2+}$

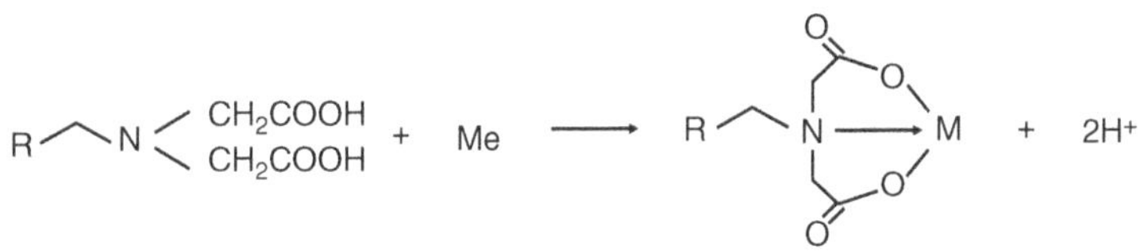

Figure 11.5 Example of chelating reaction between the resin (iminodiacetate) and ions (Me^{+2}, Ni or Co).

The chelating reaction between the resin and Me (Ni or Co) is shown in Figure 11.5. In the process, H^+ ions are released into the solution, and a strong bond occurs between the ions and the functional group.

Strauss et al. (2021) studied the separation of Co, Ni and Mn from leach solution of LIBs using chelating resin Dowex M4195 (bis-picolylamine functional group). The process was designed at pH 1.1 in columns as continuous system in which the first column separated Ni and the output fed another column for Co separation, releasing Mn in solution. Then, the metals adsorbed by the resin are desorbed using H_2SO_4 at 2.0 mol/L. As a result, highly pure solutions of Ni (99.0%) and Co (98.5%) were obtained (Strauss et al. 2021).

Virolainen et al. (2021) studied the chelating resin to separate metals from leaching solution of Li-ion batteries (LIBs). The authors evaluated four resins: Lewatit TP260 with aminomethylphosphonic acid functional group, Purolite S-930 with iminodiacetate functional group, Finex CS12GC with sulfonic acid functional group, and Finex CA16GC with carboxylic acid functional group. Purolite S-930 chelating resin was used for impurity removal, as Fe and Cu, while

Lewatit TP260 removed Fe, Al, Mn, and Cu leaving Co, Ni, and Li in solution for further recovery steps (Virolainen et al. 2021).

The use of ion exchange resins for separation of metals from PCB recycling processes has been evaluated. Nekouei et al. (2019) evaluated the chelating resins Amberlite IRA 743 (N-methylglucamine), Lewatit TP 208 (iminodiacetate), and Lewatit TP 260 (aminomethylphosphonic acid). The separation of Cu, Zn, Ni, Al, and Pb was evaluated in HNO_3 media. Results indicated that Pb may be entirely separated from the other ions by these resins, which is important due to the presence of the element in e-waste (Nekouei et al. 2019).

Neto and Soares (2018) purified a solution from PCB leaching using Dowex M4195 chelating resin to separate Cu. The leach solution is first purified by NaCl addition to recover Ag, and then Pb is precipitated using Na_2SO_4. Then, the remaining solution containing Cu and Al is treated to separate the elements in high-purity solutions (99% and 78%, respectively) (Neto and Soares 2018). In addition, Neto et al. (2016) evaluated the same resin to separate Cu directly from leach solution of PCBs by HNO_3 at 2.0 mol/L. Then, Cu ions are released from the resin using H_2SO_4 at 4.0 mol/L. The processes designed in this study occurs in columns in a continuous system (Neto et al. 2016).

The use of ion exchange resins is not limited to inorganic acids. For instance, Choi et al. (2020) studied the recovery of Au using Amberjet 4200 resin (trimethyl ammonium) and Cu using Amberlite IRC-86 resin (carboxylic acid) after bioleaching of PCBs. The authors used *Chromobacterium violaceum* bacteria, which produce cyanide for leaching metals (Choi et al. 2020). Gámez et al. (2019) tested the MTA 5011 ion exchange resin (quaternary ammonium) for Cu adsorption from thiosulfate leaching. HNO_3 was used to leach Cu, and then ammonia-thiosulfate solution was used for leaching precious metals, which were further separated using resin. About 87% of Au and 29% of remaining Cu were adsorbed. The reagent KSCN was used for desorption of precious metals from the resin (Gámez et al. 2019).

11.4.4 Electrodeposition

In order to avoid the excessive use of leaching solutions to recover the precious metals from WPCBs, staged leaching can be considered, where, for example, the base metals are usually leached at the beginning as they are present at a higher percentage in the e-waste (for instance, copper > 15%). Typically, some preliminary treatments or steps are essential for separating the precious metals that are present in the e-waste (Havlík 2008; Veit et al. 2006).

Hydrometallurgical treatments can be efficient as they selectively extract metals within the process. Recovery of copper combined with gold can be an example for efficient recovery of PMs via hydrometallurgical treatments, as depicted by Habib et al. (2013), Kim et al. (2011), Liang et al. (2013), and Birloaga et al. (2013). Nonetheless, there have been issues such as consumption of leaching reagents, rate of reactions and recovery, as well as selectivity that need to be addressed for more efficient recovery (Syed 2012).

Additionally, environmental regulations are requiring more environment friendly and economic processes to be developed. Electrochemical processes can be a solution in this regard, as they can offer advantages of environmental and efficiency benefits. Electrodeposition mainly involves the reduction of metallic ions in an electrochemical cell where an electric potential is applied within the anode and cathode materials. Hence, an external power source is required to drive the non-spontaneous reactions which later lead to the reduction of metal at the cathode surface. In general, electrodeposition is the final step in the process for metal products after leaching and solvent extraction, and the potential and current density are the primary process parameters. For efficient electrodeposition, it is important to consider production rate (higher current density expressed as A/m^2) and low specific energy consumption (expressed as kWh/kg-metal). However, there are other specific factors to consider that contribute to the energy requirements and cost of the process, such as solution conductivity and the type of electrolyte, electrode properties (i.e. distance between electrodes, composition, and surface area), temperature, pH, and rate of mixing, which also influence the process (Ashiq et al. 2019; Kaya 2018).

Electrodeposition can be useful in recovering several metals simultaneously within the same process, such as both the copper (at cathode) and lead oxide (at anode), can be recovered with a higher efficiency via using rotating cylinder electrode reactor (RCER) (Mecucci and Scott 2002). By optimizing the operating conditions, higher metal recoveries can be realized. Efforts have been made to develop a combined process of direct and mediated electrochemical oxidation, where metals (such as copper) dissolve from PCBs and are electrodeposited concurrently at the cathode material. Gold enrichment in the residue can also be achieved using iron redox couple (i.e. Fe^{3+}/Fe^{2+}) in the acidic electrolyte. It has also been reported that, by optimizing the process parameters such as current efficiency and flow rate, 99.95% copper can be recovered along with the gold and silver-enriched residues (Fogarasi et al. 2014).

Though there have been reports on successful recovery of PMs (i.e. Au and Ag) along with Cu and Zn, recovering other PMs and critical metals is still a challenge and needs further consideration. Additionally, evolution of toxic gases (e.g. Cl_2) due to the decomposition of electrolytes can be a major consideration in process design and plant infrastructure requirements.

11.5 Emerging Technologies for E-Waste Recycling

11.5.1 Ionic Liquids

Investigations into the use of ILs for the recovery and recycling of e-waste and secondary sources, as they can be employed at low temperatures relative to many molten salt electrolysis processes and can be tailored to a wide range of potential chemical separations. ILs are also less volatile and flammable compared with conventional solvent extraction systems, as reported by Kamali Ardakani et al. (2020), Zhang and Hua (2009), Dong et al. (2008), Huddleston et al. (2001), Park et al. (2014), and Hu et al. (2017). Most importantly, the insignificantly low vapor pressure has made them less hazardous and greener means of extraction process. ILs are mainly liquid organic salts at ambient temperature, made of ions and containing organic or inorganic anions, typically including a large organic cation (Table 11.5). ILs can have a unique property of being hydrophobic and hydrophilic simultaneously, as in the molecule they have both inorganic domains and ionic sites (Hagiwara and Ito 2000).

Due to the intricacy of recycling of industrial products, increasing use of metals in modern technologies, as well as the demand for recovery of precious metals from the e-waste, researchers have developed different types of task-specific ILs, such as acidic-ILs, basic-ILs, OH-group-containing ILs, and chiral-ILs, for their wider applications in the recycling industry (Yue et al. 2011; Maurice et al. 2020)

In the metal extraction processes, ILs are used as solvents that are insoluble in the metal-ion-containing aqueous solution. In the process, the two phases are generally stirred together, and the anionic group in the organic phase of the ILs adsorbs the metallic ions. After separating the two phases, the organic group that contains the metal ions is treated in the metal recovery step (Kelley et al. 2016). In IL's extraction process, usually the ion exchange and the ion-pairing mechanisms are used, where ion exchange occurs following anion or cationic exchange (Figure 11.6).

Table 11.5 Properties of ILs.

Properties	Values
Melting point	Preferably below 100 °C
Liquidus range	Often > 200 °C
Thermal stability	Usually, high
Viscosity	Normally < 100cP, workable
Dielectric constant	Implied < 30
Polarity	Moderate
Ionic conductivity	Usually < 10 mS/cm
Molar conductivity	<10 Scm^2/mol
Electrochemical window	Often > 4 V
Vapor pressure	Usually negligible

Source: Park et al. (2014/MDPI/CC by 3.0.

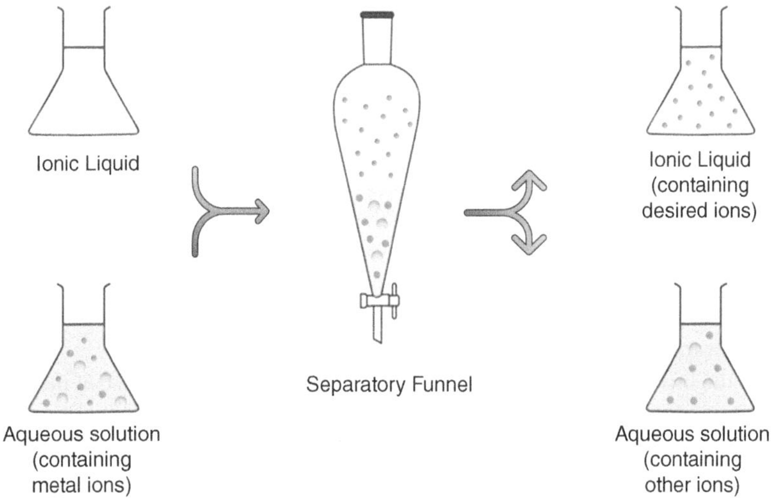

Figure 11.6 Schematic of extraction process via ILs, adapted from reference. Source: Adapted from AliAkbari et al. (2020).

There have been reports on successful leaching and recovery of precious metals and REEs from e-wastes, for instance, AliAkbari et al. (2020), Yang et al. (2019), Zante et al. (2020), Dhiman and Gupta (2019), Dhiman and Gupta (2020), Van Den Bossche et al. (2019), Micheau et al. (2019), Li et al. (2019).

ILs can exhibit some advantages over conventional two-phase solvent extraction process as they can be tailored to be more selective in separations from multi-metal-containing solutions. Also, the concentration of metal in ILs can, in principle, be higher than in the relatively dilute solvent extraction systems. Given that ILs can exhibit reasonable ionic conductivity and a wide electrochemical operating window, they can be used as electrolytes for recovering metals via electrowinning or electrorefining processes. However, the higher viscosity of some ILs can be a problem in extraction process, as solid–liquid separations are difficult. Elevated temperature and addition of diluents might be the solution to this problem of high viscosity, which again imposes other problems of higher cost, corrosion of containers and pipes, and safety/environmental problems due to the addition of higher amounts of organic diluents (Dhiman and Gupta 2019). Mixing of two different ILs can be considered for reducing the viscosity of the system (Dhiman and Gupta 2019; Łukomska et al. 2020).

There seem to be potential benefits and drawbacks for using ILs, depending on the specific application. Recyclability and build-up of impurities in the ILs need to be considered carefully in any process development (Micheau et al. 2019). Despite having some disadvantages, ILs can be used efficiently on an industrial scale due to their effectiveness in selective leaching, shorter period of leaching time, and avoiding the release of toxic metals or gases to the environment (Van Den Bossche et al. 2019). Research efforts are also being made to synthesize the biomaterial-based ILs in order to make them greener as well as lower their carbon footprints (Usmani et al. 2020). The high cost of some ILs makes their application challenging from an operating cost perspective if there are any significant losses during the process.

11.5.2 Deep Eutectic Solvents

As the newly emerged DESs and the ILs share many properties and characteristics, they are being acknowledged as the new class of ILs. However, it should be noted that they are in fact two different types of solvents. DESs are made of the eutectic mixture of Lewis acid and bases, generally contain large and nonsymmetric ions with lower melting points, whereas ILs contain primarily one type of discrete anions or cations and are made of mainly weakly coordinated ionic species that have lower melting points (i.e. <100 °C) (Smith et al. 2014). The term DES also refers to the molar ratios of components that show the lowest melting point in the system. DES are synthesized by complexing quaternary ammonium salts along with a metal salts or hydrogen bond donors (Abbott et al. 2001).

These solvents are described as $Cat^{+}X^{-}zY$, where the term Cat^{+} is the principal ammonium, sulfonium cation, or phosphonium cation, and the term X^{-} is the Lewis base. The anionic species are generally formed between X^{-} and Y, where z is the number of molecules of Y that react with the anion. Depending on the nature of complexing agents, DESs can be of four different types, as depicted in Table 11.6.

Though the chemical properties of DES are analogues to those of ILs, they have similar trends in physical properties (such as low vapor pressure, conductivity, higher thermal stability, and nonflammability) (Perna et al. 2020). Additionally, the DESs are biodegradable, as their groups come from renewable sources (for instance, urea, carbohydrates, vitamins, amino acids, polyalcohols, lactic acid, glycerol [Gly], and choline chloride [ChCl]), and they are considerably nontoxic. Hence, DES is being considered as the ideal green solvent. Besides, their properties can be designed according to the application specification. Table 11.7 summarizes some of the properties of few DESs.

DESs have potential to be used in wide range of applications, such as electrodeposition (Shivagan et al. 2007; Abbott et al. 2007), metal separation, medical industry, catalytic activity (Durand et al. 2013), synthesis through cross-coupling (Hooshmand et al. 2020), carbon capture, and production of biodiesel (Popescu et al. 2011). Among all these applications, DESs have achieved much attention in the field of metal separation and recovery. However, the mechanisms for metal and oxide dissolution have yet to be explored in detail. The leaching mechanism of DESs is believed to have the reduction of metal and the formation complexes of chloro-metalates. In fact, the oxygen acceptor, present in the eutectic, is necessary

Table 11.6 Different types of DESs and their general formulas reproduced (Smith et al. 2014).

Type	Description	General formula	Terms	Example	Application
Type I	A combination of metal and organic salts	$Cat^{+}X^{-}zMCl_x$	M = Zn (Abbott et al. 2001), Sn (Abbott et al. 2004b), Fe and Al (Hurley and Wier 1951), Ga (Xu et al. 2006), In (Yang et al. 2003)	$ZnCl_2$ +ChCl	Metal electrodeposition (Zhang et al. 2012a)
Type II	A hydrate of metal salt and organic salt	$Cat^{+}X^{-}zMCl_x \cdot yH_2O$	M = Cr (Abbott et al. 2004b), Co, Cu, Ni, Fe	$CoCl_2 \cdot 6H_2O$ + ChCl	Large scale industrial process (Smith et al. 2014; Hammond et al. 2019)
Type III	A mixture of hydrogen bond donor and organic salt	$Cat^{+}X^{-}zRZ$	Z = $CONH_2$ (Abbott et al. 2003), COOH (Abbott et al. 2004a), OH (Abbott et al. 2005)	ChCl + urea	Metal recovery, extraction, biocatalytic activity (Peeters et al. 2020)
Type IV	A mixture of metal salts in addition with hydrogen donor compound	$MCl_x + RZ = MCl_{x-1}^{+} \cdot RZ + MCl_{x+1}^{-}$	M = Al, Zn, and Z = $CONH_2$, OH		Precursor for LnDES, materials for superconductors, lasers, magnets, and electronics (Peeters et al. 2020)

Source: Padwal et al. (2022)/John Wiley & Sons/CC BY 4.0.

Table 11.7 Physicochemical and thermal properties of DES (Padwal et al. 2022).

DESs	Molar ratio	Melting point (°C)	Freezing point (°C)	Density (g/cm)	Viscosity (cP)	Ionic conductivity (mS/cm)	Surface tension (mN/m)
Urea:ChCl (reline)	2:1	134	12	1.25	632–750	0.199 (40 °C)	52
Ethylene glycol:ChCl (ethalene)	2:1	−12.9	−66	1.12	36–52	7.61 (20 °C)	49
Glycerol:ChCl (glyceline)	2:1	17.8	−40	1.18	259–376	0.75	55.8
Thiourea:ChCl	2:1	175	69	–	–	–	–
Malonic acid:ChCl	1:2	135	85	1.25	1124	0.55	65.7
Oxalic acid:ChCl	1:1	190	34	–	1500	–	–
Imidazole:ChCl	3:7	89	56	–	15	12	–
Benzamide:ChCl	1:2	129	92	–	–	–	–
Citric acid:ChCl	1:2	149	69	–	–	–	–
Adipic acid:ChCl	1:1	153	85	–	–	–	–
$ZnCl_2$:ChCl	1:2	–	–	–	85,000	0.06	–
DeTAB:HFIP	1:2	–	25	1.28	40	–	–
Benzoic acid:ChCl	1:1	122	95	–	–	–	–
2,2,2-Triflouroacetamide:ChCl	1:2	72	−45	1.342	77	0.286	35.9
Acetamide:ChCl	1:2	80	51	–	–	–	–
Glucose:ChCl	1:1	–	–	–	34,000	–	–
Phenyllactic acid:ChCl	1:1	77	25	–	–	–	–
Phenylpropionic acid:ChCl	1:1	48	20	–	–	–	–
Tricarballylic acid:ChCl	1:1	159	90	–	–	–	–
Levulinic acid:ChCl	1:2	RT	–	–	–	–	–
Succinnic acid:ChCl	1:1	185	71	–	–	–	–
$EtNH_3Cl$:TFA	1:1.5	–	–	1.041	64	0.688	46.3
$EtNH_3Cl$:Urea	1:1.5	–	–	1.14	128	0.348	52.9
EMIM:TFSI	–	−16	–	1.52	36	9	–

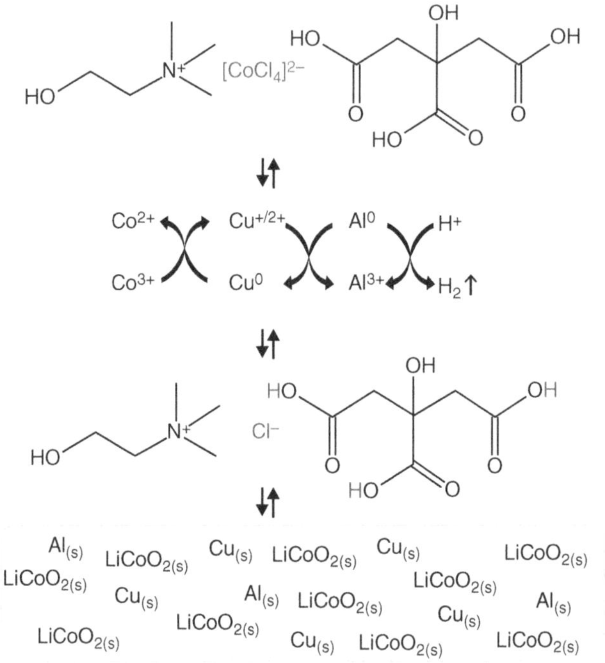

Figure 11.7 Schematic representation of leaching of $LiCoO_2$ using ChCl:citric acid DES, in the presence of aluminium and copper metal (Peeters et al. 2020; Padwal et al. 2022).

for the oxide dissolution, as it weakens the metal-oxygen bond (Abbott et al. 2006). In an effort to recycle cobalt using a choline chloride–citric acid deep-eutectic solvent as lixiviant from the end-of-life lithium ion batteries (i.e. $LiCoO_2$ cathode material), the presence of tetrachlorocobaltate (II) anion complex was observed as dissolved species, along with the concurrent reduction of Co(III) to Co(II) (Peeters et al. 2020).

This process of redox reaction and complexation formation proposes that the $LiCoO_2$ dissolution in ChCl:EG requires simultaneous reduction of cobalt and oxidation of ethylene glycol, where ethylene glycol plays the role of common reducing agent in the system (Figure 11.7). However, higher temperatures (up to 220 °C) and longer reaction period are necessary for the extraction process as the reducing potential of ethylene glycol is poor. Hence, to improve the reducibility of DESs and accelerate the leaching reaction, other approaches such as the use of other functional groups (ChCl:urea or ChCl:citric acid-based DESs) (Padwal et al. 2022) or addition of other metals (copper and aluminum) (Peeters et al. 2020) have been embraced. In the case of metal addition (e.g. Cu and Al) to the system, Co (III) is reduced to Co(II) while oxidizing copper to Cu(I/II), followed by formation of chloro-complexes. Copper(I/II) complexes are again reduced to generate metallic copper via cementation process by metallic aluminum. Aluminum also reduces protons in the system, which enables hydrogen gas evolution. Briefly, the leaching or dissolution of metal oxides by DESs can be demonstrated in three steps: (1) reducing agent reduces the high-valance metal, (2) chloride anions generate chloro-metalates and finally, (3) water molecules are produced via reaction between protons and oxide ions.

DESs are being considered as potential alternatives to the conventional inorganic or organic acid leaching for metal recovery processes. However, the full potential of DESs in metal recovery has yet to be discovered. In many cases, especially at the industrial scale or in continuous-flow reactions, the use of DESs is limited by their high densities and viscosities, reduced conductivity, and moisture sensitivity compared to those of other electrolytes (Yadav and Pandey 2014).

11.5.3 Supercritical Fluids

Supercritical fluids have been studied in the past few years due to their potential applications in industry, such as recycling, dry cleaning, high-pressure sterilization, and separation in fermentation broths (Bertuol et al. 2020). The process occurs above critical pressure (Pc) and temperature (Tc), in the zone called supercritical. Among the main proprieties, highlighted are high density, diffusivity, solubility, reactivity, and low viscosity (Botelho Junior et al. 2022). High diffusivities are observed in supercritical zone close to the gaseous state, which turns the fluid a potential solvent for higher reaction rates. Among the fluids, it is highlighted CO_2, C_3H_8, C_2H_6, and C_2H_4, which the supercritical zone is near to ambient conditions (Chaudhary et al. 2021).

The reaction has been used for recycling of different e-waste. For instance, the technique is used for electrolyte and polymer recovery and extraction of metals from batteries (Botelho Junior et al. 2022; Li and Xu 2019). It is reported that the use of supercritical fluids may decrease the consumption of reagents, have a low environmental impact, and have fast kinetics, making them attractive for environmental and economic aspects (Chaudhary et al. 2021; Chen et al. 2022).

Grützke et al. (2014) evaluated the supercritical CO_2 (helium head pressure carbon dioxide – sc $HHPCO_2$) to recover electrolyte of LIBs. The process occurred at room temperature (25 °C) and 90–120 bar. According to the authors, it was possible to obtain DEC (diethyl carbonate), DMDOHC (dimethyl-2,5-dioxahexane dicarboxylate), EMDOHC (ethylmethyl-2,5-dioxahexane dicarboxylate), and DEDOHC (diethyl-2,5-dioxahexane dicarboxylate) from aging of electrolyte, and traces of $LiPF_6$ (Grützke et al. 2014).

LIBs have Al and Cu foils bonded to cathode and anode materials, respectively, and the most challenging in recycling is the separation before leaching step (Martins et al. 2021b; Guimarães et al. 2022). Several studies used organic solvents to dissolve the binder (mainly polyvinylidene fluoride – PVDF), such as dimethyl sulfoxide (DMSO), propylene carbonate ($C_4H_6O_3$), and ethylene carbonate ($(CH_2O)_2CO$) (Raj et al. 2022), or even pyrolysis (Zhang et al. 2018). Fu et al. (2021) evaluated the supercritical CO_2, where at 70 °C and 80 bar within 13 min, about 98.5% of PVDF was dissolved, which can be recovered and reused.

For extraction of metals, Bertuol et al. (2016) compared the recovery of Co from LIBs using supercritical CO_2 and H_2SO_4 leaching combined with H_2O_2. About 95% of Co was extracted from LIBs in 5 min, while the same extraction efficiency was obtained by leaching within 60 min. As stated by the authors, despite the high energy consumption to keep the CO_2 in supercritical conditions, the main advantage is the faster reaction than acid leaching and less consumption of reagents (Bertuol et al. 2016). Yao et al. (2018) used supercritical CO_2 for recycling of NiMH, recovering up to 90% of REEs at 35 °C and 31 MPa within 2 h. These batteries are used in electronic equipment and hybrid vehicles as well. According to the authors, the process has less energy consumption than pyrometallurgical route (Yao et al. 2018).

Supercritical fluids may also be used for the recovery of critical metals from PCBs. Liu et al. (2016) tested supercritical H_2O and supercritical CO_2 extraction to obtain Pd (93.7% efficiency) and Ag (96.4% efficiency). The economic evaluation showed that the process is feasible allied to the degradation of organic polymers, organic halogens, and recycling of metals (Liu et al. 2016).

The recovery of Cu may be carried out by supercritical CO_2 mixed with solvents such as methanol, ethanol, and propanol, along with complexing agents such as cyanide 301, 302, and tri-butyl phosphate, to increase the Cu solubility in the supercritical fluid. About 97% of Cu with 98.7% purity was obtained at 60 °C and 200 bar within 30 min (Golzary and Abdoli 2020). Hsu et al. (2021) evaluated the combination of supercritical CO_2, H_2SO_4, and H_2O_2 (2.0 mol/L and 0.2 mol/L, respectively) for extraction of Cu from PCBs. At 120 °C and 148 atm during 30 min, the extraction of Cu achieved 82%, which was improved due to the action of supercritical fluid creating pores and weakening the structure of the e-waste. In comparison, direct leaching at same conditions (without supercritical fluid) produces similar results within 6 h (Correa et al. 2018).

Zhan et al. (2020) evaluated subcritical H_2O+ethanol (C_2H_5OH) and supercritical C_2H_5OH to recover critical metals from LEDs. In this case, the subcritical critical fluid was important to dissolve encapsulation material to release LED chips and Ag line (Chapter 2). The recovery of Ga, In, As, and Ag was achieved at up to 90% at 250–300 °C and 7.5 MPa in 240 min (Zhan et al. 2020). The recovery of REEs from e-waste may also be carried out by supercritical fluids, as reported by Zhang et al. (2022a). In their study, the authors mixed supercritical CO_2 with TBP-HNO_3 as chelating agent for recycling of fluorescent lamps. At 40 °C, 31 MPa, and 3 h, the extraction achieved 70% for Y and Eu and 50% for Tb (Zhang et al. 2022a).

11.5.4 Nanohydrometallurgy

The nanotechnology is considered the key technology of 21st century providing high-performance materials for different applications, including better or new products and manufacturing (Botelho Junior et al. 2022; Toma 2013). In recycling of e-waste, the nanotechnology has been applied for separation of metals after leaching due to its high surface area-to-volume ratio, strong sorption, and high affinity toward the target metal (Botelho Junior et al. 2022). The use of nanomaterials in hydrometallurgical processes has been denoted as *nanohydrometallurgy* (Condomitti et al. 2012; Toma 2015).

The are several types of nanomaterials used for separation of metals from solutions, such as nanofibrillated cellulose and polymer-derived or modified nanoadsorbents (Maatar and Boufi 2015), zinc oxides (Dhiman and Kondal 2021), carbon nanotubes (Rao et al. 2007), and magnetic, superparamagnetic, and carbon-based nanoparticles (Melo et al. 2019). The literature reports several examples that can be used in hydrometallurgical processes for e-waste.

Chaves et al. (2022) evaluated the recovery of Co using nanofibers functionalized with Cyanex 272. The nanomaterial was prepared by mixing Nylon 6 and Cyanex 272 in formic acid under stirring at 40 °C for 4 h. The solid–liquid separation achieved close to 100% at pH 8.0 with 25% of Cyanex 272 in 60 min. The authors stated that the recovery of Co remained over 85% after several cycles (Chaves et al. 2022). These results are similar to those found for ion exchange resins (Botelho Junior et al. 2018; Perez et al. 2020). For recovery of Ga, Segala et al. (2022) developed a polyacrylonitrile nanofibers functionalized with Cyanex 272, which may be applied in recycling of LED lamps (Table 11.1). The fibers were mixed with the extractant in dimethylformamide (DMF) solvent under stirring at 60 °C for 4 h. As reported by Chaves et al. (2022), the separation of Ga slightly decreased after reuse of nanofibers from 80% to 70% in 5 cycles at pH 2.5 (Segala et al. 2022).

Magnetic nanoparticles are usually tested for separation of metals from solutions. In addition, the materials are functionalized to improve the adsorption and selectivity rates. Then, the solid–liquid separation occurs by applying an external magnetic field (Condomitti et al. 2012). For instance, the co-precipitation method is commonly reported as a synthesis method. Almeida and Toma (2016) used $FeCl_3$ and $FeSO_4$ with NH_4OH to produce the nanoparticles, and the resulting material was treated with tetramethylammonium hydroxide (TMAOH). The material was further coated with SiO_2 using tetraethylorthosilicate (TEOS) and then functionalized with DTPA (ethylenediaminepropylsilane/diethylenetriaminepentaacetic acid) (Almeida and Toma 2016). The authors evaluated the separation of Nd and La, both found in lamp waste and permanent magnets (Table 11.1). Melo et al. (2019) achieved similar results to Almeida and Toma (2016) for adsorption of La, Mn, Co, Ni, and Cu (Melo et al. 2019).

Da Silva et al. (2016) evaluated the removal of Ni, Co, Pb, Hg, and Cu using magnetic nanoparticles functionalized with humic acid. These metals can be found in many e-wastes, such as PCBs and LED lamps (Table 11.1). In the case of Hg, it may be found in fluorescent lamps (Dhawan and Tanvar 2022). The contact between the nanoparticles and metals occurs in multielementary solution. After adsorption, the magnetic nanoparticles were separated from the solution using a permanent magnet (NdFeB). The nanomaterial adsorbed over 94% of Ni, Cu, Co and Pb, and at least 62% of Hg in batch process. The nanoparticles were recycled using concentrated ammonia solution for metals desorption, which is important for reuse in industrial applications (Da Silva et al. 2016).

For separation of Co, Lobato et al. (2016) evaluated magnetic nanoparticles coated with oleic acid. The metal is considered crucial, as well as Li, for batter market (Table 11.1). In this study, the authors synthesized the magnetic nanoparticles and functionalized them with Cyanex 272 (Chapter 4.2). Compared to solvent extraction techniques, the separation of Co was three times faster using nanoparticles – 51s (Lobato et al. 2016). If it is considered that most of solvent extraction reactions achieve the equilibrium in 5 min, the magnetic nanoparticles were at least 5 times faster (Botelho Junior et al. 2021b). As reported by Lobato et al. (2016), the Co may be eluted using H_2SO_4 at 1.0 mol/L and the nanoparticles reused.

11.6 Conclusion and Futures Perspectives

The advancement of green technologies has pushed the boundaries of raw materials market. Among those, the most crucial metals for the production of strategic technologies such as electric cars, LED lamps, and wind and solar energy, are classified as critical. In addition, the market of these metals is controlled by a few countries, and the risk of supply chain interruption is high (Bobba et al. 2020; U.S. Department of Energy 2010; MCTIC 2018). Among them, Co and Li are important for Li-ion batteries used in electric vehicles, where Co production is controlled by D. R. of Congo (van den Brink et al. 2020; Bridge and Faigen 2022). PCBs contain Ag and Au in concentrations 100–1000 times higher than in natural resources (Dutta et al. 2018; Martins et al. 2021a), while LED lamps and Fe-Nd-B permanent magnets contain REEs, where 95% of global production occurs in China (München et al. 2018; Gabriel et al. 2020; Cenci et al. 2020b; Vekasi 2019). Indeed, selection of system for precious metal recovery is important due to their economic value and criticality.

Allied to it, the growth in e-waste generation containing critical metals brings the recycling as an opportunity to reduce dependence on primary sources as well as a destination for these wastes. It is expected to increase from 53.6Mt of e-waste generated to 74.7Mt between 2019 and 2030 (Forti et al. 2020), and the impacts of COVID-19 pandemic could intensify (Zhang et al. 2022d).

The present chapter presents the technologies for e-waste recycling by hydrometallurgical processes. The literature review about the composition of main e-waste demonstrated the high concentration of critical and valuable metals. The first step of hydrometallurgical process is the leaching to solubilize the metals present in e-waste, where the techniques used are acid, alkaline, cyanide, and thiosulfate/thiourea. Most of metals are soluble in acid media, and for this reason H_2SO_4, HCl, HNO_3, and H_3PO_4 are evaluated. In order to design sustainable processes, organic acids have been explored such as citric acid, DL-malic acid, glycine, and ascorbic acid. Alkaline leaching is used to remove Al before the acid leaching, as the metal is soluble in alkali conditions. Cyanide is still used for Ag, Au, and PMG leaching, but it has been replaced by thiosulfate and thiourea due to environmental and health risks. Indeed, multistage leaching needs to be considered to improve kinetic reaction within several hours (~24 h).

After leaching, the metals present in leach liquor must be separated to obtain high-purity products. Precipitation is a common technique to convert the metallic ions into an insoluble compound, such as carbonate, sulfate, hydroxide, or sulfide. The technique is mostly used to obtain the final product since for separation of metallic ions, the loss by co-precipitation is considered too high. Solvent extraction is a classic liquid–liquid separation involving an aqueous phase (leach solution) and an organic phase (organic extractant), and the selectivity on separation among metallic ions is considered high. However, the use of volatile and flammable compounds allied to the high cost are drawbacks to the process. Separation using ion exchange resins (mainly chelating resins) is a solid–liquid separation using polymeric resins with functional groups, and the literature reports its potential use for Ni and Co recovery as well as REEs. However, there is a lack in literature regarding e-waste recycling due to its low kinetics and high costs for industrial applications.

Emerging technologies have been explored to improve the kinetics and efficiency of recycling, as well as turning it more sustainable and greener. Among those, there is an increase in interest in application of ILs, DESs, supercritical fluids, and nanotechnology for leaching and separation of metals in e-waste processing. These technologies are still incipient for pilot or industrial scale with high-cost investment. Despite that, the literature has demonstrated great potential, and future research might focus on development of greener processes allied to the circularity of wastes and effluents. Streamlined separation and purification required using SX/IX or other emerging techniques is important to economically recover metals or metal compound products.

Acknowledgments

To the University of Sao Paulo (USP) and the University of Queensland to support this project. To Fundação de Amparo à Pesquisa do Estado de São Paulo and Capes

(2019/11866-5, 2020/00493-0, 2021/14842-0) for the financial support. This project was developed with the support of SemeAd (FEAUSP), FIA Fundação Instituto de Administração and Cactvs Instituto de Pagamento S.A. through the granting of assistance to a research project Bolsa SemeAd PQ Jr (Public Notice 2021.01).

References

Abbott, A.P., Capper, G., Davies, D.L. et al. (2001). Preparation of novel, moisture-stable, lewis-acidic ionic liquids containing quaternary ammonium salts with functional side chains. *Chemical Communications* 1: 2010–2011. https://doi.org/10.1039/b106357j.

Abbott, A.P., Capper, G., Davies, D.L. et al. (2003). Novel solvent properties of choline chloride/urea mixtures. Electronic supplementary information (ESI) available: spectroscopic data. See http://www.rsc.org/suppdata/cc/b2/b210714g/. Chem. Commun., 70–71. https://doi.org/10.1039/b210714g.

Abbott, A.P., Boothby, D., Capper, G. et al. (2004a). Deep eutectic solvents formed between choline chloride and carboxylic acids: versatile alternatives to ionic liquids. *Journal of the American Chemical Society* 126: 9142–9147. https://doi.org/10.1021/ja048266j.

Abbott, A.P., Capper, G., Davies, D.L., and Rasheed, R. (2004b). Ionic liquids based upon metal halide/substituted quaternary ammonium salt mixtures. *Inorganic Chemistry* 43: 3447–3452. https://doi.org/10.1021/ic049931s.

Abbott, A.P., Capper, G., Swain, B.G., and Wheeler, D.A. (2005). Electropolishing of stainless steel in an ionic liquid. *Transactions of the Institute of Metal Finishing* 83: 51–53. https://doi.org/10.1179/002029605X17657.

Abbott, A.P., Capper, G., Davies, D.L. et al. (2006). Solubility of metal oxides in deep eutectic solvents based on choline chloride. *Journal of Chemical & Engineering Data* 51: 1280–1282. https://doi.org/10.1021/je060038c.

Abbott, A.P., Capper, G., McKenzie, K.J., and Ryder, K.S. (2007). Electrodeposition of zinc-tin alloys from deep eutectic solvents based on choline chloride. *Journal of Electroanalytical Chemistry* 599: 288–294. https://doi.org/10.1016/j.jelechem.2006.04.024.

Abbruzzese, C., Fornari, P., Massidda, R. et al. (1995). Thiosulphate leaching for gold hydrometallurgy. *Hydrometallurgy* 39: 265–276. https://doi.org/10.1016/0304-386X(95)00035-F.

Addison, T. and Roe, A. (ed.) (2018). *Extractive Industries: The Management of Resources as a Driver of Sustainable Development. First*. Oxford: Oxford University Press.

Ahamed, M.I. and Asiri, A.M. (2019). Applications of Ion Exchange Materials in the Environment. https://doi.org/10.1007/978-3-030-10430-6.

Akbari, S. and Ahmadi, A. (2019). Recovery of copper from a mixture of printed circuit boards (PCBs) and sulphidic tailings using bioleaching and solvent extraction processes. *Chemical Engineering and Processing Process Intensification* 142: 107584. https://doi.org/10.1016/j.cep.2019.107584.

Akcil, A., Erust, C., Gahan, C.S. et al. (2015). Precious metal recovery from waste printed circuit boards using cyanide and non-cyanide lixiviants – A review. *Waste Management* 45: 258–271. https://doi.org/10.1016/j.wasman.2015.01.017.

AliAkbari, R., Marfavi, Y., Kowsari, E., and Ramakrishna, S. (2020). Recent studies on ionic liquids in metal recovery from e-waste and secondary sources by liquid-liquid extraction and electrodeposition: a review. *Materials Circular Economy* 2. https://doi.org/10.1007/s42824-020-00010-2.

Aliprandini, P., Veiga, M.M., Marshall, B.G. et al. (2020). Investigation of mercury cyanide adsorption from synthetic wastewater aqueous solution on granular activated carbon. *Journal of Water Process Engineering* 34: 101154. https://doi.org/10.1016/j.jwpe.2020.101154.

Almeida, S.D.N. and Toma, H.E. (2016). Neodymium(III) and lanthanum(III) separation by magnetic nanohydrometallurgy using DTPA functionalized magnetite nanoparticles. *Hydrometallurgy* 161: 22–28. https://doi.org/10.1016/j.hydromet.2016.01.009.

Arima, H., Fujita, T., and Yen, W.T. (2003). Gold recovery from nickel catalyzed ammonium thiosulfate solution by strongly basic anion exchange resin. *Materials Transactions* 44: 2099–2107. https://doi.org/10.2320/matertrans.44.2099.

Arima, H., Fujita, T., and Yen, W.T. (2004). Using nickel as a catalyst in ammonium thiosulfate leaching for gold extraction. *Materials Transactions* 45: 516–526. https://doi.org/10.2320/matertrans.45.516.

Ashiq, A., Cooray, A., Srivatsa, S.C., and Vithanage, M. (2019). Electrochemical enhanced metal extraction from E-waste. https://doi.org/10.1016/B978-0-12-817030-4.00004-8.

Assefi, M., Maroufi, S., Yamauchi, Y., and Sahajwalla, V. (2020). Pyrometallurgical recycling of Li-ion, Ni–Cd and Ni–MH batteries: a minireview. *Current Opinion in Green and Sustainable Chemistry* 24: 26–31. https://doi.org/10.1016/j.cogsc.2020.01.005.

Awadalla, F.T. and Ritcey, G.M. (1991). Recovery of gold from thiourea, thiocyanate, or thiosulfate solutions by reduction-precipitation with a stabilized form of sodium borohydride. *Separation Science and Technology* 26: 1207–1228. https://doi.org/10.1080/01496399108050525.

Aylmore, M.G. (2001). Treatment of a refractory gold-copper sulfide concentrate by copper ammoniacal thiosulfate leaching. *Minerals Engineering* 14: 615–637. https://doi.org/10.1016/S0892-6875(01)00057-7.

Bertuol, D.A., Machado, C.M., Silva, M.L. et al. (2016). Recovery of cobalt from spent lithium-ion batteries using supercritical carbon dioxide extraction. *Waste Management* 51: 245–251. https://doi.org/10.1016/j.wasman.2016.03.009.

Bertuol, D.A., Amado, F.R., Cruz, E.D., and Tanabe, E.H. (2020). Metal recovery using supercritical carbon dioxide. In: *Green Sustain. Process Chem. Environ. Eng. Sci.*, 85–103. Elsevier https://doi.org/10.1016/B978-0-12-817388-6.00005-2.

Birloaga, I., De Michelis, I., Ferella, F. et al. (2013). Study on the influence of various factors in the hydrometallurgical processing of waste printed circuit boards for copper and gold recovery. *Waste Management* 33: 935–941. https://doi.org/10.1016/j.wasman.2013.01.003.

Bisceglie, F., Civati, D., Bonati, B., and Faraci, F.D. (2017). Reduction of potassium cyanide usage in a consolidated industrial process for gold recovery from wastes and scraps. *Journal of Cleaner Production* 142: 1810–1818. https://doi.org/10.1016/j.jclepro.2016.11.103.

Bobba, S., Carrara, S., Huisman, J. et al. (2020). *Critical Raw Materials for Strategic Technologies and Sectors in the EU – a Foresight Study*. European Union https://doi.org/10.2873/58081.

Botelho Junior, A.B., Dreisinger, D.B., Espinosa, D.C.R., and Tenório, J.A.S. (2018). Pre-reducing process kinetics to recover metals from nickel leach waste using chelating resins. *International Journal of Chemical Engineering* 2018: 1–7. https://doi.org/10.1155/2018/9161323.

Botelho Junior, A.B., Espinosa, D.C.R., Dreisinger, D., and Tenório, J.A.S. (2019). Recovery of nickel and cobalt from nickel laterite leach solution using chelating resins and pre-reducing process. *Canadian Journal of Chemical Engineering* 97: 1181–1190. https://doi.org/10.1002/cjce.23359.

Botelho Junior, A.B., Espinosa, D.C.R., and Tenório, J.A.S. (2021a). The use of computational thermodynamic for yttrium recovery from rare earth elements-bearing residue. *Journal of Rare Earths* 39: 201–207. https://doi.org/10.1016/j.jre.2020.02.019.

Botelho Junior, A.B., Espinosa, D.C.R., and Tenório, J.A.S. (2021b). Selective separation of Sc(III) and Zr(IV) from the leaching of bauxite residue using trialkylphosphine acids, tertiary amine, tri-butyl phosphate and their mixtures. *Separation and Purification Technology* 279: 119798. https://doi.org/10.1016/j.seppur.2021.119798.

Botelho Junior, A.B., Stopic, S., Friedrich, B. et al. (2021c). Cobalt recovery from Li-ion battery recycling: a critical review. *Metals (Basel)* 11: 1999. https://doi.org/10.3390/met11121999.

Botelho Junior, A.B., Pavoski, G., da Silva, C.R. et al. (2022). Promising technologies under development for recycling, remanufacturing, and reusing batteries: an introduction. In: *Nano Technol. Batter. Recycl. Remanufacturing, Reusing*, 79–104. Elsevier https://doi.org/10.1016/B978-0-323-91134-4.00006-6.

Breuer, P.L. and Jeffrey, M.I. (2002). An electrochemical study of gold leaching in thiosulfate solutions containing copper and ammonia. *Hydrometallurgy* 65: 145–157. https://doi.org/10.1016/S0304-386X(02)00086-5.

Bridge, G. and Faigen, E. (2022). Energy Research & Social Science Towards the lithium-ion battery production network: thinking beyond mineral supply chains. *Energy Research and Social Science* 89: 102659. https://doi.org/10.1016/j.erss.2022.102659.

van den Brink, S., Kleijn, R., Sprecher, B., and Tukker, A. (2020). Identifying supply risks by mapping the cobalt supply chain. *Resources, Conservation and Recycling* 156: 104743. https://doi.org/10.1016/j.resconrec.2020.104743.

Byerley, J.J., Fouda, S.A., and Rempel, G.L. (1973a). Kinetics and mechanism of the oxidation of thiosulphate ions by copper(II) ions in aqueous ammonia solution. *Journal of the Chemical Society, Dalton Transactions* 889: 889–893. https://doi.org/10.1039/DT9730000889.

Byerley, J.J., Fouda, S.A., and Rempel, G.L. (1973b). The oxidation of thiosulfate in aqueous ammonia by copper (II) oxygen complexes. *Inorganic and Nuclear Chemistry Letters* 9: 879–883. https://doi.org/10.1016/0020-1650(73)90041-X.

Byerley, J.J., Fouda, S.A., and Rempel, G.L. (1975). Activation of copper(II) ammine complexes by molecular oxygen for the oxidation of thiosulphate ions. *Journal of the Chemical Society, Dalton Transactions* 29: 1329–1338. https://doi.org/10.1039/DT9750001329.

Cenci, M.P., Dal Berto, F.C., Castillo, B.W., and Veit, H.M. (2020a). Precious and critical metals from wasted LED lamps: characterization and evaluation. *Environmental Technology* 0: 1–12. https://doi.org/10.1080/09593330.2020.1856939.

Cenci, M.P., Dal Berto, F.C., Schneider, E.L., and Veit, H.M. (2020b). Assessment of LED lamps components and materials for a recycling perspective. *Waste Management* 107: 285–293. https://doi.org/10.1016/j.wasman.2020.04.028.

Cenci, M.P., Dal Berto, F.C., Camargo, P.S.S., and Veit, H.M. (2021). Separation and concentration of valuable and critical materials from wasted LEDs by physical processes. *Waste Management* 120: 136–145. https://doi.org/10.1016/j.wasman.2020.11.023.

Chaudhary, A., Dwivedi, A., and Upadhyayula, S. (2021). Supercritical fluids as green solvents. In: *Handbook of Greener Synthesis of Nanomaterials and Compounds*, 26. Elsevier https://doi.org/10.1016/C2019-0-03121-4.

Chaves, R.M., Power, N.P., Collinson, S.R. et al. (2022). Development of Nylon 6 nanofibers modified with Cyanex-272 for cobalt recovery. *Environmental Technology (United Kingdom)* 1–13. https://doi.org/10.1080/09593330.2022.2047111.

Chen, W.S. and Ho, H.J. (2018). Recovery of valuable metals from lithium-ion batteries NMC cathode waste materials by hydrometallurgical methods. *Metals (Basel)* 8. https://doi.org/10.3390/met8050321.

Chen, M., Wang, R., Qi, Y. et al. (2021). Cobalt and lithium leaching from waste lithium ion batteries by glycine. *Journal of Power Sources* 482: 228942. https://doi.org/10.1016/j.jpowsour.2020.228942.

Chen, J., Meng, T., Leng, E., and E J. (2022). Review on metal dissolution characteristics and harmful metals recovery from electronic wastes by supercritical water. *Journal of Hazardous Materials* 424: 127693. https://doi.org/10.1016/j.jhazmat.2021.127693.

Cheng, S., Huang, C.M., and Pecht, M. (2017). A review of lead-free solders for electronics applications. *Microelectronics and Reliability* 75: 77–95. https://doi.org/10.1016/j.microrel.2017.06.016.

Choi, J.W., Song, M.H., Bediako, J.K., and Yun, Y.S. (2020). Sequential recovery of gold and copper from bioleached wastewater using ion exchange resins. *Environmental Pollution* 266: 115167. https://doi.org/10.1016/j.envpol.2020.115167.

Condomitti, U., Zuin, A., Silveira, A.T. et al. (2012). Magnetic nanohydrometallurgy: a promising nanotechnological approach for metal production and recovery using functionalized superparamagnetic nanoparticles. *Hydrometallurgy* 125–126: 148–151. https://doi.org/10.1016/j.hydromet.2012.06.005.

Correa, M.M.J., Silvas, F.P.C., Aliprandini, P. et al. (2018). Separation of copper from a leaching solution of printed circuit boards by using solvent extraction with D2EHPA. *Brazilian Journal of Chemical Engineering* 35: 919–930. https://doi.org/10.1590/0104-6632.20180353s20170144.

Crundwell, F.K., Moats, M.S., Ramachandran, V. et al. (2011). *Extractive Metallurgy of Nickel, Cobalt and Platinum-Group Metals. First*. Oxford: Elsevier.

Da Silva, D.G., De Melo, F.M., Silveira, A.T. et al. (2016). Introducing environmental and sustainable chemistry topics using a nanotechnology approach: removing hazardous metal ions by means of humic-acid-modified superparamagnetic nanoparticles. *Journal of Chemical Education* 93: 1929–1934. https://doi.org/10.1021/acs.jchemed.6b00172.

Deschenes, G. and Ritcey, G.M. (1990). Recovery of gold from aqueous solutions. US4913730A.

Dhawan, N. and Tanvar, H. (2022). A critical review of end-of-life fluorescent lamps recycling for recovery of rare earth values. Sustain. *Materials and Technologies* e00401. https://doi.org/10.1016/j.susmat.2022.e00401.

Dhiman, S. and Gupta, B. (2019). Partition studies on cobalt and recycling of valuable metals from waste Li-ion batteries via solvent extraction and chemical precipitation. *Journal of Cleaner Production* 225: 820–832. https://doi.org/10.1016/j.jclepro.2019.04.004.

Dhiman, S. and Gupta, B. (2020). Cyphos IL 104 assisted extraction of indium and recycling of indium, tin and zinc from discarded LCD screen. *Separation and Purification Technology* 237: 116407. https://doi.org/10.1016/j.seppur.2019.116407.

Dhiman, V. and Kondal, N. (2021). ZnO nanoadsorbents: a potent material for removal of heavy metal ions from wastewater. *Colloid and Interface Science Communications* 41: 100380. https://doi.org/10.1016/j.colcom.2021.100380.

Dias, P., Javimczik, S., Benevit, M. et al. (2016). Recycling WEEE: extraction and concentration of silver from waste crystalline silicon photovoltaic modules. *Waste Management* 57: 220–225. https://doi.org/10.1016/j.wasman.2016.03.016.

Dias, P., Schmidt, L., Monteiro Lunardi, M. et al. (2021). Comprehensive recycling of silicon photovoltaic modules incorporating organic solvent delamination – technical, environmental and economic analyses. *Resources, Conservation and Recycling* 165: 105241. https://doi.org/10.1016/j.resconrec.2020.105241.

Dong, B., Zhao, X., Zheng, L. et al. (2008). Aggregation behavior of long-chain imidazolium ionic liquids in aqueous solution: micellization and characterization of micelle microenvironment. *Colloids and Surfaces A: Physicochemical and Engineering Aspects* 317: 666–672. https://doi.org/10.1016/j.colsurfa.2007.12.001.

Drace, K., Kiefer, A.M., and Veiga, M.M. (2016). Cyanidation of mercury-contaminated tailings: potential health effects and environmental justice. *Current Environmental Health Reports* 3: 443–449. https://doi.org/10.1007/s40572-016-0113-0.

Durand, E., Lecomte, J., and Villeneuve, P. (2013). Deep eutectic solvents: synthesis, application, and focus on lipase-catalyzed reactions. *European Journal of Lipid Science and Technology* 115: 379–385. https://doi.org/10.1002/ejlt.201200416.

Dutta, D., Panda, R., Kumari, A. et al. (2018). Sustainable recycling process for metals recovery from used printed circuit boards (PCBs). *Sustainable Materials and Technologies* 17: e00066. https://doi.org/10.1016/j.susmat.2018.e00066.

Espinosa, D.C.R. and Mansur, M.B. (2019). *Recycling Batteries*. Elsevier Ltd https://doi.org/10.1016/B978-0-08-102158-3.00014-8.

Feng, D. and van Deventer, S.J. (2010). Thiosulphate leaching of gold in the presence of ethylenediaminetetraacetic acid (EDTA). *Minerals Engineering* 23: 143–150. https://doi.org/10.1016/j.mineng.2009.11.009.

Feng, D. and Van Deventer, J.S.J. (2011). The role of amino acids in the thiosulphate leaching of gold. *Minerals Engineering* 24: 1022–1024. https://doi.org/10.1016/j.mineng.2011.04.017.

Fogarasi, S., Imre-lucaci, F., Imre-lucaci, Á., and Ilea, P. (2014). Copper recovery and gold enrichment from waste printed circuit boards by mediated electrochemical oxidation. *Journal of Hazardous Materials* 273: 215–221. https://doi.org/10.1016/j.jhazmat.2014.03.043.

Forti, V., Baldé, C.P., Kuehr, R., and Bel, G. (2020). The Global E-waste Monitor 2020. Global E-Waste 2020. https://www.itu.int/myitu/-/media/Publications/2020-Publications/Global-E-waste-Monitor-2020.pdf (accessed 22 November 2021).

Fu, Y., Schuster, J., Petranikova, M., and Ebin, B. (2021). Innovative recycling of organic binders from electric vehicle lithium-ion batteries by supercritical carbon dioxide extraction. *Resources, Conservation and Recycling* 172: 105666. https://doi.org/10.1016/j.resconrec.2021.105666.

Gabriel, A.P., Kasper, A.C., and Veit, H.M. (2020). Acid leaching of indium from the screens of obsolete LCD monitors. *Journal of Environmental Chemical Engineering* 8: 103758. https://doi.org/10.1016/j.jece.2020.103758.

Gámez, S., Garcés, K., de la Torre, E., and Guevara, A. (2019). Precious metals recovery from waste printed circuit boards using thiosulfate leaching and ion exchange resin. *Hydrometallurgy* 186: 1–11. https://doi.org/10.1016/j.hydromet.2019.03.004.

García, O., Veiga, M.M., Cordy, P. et al. (2015). Artisanal gold mining in Antioquia, Colombia: a successful case of mercury reduction. *Journal of Cleaner Production* 90: 244–252. https://doi.org/10.1016/j.jclepro.2014.11.032.

Golmohammadzadeh, R., Rashchi, F., and Vahidi, E. (2017). Recovery of lithium and cobalt from spent lithium-ion batteries using organic acids: Process optimization and kinetic aspects. *Waste Management* 64: 244–254. https://doi.org/10.1016/j.wasman.2017.03.037.

Golzary, A. and Abdoli, M.A. (2020). Recycling of copper from waste printed circuit boards by modified supercritical carbon dioxide combined with supercritical water pretreatment. *Journal of CO_2 Utilization* 41: 101265. https://doi.org/10.1016/j.jcou.2020.101265.

Gönen, N., Kabasakal, O.S., and Özdil, G. (2004). Recovery of cyanide in gold leach waste solution by volatilization and absorption. *Journal of Hazardous Materials* 113: 231–236. https://doi.org/10.1016/j.jhazmat.2004.06.029.

Groenewald, T. and Jones, B.M. (1971). Determination of gold in solutions of thiourea. *Analytical Chemistry* 43: 1689–1691. https://doi.org/10.1021/ac60306a033.

Grosse, A.C., Dicinoski, G.W., Shaw, M.J., and Haddad, P.R. (2003). Leaching and recovery of gold using ammoniacal thiosulfate leach liquors (a review). *Hydrometallurgy* 69: 1–21. https://doi.org/10.1016/S0304-386X(02)00169-X.

Grützke, M., Kraft, V., Weber, W. et al. (2014). Supercritical carbon dioxide extraction of lithium-ion battery electrolytes. *Journal of Supercritical Fluids* 94: 216–222. https://doi.org/10.1016/j.supflu.2014.07.014.

Gui, Q., Ibrar, M., Wang, S., and Zhang, L. (2020). Hydrometallurgy The ultrasound leaching kinetics of gold in the thiosulfate leaching process catalysed by cobalt ammonia. *Hydrometallurgy* 196: 105426. https://doi.org/10.1016/j.hydromet.2020.105426.

Guimarães, L.F., Botelho Junior, A.B., and Espinosa, D.C.R. (2022). Sulfuric acid leaching of metals from waste Li-ion batteries without using reducing agent. *Minerals Engineering* 183: 107597. https://doi.org/10.1016/j.mineng.2022.107597.

Gupta CK, Mukherjee TK. *Hydrometallurgy in Extraction Processes*. 1990.

Ha, V.H., Lee, J., Jeong, J. et al. (2010). Thiosulfate leaching of gold from waste mobile phones. *Journal of Hazardous Materials* 178: 1115–1119. https://doi.org/10.1016/j.jhazmat.2010.01.099.

Habashi, F. (2005). A short history of hydrometallurgy. *Hydrometallurgy* 79: 15–22. https://doi.org/10.1016/j.hydromet.2004.01.008.

Habib, M., Miles, N.J., and Hall, P. (2013). Recovering metallic fractions from waste electrical and electronic equipment by a novel vibration system. *Waste Management* 33: 722–729. https://doi.org/10.1016/j.wasman.2012.11.017.

Hagiwara, R. and Ito, Y. (2000). Room temperature ionic liquids of alkylimidazolium cations and fluoroanions. *Journal of Fluorine Chemistry* 105: 221–227. https://doi.org/10.1016/S0022-1139(99)00267-5.

Hammond, O.S., Bowron, D.T., and Edler, K.J. (2019). Structure and properties of "type IV" lanthanide nitrate hydrate:urea deep eutectic solvents. *ACS Sustainable Chemistry & Engineering* 7: 4932–4940. https://doi.org/10.1021/acssuschemeng.8b05548.

Han, K.N. (2002). *Fundamentals of Aqueous Metallurgy*, 1ste, vol. 1. Littleton: Society for Mining, Metallurgy, and Exploration, Inc.

Havlík, T. (2008). *Hydrometallurgy: Principles and Application*, vol. 61. Cambridge: Cambridge International Science Publishing Limited https://doi.org/10.1016/S0304-386X(01)00178-5.

Hernandez, J., Patino, F., Rivera, I. et al. (2014). Leaching kinetics in cyanide media of Ag contained in the industrial mining-metallurgical wastes in the state of Hidalgo, Mexico. *International Journal of Mining Science and Technology* 24: 689–694. https://doi.org/10.1016/j.ijmst.2014.07.003.

Hooshmand, S.E., Afshari, R., Ramón, D.J., and Varma, R.S. (2020). Deep eutectic solvents: cutting-edge applications in cross-coupling reactions. *Green Chemistry* 22: 3668–3692. https://doi.org/10.1039/d0gc01494j.

Hsu, E., Durning, C.J., West, A.C., and Park, A.H.A. (2021). Enhanced extraction of copper from electronic waste via induced morphological changes using supercritical CO2. *Resources, Conservation and Recycling* 168: 105296. https://doi.org/10.1016/j.resconrec.2020.105296.

Hu, J., Tian, G., Zi, F., and Hu, X. (2017). Leaching of chalcopyrite with hydrogen peroxide in 1-hexyl-3-methyl-imidazolium hydrogen sulfate ionic liquid aqueous solution. *Hydrometallurgy* 169: 1–8. https://doi.org/10.1016/j.hydromet.2016.12.001.

Huddleston, J.G., Visser, A.E., Reichert, W.M. et al. (2001). Characterization and comparison of hydrophilic and hydrophobic room temperature ionic liquids incorporating the imidazolium cation. *Green Chemistry* 3: 156–164. https://doi.org/10.1039/b103275p.

Hurley, F.H. and Wier, T.P. (1951). Electrodeposition of metals from fused quaternary ammonium salts. *Journal of the Electrochemical Society* 98: 203. https://doi.org/10.1149/1.2778132.

Ibn-Mohammed, T., Mustapha, K.B., Godsell, J. et al. (2021). A critical review of the impacts of COVID-19 on the global economy and ecosystems and opportunities for circular economy strategies. *Resources, Conservation and Recycling* 164: 105169. https://doi.org/10.1016/j.resconrec.2020.105169.

Inamuddin, M.L. (2012). *Ion Exchange Technology I*. Dordrecht: Springer Netherlands https://doi.org/10.1007/978-94-007-1700-8.

Innocenzi, V., Ippolito, N.M., Pietrelli, L. et al. (2018). Application of solvent extraction operation to recover rare earths from fluorescent lamps. *Journal of Cleaner Production* 172: 2840–2852. https://doi.org/10.1016/j.jclepro.2017.11.129.

Jadhao, P.R., Pandey, A., Pant, K.K., and Nigam, K.D.P. (2021). Efficient recovery of Cu and Ni from WPCB via alkali leaching approach. *Journal of Environmental Management* 296: 113154. https://doi.org/10.1016/j.jenvman.2021.113154.

Jadhav, U. and Hocheng, H. (2015). Hydrometallurgical recovery of metals from large printed circuit board pieces. *Scientific Reports* 5. https://doi.org/10.1038/srep14574.

Jeffrey, M.I., Breuer, P.L., and Choo, W.L. (2001). A kinetic study that compares the leaching of gold in the cyanide, thiosulfate, and chloride systems. *Metallurgical and Materials Transactions B: Process Metallurgy and Materials Processing Science* 32: 979–986. https://doi.org/10.1007/s11663-001-0086-7.

Jing-ying, L., Xiu-li, X., and Wen-quan, L. (2012). Thiourea leaching gold and silver from the printed circuit boards of waste mobile phones. *Waste Management* 32: 1209–1212. https://doi.org/10.1016/j.wasman.2012.01.026.

Jowitt, S.M., Mudd, G.M., Werner, T.T. et al. (2018). The critical metals: an overview and opportunities and concerns for the future. In: *Met. Miner. Soc.*, 1ste (ed. A. Arribas, D. Broughton, and J. Mauk), 25–38. Lawrence: Society of Economic Geologists https://doi.org/10.5382/SP.21.02.

Kamali Ardakani, E., Kowsari, E., and Ehsani, A. (2020). Imidazolium-derived polymeric ionic liquid as a green inhibitor for corrosion inhibition of mild steel in 1.0 M HCl: experimental and computational study. *Colloids and Surfaces A: Physicochemical and Engineering Aspects* 586: 124195. https://doi.org/10.1016/j.colsurfa.2019.124195.

Kaya, M. (2018). *Current WEEE Recycling Solutions*. Elsevier Ltd https://doi.org/10.1016/B978-0-08-102057-9.00003-2.

Kaya M. *Electronic Waste and Printed Circuit Board Recycling Technologies*. 2019. https://doi.org/10.1007/978-3-030-26593-9.

Kelley, Steven P, J Seth Nuss and RDR. *Application of Ionic Liquids on Rare Earth Green Separation and Utilization*, 2016.

Kim, E.Y., Kim, M.S., Lee, J.C., and Pandey, B.D. (2011). Selective recovery of gold from waste mobile phone PCBs by hydrometallurgical process. *Journal of Hazardous Materials* 198: 206–215. https://doi.org/10.1016/j.jhazmat.2011.10.034.

Kislik, V.S. (2014). Solvent Extraction: Classical and Novel Approaches. https://doi.org/10.1007/s13398-014-0173-7.2.

Kołodyńska, D., Fila, D., Gajda, B. et al. (2019). Rare earth elements – Separation methods yesterday and today. In: *Appl. Ion Exch. Mater. Environ.*, 161–185. Cham: Springer International Publishing https://doi.org/10.1007/978-3-030-10430-6_8.

Krzewska, S. and Podsiadły, H. (1980). Studies on the reaction of copper(II) with thiourea – I. Silver-silver thiourea electrode for determination of free thiourea concentration in HClO4 medium. *Journal of Inorganic and Nuclear Chemistry* 42: 83–85. https://doi.org/10.1016/0022-1902(80)80048-0.

Kumar, A., Holuszko, M., and Espinosa, D.C.R. (2017). E-waste: an overview on generation, collection, legislation and recycling practices. *Resources, Conservation and Recycling* 122: 32–42. https://doi.org/10.1016/j.resconrec.2017.01.018.

Kumar, J., Shen, X., Li, B. et al. (2020). Selective recovery of Li and FePO4 from spent LiFePO4 cathode scraps by organic acids and the properties of the regenerated LiFePO4. *Waste Management* 113: 32–40. https://doi.org/10.1016/j.wasman.2020.05.046.

Lai, X., Huang, Y., Gu, H. et al. (2021). Turning waste into wealth: a systematic review on echelon utilization and material recycling of retired lithium-ion batteries. *Energy Storage Materials* 40: 96–123. https://doi.org/10.1016/j.ensm.2021.05.010.

Langhans, J.W., Lei, K.P.V., and Carnahan, T.G. (1992). Copper-catalyzed thiosulfate leaching of low-grade gold ores. *Hydrometallurgy* 29: 191–203. https://doi.org/10.1016/0304-386X(92)90013-P.

Li, K. and Xu, Z. (2019). A review of current progress of supercritical fluid technologies for e-waste treatment. *Journal of Cleaner Production* 227: 794–809. https://doi.org/10.1016/j.jclepro.2019.04.104.

Li, L., Ge, J., Chen, R. et al. (2010a). Environmental friendly leaching reagent for cobalt and lithium recovery from spent lithium-ion batteries. *Waste Management* 30: 2615–2621. https://doi.org/10.1016/j.wasman.2010.08.008.

Li, L., Ge, J., Wu, F. et al. (2010b). Recovery of cobalt and lithium from spent lithium ion batteries using organic citric acid as leachant. *Journal of Hazardous Materials* 176: 288–293. https://doi.org/10.1016/j.jhazmat.2009.11.026.

Li, L., Lu, J., Ren, Y. et al. (2012). Ascorbic-acid-assisted recovery of cobalt and lithium from spent Li-ion batteries. *Journal of Power Sources* 218: 21–27. https://doi.org/10.1016/j.jpowsour.2012.06.068.

Li, L., Dunn, J.B., Zhang, X.X. et al. (2013). Recovery of metals from spent lithium-ion batteries with organic acids as leaching reagents and environmental assessment. *Journal of Power Sources* 233: 180–189. https://doi.org/10.1016/j.jpowsour.2012.12.089.

Li, X., Li, Z., Orefice, M., and Binnemans, K. (2019). Metal recovery from spent samarium-cobalt magnets using a trichloride ionic liquid. *ACS Sustainable Chemistry & Engineering* 7: 2578–2584. https://doi.org/10.1021/acssuschemeng.8b05604.

Li, H., Oraby, E., and Eksteen, J. (2020). Extraction of copper and the co-leaching behaviour of other metals from waste printed circuit boards using alkaline glycine solutions. *Resources, Conservation and Recycling* 154: 104624. https://doi.org/10.1016/j.resconrec.2019.104624.

Liang, G., Tang, J., Liu, W., and Zhou, Q. (2013). Optimizing mixed culture of two acidophiles to improve copper recovery from printed circuit boards (PCBs). *Journal of Hazardous Materials* 250–251: 238–245. https://doi.org/10.1016/j.jhazmat.2013.01.077.

Lie, J. and Liu, J.C. (2021). Closed-vessel microwave leaching of valuable metals from spent lithium-ion batteries (LIBs) using dual-function leaching agent: ascorbic acid. *Separation and Purification Technology* 266. https://doi.org/10.1016/j.seppur.2021.118458.

Liu, K., Zhang, Z., and Zhang, F.S. (2016). Direct extraction of palladium and silver from waste printed circuit boards powder by supercritical fluids oxidation-extraction process. *Journal of Hazardous Materials* 318: 216–223. https://doi.org/10.1016/j.jhazmat.2016.07.005.

Liu, X., Jiang, T., Xu, B. et al. (2020). Thiosulphate leaching of gold in the Cu–NH3–S2O32−H2O system: an updated thermodynamic analysis using predominance area and species distribution diagrams. *Minerals Engineering* 151: 106336. https://doi.org/10.1016/j.mineng.2020.106336.

Lobato, N.C.C., Ferreira, Â.D.M., and Mansur, M.B. (2016). Evaluation of magnetic nanoparticles coated by oleic acid applied to solvent extraction processes. *Separation and Purification Technology* 168: 93–100. https://doi.org/10.1016/j.seppur.2016.05.027.

Łukomska, A., Wiśniewska, A., Dąbrowski, Z., and Domańska, U. (2020). Liquid-liquid extraction of cobalt(II) and zinc(II) from aqueous solutions using novel ionic liquids as an extractants. *Journal of Molecular Liquids* 307. https://doi.org/10.1016/j.molliq.2020.112955.

Łukomska, A., Wiśniewska, A., Dąbrowski, Z. et al. (2021). Recovery of zinc and manganese from "black mass" of waste Zn-MnO2 alkaline batteries by solvent extraction technique with ionic liquids, DESs and organophosphorous-based acids. *Journal of Molecular Liquids* 338. https://doi.org/10.1016/j.molliq.2021.116590.

Ma, E. (2019). Recovery of waste printed circuit boards through pyrometallurgy. *Electronic Waste Management and Treatment Technology* 247–267. https://doi.org/10.1016/B978-0-12-816190-6.00011-X.

Maatar, W. and Boufi, S. (2015). Poly(methacylic acid-co-maleic acid) grafted nanofibrillated cellulose as a reusable novel heavy metal ions adsorbent. *Carbohydrate Polymers* 126: 199–207. https://doi.org/10.1016/j.carbpol.2015.03.015.

Marchbank, A.R., Thomas, K.G., Dreisinger, D., and Fleming, C. (1998). Gold recovery from refractory carbonaceous ores by pressure oxidation, thiosulfate leaching and resin-in-pulp adsorption. US5785736A.

Marsden, J.O. and House, I. (2006). *The Chemistry of Gold Extraction*. SME.

Martell, A.E. and Smith, R.M. (1989). Critical Stability Constants., vol. 6. https://doi.org/10.1007/978-1-4615-6764-6.

Martins, T.A.G., Caldas, M.P.K., de Moraes, V.T. et al. (2021a). Recovering metals from motherboard and memory board waste through sulfuric leaching. *Journal of Environmental Chemical Engineering* 9. https://doi.org/10.1016/j.jece.2021.106789.

Martins, L.S., Guimarães, L.F., Botelho Junior, A.B. et al. (2021b). Electric car battery: an overview on global demand, recycling and future approaches towards sustainability. *Journal of Environmental Management* 295: 113091. https://doi.org/10.1016/j.jenvman.2021.113091.

Maurice, A., Theisen, J., and Gabriel, J.C.P. (2020). Microfluidic lab-on-chip advances for liquid–liquid extraction process studies. *Current Opinion in Colloid & Interface Science* 46: 20–35. https://doi.org/10.1016/j.cocis.2020.03.001.

MCTIC. Plano de ciência, tecnologia e inovação para minerais estratéticos 2018: 50. https://www.inova.rs.gov.br/upload/arquivos/202006/16181825-plano-de-ciencia-tecnologia-e-inovacao-para-minerais-estrategicos.pdf (accessed 25 August 2020).

Mecucci, A. and Scott, K. (2002). Leaching and electrochemical recovery of copper, lead and tin from scrap printed circuit boards. *Journal of Chemical Technology and Biotechnology* 77: 449–457. https://doi.org/10.1002/jctb.575.

Melo, F.M., Silveira, A.T., Quartarolli, L.F. et al. (2019). Magnetic behavior of superparamagnetic nanoparticles containing chelated transition metal ions. *Journal of Magnetism and Magnetic Materials* 487: 165324. https://doi.org/10.1016/j.jmmm.2019.165324.

Merli, G., Becci, A., and Amato, A. (2022). Recovery of precious metals from printed circuit boards by cyanogenic bacteria: optimization of cyanide production by statistical analysis. *Journal of Environmental Chemical Engineering* 10: 107495. https://doi.org/10.1016/j.jece.2022.107495.

Micheau, C., Lejeune, M., Arrachart, G. et al. (2019). Recovery of tantalum from synthetic sulfuric leach solutions by solvent extraction with phosphonate functionalized ionic liquids. *Hydrometallurgy* 189: 105107. https://doi.org/10.1016/j.hydromet.2019.105107.

Moldoveanu, S., David, V., Comments, G. et al. (2015). Solvent Extraction. In: *Mod. Sample Prep. Chromatogr.*, 131–189. Elsevier https://doi.org/10.1016/B978-0-444-54319-6.00006-2.

Muir, D.M., La Brooy, S.R., and Cao, C. (1993). Recovery of gold from copper-bearing ores. *World Gold* 89: 363–374. https://doi.org/10.2307/j.ctv1tbhrfw.49.

München, D.D., Bernardes, A.M., and Veit, H.M. (2018). Evaluation of neodymium and praseodymium leaching efficiency from post-consumer NdFeB magnets. *Journal of Sustainable Metallurgy* 4: 288–294. https://doi.org/10.1007/s40831-018-0180-6.

Nadimi, H. and Karazmoudeh, N.J. (2021). Selective separation and purification of Mn from Co and Ni in waste mobile phone lithium-ion batteries using D2EHAP via solvent extraction method. *Journal of Sustainable Metallurgy* 7: 653–663. https://doi.org/10.1007/s40831-021-00371-1.

Naughton, C.C. (2020). Will the COVID-19 pandemic change waste generation and composition?: The need for more real-time waste management data and systems thinking. *Resources, Conservation and Recycling* 162: 105050. https://doi.org/10.1016/j.resconrec.2020.105050.

Nekouei, R.K., Pahlevani, F., Assefi, M. et al. (2019). Selective isolation of heavy metals from spent electronic waste solution by macroporous ion-exchange resins. *Journal of Hazardous Materials* 371: 389–396. https://doi.org/10.1016/j.jhazmat.2019.03.013.

Neto, I.F.F. and Soares, H.M.V.M. (2018). Sequential separation of Ag, Al, Cu and Pb from a multi-metal leached solution using a zero waste technology. *Separation Science and Technology* 53: 2961–2970. https://doi.org/10.1080/01496395.2018.1482342.

Neto, I.F.F., Sousa, C.A., Brito, M.S.C.A. et al. (2016). A simple and nearly-closed cycle process for recycling copper with high purity from end life printed circuit boards. *Separation and Purification Technology* 164: 19–27. https://doi.org/10.1016/j.seppur.2016.03.007.

Oliveira, R.P., Botelho Junior, A.B., and DCR, E. (2020). Characterization of wasted LEDs from tubular lamps focused on recycling process by hydrometallurgy. In: *Energy Technology 2020: Recycling, Carbon Dioxide Manageent and Other Technologies*, 317–325. Springer International Publishing https://doi.org/10.1007/978-3-030-36830-2_30.

de Oliveira, R.P., Benvenuti, J., and Espinosa, D.C.R. (2021). A review of the current progress in recycling technologies for gallium and rare earth elements from light-emitting diodes. *Renewable and Sustainable Energy Reviews* 145: 111090. https://doi.org/10.1016/j.rser.2021.111090.

Osibanjo, O. and Nnorom, I.C. (2008). Material flows of mobile phones and accessories in Nigeria: environmental implications and sound end-of-life management options. *Environmental Impact Assessment Review* 28: 198–213. https://doi.org/10.1016/j.eiar.2007.06.002.

Padwal, C., Pham, H.D., Jadhav, S. et al. (2022). Deep eutectic solvents: green approach for cathode recycling of Li-ion batteries. *Advanced Energy and Sustainability Research* 3: 2100133. https://doi.org/10.1002/aesr.202100133.

Park, J., Jung, Y., Kusumah, P. et al. (2014). Application of ionic liquids in hydrometallurgy. *International Journal of Molecular Sciences* 15: 15320–15343. https://doi.org/10.3390/ijms150915320.

Patil, A.B., Tarik, M., Struis, R.P.W.J., and Ludwig, C. (2021). Exploiting end-of-life lamps fluorescent powder e-waste as a secondary resource for critical rare earth metals. *Resources, Conservation and Recycling* 164: 105153. https://doi.org/10.1016/j.resconrec.2020.105153.

Pavón, S., Lapo, B., Fortuny, A. et al. (2021). *Recycling of rare earths from fluorescent lamp waste by the integration of solid-state chlorination, leaching and solvent extraction processes*, 272. Sep Purif Technol https://doi.org/10.1016/j.seppur.2021.118879.

Peeters, N., Binnemans, K., and Riaño, S. (2020). Solvometallurgical recovery of cobalt from lithium-ion battery cathode materials using deep-eutectic solvents. *Green Chemistry* 22: 4210–4221. https://doi.org/10.1039/d0gc00940g.

Perea, C.G., Baena, O.J.R., Ihle, C.F., and Estay, H. (2021). Copper leaching from wastes electrical and electronic equipment (WEEE) using alkaline monosodium glutamate: thermodynamics and dissolution tests. *Cleaner Engineering and Technology* 5: 100312. https://doi.org/10.1016/j.clet.2021.100312.

Perez, I.D., Anes, I.A., Botelho Junior, A.B., and Espinosa, D.C.R. (2020). Comparative study of selective copper recovery techniques from nickel laterite leach waste towards a competitive sustainable extractive process. *Cleaner Engineering and Technology* 1: 100031. https://doi.org/10.1016/j.clet.2020.100031.

Perna, F.M., Vitale, P., and Capriati, V. (2020). Deep eutectic solvents and their applications as green solvents. *Current Opinion in Green and Sustainable Chemistry* 21: 27–33. https://doi.org/10.1016/j.cogsc.2019.09.004.

Popescu, A.M., Constantin, V., Cojocaru, A., and Olteanu, M. (2011). Electrochemical behaviour of copper (II) chloride in choline chloride-urea deep eutectic solvent. *Revista de Chimie* 62: 206–211.

Pourbaix, M. (1974). *Atlas of Electrochemical Equilibria in Aqueous Solutions*, seconde. Houston: National Association of Corrosion Engineers.

Puente-Siller, D.M., Fuentes-Aceituno, J.C., and Nava-Alonso, F. (2013). A kinetic-thermodynamic study of silver leaching in thiosulfate-copper-ammonia-EDTA solutions. *Hydrometallurgy* 134–135: 124–131. https://doi.org/10.1016/j.hydromet.2013.02.010.

Qi, D. (2018). Hydrometallurgy of Rare Earths. https://doi.org/10.1016/c2016-0-05328-7.

Raj, T., Chandrasekhar, K., Kumar, A.N. et al. (2022). Recycling of cathode material from spent lithium-ion batteries: challenges and future perspectives. *Journal of Hazardous Materials* 429: 128312. https://doi.org/10.1016/j.jhazmat.2022.128312.

Rao, G.P., Lu, C., and Su, F. (2007). Sorption of divalent metal ions from aqueous solution by carbon nanotubes: a review. *Separation and Purification Technology* 58: 224–231. https://doi.org/10.1016/j.seppur.2006.12.006.

Rao, M.D., Singh, K.K., Morrison, C.A., and Love, J.B. (2021). Recycling copper and gold from e-waste by a two-stage leaching and solvent extraction process. *Separation and Purification Technology* 263: 118400. https://doi.org/10.1016/j.seppur.2021.118400.

Rao, M.D., Singh, K.K., Morrison, C.A., and Love, J.B. (2022). Selective recovery of nickel from obsolete mobile phone PCBs. *Hydrometallurgy* 210: 105843. https://doi.org/10.1016/j.hydromet.2022.105843.

Santanilla, A.J.M., Aliprandini, P., Benvenuti, J. et al. (2021). Structure investigation for nickel and cobalt complexes formed during solvent extraction with the extractants Cyanex 272, Versatic 10 and their mixtures. *Minerals Engineering* 160: 106691. https://doi.org/10.1016/j.mineng.2020.106691.

Schlesinger, M.E., King, M.J., Sole, K.C., and Davenport, W.G. (2011). Solvent extraction. In: *Extr. Metall. Copp.*, 323–347. Elsevier https://doi.org/10.1016/B978-0-08-096789-9.10016-2.

Segala, B.N., Bertuol, D.A., and Tanabe, E.H. (2022). Production of polyacrylonitrile nanofibres modified with Cyanex 272 for recovery of gallium from solution. *Environ Technol (United Kingdom)* 43: 737–750. https://doi.org/10.1080/09593330.2020.1803991.

Senćanski, J., Bajuk-Bogdanović, D., Majstorović, D. et al. (2017). The synthesis of Li(Co Mn Ni)O2 cathode material from spent-Li ion batteries and the proof of its functionality in aqueous lithium and sodium electrolytic solutions. *Journal of Power Sources* 342: 690–703. https://doi.org/10.1016/j.jpowsour.2016.12.115.

Shivagan, D.D., Dale, P.J., Samantilleke, A.P., and Peter, L.M. (2007). Electrodeposition of chalcopyrite films from ionic liquid electrolytes. *Thin Solid Films* 515: 5899–5903. https://doi.org/10.1016/j.tsf.2006.12.092.

Smith, N.M. (2019). "Our gold is dirty, but we want to improve": challenges to addressing mercury use in artisanal and small-scale gold mining in Peru. *Journal of Cleaner Production* 222: 646–654. https://doi.org/10.1016/j.jclepro.2019.03.076.

Smith, E.L., Abbott, A.P., and Ryder, K.S. (2014). Deep eutectic solvents (DESs) and their applications. *Chemical Reviews* 114: 11060–11082. https://doi.org/10.1021/cr300162p.

Sole, K.C., Mooiman, M.B., and Hardwick, E. (2018). Ion exchange in hydrometallurgical processing: an overview and selected applications. *Separation and Purification Reviews* 47: 159–178. https://doi.org/10.1080/15422119.2017.1354304.

Strauss, M.L., Diaz, L.A., McNally, J. et al. (2021). Separation of cobalt, nickel, and manganese in leach solutions of waste lithium-ion batteries using Dowex M4195 ion exchange resin. *Hydrometallurgy* 206: 105757. https://doi.org/10.1016/j.hydromet.2021.105757.

Suzuki, T., Nakamura, T., Inoue, Y. et al. (2012). A hydrometallurgical process for the separation of aluminum, cobalt, copper and lithium in acidic sulfate media. *Separation and Purification Technology* 98: 396–401. https://doi.org/10.1016/j.seppur.2012.06.034.

Syed, S. (2012). Recovery of gold from secondary sources-A review. *Hydrometallurgy* 115–116: 30–51. https://doi.org/10.1016/j.hydromet.2011.12.012.

Takahashi, V.C.I., Botelho Junior, A.B., Espinosa, D.C.R., and Tenório, J.A.S. (2020). Enhancing cobalt recovery from Li-ion batteries using grinding treatment prior to the leaching and solvent extraction process. *Journal of Environmental Chemical Engineering* 8: 103801. https://doi.org/10.1016/j.jece.2020.103801.

Toma, H.E. (2013). Developing nanotechnological strategies for green industrial processes. *Pure and Applied Chemistry* 85: 1655–1669. https://doi.org/10.1351/PAC-CON-12-12-02.

Toma, H.E. (2015). Magnetic nanohydrometallurgy: a nanotechnological approach to elemental sustainability. *Green Chemistry* 17: 2027–2041. https://doi.org/10.1039/c5gc00066a.

Tuncuk, A., Stazi, V., Akcil, A. et al. (2012). Aqueous metal recovery techniques from e-scrap: hydrometallurgy in recycling. *Minerals Engineering* 25: 28–37. https://doi.org/10.1016/j.mineng.2011.09.019.

Tunsu, C., Petranikova, M., Ekberg, C., and Retegan, T. (2016). A hydrometallurgical process for the recovery of rare earth elements from fluorescent lamp waste fractions. *Separation and Purification Technology* 161: 172–186. https://doi.org/10.1016/j.seppur.2016.01.048.

U.S. Department of Energy. Critical materials strategy. 2010. https://doi.org/10.1109/DSP-SPE.2013.6642605.

Ubaldini, S., Vegliò, F., Fornari, P., and Abbruzzese, C. (2000). Process flow-sheet for gold and antimony recovery from stibnite. *Hydrometallurgy* 57: 187–199. https://doi.org/10.1016/S0304-386X(00)00107-9.

UNEP. Environmental risks and challenge of anthropogenic metlas flows and cycles. 2013.

Usmani, Z., Sharma, M., Gupta, P. et al. (2020). Ionic liquid based pretreatment of lignocellulosic biomass for enhanced bioconversion. *Bioresource Technology* 304: 123003. https://doi.org/10.1016/j.biortech.2020.123003.

Van Den Bossche, A., Vereycken, W., Vander Hoogerstraete, T. et al. (2019). Recovery of gallium, indium, and arsenic from semiconductors using tribromide ionic liquids. *ACS Sustainable Chemistry & Engineering* 7: 14451–14459. https://doi.org/10.1021/acssuschemeng.9b01724.

Veit, H.M., Bernardes, A.M., Ferreira, J.Z. et al. (2006). Recovery of copper from printed circuit boards scraps by mechanical processing and electrometallurgy. *Journal of Hazardous Materials* 137: 1704–1709. https://doi.org/10.1016/j.jhazmat.2006.05.010.

Vekasi, K. (2019). Politics, markets, and rare commodities: responses to Chinese rare earth policy. *Japanese Journal of Political Science* 20: 2–20. https://doi.org/10.1017/s1468109918000385.

Velázquez-Martínez, V., Santasalo-Aarnio, R., and Serna-Guerrero. (2019). A critical review of lithium-ion battery recycling processes from a circular economy perspective. *Batteries* 5: 68. https://doi.org/10.3390/batteries5040068.

Vieceli, N., Reinhardt, N., Ekberg, C., and Petranikova, M. (2020). Optimization of manganese recovery from a solution based on lithium-ion batteries by solvent extraction with D2EHPA. *Metals (Basel)* 11: 54. https://doi.org/10.3390/met11010054.

Vignes, A. (2013). *Extractive Metallurgy 3 – Processing Operations and Routes*. Hoboken, NJ: John Wiley & Sons, Inc. https://doi.org/10.1002/9781118617106.

Virolainen, S., Wesselborg, T., Kaukinen, A., and Sainio, T. (2021). Removal of iron, aluminium, manganese and copper from leach solutions of lithium-ion battery waste using ion exchange. *Hydrometallurgy* 202: 105602. https://doi.org/10.1016/j.hydromet.2021.105602.

Wang, Q., Hu, X., Zi, F. et al. (2019). Environmentally friendly extraction of gold from refractory concentrate using a copper e ethylenediamine e thiosulfate solution. *Journal of Cleaner Production* 214: 860–872. https://doi.org/10.1016/j.jclepro.2019.01.007.

Wang, J., Xie, F., Wang, W. et al. (2020). Eco-friendly leaching of gold from a carbonaceous gold concentrate in copper-citrate-thiosulfate solutions. *Hydrometallurgy* 191: 105204. https://doi.org/10.1016/j.hydromet.2019.105204.

Wang, J., Faraji, F., Ramsay, J., and Ghahreman, A. (2021). A review of biocyanidation as a sustainable route for gold recovery from primary and secondary low-grade resources. *Journal of Cleaner Production* 296: 126457. https://doi.org/10.1016/j.jclepro.2021.126457.

Werner, T.T., Mudd, G.M., and Jowitt, S.M. (2015). Indium: key issues in assessing mineral resources and long-term supply from recycling. *Transactions of the Institutions of Mining and Metallurgy: Section B Applied Earth Science* 124: 213–226. https://doi.org/10.1179/1743275815Y.0000000007.

Wilkinson, G. and Gillard, R.D. (1987). Comprehensive coordination chemistry The synthesis, reactions, properties and applications of coordination compounds V3 Main group and early transition elements., 19.

Xu, W.-G., Lü, X.-M., Zhang, Q.-G. et al. (2006). Studies on the thermodynamic properties of the ionic liquid BMIGaCl4. *Chinese Journal of Chemistry* 24: 331–335. https://doi.org/10.1002/cjoc.200690063.

Xu, B., Yang, Y., Li, Q. et al. (2016). Stage leaching of a complex polymetallic sulfide concentrate: focus on the extraction of Ag and Au. *Hydrometallurgy* 159: 87–94. https://doi.org/10.1016/j.hydromet.2015.10.008.

Xu, B., Li, K., Dong, Z. et al. (2019). Eco-friendly and economical gold extraction by nickel catalyzed ammoniacal thiosulfate leaching-resin adsorption recovery. *Journal of Cleaner Production* 233: 1475–1485. https://doi.org/10.1016/j.jclepro.2019.06.182.

Xuan, W., de Souza, B.A., Korbel, C., and Chagnes, A. (2021). New insights in the leaching kinetics of cathodic materials in acidic chloride media for lithium-ion battery recycling. *Hydrometallurgy* 204. https://doi.org/10.1016/j.hydromet.2021.105705.

Yadav, A. and Pandey, S. (2014). Densities and viscosities of (choline chloride + urea) deep eutectic solvent and its aqueous mixtures in the temperature range 293.15 K to 363.15 K. *Journal of Chemical & Engineering Data* 59: 2221–2229. https://doi.org/10.1021/je5001796.

Yang, J.Z., Tian, P., He, L.L., and Xu, W.G. (2003). Studies on room temperature ionic liquid InCl3-EMIC. *Fluid Phase Equilibria* 204: 295–302. https://doi.org/10.1016/S0378-3812(02)00265-0.

Yang, S., Liu, G., Wang, J. et al. (2019). Recovery of lithium from alkaline brine by solvent extraction with functionalized ionic liquid. *Fluid Phase Equilibria* 493: 129–136. https://doi.org/10.1016/j.fluid.2019.04.015.

Yao, Z., Xie, Z., and Tang, J. (2016). A typical e-waste – cathode ray tube glass: alkaline leaching in the sulfur-containing medium. *Procedia Environmental Sciences* 31: 880–886. https://doi.org/10.1016/j.proenv.2016.02.104.

Yao, Y., Farac, N.F., and Azimi, G. (2018). Supercritical fluid extraction of rare earth elements from nickel metal hydride battery. *ACS Sustainable Chemistry & Engineering* 6: 1417–1426. https://doi.org/10.1021/acssuschemeng.7b03803.

Yi, C., Zhou, L., Wu, X. et al. (2021). Technology for recycling and regenerating graphite from spent lithium-ion batteries. *Chinese Journal of Chemical Engineering* 39: 37–50. https://doi.org/10.1016/j.cjche.2021.09.014.

Yu, H., Zi, F., Hu, X. et al. (2014). The copper-ethanediamine-thiosulphate leaching of gold ore containing limonite with cetyltrimethyl ammonium bromide as the synergist. *Hydrometallurgy* 150: 178–183. https://doi.org/10.1016/j.hydromet.2014.10.008.

Yuan, Z., Ruan, J., Li, Y., and Qiu, R. (2018). A new model for simulating microbial cyanide production and optimizing the medium parameters for recovering precious metals from waste printed circuit boards. *Journal of Hazardous Materials* 353: 135–141. https://doi.org/10.1016/j.jhazmat.2018.04.007.

Yue, C., Fang, D., Liu, L., and Yi, T.F. (2011). Synthesis and application of task-specific ionic liquids used as catalysts and/or solvents in organic unit reactions. *Journal of Molecular Liquids* 163: 99–121. https://doi.org/10.1016/j.molliq.2011.09.001.

Zagorodni, A.A. (2012). Ion Exchnange Materials: Properties and Application., 1ste, vol. XXXIII. Stockholm: Elsevier https://doi.org/10.1007/s13398-014-0173-7.2.

Zante, G., Masmoudi, A., Barillon, R. et al. (2020). Separation of lithium, cobalt and nickel from spent lithium-ion batteries using TBP and imidazolium-based ionic liquids. *Journal of Industrial and Engineering Chemistry* 82: 269–277. https://doi.org/10.1016/j.jiec.2019.10.023.

Zhan, L., Wang, Z., Zhang, Y., and Xu, Z. (2020). Recycling of metals (Ga, In, As and Ag) from waste light-emitting diodes in sub/supercritical ethanol. *Resources, Conservation and Recycling* 155: 104695. https://doi.org/10.1016/j.resconrec.2020.104695.

Zhang, Q.B. and Hua, Y.X. (2009). Corrosion inhibition of mild steel by alkylimidazolium ionic liquids in hydrochloric acid. *Electrochimica Acta* 54: 1881–1887. https://doi.org/10.1016/j.electacta.2008.10.025.

Zhang, Q., De Oliveira, V.K., Royer, S., and Jérôme, F. (2012a). Deep eutectic solvents: syntheses, properties and applications. *Chemical Society Reviews* 41: 7108. https://doi.org/10.1039/c2cs35178a.

Zhang, Y., Liu, S., Xie, H. et al. (2012b). Current status on leaching precious metals from waste printed circuit boards. *Procedia Environmental Sciences* 16: 560–568. https://doi.org/10.1016/j.proenv.2012.10.077.

Zhang, C., Zhuang, L., Yuan, W. et al. (2016). Extraction of lead from spent leaded glass in alkaline solution by mechanochemical reduction. *Hydrometallurgy* 165: 312–317. https://doi.org/10.1016/j.hydromet.2016.01.017.

Zhang, G., He, Y., Feng, Y. et al. (2018). Pyrolysis-ultrasonic-assisted flotation technology for recovering graphite and LiCoO2 from spent lithium-ion batteries. *ACS Sustainable Chemistry & Engineering* 6: 10896–10904. https://doi.org/10.1021/acssuschemeng.8b02186.

Zhang, C., Hu, M., Yang, X. et al. (2020). Upgrading construction and demolition waste management from downcycling to recycling in the Netherlands. *Journal of Cleaner Production* 266: 121718. https://doi.org/10.1016/j.jclepro.2020.121718.

Zhang, Y., Xu, B., Cui, M. et al. (2021). Thiosulfate leaching of gold catalyzed by hexaamminecobalt(III): Electrochemical behavior and mechanisms. *Electrochimica Acta* 399: 139393. https://doi.org/10.1016/j.electacta.2021.139393.

Zhang, J., Anawati, J., and Azimi, G. (2022a). Urban mining of terbium, europium, and yttrium from real fluorescent lamp waste using supercritical fluid extraction: Process development and mechanistic investigation. *Waste Management* 139: 168–178. https://doi.org/10.1016/j.wasman.2021.12.033.

Zhang, Y., Cui, M., Wang, J. et al. (2022b). A review of gold extraction using alternatives to cyanide: Focus on current status and future prospects of the novel eco-friendly synthetic gold lixiviants. *Minerals Engineering* 176: 107336. https://doi.org/10.1016/j.mineng.2021.107336.

Zhang, Y., Li, Q., Liu, X., and Jiang, T. (2022c). A thermodynamic analysis on thiosulfate leaching of gold under the catalysis of Fe3+/Fe2+ complexes. *Minerals Engineering* 180: 107511. https://doi.org/10.1016/j.mineng.2022.107511.

Zhang, Z., Malik, M.Z., Khan, A. et al. (2022d). Environmental impacts of hazardous waste, and management strategies to reconcile circular economy and eco-sustainability. *Science of The Total Environment* 807: 150856. https://doi.org/10.1016/j.scitotenv.2021.150856.

Zhao, J., Wu, Z., and Chen, J. (1998). Extraction of gold from thiosulfate solutions using amine mixed with neutral donor reagents. *Hydrometallurgy* 48: 133–144. https://doi.org/10.1016/S0304-386X(97)00059-5.

Zipperian, D., Raghavan, S., and Wilson, J.P. (1988). Gold and silver extraction by ammoniacal thiosulfate leaching from a rhyolite ore. *Hydrometallurgy* 19: 361–375. https://doi.org/10.1016/0304-386X(88)90041-2.

12

Microbiology Behind Biological Metal Extraction

Mishra Bhawana and Pant Deepak

Department of Environmental Sciences, Central University of Himachal Pradesh (CUHP), Shahpur, Dharamshala, Kangra, Himachal Pradesh, India

Abstract

E-waste includes both base metals and precious metals. Traditional e-waste metal recovery processes, such as hydrometallurgical and pyrometallurgical operations, are rapid and effective, but they harm the environment and are not cost-effective. Due to the limitations of traditional procedures, a transition to bio-metallurgical processes including eco-friendly microbial leaching of metals from e-waste has occurred. Biological heavy metal remediation employs organisms such as diverse microbes, plants, and their derivatives or products as a gentle and ecologically acceptable approach of decontaminating the environment. Bioleaching is the process of extracting metals using either metabolic chemicals (indirect bioleaching) or microorganisms (direct bioleaching). This chapter discusses microorganism-based e-waste remediation, as well as processes and chembio engineering techniques that have previously been used in waste remediation and may be used in bioremediation. A discussion of technological breakthroughs that have permitted this study path, and how these technologies could be applied in future research, will also be included.

Keywords *E-waste; microbial participation; heavy metal extraction; bioleaching; microbial remediation; sustainable environment*

12.1 Background

The problem of e-waste is of global concern due to its productive nature and disposal in a globalized world. For decades, the e-waste problem has gotten worse (Balde et al. 2015). India is the second-largest mobile market in the world and also the fifth-largest producer among the top five countries in e-waste production. India discards about 18.5 lakh tonnes of e-waste each year, and there is an

Management of Electronic Waste: Resource Recovery, Technology and Regulation, First Edition. Edited by Anshu Priya.

estimate that about three million tonnes of e-waste production will occur by the end of 2018 (Breivik et al. 2014). The possibility of recovering valuable elements from e-waste such as gold, aluminum, iron, silver, copper, and rare earth metals, has made it a worldwide recognized resource. Despite the fact that recovering these precious materials might result in economic benefits, only about 15% of worldwide e-waste is properly recycled (Modak 2011). Though, in developing countries, e-waste handling and disposal are frequently dangerous, resulting in contaminated surroundings. About 2 million tonnes per annum (TPA) of e-waste are produced in India, while only 4,38,085 TPA are recycled, out of which 5% of total electronic waste gets recycled only via deprived infrastructure, legislation, and framework. Over 95% of e-waste is handled by scrap dealers and unorganized sector, which take pieces of these e-waste products instead of recycling them. The growing number of reports of negative health impacts from e-waste sites necessitates environmental and human health protection against e-waste pollution. At the national, local, and global levels, efforts must be made to build secure recycling operations that take into account a wide range of security concerns for individuals who rely on the processing of e-waste for their livelihood. Therefore, a challenge is emerging to develop cost-effective and innovative approach to remediate pollutants from surroundings.

Recently, various bioremediation strategies, including microbial metal extraction from e-waste have gained lots of attention. This approach includes biosorption, bioadsorption, biomineralization, bioaccumulation, and enzyme catalyzation that utilize various microorganisms (either alone or in combination with plants) that speed up the recycling and metal extraction methods (Gadd 2009; Krishnamoorthy et al. 2017; Lebeau et al. 2008; Ojo et al. 2018; Shah and Daverey 2020).

Metal bio-extraction from e-waste using various microorganisms (bacteria and fungus) is becoming progressively more popular in order to congregate the dual goals of assisting pollution reduction and recycling e-waste. Despite known hazards, in several regions of the world, the rates of e-waste heavy metal deposition in soil continue to increase, as well as human exposure to them (Farooq et al. 2016; Lebeau et al. 2008; Schippers et al. 2014). Phytoextraction is a major bioremediation strategy studied for heavy metal remediation from e-waste-deposited soil (Mulligan et al. 2001; Naila et al. 2019). When plant material is harvested, the metal is permanently removed from the contaminated area. Much of the research in this area focuses on a special plant called the heavy metal hyperaccumulator (HMH). It can reach 100–1000 times higher ground tissue heavy metal concentrations than are normally found in non-HMH (Rascio and Navari-Izzo 2011; Wood et al. 2016).

Bioremediation or microbial collaboration can help make the e-waste treatment process more environmentally friendly. The inorganic and organic components of e-waste are the focus of every management approach. The organic component is made up of a variety of thermal and thermosetting plastics, as well as halogenated materials. In numerous ways, microorganisms play a role in the dehalogenation process. Microorganisms can handle the oozing of e-wastes' inorganic fraction includes, which includes metallic as well as nonmetallic elements (Pant et al. 2018). Plant biomass, algal, bacterial, and fungal organisms have all been proposed as appropriate organisms for waste remediation techniques. Bioremediation has been proven to correctly take away lots of heavy metals from e-waste, which includes incredibly poisonous metallic ions of Cd, Cr, Pb, and Hg (Dolker and Pant 2019; Gadd 2009; Krishnamoorthy et al. 2017). The achievement of bioremediation is based on the variety of molecular cell wall components (such as nucleic acids, chitin, proteins, and polysaccharides) and molecular wall structures, which offer chemically numerous binding companions for extraordinary heavy metallic compounds and ions (Ahluwalia and Goyal 2007). BioSolutions generally use metabolically inactive organic substances, while in-situ bioremediation makes utilization of catabolic-decreasing microbes to precipitate heavy metals (Aryal and Liakopoulou-Kyriakides 2014; Gadd 2004). In this chapter, we will discuss how and which microorganisms are used and the strategy/mechanism for their utilization in heavy metal extraction from e-waste to show how the microbial community engineering strategies are used in heavy metal remediation from e-waste-contaminated environments. This chapter explains how microorganisms are utilized in metal extraction from e-waste and also gives bright information about how microbial communities are engineered to improve the metal remediation tendency of microbes from e-waste and heavy metal-contaminated soil, which was previously used in waste remediation. Further, a discussion will be made about technological advancements that have enabled this study direction as well as how these technologies may be used in future research.

12.2 Overview of E-Waste: A Global Hazard

Globally, enormous quantities of e-waste are generated and redistributed as well. Electronic garbage (e-waste) is created in massive numbers, with 41.8 million tonnes estimated to have been produced globally in 2014. According to 2014 estimation, the largest e-waste producer, the United States, generates 7.1 million tonnes (Mt) of e-waste, followed by roughly 6.0 Mt of e-waste produced in China (Balde et al. 2015). Some selected African countries and China receive up to 80% of the world's e-waste (Lundgren 2012). Therefore, in many countries throughout the world, the management of electronic and electrical waste has become a huge concern. Developing countries, in particular, confront a lot of challenges when it comes to e-waste transboundary movement, management, and generation. E-waste management that is environmentally sound (ESM) is either nonexistent or very limited in underdeveloped countries (Widmer et al. 2005). Now, a number of countries throughout the world are grappling with how to respond to this new hazard. Even though the present focus is on management activities of e-waste end-of-life like reuse, remanufacturing disposal, recycling, service, and upstream e-waste generation reduction via cleaner production and green designing. Globally, 20–50 million tonnes per year of e-waste production were estimated, with the majority of it coming from Asian countries. Because of its perilous components, inappropriate management of electronic waste can affect the environment as well as human health (Widmer et al. 2005; Herat and Agamuthu 2012).

12.3 E-Waste Categories and Classification

12.3.1 E-Waste Categories

There are several categories for e-waste such as small appliances (including irons, fryers, vacuum cleaners, and blenders), major appliances (including dryers washing machines, and refrigerators), consumer electronics (including musical instruments, audio, and video equipment), telecommunication and computer appliances (such as mobile phones, PCs, laptops, and telephones), electrical toys and e-tools (such as saws, gardening devices, and drills), and recreational activities (models, electronic toys, sports equipment), medical gadgets (all medical equipment with the exception of implants), Vending machines, monitoring devices (such as thermostats, detectors, and laboratory equipment) and lighting devices (such as fluorescent tubes, incandescent light bulbs, and gas-discharge lamps).

12.3.2 Physical and Chemical Composition of E-Waste

Arsenic, lead, nickel, chromium, cadmium, mercury, manganese, copper, zinc, iron, and aluminum are among the 60 chemical elements commonly found in complex electronics such as semiconductor chips, coatings, circuit boards, batteries, and cathode ray tubes, and many of these metals are potentially hazardous, as listed in Table 12.1 (Grant et al. 2013).

Recently, the utilization of electronic tools has expanded drastically in recent decades, and the quantity of electronic apparatus discarded, such as entertainment gadgets (play stations, TVs, and air conditioners), mobile phones, laptops, and PCs, is rapidly expanding around the world. In 1994, there was a prediction that over 20 million computers (weighing around 7 million tonnes) would be obsolete. This data would have climbed by 2004 to almost 100 million PCs. ~2,872,000 tonne of plastics, 1,363 tonne of cadmium, 7,18,000 tonne of lead, and 287 tonne of mercury are present in 500 million PCs (Puckett and Smith 2002).

Table 12.1 E-waste sources, components, and their hazardous effect on the environment and human health.

Component of E-waste	Source of E-waste	Effect on human health	Effect of environment
Polychlorinated Biphenyls (PCBs)	Paints, electrical transformers, capacitors, printing inks, and PVC and polymer	Immune system suppression, damage to the liver, neurological, and reproductive system	Extreme pollution from production, harmful chemical exposure during use, and fire hazards are only a few examples
Mercury	Medical and housing equipment, batteries	Muscle tumors, cerebral palsy, renal toxicity, and mental retardation	Glass dust is emitted into the atmosphere as well as discharged into waterways, resulting in air and water pollution
Barium, lead, and other heavy metals	PVC cables, Batteries, Paints, CRTs (used in video cameras, ATMs, TVs, monitors, etc.)	Insomnia, anemia, and renal toxicity	These metals seep into groundwater, releasing harmful phosphorus
Plasticizers DEHP, DBP, and BPA (bisphenol A), plastic compounds well-known as phthalates	Plastics from monitors, keyboards, and printers	Obesity, reproductive diseases, and cardiac troubles	Chlorinated plastics leak hazardous chemicals into the surrounding soil, which leaks into groundwater and other nearby water sources, causing major harm to the species that consume it

12.4 Environmental Hazards Due to E-Waste Composition

Electronic waste has harmful consequences on the environment as well as on human health, as listed in Table 12.1. In many regions, the dumping of e-waste is a special issue. Toxic leachates are produced by landfilled computers that eventually damage groundwater. Melting of computer chips, dumped on the ground, results in the acidification of soil due to the discharge of sludge and acids into the soil. To meet the population's demands, water is now being carried from faraway towns. Hazardous gases and fumes can be emitted into the environment as a result of the burning of e-waste, which can affect the environment, as well as landfills that aren't appropriately monitored. When broken Pb-containing glassware (cone glass from cathode ray tubes) combines with acid waters in landfills, substantial amounts of lead ions are released. Cadmium and polybrominated diphenyl ethers (PBDE) leach into the soil and groundwater when cadmium-containing plastics or brominated flame retardant polymers are used. Mercury and PCBs (polychlorinated biphenyls) seep into the soil when certain electronic components, such as circuit breakers and condensers, are broken (Ramachandra and Saira 2004). The source of e-waste and its environmental hazards are summarized in Table 12.1.

12.5 Health Risks from E-Waste Exposure

E-waste components have an impact on public health in two ways: through direct exposure to toxic substances or via food chain contamination. Because there are several sources and ways, time period of e-waste exposure, and various e-waste exposure effects such as inhibitory, additive, and synergistic effects, e-waste exposure is considered as a complicated process (Grant et al. 2013). The quality, type, amount, and exposure time of e-waste affect human health, especially in children and pregnant women. The contamination method, exposure source of e-waste, and hazardous compounds are continuously discarded EEE to the surroundings and affect the environment accordingly. Humans get exposed to these toxicants through contaminated food, water, air, and soil and show various toxic effects on health as listed in Table 12.1 (Gao et al. 2019; Perkins et al. 2014).

12.6 Bioremediation Techniques for E-Waste Management

Bioremediation is a comprehensive perception that encompasses all activities that occur to biotransform an environment that has been contaminated by pollutants back to its natural state (Figure 12.1).

Bioremediation entailed microbial collaboration, which could make the e-waste disposal process more environmentally friendly. Organic and inorganic are the main components of e-waste that are the focus of every management approach. The organic portion is made up of a variety of thermal and thermosetting plastics, as well as halogenated materials. In numerous ways, microbes play a role in the dehalogenation process. Microbes can handle the inorganic fraction of e-waste, which includes both metallic and nonmetallic components, by leaching (Pant et al. 2018). The biological procedures, which have a higher elimination efficiency and are more successful, are more effective than the thermal or physiochemical techniques. Bioremediation is now established as a widely recognized method for clearing polluted soils and aquifers. In the process of recycling relevant materials from various elements of electronic and electrical devices, hazardous

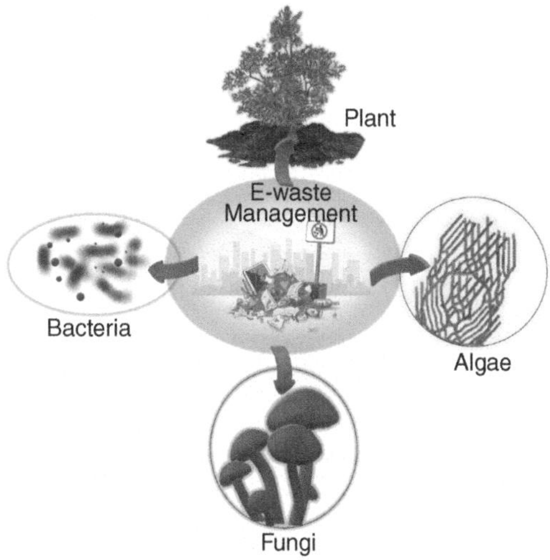

Figure 12.1 Different strategies for heavy metal extraction from e-waste.

organics (represented by polybrominated diphenyl ethers, polycyclic aromatic hydrocarbons, and PCBs) and heavy metals were produced. Due to its low cost and environmental friendliness, bioremediation technology has enhanced systematization, extensivity, and depth of study and has made significant progress in the field of organic or heavy metal-contaminated soil. Metal extraction biological techniques have not been extensively applied up to this point. The two main bioremediation methods are phytoremediation and microbial bioremediation (bioleaching) (Figure 12.2). Microbial remediation technique utilized various microorganisms that developed strong metal tolerance abilities such as *Cyanobacteria microcystis, Aphanothece halophytica, Nostoc calcicola* HH-12, *Chroococcus Xanthofacter, Actinobacter, Mycococcus, Acinethobacter, Acaligenes, Penicillium, Serratio, Nitrosomonas, Bacillins, Arthrobacter, Pseudomonas, Flavobacterium, Nocardia, Methylosinus, Berijerinckia, Mycrobacterium, Trametes, Phanerochaete,* and *Rhizoctomia* (Abatenh et al. 2017; Sharma 2012; Singh et al. 2014). Phytoremediation is the most sustainable (both economically and environmentally) technique (Figure 12.1). Plant remediation is emerging as a powerful technology that remediates the heavy metals from the surroundings/environment using plants (Figure 12.2). Numbers of plants such as *Polygonum sachalase, Chenopodium, Urtica, Thlaspi,* and *Alyssim* have the capability to accumulate and remediate heavy metals such as Pb, Cd, Cu, Ni, and Zn from soil and might be a potential candidate for treating heavy metal-contaminated soils (Morin et al. 2006; Mulligan et al. 2001).

Biological heavy metal remediation employs organisms such as diverse microbes and plants, as well as their derivatives or products, as a gentle and ecologically acceptable approach to decontaminating the environment (Figure 12.1) (Wood et al. 2016). Biological techniques, unlike other remediation

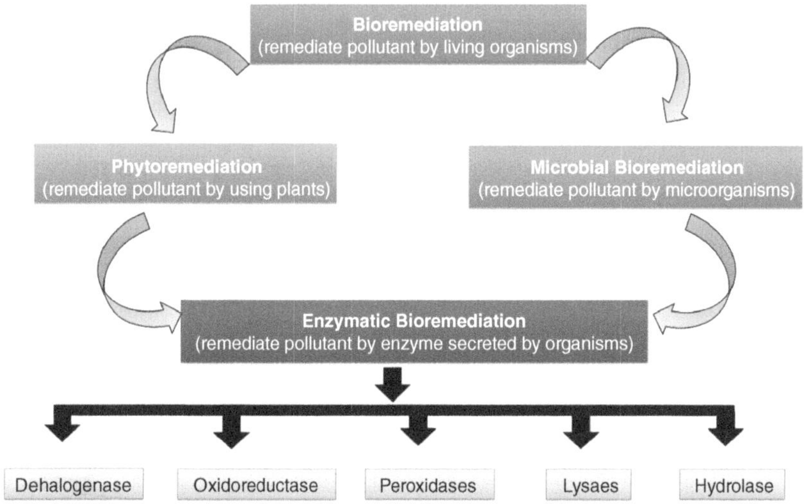

Figure 12.2 Types of bioremediation techniques and mechanisms.

methods, are distinct in that they are ecologically benign and acceptable, the species diversity involved is vast and has not yet been fully explored, and they are also amenable to genetic manipulation for quicker bioremediation. The harsh, poisonous situations wherein metals are extracted from the earth might not produce hospitable surroundings for humans; however, for a few microorganisms, those surroundings are a comfortable bath. Reveling in slurries, which can be frequently extraordinarily acidic or basic, those microbes swim around, deriving power from oxidation and discount reactions that launch economically precious metals from minerals (Adebayo and Obiekezie 2018; Mondal and Palit 2019; Srivastav et al. 2018).

12.7 Why Biological Methods for Metal Extraction from E-Waste

Applications of microbiology to mining operations have been surveyed and evaluated. This has been performed to be able to search for guidelines wherein genetic and molecular organic strategies can be implemented to optimize "bio-mining" processes. Remediation of websites infected with poisonous metals is especially challenging. Unlike natural compounds, metals released from e-waste cannot be degraded, and the cleanup commonly calls for their removal. However, this power-extensive technique may be prohibitively expensive (Lebeau et al. 2008; Usman et al. 2020). Phytoremediation has been proposed as a cost-effective, environmentally pleasant opportunity generation method. A first-rate deal of studies shows that flora have the genetic capability to dispose of many poisonous metals from the soil. Despite this capability, phytoremediation is but to come to be a commercially to-be-had generation. There is a limited information about the comprehension of complex linkages with the rhizosphere and plant-primarily based mechanisms that allow metal transmission and accumulation in flora (Gudeppu et al. 2019; Verma et al. 2017).

The metal pollution management associated with e-waste is widespread worldwide. Nowadays, biological or chemical leaching mechanisms have their own limitations. The biological leaching process is comparatively cost-effective but time-consuming, and complete metal leaching is not possible, while the chemical leaching process is more efficient and rapid but has its own environmental consequences (Pant et al. 2012). However, most of these methods are often expensive, not convenient for soil ecology, and cannot be used over a large area. For that reason, to conquer these problems, biological treatment methods are needed. Bioremediation of heavy metal toxicity, in a general view, is to make fallow land arable again, using different methods such as ion exchange, ultrafiltration, electrodialysis, coagulation, recovery evaporation, chemical precipitation, reverse osmosis computation, absorption, and nano-annealing as well as physicochemical methods such as leaching, immobilization, extraction, and stabilization from landfill soils and mining excavation. In general terms, bioremediation is "a process used to treat contaminated environments, including water, soil, and underground materials, by altering environmental conditions to stimulate growth" of microorganisms and degrading target pollutants," and have been classified into two categories: (a) ex situ: transport contaminated soil (b) in situ bioremediation: onsite xenobiotic treatment and then treating it (Adebayo and Obiekezie 2018; Rizvi et al. 2020; Stingu et al. 2012).

12.7.1 Leaching Mechanisms of Heavy Metals from E-Waste

The key problem with e-waste dumped in the environment is the abundance of a variety of hazardous and toxic metals that accumulated in the soil in nondegradable form and showed major concern due to their complexation characteristics and intrinsic properties such as oxidation-reduction potentials and solubility at different pHs. The waste that's comes from the industrial, domestic, or commercial sectors contained organic fraction (easily degradable) as well as inorganic fraction that is not decomposed or undergoes transformation (Priya and Hait 2017). This mobilization and dissolution of metals from minerals and wastes are usually mediated by microbial action known as bioleaching, which is carried by various micro-organisms that produce various metabolic products. Several cyanobacteria, including *C. Microcystis, A. halophytica, N. calcicola* HH-12, *Chroococcus*, thermophilic bacteria like *Thermothrix thioparus, Thiobacillus thermophilica, Sulfolobus acidocaldarius* and *Thiobacillus TH1,* and other microorganisms such as *Pseudomonas florescens, Thiobacillus concretivorus, Pseudomonas putida, Bacillus licheniformis*, Achromobacter, *B. luteus, B. cereus, Bacillus megaterium, Bacillus polymyxa* are used for the bioleaching process (Ilyas et al. 2013; Sand et al. 2001; Wang et al. 2009). There are two types of bioleaching procedures:

12.7.2 Direct Bacterial Leaching

The direct bacterial leaching process includes the physical contact between the mineral sulfide surface and the bacterial cell. Several enzymatic oxidation reactions occur for sulfates such as non-iron metal sulfides such as covellite (CuS), millerite (NiS), galena (PbS), and chalcocite (Cu_2S) can be oxidized by *Thiobacillus ferrooxidans*. However, for MeS (metal sulfide), metal solubilization occurs due to the intimate contact between the mineral surface, the whole mechanism behind the attachment is not completely understood. Perceptibly, there is some specific site for the electrochemical interactions for the attachment of bacteria to the mineral surface to solubilize the metal, as microbes do not append to the entire surface of the mineral (Priya and Hait 2017; Sand et al. 2001; Yazdi et al. 2014). This can be expressed in terms of the following reaction equation:

$$\mathrm{MS} + 2\mathrm{O}2 \xrightarrow{A.ferrooxidans;\ A.thiooxidans} \mathrm{MSO4}$$

12.7.3 Indirect Bacterial Leaching

The indirect bioleaching technique uses sulphide minerals that are chemically oxidized by a lixiviant and produced by bacteria. There is no direct involvement of microorganisms or require to be mineral surface contact, but they are involved indirectly as the product produced by them is responsible for mineral oxidation (Priya and Hait 2017; Sand et al. 2001; Yazdi et al. 2014).

$$4\mathrm{FeS2} + 15\mathrm{O2} + 2\mathrm{H2O} \xrightarrow{A.ferrooxidans} 2\mathrm{Fe2}\ (\mathrm{SO4})\ 3 + 2\mathrm{H2SO4}$$

12.7.4 Role of Microbes in Metal Leaching Process from E-Waste

The most appropriate soil and sediment treatment technique is determined by the area's features, the concentration, the kind of pollutant to be eliminated, and the final use of the contaminated medium. Isolation, immobilization, decreased toxicity, separation, and physical extraction are among approaches. Many of these technologies are already in widespread use (Mulligan et al. 2001). Several chemical, biological, and physical processes are being used to remediate heavy metals from contaminated soil.

Methods such as solidification/stabilization remediate contaminants in an area by injecting or mixing the substances. The solidification process includes the encapsulation of contaminants in a solid matrix, while the chemical bond formation occurs in stabilization technique to reduce contaminant mobility. Sizing processes such as removal and cleaning of the larger particle from smaller or contaminated particles are another approach. To do this, several methods are being utilized such as flotation, fluidized bed separation, and hydrocyclones. The addition of special chemicals and aeration in the second case causes these pollutant particles to emerge. The electrodynamic process consists of a fleeting electric current with low intensity between the anode and cathode embedded in contaminated soil. Small charge carriers and ions, in addition to water, are transported between the electrodes. The technology has been demonstrated in the United States on a limited, large scale, but in Europe, it is exploited for various heavy metals such as zinc, copper, lead, cadmium, nickel, arsenic, and chromium. The length of time the electrode stays in the soil and the distance are site-specific (Mulligan et al. 2001; Ni et al. 2019; Tian et al. 2018).

Metal extraction biological techniques have not been extensively applied up to this point. The two main bioremediation methods are phytoremediation and microbial bioremediation (bioleaching) (Figure 12.2). Microbial remediation technique utilized various microorganisms that developed strong metal tolerance abilities such as *cyanobacteria*, *C. Microcystis A. halophytica*, *N. calcicola* HH-12, *Chroococcus*, *Xanthofacter*, *Actinobacter*, *Mycococcus*, *Acinethobacter*, *Acaligenes*, *Penicillium*, *Serratio*, *Nitrosomonas*, *Bacillins*, *Arthrobacter*, *Pseudomonas*, *Flavobacterium*, *Nocardia*, *Methylosinus*, *Berijerinckia*, *Mycrobacterium*, *Trametes Phanerochaete*, and *Rhizoctomia* (Abatenh et al. 2017; Sharma 2012; Silver and Phung 1996; Singh et al. 2014).

Biological heavy metal remediation employs organisms such as diverse microbes and plants, as well as their derivatives or products, as a gentle and ecologically acceptable approach to decontaminating the environment (Table 12.2) (Wood et al. 2016). The harsh, poisonous situations wherein metals are extracted from the earth might not produce hospitable surroundings for humans; however, for a few microorganisms, those surroundings are a comfortable bath. Reveling in slurries, which can be frequently extraordinarily acidic or basic, those microbes swim around, deriving power from oxidation and discount reactions that launch economically precious metals from minerals (Adebayo and Obiekezie 2018; Mondal and Palit 2019; Srivastav et al. 2018).

12.7.5 Major Microorganisms Involved in Metal Leaching

Several microorganisms are being used to remove metals from e-waste listed in Table 12.3. Microorganisms have a clear advantage over it, since they now not only give a cost-effective alternative, but are also smooth technology (Gadd

Table 12.2 Detoxifying strategies for different metals adopted by metal-tolerant bacteria used in metal remediation from e-waste.

Different metal detoxifying strategies	Used microorganism	Remediated heavy metal	References
Bioaccumulation	*Pseudomonas, Bacillus, Staphylococcus, Deinococcus radiodurans, Delftia, Desulfovibrio* sp.	Cu, Zn, Hg, Cd, Ni, Pb, As, Ag, Cr, and Ag	Ahemad and Malik (2011), Annamalai and Gurumurthy (2019), Banerjee et al. (2015), Beolchini et al. (2010), Limcharoensuk et al. (2015), Liu et al. (2018), Velásquez and Dussan (2009)
Biosorption	*Bacillus, Microbacterium*, Ochrobactrum, Lysinibacillus, Serratia, *Desulfovibrio: Pseudomonas, Micrococcus, Enterobacter, Zoogloea, Comamonas, Aphanothece halophytica*	Cr, Cd, Zn, Ni, Hg, As, Hg, Cu, Pb, Pt, and Pd	Abbas et al. (2014), Aryal and Liakopoulou-Kyriakides (2014), de Vargas et al. (2004), Fabienne et al. (2012), Gadd (2009), Iyer et al. (2005), Limcharoensuk et al. (2015), Mohapatra et al. (2019), Rani et al. (2010), Vijayaraghavan and Yun (2008)
Biotransformation	*Enterobacteriaceae, Pseudomonas*, Cyanobacteria, *Cornebaterium, Xanthomonas, Acinetobacter, Flavimonas, Micrococcus, Alkaliphilus, Shewanella, Sinorhizobium, Bacillus, Delftia, Klebsiella, Brevibacillus*	As, Cd, Cr, Cu, Hg, Ur, Ni, and Pb	Garavaglia et al. (2010), Govarthanan et al. (2015), Jobby et al. (2019), Nagvenkar and Ramaiah (2010), Paul et al. (2015), Srivastava and Thakur (2012), Tian et al. (2015)
Bioadsorption	*Pseudomonas* sp.	Cu, Cd, Zn, Pb, and Ni	Al-qodah and Al-shannag (2017), Ghaima and Mohamed (2017), González et al. (2017), Olguín and Sánchez-Galván (2012), Vijayaraghavan and Yun (2008)
Bioreduction	*Pseudomonas* sp., *Azotobacter chroococcum, Bacillus* sp., *Shewanella oneidensis, Stenotrophomonas maltophilia*, and *Acinetobacter* sp., *Sporolactobacillus, Clostridium,* Exiguobacterium sp., *Sporosarcina saromensis, Bacillus* sp.	Cr, Hg, Cd, Pb, Zn, Ag, Pt, Cu, Ni, and Co	Andreazza et al. (2012), Cheng and Li (2009), Huang et al. (2021), Kumaresan Sarankumar et al. (2020), Lin et al. (2020), Mal et al. (2016), Masaki et al. (2015), Rashamuse and Whiteley (2007), Rizvi et al. (2019), Sable et al. (2020), Tarhriz et al. (2019)

Table 12.3 Microorganisms used for heavy metal remediation from e-waste (Publication onwards 2010).

Targeted metals	Micro-organisms utilized	References
Au (gold)	*Ammonium thiosulfate* (AT), *Lactobacillus acidophilus* (LA), and *Chromobacterium violaceum*	Sheel and Pant (2018), Tay et al. (2013)
Cu	*Acidithiobacillus thiooxidans*	Hong and Valix (2014)
Zn, Au, Ag, Fe, and Cu	*Pseudomonas aeruginosa C. violaceum,* and *Pseudomonas fluorescens*	Pradhan and Kumar (2012)
Heavy metals, Cd	*P. fluorescens* and *P. aeruginosa*	Chellaiah (2018), Tang et al. (2018), Wasi et al. (2011)
Cd	*Cunninghamella elegans*	De Lima et al. (2013)
Heavy metals and gold	*Lysinibacillus sphaericus* CBAM5	Nanda et al. (2019), Páez-Vélez et al. (2019)
Cd	*Aspergillus sp, Paecilomyces* sp., *Microsporum* sp., *Terichoderma* sp., and *Cladosporium* sp.	Mohammadian Fazli et al. (2015)
Fe	*Microbacterium profundi* strain Shh49T	Gupta et al. (2016)
Fe (III), U (VI)	*Geobacter* spp	Iwahori et al. (2014)
Cd	*Bacillus safensis* (JX126862)	Murugan and Raj (2014)
Ni and V	*Microbacterium arabinogalactanolyticum* *Bacillus megaterium*	Fierros-Romero et al. (2016)
Cr(IV)	*Bacillus subtilis, Enterobacter cloacae, and Pseudomonas putida*	Balamurugan et al. (2014), Rahman et al. (2015)
Cd, Zn, Cu	*Pseudomonas veronii*	Vullo et al. (2008)
Cr(IV)	*Bacillus cereus*	Chen et al. (2012), Kanmani et al. (2012)
Fe(III)	*Geobacter* spp.	Singer et al. (2011)
Cu, Ni, Cr	*P. aeruginosa*	Sinha et al. (2011)
As	*Pseudomonas* sp. *P1III2, Bacillus* sp. *MPV12, Delftia* sp. *P2III5, Pseudoxanthomonas* sp. *P4V6, and Variovorax* sp. *P4III4,*	Lampis et al. (2015)
Pb, Cr, Cd	*Rhodopseudomonas palustris, Aerococcus* sp.	Sinha and Biswas (2014), Sinha and Paul (2013)
Cr, Cu, Pb, Co	*L. sphaericus*	Peña-Montenegro et al. (2015)

(Continued)

Table 12.3 (Continued)

Targeted metals	Micro-organisms utilized	References
Fe	*M. profundi*	Wu et al. (2020)
Ni and V	*B. megateriumSR28C*	Fierros Romero et al. (2016), Rajkumar et al. (2013)
Pb, Cu	*Spirogyra, Cladophora*	Lee and Chang (2011)
Ni, Cu, Pb	*Aspergillus versicolor, Aspergillus fumigatus*	Taştan et al. (2010), Kumar Ramasamy et al. (2011)
As(III)	*Sporosarcina ginsengisoli*	Achal et al. (2012)
As	*Hydrodictyon, Oedogonium*	Coelho et al. (2015)
Cd	*P. putida, Comamonas aquatica*	Koul and Taak (2018)
Ni	*Pseudomonas* sp.*A3R3*	Ma et al. (2011)
Zn	*Microbacterium saperdae BJ1, Pseudomonas monteilii BJ5, Enterobacter cancerogenus BJ 10]*	Whiting et al. (2001)
Cd	*Bacillus mycoides, Micrococcus roseus*	Malekzadeh et al. (2012)
Cd, Zn	*Enterobacter intermedius MH8b*	Płociniczak et al. (2013)
Zn^{2+}, Cu^{2+}, and Cd^{2+}	*D. desulfuricans*	Yue et al. (2015)
Cu	*B. subtilis* and *P. putida*	Fang et al. (2011)
Cu(II), Pb(II), and Cd (II)	*Aspergillus fumigates*	Yin et al. (2013)
Cr and Ni	*Micrococcus* sp. and *Aspergillus* sp.	Congeevaram et al. (2007)
Hg, Cd, and Pb	*Alcaligenes faecalis, Bacillus pumilus, P. aeruginosa, and Brevibacterium iodinium*	De et al. (2008)

2009; Stingu et al. 2012). Metals can be detoxified by microorganisms via extracellular chemical precipitation, volatilization, and valence change (Aryal and Liakopoulou-Kyriakides 2014; Gadd 2004, 2009; Lovley et al. 2004). Mineral oxidizing microorganisms such as iron-oxidizing (*Leptospirillum ferri* and *Leptospirillum ferrooxidans),* the sulfur-oxidizing (*Acidithiobacillus caldus* and *Acidithiobacillus thiooxidans*), and iron- and sulfur-oxidizing *Acidithiobacillus ferrooxidans* have been identified and can easily oxidize sulfur-containing minerals and iron. Several fungus species might be employed for biomining (Falco et al. 2003; Vera et al. 2013; Zhu et al. 2012). Some of the examples of the metal-extracting bacteria in the form of metal leaching, phytoextraction, and plant development connected with microbes and heavy metal buildup to promote phytoextraction from polluted soils are as follows (Table 12.3).

12.7.5.1 Acidophiles

A. ferrooxidans are well understood in terms of iron oxidation in acidophiles. Acidic environments are frequently rich in extreme acidophiles (e.g. the *Leptospirillum* genus) and soluble iron and have far fewer iron absorption structures than neutrophils (Falco et al. 2003; Sajjad et al. 2019; Vera et al. 2013; Zhu et al. 2012). Copper leaching performance is improved by using microorganisms linked to chalcopyrite. Bioremediation has been accomplished by the adhesion of acidophiles leaching microorganisms: *A. thiooxidans, A. ferrooxidans*, and *L. ferrooxidans* (Falco et al. 2003; Sajjad et al. 2019; Zhu et al. 2012).

12.7.5.2 Cynobacteria

Cyanobacteria is a suitable candidate for biosorbent materials due to its easy cultivation. Cyanobacteria have the potential to endure processing into biosorbent materials. *Anabaena is* reported to be biosorption of nickel (Corder and Reeves 1994). *Aphanothece halophytica* was investigated for zinc biosorption and the result suggested that Zinc biosorption was modest at pH 4–6, but increased as pH 6.5 and 7 approached in the presence of *A. halophytica* (Incharoensakdi and Kitjaharn 2002). *N. calcicola* HH-12 and *Chroococcus* sp. HH-11 were discovered to be more suited for the production of an effective Cr-removing biosorbent (VI) (Anjana et al. 2007). Microcystis from cyanobacteria blooms reported to have the potential to remediate Antimony Sb (V) from effluents (Sun et al. 2014).

12.7.5.3 Thiobacillus

The microorganism maximum energy in bioleaching belongs to the genus *Thiobacillus*. These are nonspore-forming rods, gram-negative, that grow in anaerobic environments. The majority of *thiobacilli* are chemolithoautotrophic organisms that utilize carbon dioxide from the environment to synthesize new cell material. (Bosecker 1997; Priya and Hait 2017). *Thiobacillus thiooxidans* and *Thiobacillus ferrooxidans* are crucial. Other *thiobacilli* are also capable of oxidizing sulfur and sulfides, but they develop more effectively at higher pH levels when metallic ions no longer remain in solution (Anders and Colin 1995; Xiang et al. 2010). *T. ferrooxidans*, on the other hand, performs the most important job in bacterial leaching (Ewart and Hughes 1991; Mulligan et al. 2001; Priya and Hait 2017).

12.7.5.4 Thermophilic Bacteria

Thiobacillus-like microorganisms, often known as Th-microorganisms, are moderately thermophilic microorganisms that grow on chalcopyrite, pentlandite, and pyrite at temperatures of about 50 °C (Willner and Fornalczyk 2013). The strength source is ferrous iron; however, growth is detected best in the presence of yeast extract. Brierley and his anco-worker have isolated Thermophilic bacteria that grow above 60 °C (Brierley and Brierley 2001; Priya and Hait 2017). *Acidianus brierleyi* is a chemolithoautotrophic, facultatively aerobic, acidophilic Archae that grows on metallic sulfides, elemental sulfur, and ferrous iron. It was originally associated with the genus Sulfolobus (Samadzadeh Yazdi et al. 2020; Yazdi et al. 2014; Zeldes et al. 2019). In anaerobic environments, elemental sulfur is employed as an electron acceptor and is reduced to H_2S. Sulfolobus members are aerobic, facultatively chemolithotrophic microorganisms that oxidize elemental sulfur, sulfide minerals, and ferrous iron (Priya and Hait 2017). *Sulfobacillus thermosulfidooxidans*, a spore-forming facultatively autotrophic bacteria, uses the same chemicals as an energy source. Growth, on the other hand, will occur most readily in the presence of yeast extract, which is reported to be used for Cu bioleaching (Ilyas et al. 2022).

12.7.5.5 Siderophores

Siderophores are natural chemicals created by bacteria to harvest iron in times of low iron availability (Vakilchap et al. 2020; Yi et al. 2022). Pseudomonads (such as *Pseudomonas azotoformans, Pseudomonas putida*, and *P. fluorescens*) generate organic siderophores, which were investigated for metal extraction. This study provides significant knowledge on siderophore-metal interactions (Williamson et al. 2021).

12.7.5.6 Heterotrophic Microorganisms

Heterotrophic microorganisms and fungi that require natural organic supplements for development and energy contribute to metal leaching as well. Metal leaching has little effect on heterotrophic bacteria. Individuals from the genus Bacillus are the best choice for metal solubilization among microorganisms, while those from the fungal genera Aspergillus and Penicillium are the most important (Mondal and Palit 2019). Metal solubilization, like manganese leaching, can be caused by enzymatic oxidation of highly oxidized metal complexes or by the production of organic acids and compounds with hydrophilic reactive groups (Priya and Hait 2017).

12.8 Types of Bioremediation

Bioremediation techniques include various methods such as bioattenuation, biosimulation, bioventing, bioaugmentation, and biopiles (Figure 12.3). Biosimulation is a simulation process of natural or indigenous microbes via supplying specific nutrients, minerals, or fertilizer at the specific site. Bioattenuation remediates the pollutant from its surrounding via various biological processes including physical phenomena (volatilization, dispersion, advection, diffusion, dilution, sorption/desorption), aerobic/anaerobic biodegradation, and chemical reactions (complexation, abiotic transformation, ion exchange). Bioaugmentation is one of the biodegradation mechanisms that includes the supplementation

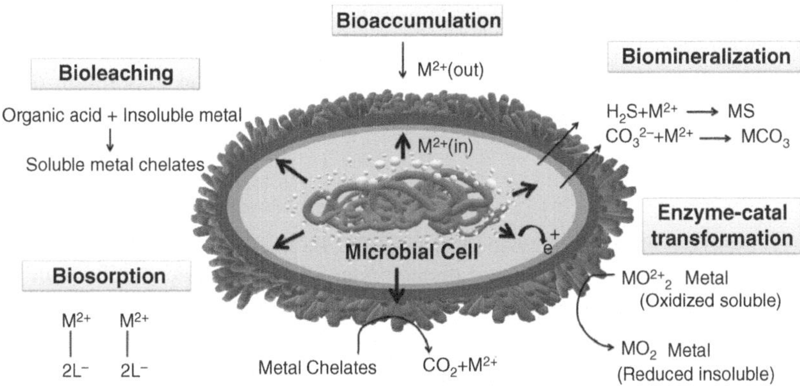

Figure 12.3 Different Metal detoxification strategies adopted by metal-tolerant bacteria.

of pollutant-degrading natural microorganisms or engineered microorganisms. The bioventing technique is used to stimulate the growth of specific microbes including fungi and bacteria by venting oxygen through the soil. Biopiles is a technique to aerobically remediate petroleum or hydrocarbons pollutants in excavated soils, also known as bioheaps, biocells, compost, and biomounds piles (Abatenh et al. 2017; Sharma 2012; Singh et al. 2014).

A recent study aimed to establish the links between circular bio-economy (CirBioeco) and electronic (e-)waste recycling using microbial activity rather than smelting and chemical technologies (Khalid et al. 2021). This research could pave the way for environmentally friendly biological e-waste recycling (Khalid et al. 2021). A case study demonstrates the circular bioeconomy and environmental friendliness technique for copper and gold restoration via microbial recycling of e-waste (Ilyas et al. 2021). The open burning of e-waste resulted in a noticeable heat effect and enriched thermophilic/mesophilic microorganisms in the surrounding region. The enrichment of acidophilic bacteria may occur as a result of acid washing during the e-waste recycling process (Pant et al. 2018).

12.9 Factors Influencing Microbial Metal Leaching

Both biotic factors such as microbial culture, growth, and efficiency of microorganisms, and abiotic factors such as composition, pH, nutrient concentration, temperature, and aeration affect the microbial-based metal leaching/extraction mechanism (Xiang et al. 2010; Ilyas et al. 2013).

12.9.1 Availability of Nutrients

Inorganic compounds are required for the growth of microorganisms used for metal extraction. For the optimal growth of microorganisms, phosphate, sulfur, magnesium, iron, and ammonium compounds are supplemented in leaching medium for great significance. These nutrient supplements in media significantly affect or influence microbial metabolism, growth, and metal dissolution and recovery by altering cell growth (Xiang et al. 2010; Ilyas et al. 2013).

12.9.2 Aeration

Microorganisms used for bioleaching require plenty of air supply (aerobic in nature) for their better growth and metabolism. So, an adequate supply of air is required for better metal leaching. Higher as well as lower supply of air causes stress in bacteria, resulting in minimal metal extraction rates. Therefore, sufficient O_2 supply is required for better microbial growth and metal bioleaching property (Mulligan et al. 2001; Beolchini et al. 2010; Ilyas et al. 2022).

12.9.3 Substrate

The composition of substrate and e-waste has an important role in the metal extraction process, such as high metallic content in e-waste possess toxicity to microorganisms, or high PCB content minimizes the metal leaching efficiency (Brandl 2001). The size of particles in e-waste also has an impact on metal dissolution (optimum size range: 40–200 μm) (Mulligan et al. 2001; Beolchini et al. 2010; Ilyas et al. 2022).

12.9.4 Surfactant, Chelators, and Complexing Agents

Metal dissolution rate is also affected by the addition of chelators, surfactants, and complexing agents with bioleaching microbe such as supplementation of citric acid enhanced 37–80% Cu leaching. There is an increment in metal dissolution rate after the addition of mixture of biosurfactant-producing bacteria and sulfur oxidizing bacteria in comparison to sulfur-oxidizing bacteria's' pure culture (Ilyas et al. 2022; Pant et al. 2012).

12.9.5 Temperature

The temperature that is favorable for the higher metal dissolution rate and leaching microorganism growth is considered as optimum temperature, which changes with the growth constraints of different microbial cultures. For example, optimum temperature for thermophiles is 40–45 °C and for mesophilic acidophiles is 25–30 °C. Lower temperature decreases the metal extraction, while in some cases zinc, nickel, cobalt, and copper soulzation were observed at 4 °C and high temperature like 50–80 °C is required for the thermophilic leaching bacteria (Xiang et al. 2010; Ilyas et al. 2013).

12.9.6 Genomic and Metagenomic Challenges

High-throughput genomic technologies are speeding up research into the diversity of microbial life in a variety of situations. The procedures and mechanisms underpinning microbial successions, as well as the impact of community composition on ecosystem functioning, can be studied in bioleaching heaps used for metal recovery. Advances in DNA sequencing technology have made it possible to gather unprecedented amounts of information on the genomes of bioleaching bacteria, enabling the construction of predictive models of metabolic potential and ecosystem-level interactions. Many species that are resistant to genetic procedures or are unculturable can now be phenotyped in advance using these methods. Genomic information is being used to generate metabolic models, determine genetic diversity, predict the interaction of microbes, and explore microbial evolution via metagenomics and genomics approaches (Cárdenas et al. 2016).

12.10 Conclusion

In recent years, metal contamination in soil has risen dramatically. E-waste contains both base and precious heavy metals released by electronic equipment. Due to the limits of conventional approaches, a shift to biometallurgical technologies has begun, including environmentally friendly microbial leaching of metals from e-waste. The current chapter dealt with the new paradigm shift to bioremediation methods from chemical-based. Bioremediation approaches such as phytoextraction technology for sequestering different hazardous heavy metals in plant tissues from extensively contaminated/polluted soils are reviewed in length. New approaches for increasing phytoextraction employing microorganisms are also being considered. Cyanogenic bacteria, Chemolithotrophs, Acidophiles, Leptospirillum, and other major extremophiles are now being used for metal remediation. The chapter examines technical advancements and emerging technologies, as well as future research directions.

12.11 Future Prospects

New bioremediation solutions will become accessible when new biotransformations are better known at the ecological, biochemical, and genetic levels. Biological extraction of metals from e-waste is being pushed to new heights thanks to cutting-edge technical solutions for delivering ingredients that excite microorganisms. Emerging biological heavy metal remediation applications will be aided by research. Microbial bioengineering innovations are being researched to enhance the availability of pollutants to microbes. These advances, if successfully implemented, would boost the efficacy of metal remediation from e-waste remediation.

References

Abatenh, E., Gizaw, B., Tsegaye, Z., and Wassie, M. (2017). The role of microorganisms in bioremediation—a review. *Open Journal of Environemental Biology* 2: 38–46.

Abbas, S.H., Ismail, I.M., Mostafa, T.M., and Sulaymon, A.H. (2014). Biosorption of heavy metals: a review. *Journal of Chemical Science and Technology* 3 (4): 74–102.

Achal, V., Pan, X., Fu, Q., and Zhang, D. (2012). Biomineralization based remediation of As(III) contaminated soil by *Sporosarcina ginsengisoli*. *Journal of Hazardous Materials* 201–202: 178–184. https://doi.org/10.1016/j.jhazmat.2011.11.067.

Adebayo, F. and Obiekezie, S. (2018). Microorganisms in waste management. *Research Journal of Science and Technology* 10: 28–39. https://doi.org/10.5958/2349-2988.2018.00005.0.

Ahemad, M. and Malik, A. (2011). Bioaccumulation of heavy metals by zinc resistant bacteria isolated from agricultural soils irrigated with wastewater. *Bacteriology Journal* 2 (1): 12–21.

Ahluwalia, S.S. and Goyal, D. (2007). Microbial and plant derived biomass for removal of heavy metals from wastewater. *Bioresource Technology* 98: 2243–2257. https://doi.org/10.1016/j.biortech.2005.12.006.

Al-qodah, Z. and Al-shannag, M. (2017). On the performance of bioadsorption processes for heavy metal ions removal by low-cost agricultural and natural by-products bioadsorbent: a review. *Desalination and Water Treatment* 85: 339–357. https://doi.org/10.5004/dwt.2017.21256.

Anders, B. and Colin, W. (1995). Ferric sulphate oxidation using *Thiobacillus ferrooxidans*: a review. *Process Biochemistry* 30: 225–236.

Andreazza, R., Okeke, B.C., Pieniz, S., and Camargo, F.A.O. (2012). Characterization of copper-resistant rhizosphere bacteria from avena sativa and plantago lanceolata for copper bioreduction and biosorption. *Biological Trace Element Research* 146: 107–115. https://doi.org/10.1007/s12011-011-9228-1.

Anjana, K., Kaushik, A., Kiran, B., and Nisha, R. (2007). Biosorption of Cr(VI) by immobilized biomass of two indigenous strains of cyanobacteria isolated from metal contaminated soil. *Journal of Hazardous Materials* 148: 383–386. https://doi.org/10.1016/j.jhazmat.2007.02.051.

Annamalai, M. and Gurumurthy, K. (2019). Enhanced bioleaching of copper from circuit boards of computer waste by *Acidithiobacillus ferrooxidans*. *Environmental Chemistry Letters* 17: 1873–1879.

Annamalai, M. and Gurumurthy, K. (2019). Microbiological leaching of metals and its recovery from waste electrical and electronic equipment: a review. *World Review of Science, Technology and Sustainable Development* 15 (1): 1–16.

Aryal, M. and Liakopoulou-Kyriakides, M. (2014). Bioremoval of heavy metals by bacterial biomass. *Environmental Monitoring and Assessment* 187: 4173. https://doi.org/10.1007/s10661-014-4173-z.

Balamurugan, D., Udayasooriyan, C., and Kamaladevi, B. (2014). Chromium (VI) reduction by *Pseudomonas putida* and *Bacillus subtilis* isolated from contaminated soils. *International Journal of Environmental Sciences* 5: 522–529.

Balde, C., Kuehr, R., Blumenthal, K. et al. (2015). *E-Waste Statistics-Guidelines on Classification, Reporting and Indicators*, 1–51. Bonn: United Nations Univ. IAS-SCYCLE.

Banerjee, G., Pandey, S., Ray, A.K., and Kumar, R. (2015). Bioremediation of heavy metals by a novel bacterial strain *Enterobacter cloacae* and its antioxidant enzyme activity, flocculant production, and protein expression in presence of lead, cadmium, and nickel. *Water, Air, & Soil Pollution* 226: 1–9.

Beolchini, F., Fonti, V., Ferella, F., and Vegliò, F. (2010). Metal recovery from spent refinery catalysts by means of biotechnological strategies. *Journal of Hazardous Materials* 178: 529–534. https://doi.org/10.1016/j.jhazmat.2010.01.114.

Bosecker, K. (1997). Bioleaching: metal solubilization by microorganisms. *FEMS Microbiology Reviews* 20: 591–604. https://doi.org/10.1111/j.1574-6976.1997.tb00340.x.

Brandl, H. (2001). Microbial leaching of metals. *Biotechnology* 10: 191–224.

Breivik, K., Armitage, J.M., Wania, F., and Jones, K.C. (2014). Tracking the global generation and exports of e-waste. Do existing estimates add up? *Environmental Science & Technology* 48: 8735–8743. https://doi.org/10.1021/es5021313.

Brierley, J. and Brierley, C. (2001). Present and future commercial applications of biohyudrometallurgy. *Hydrometallurgy* 59: 233–239.

Cárdenas, J.P., Quatrini, R., and Holmes, D.S. (2016). Genomic and metagenomic challenges and opportunities for bioleaching: a mini-review. *Research in Microbiology* 167: 529–538. https://doi.org/10.1016/j.resmic.2016.06.007.

Chellaiah, E.R. (2018). Cadmium (heavy metals) bioremediation by *Pseudomonas aeruginosa*: a minireview. *Applied Water Science* 8: 154. https://doi.org/10.1007/s13201-018-0796-5.

Chen, Z., Huang, Z., Cheng, Y. et al. (2012). Cr(VI) uptake mechanism of *Bacillus cereus*. *Chemosphere* 87: 211–216. https://doi.org/10.1016/j.chemosphere.2011.12.050.

Cheng, G. and Li, X. (2009). Bioreduction of chromium (VI) by *Bacillus* sp. isolated from soils of iron mineral area. *European Journal of Soil Biology* 45: 483–487. https://doi.org/10.1016/j.ejsobi.2009.06.009.

Coelho, L., Rezende, H., Coelho, L. et al. (2015). Bioremediation of polluted waters using microorganisms. *Advances in Bioremediation of Wastewater and Polluted Soil* 10: 60770.

Congeevaram, S., Dhanarani, S., Park, J. et al. (2007). Biosorption of chromium and nickel by heavy metal resistant fungal and bacterial isolates. *Journal of Hazardous Materials* 146: 270–277. https://doi.org/10.1016/j.jhazmat.2006.12.017.

Corder, S.L. and Reeves, M. (1994). Biosorption of nickel in complex aqueous waste streams by cyanobacteria. *Applied Biochemistry and Biotechnology* 45: 847–859. https://doi.org/10.1007/BF02941854.

De Lima, M.A.B., Franco, L.D.O., De Souza, P.M. et al. (2013). Cadmium tolerance and removal from *Cunninghamella elegans* related to the polyphosphate metabolism. *International Journal of Molecular Sciences* 14 (4): 7180–7192. https://doi.org/10.3390/ijms14047180.

De, J., Ramaiah, N., and Vardanyan, L. (2008). Detoxification of toxic heavy metals by marine bacteria highly resistant to mercury. *Marine Biotechnology* 10: 471–477. https://doi.org/10.1007/s10126-008-9083-z.

Dolker, T. and Pant, D. (2019). Chemical-biological hybrid systems for the metal recovery from waste lithium ion battery. *Journal of Environmental Management* 248: 109270. https://doi.org/10.1016/j.jenvman.2019.109270.

Ewart, D.K. and Hughes, M.N. (1991). *The Extraction of Metals from Ores Using Bacteria* (ed. I.C. Sykes), 103–135. Academic Press. https://doi.org/10.1016/S0898-8838(08)60038-0.

Fabienne, F., Carine, L., Jean-Michel, G. et al. (2012). Isolation and characterization of environmental bacteria capable of extracellular biosorption of mercury. *Applied and Environmental Microbiology* 78: 1097–1106. https://doi.org/10.1128/AEM.06522-11.

Falco, L., Pogliani, C., Curutchet, G., and Donati, E. (2003). A comparison of bioleaching of covellite using pure cultures of *Acidithiobacillus ferrooxidans* and *Acidithiobacillus thiooxidans* or a mixed culture of *Leptospirillum ferrooxidans* and *Acidithiobacillus thiooxidans*. *Hydrometallurgy* 71: 31–36. https://doi.org/10.1016/S0304-386X(03)00170-1.

Fang, D., Zhang, R., Zhou, L., and Li, J. (2011). A combination of bioleaching and bioprecipitation for deep removal of contaminating metals from dredged sediment. *Journal of Hazardous Materials* 192 (1): 226–233.

Farooq, M.A., Islam, F., Ali, B. et al. (2016). Arsenic toxicity in plants: cellular and molecular mechanisms of its transport and metabolism. *Environmental and Experimental Botany* 132: 42–52. https://doi.org/10.1016/j.envexpbot.2016.08.004.

Fierros Romero, G., Rivas Castillo, A., Gómez Ramírez, M. et al. (2016). Expression analysis of Ni- and V-associated resistance genes in a *Bacillus megaterium* strain isolated from a mining site. *Current Microbiology* 73: 165–171. https://doi.org/10.1007/s00284-016-1044-6.

Fierros-Romero, G., Gómez-Ramírez, M., Arenas-Isaac, G.E. et al. (2016). Identification of *Bacillus megaterium* and *Microbacterium liquefaciens* genes involved in metal resistance and metal removal. *Canadian Journal of Microbiology* 62: 505–513. https://doi.org/10.1139/cjm-2015-0507.

Gadd, G.M. (2004). Microbial influence on metal mobility and application for bioremediation. *Geoderma* 122: 109–119. https://doi.org/10.1016/j.geoderma.2004.01.002.

Gadd, G.M. (2009). Biosorption: critical review of scientific rationale, environmental importance and significance for pollution treatment. *Journal of Chemical Technology and Biotechnology* 84: 13–28. https://doi.org/10.1002/jctb.1999.

Gao, Y., Ge, L., Shi, S. et al. (2019). Global trends and future prospects of e-waste research: a bibliometric analysis. *Environmental Science and Pollution Research* 26: 17809–17820. https://doi.org/10.1007/s11356-019-05071-8.

Garavaglia, L., Cerdeira, S.B., and Vullo, D.L. (2010). Chromium (VI) biotransformation by β- and γ-Proteobacteria from natural polluted environments: a combined biological and chemical treatment for industrial wastes. *Journal of Hazardous Materials* 175: 104–110. https://doi.org/10.1016/j.jhazmat.2009.09.134.

Ghaima, K.K. and Mohamed, A.I. (2017). Resistance and bioadsorption of cadmium by *Pseudomonas aeruginosa* isolated from agricultural soil. *International Journal of Applied Environmental Sciences* 12 (9): 1649–1660.

González, A.G., Pokrovsky, O.S., Santana-Casiano, J.M., and González-Dávila, M. (2017). Bioadsorption of heavy metals. In: *Prospects and Challenges in Algal Biotechnology* (ed. B.N. Tripathi and D. Kumar), 233–255. Singapore: Springer Singapore. https://doi.org/10.1007/978-981-10-1950-0_8.

Govarthanan, M., Park, S.-H., Park, Y.-J. et al. (2015). Lead biotransformation potential of allochthonous *Bacillus* sp. SKK11 with sesame oil cake extract in mine soil. *RSC Advances* 5: 54564–54570. https://doi.org/10.1039/C5RA06945A.

Grant, K., Goldizen, F.C., Sly, P.D. et al. (2013). Health consequences of exposure to e-waste: a systematic review. *The Lancet Global Health* 1 (6): e350–e361.

Gudeppu, M., Varier, K.M., Chinnasamy, A. et al. (2019). Nanobiotechnology approach for the remediation of environmental hazards generated from industrial waste. *Emerging Nanostructured Materials for Energy and Environmental Science* 1: 531–561. https://doi.org/10.1007/978-3-030-04474-9_13.

Gupta, A., Joia, J., Sood, A. et al. (2016). Microbial & biochemical technology microbes as potential tool for remediation of heavy metals: a review. *Journal of Microbial & Biochemical Technology* 8: 364–372. https://doi.org/10.4172/1948-5948.1000310.

Herat, S. and Agamuthu, P. (2012). E-waste: a problem or an opportunity? Review of issues, challenges and solutions in Asian countries. *Waste Management & Research* 30 (11): 1113–1129.

Hong, Y. and Valix, M. (2014). Bioleaching of electronic waste using acidophilic sulfur oxidising bacteria. *Journal of Cleaner Production* 65: 465–472. https://doi.org/10.1016/j.jclepro.2013.08.043.

Huang, Y., Zeng, Q., Hu, L. et al. (2021). Bioreduction performances and mechanisms of Cr(VI) by *Sporosarcina saromensis* W5, a novel Cr(VI)-reducing facultative anaerobic bacteria. *Journal of Hazardous Materials* 413: 125411. https://doi.org/10.1016/j.jhazmat.2021.125411.

Ilyas, S., Lee, J., and Chi, R. (2013). Bioleaching of metals from electronic scrap and its potential for commercial exploitation. *Hydrometallurgy* 131–132: 138–143. https://doi.org/10.1016/j.hydromet.2012.11.010.

Ilyas, S., Srivastava, R.R., Kim, H. et al. (2021). Circular bioeconomy and environmental benignness through microbial recycling of e-waste: a case study on copper and gold restoration. *Waste Management* 121: 175–185. https://doi.org/10.1016/j.wasman.2020.12.013.

Ilyas, S., Srivastava, R.R., Kim, H., and Ilyas, N. (2022). Biotechnological recycling of hazardous waste PCBs using *Sulfobacillus thermosulfidooxidans* through pretreatment of toxicant metals: process optimization and kinetic studies. *Chemosphere* 286: 131978. https://doi.org/10.1016/j.chemosphere.2021.131978.

Incharoensakdi, A. and Kitjaharn, P. (2002). Zinc biosorption from aqueous solution by a halotolerant cyanobacterium *Aphanothece halophytica*. *Current Microbiology* 45: 261–264. https://doi.org/10.1007/s00284-002-3747-0.

Iwahori, K., Watanabe, J., Tani, Y. et al. (2014). Removal of heavy metal cations by biogenic magnetite nanoparticles produced in Fe(III)-reducing microbial enrichment cultures. *Journal of Bioscience and Bioengineering* 117: 333–335. https://doi.org/10.1016/j.jbiosc.2013.08.013.

Iyer, A., Mody, K., and Jha, B. (2005). Biosorption of heavy metals by a marine bacterium. *Marine Pollution Bulletin* 50: 340–343. https://doi.org/10.1016/j.marpolbul.2004.11.012.

Jobby, R., Jha, P., Gupta, A. et al. (2019). Biotransformation of chromium by root nodule bacteria *Sinorhizobium* sp. SAR1. *PLoS ONE* 14: e0219387.

Kanmani, P., Aravind, J., and Preston, D. (2012). Remediation of chromium contaminants using bacteria. *International Journal of Environmental Science and Technology* 9: 183–193. https://doi.org/10.1007/s13762-011-0013-7.

Khalid, F., Hashmi, M.Z., Jamil, N. et al. (2021). Microbial and enzymatic degradation of PCBs from e-waste-contaminated sites: a review. *Environmental Science and Pollution Research* 28: 10474–10487. https://doi.org/10.1007/s11356-020-11996-2.

Koul, B. and Taak, P. (2018). Soil remediation through microbes. In: *Biotechnological Strategies for Effective Remediation of Polluted Soils* (ed. B. Koul and P. Taak), 101–128. Singapore: Springer Singapore https://doi.org/10.1007/978-981-13-2420-8_6.

Krishnamoorthy, R., Venkateswaran, V., Senthilkumar, M. et al. (2017). Potential microbiological approaches for the remediation of heavy metal-contaminated soils. In: *Plant-Microbe Interactions in Agro-Ecological Perspectives: Volume 2: Microbial Interactions and Agro-Ecological Impacts* (ed. D.P. Singh, H.B. Singh, and R. Prabha), 341–366. Singapore: Springer Singapore. https://doi.org/10.1007/978-981-10-6593-4_14.

Kumar Ramasamy, R., Congeevaram, S., and Thamaraiselvi, K. (2011). Evaluation of isolated fungal strain from e-waste recycling facility for effective sorption of toxic heavy metal Pb (II) ions and fungal protein molecular characterization—a mycoremediation approach. *Asian Journal of Experimental Biological Sciences* 2: 342–347.

Lampis, S., Santi, C., Ciurli, A. et al. (2015). Promotion of arsenic phytoextraction efficiency in the fern *Pteris vittata* by the inoculation of As-resistant bacteria: a soil bioremediation perspective. *Frontiers in Plant Science* 6: 80.

Lebeau, T., Braud, A., and Jézéquel, K. (2008). Performance of bioaugmentation-assisted phytoextraction applied to metal contaminated soils: a review. *Environmental Pollution* 153: 497–522. https://doi.org/10.1016/j.envpol.2007.09.015.

Lee, Y.-C. and Chang, S.-P. (2011). The biosorption of heavy metals from aqueous solution by *Spirogyra* and *Cladophora filamentous* macroalgae. *Bioresource Technology* 102: 5297–5304. https://doi.org/10.1016/j.biortech.2010.12.103.

Limcharoensuk, T., Sooksawat, N., Sumarnrote, A. et al. (2015). Bioaccumulation and biosorption of Cd^{2+} and Zn^{2+} by bacteria isolated from a zinc mine in Thailand. *Ecotoxicology and Environmental Safety* 122: 322–330. https://doi.org/10.1016/j.ecoenv.2015.08.013.

Lin, W.-H., Chen, S.-C., Chien, C.-C. et al. (2020). Application of enhanced bioreduction for hexavalent chromium-polluted groundwater cleanup: microcosm and microbial diversity studies. *Environmental Research* 184: 109296. https://doi.org/10.1016/j.envres.2020.109296.

Liu, Q., Li, Q., Wang, N. et al. (2018). Bioremediation of petroleum-contaminated soil using aged refuse from landfills. *Waste Management* 77: 576–585.

Lovley, D., Holmes, D., and Nevin, K. (2004). Dissimilatory Fe(III) and Mn(IV) reduction. *Advances in Microbial Physiology* 49: 219–226.

Lundgren, K. (2012). The global impact of e-waste: addressing the challenge. International Labour Organization.

Ma, Y., Rajkumar, M., Luo, Y., and Freitas, H. (2011). Inoculation of endophytic bacteria on host and non-host plants—effects on plant growth and Ni uptake. *Journal of Hazardous Materials* 195: 230–237. https://doi.org/10.1016/j.jhazmat.2011.08.034.

Mal, J., Nancharaiah, Y.V., van Hullebusch, E.D., and Lens, P.N.L. (2016). Effect of heavy metal co-contaminants on selenite bioreduction by anaerobic granular sludge. *Bioresource Technology* 206: 1–8. https://doi.org/10.1016/j.biortech.2016.01.064.

Malekzadeh, E., Alikhani, H.A., Savaghebi-Firoozabadi, G.R., and Zarei, M. (2012). Bioremediation of cadmium-contaminated soil through cultivation of maize inoculated with plant growth–promoting rhizobacteria. *Bioremediation Journal* 16: 204–211. https://doi.org/10.1080/10889868.2012.703258.

Masaki, Y., Hirajima, T., Sasaki, K., and Okibe, N. (2015). Bioreduction and immobilization of hexavalent chromium by the extremely acidophilic Fe(III)-reducing bacterium *Acidocella aromatica* strain PFBC. *Extremophiles* 19: 495–503. https://doi.org/10.1007/s00792-015-0733-6.

Modak, P. (2011). Waste: investing in energy and resource efficiency. *Towards a Green Economy Pathways to Sustainability Development Poverty Eradication* 2: 1–20.

Mohammadian Fazli, M., Soleimani, N., Mehrasbi, M. et al. (2015). Highly cadmium tolerant fungi: their tolerance and removal potential. *Journal of Environmental Health Science and Engineering* 13: 19. https://doi.org/10.1186/s40201-015-0176-0.

Mohapatra, R.K., Parhi, P.K., Pandey, S. et al. (2019). Active and passive biosorption of Pb(II)using live and dead biomass of marine bacterium *Bacillus xiamenensis* PbRPSD202: kinetics and isotherm studies. *Journal of Environmental Management* 247: 121–134. https://doi.org/10.1016/j.jenvman.2019.06.073.

Mondal, S. and Palit, D. (2019). Effective role of microorganism in waste management and environmental sustainability. *Sustainable Agriculture for Environment* 1: 485–515.

Morin, D., Lips, A., Pinches, T. et al. (2006). BioMinE–integrated project for the development of biotechnology for metal-bearing materials in Europe. *Hydrometallurgy* 83: 69–76. https://doi.org/10.1016/j.hydromet.2006.03.047.

Mulligan, C.N., Yong, R.N., and Gibbs, B.F. (2001). Remediation technologies for metal-contaminated soils and groundwater: an evaluation. *The Engineering Geologist* 60: 193–207. https://doi.org/10.1016/S0013-7952(00)00101-0.

Nagvenkar, G.S. and Ramaiah, N. (2010). Arsenite tolerance and biotransformation potential in estuarine bacteria. *Ecotoxicology* 19: 604–613. https://doi.org/10.1007/s10646-009-0429-8.

Naila, A., Meerdink, G., Jayasena, V. et al. (2019). A review on global metal accumulators—mechanism, enhancement, commercial application, and research trend. *Environmental Science and Pollution Research* 26: 26449–26471. https://doi.org/10.1007/s11356-019-05992-4.

Nanda, M., Kumar, V., and Sharma, D.K. (2019). Multimetal tolerance mechanisms in bacteria: the resistance strategies acquired by bacteria that can be exploited to 'clean-up' heavy metal contaminants from water. *Aquatic Toxicology* 212: 1–10. https://doi.org/10.1016/j.aquatox.2019.04.011.

Ni, L., Tian, J., Song, T. et al. (2019). Optimizing geometric parameters in hydrocyclones for enhanced separations: a review and perspective. *Separation and Purification Reviews* 48: 30–51. https://doi.org/10.1080/15422119.2017.1421558.

Ojo, F. O., Tukura, B. W., and Madu, P. C. (2018). Comparative analysis of heavy metal content in soil and their transfer to *Ocimum gratissimum* and *Moringa oleifera* vegetables planted in farms by Bmuko quarry sites. 3, 206–225.

Olguín, E.J. and Sánchez-Galván, G. (2012). Heavy metal removal in phytofiltration and phycoremediation: the need to differentiate between bioadsorption and bioaccumulation. *New Biotechnology* 30: 3–8. https://doi.org/10.1016/j.nbt.2012.05.020.

Páez-Vélez, C., Rivas, R.E., Dussán, J., 2019. Enhanced gold biosorption of *Lysinibacillus sphaericus* CBAM5 by encapsulation of bacteria in an alginate matrix. *Metals* 9(8), 818. https://doi.org/10.3390/met9080818

Pant, D., Joshi, D., Upreti, M.K., and Kotnala, R.K. (2012). Chemical and biological extraction of metals present in E waste: a hybrid technology. *Waste Management* 32: 979–990. https://doi.org/10.1016/j.wasman.2011.12.002.

Pant, D., Giri, A., and Dhiman, V. (2018). Bioremediation techniques for E-waste management. In: *Waste Bioremediation. Energy, Environment, and Sustainability* (ed. S.J. Varjani, E. Gnansounou, B. Gurunathan, et al.), 105–125. Singapore: Springer Singapore. https://doi.org/10.1007/978-981-10-7413-4_5.

Paul, D., Kazy, S.K., Banerjee, T.D. et al. (2015). Arsenic biotransformation and release by bacteria indigenous to arsenic contaminated groundwater. *Bioresource Technology* 188: 14–23. https://doi.org/10.1016/j.biortech.2015.02.039.

Peña-Montenegro, T.D., Lozano, L., and Dussán, J. (2015). Genome sequence and description of the mosquitocidal and heavy metal tolerant strain *Lysinibacillus sphaericus* CBAM5. *Standards in Genomic Sciences* 10 (1): 1–10.

Perkins, D.N., Brune Drisse, M.-N., Nxele, T., and Sly, P.D. (2014). E-waste: a global hazard. *Annals of Global Health* 80: 286–295. https://doi.org/10.1016/j.aogh.2014.10.001.

Płociniczak, T., Sinkkonen, A., Romantschuk, M., and Piotrowska-Seget, Z. (2013). Characterization of *Enterobacter intermedius* MH8b and its use for the enhancement of heavy metals uptake by *Sinapis alba* L. *Applied Soil Ecology* 63: 1–7. https://doi.org/10.1016/j.apsoil.2012.09.009.

Pradhan, J.K. and Kumar, S. (2012). Metals bioleaching from electronic waste by *Chromobacterium violaceum* and *Pseudomonads* sp. *Waste Management & Research* 30: 1151–1159. https://doi.org/10.1177/0734242X12437565.

Priya, A. and Hait, S. (2017). Comparative assessment of metallurgical recovery of metals from electronic waste with special emphasis on bioleaching. *Environmental Science and Pollution Research* 24: 6989–7008. https://doi.org/10.1007/s11356-016-8313-6.

Puckett, J. and Smith, T. (2002). *Exporting Harm: The High-Tech Trashing of Asia The Basel Action Network*. Seattle, WA: Silicon Valley Toxics Coalition.

Rahman, A., Nahar, N., Nawani, N.N. et al. (2015). Bioremediation of hexavalent chromium (VI) by a soil-borne bacterium, *Enterobacter cloacae* B2-DHA. *Journal of Environmental Science and Health, Part A* 50: 1136–1147. https://doi.org/10.1080/10934529.2015.1047670.

Rajesh, P., Athiappan, M., Paul, R., and Raj, K.D. (2014). Bioremediation of cadmium by Bacillus safensis (JX126862), a marine bacterium isolated from mangrove sediments. *International Journal of Current Microbiology and Applied Sciences* 3 (12): 326–335.

Rajkumar, M., Ma, Y., and Freitas, H. (2013). Improvement of Ni phytostabilization by inoculation of Ni resistant *Bacillus megaterium* SR28C. *Journal of Environmental Management* 128: 973–980. https://doi.org/10.1016/j.jenvman.2013.07.001.

Ramachandra, T. and Saira, V. (2004). Environmentally sound options for e-wastes management. *Envis Journal of Human Settlements* 5: 1–5.

Rani, M.J., Hemambika, B., Hemapriya, J., and Kannan, V.R. (2010). Comparative assessment of heavy metal removal by immobilized and dead bacterial cells: a biosorption approach. *African Journal of Environmental Science and Technology* 4: 77–83.

Rascio, N. and Navari-Izzo, F. (2011). Heavy metal hyperaccumulating plants: how and why do they do it? And what makes them so interesting? *Plant Science* 180: 169–181. https://doi.org/10.1016/j.plantsci.2010.08.016.

Rashamuse, K.J. and Whiteley, C.G. (2007). Bioreduction of Pt (IV) from aqueous solution using sulphate-reducing bacteria. *Applied Microbiology and Biotechnology* 75: 1429–1435. https://doi.org/10.1007/s00253-007-0963-3.

Rizvi, A., Ahmed, B., Zaidi, A., and Khan, M.S. (2019). Bioreduction of toxicity influenced by bioactive molecules secreted under metal stress by *Azotobacter chroococcum*. *Ecotoxicology* 28: 302–322. https://doi.org/10.1007/s10646-019-02023-3.

Rizvi, A., Zaidi, A., Ameen, F. et al. (2020). Heavy metal induced stress on wheat: phytotoxicity and microbiological management. *Royal Society* 10: 38379–38403. https://doi.org/10.1039/d0ra05610c.

Sable, S.V., Kawade, S., Ranade, S., and Joshi, S. (2020). Bioreduction mechanism of silver nanoparticles. *Materials Science and Engineering: C* 107: 110299. https://doi.org/10.1016/j.msec.2019.110299.

Sajjad, W., Zheng, G., Din, G. et al. (2019). Metals extraction from sulfide ores with microorganisms: the bioleaching technology and recent developments. *Transactions of the Indian Institute of Metals* 72: 559–579. https://doi.org/10.1007/s12666-018-1516-4.

Samadzadeh Yazdi, M.R., Abdollahi, M., Mousavi, S.M., and Khodadadi Darban, A. (2020). Comparison of copper dissolution in chalcopyrite concentrate bioleaching with *Acidianus brierleyi* in different initial pH values. *Journal of Mining Environment* 11: 753–764. https://doi.org/10.22044/jme.2020.9447.1853.

Sand, W., Gehrke, T., Jozsa, P.-G., and Schippers, A. (2001). (Bio)chemistry of bacterial leaching—direct vs. indirect bioleaching. *Hydrometallurgy* 59: 159–175. https://doi.org/10.1016/S0304-386X(00)00180-8.

Sarankumar, R.K., Arulprakash, A., Devanesan, S. et al. (2020). Bioreduction of hexavalent chromium by chromium resistant alkalophilic bacteria isolated from tannery effluent. *Journal of King Saud University - Science* 32: 1969–1977. https://doi.org/10.1016/j.jksus.2020.02.010.

Schippers, A., Hedrich, S., Vasters, J. et al. (2014). Biomining: metal recovery from ores with microorganisms. *Advances in Biochemical Engineering/Biotechnology* 141: 1–47. https://doi.org/10.1007/10_2013_216.

Shah, V. and Daverey, A. (2020). Phytoremediation: a multidisciplinary approach to clean up heavy metal contaminated soil. *Environmental Technology and Innovation* 18: 100774. https://doi.org/10.1016/j.eti.2020.100774.

Sharma, S. (2012). Bioremediation: features. *Strategies and Applications* 2: 202–213.

Sheel, A. and Pant, D. (2018). Recovery of gold from electronic waste using chemical assisted microbial biosorption (hybrid) technique. *Bioresource Technology* 247: 1189–1192. https://doi.org/10.1016/j.biortech.2017.08.212.

Silver, S. and Phung, L.T. (1996). Bacterial heavy metal resistance: new surprises. *Annual Review of Microbiology* 50: 753–789. https://doi.org/10.1146/annurev.micro.50.1.753.

Singh, R., Gandhi, M., Gramoday, C. et al. (2014). Microorganism as a tool of bioremediation technology for cleaning environment: A review. *Proceedings of the International Academy of Ecology and Environmental Sciences* 4 (1): 1–6.

Sinha, S.N. and Biswas, K. (2014). Bioremediation of lead from river water through lead-resistant purple-nonsulfur bacteria. *Global Journal of Microbiology and Biotechnology* 2 (1): 11–18.

Sinha, S.N. and Paul, D. (2013). Heavy metal tolerance and accumulation by bacterial strains isolated from waste water. *Journal of Chemical, Biological and Physical Sciences (JCBPS)* 4 (1): 812.

Sinha, A., Singh, V.N., Mehta, B.R., and Khare, S.K. (2011). Synthesis and characterization of monodispersed orthorhombic manganese oxide nanoparticles produced by *Bacillus* sp. cells simultaneous to its bioremediation. *Journal of Hazardous Materials* 192 (2): 620–627.

Singer, E., Emerson, D., Webb, E.A. et al. (2011). Mariprofundus ferrooxydans PV-1 the first genome of a marine Fe (II) oxidizing Zetaproteobacterium. *PloS one* 6 (9): e25386.

Srivastav, A., Yadav, K.K., Yadav, S., and Gupta, N. (2018). Nano-phytoremediation of pollutants from contaminated soil environment: current scenario and future prospects. In: *Phytoremediation* (ed. A. Ansari, S. Gill, R. Gill, et al.), 383–403. Cham: Springer. https://doi.org/10.1007/978-3-319-99651-6.

Srivastava, S. and Thakur, I.S. (2012). Biosorption and biotransformation of chromium by *Serratia* sp. isolated from tannery effluent. *Environmental Technology* 33: 113–122. https://doi.org/10.1080/09593330.2011.551842.

Stingu, A., Volf, I., Popa, V.I., and Gostin, I. (2012). New approaches concerning the utilization of natural amendments in cadmium phytoremediation. *Industrial Crops and Products* 35: 53–60. https://doi.org/10.1016/j.indcrop.2011.06.005.

Sun, F., Yan, Y., Liao, H. et al. (2014). Biosorption of antimony(V) by freshwater cyanobacteria *Microcystis* from Lake Taihu, China: effects of pH and competitive ions. *Environmental Science and Pollution Research* 21: 5836–5848. https://doi.org/10.1007/s11356-014-2522-7.

Tang, X., Zeng, G., Fan, C. et al. (2018). Chromosomal expression of CadR on *Pseudomonas aeruginosa* for the removal of Cd(II) from aqueous solutions. *Science of the Total Environment* 636: 1355–1361. https://doi.org/10.1016/j.scitotenv.2018.04.229.

Tarhriz, V., Akbari, Z., Dilmaghani, A. et al. (2019). Bioreduction of iron and biosorption of heavy metals (Ni^{2+}, Co^{2+}, Pb^{2+}) by a novel environmental bacterium,

Tabrizicola aquatica RCRI19^T. *Asian Journal of Water, Environment and Pollution* 16: 73–81. https://doi.org/10.3233/AJW190035.
Taştan, B.E., Ertuğrul, S., and Dönmez, G. (2010). Effective bioremoval of reactive dye and heavy metals by *Aspergillus versicolor*. *Bioresource Technology* 101 (3): 870–876.
Tay, S.B., Natarajan, G., Rahim, M.N.A. et al. (2013). Enhancing gold recovery from electronic waste via lixiviant metabolic engineering in *Chromobacterium violaceum*. *Scientific Reports* 3: 2236. https://doi.org/10.1038/srep02236.
Tian, H., Shi, Q., and Jing, C. (2015). Arsenic biotransformation in solid waste residue: comparison of contributions from bacteria with arsenate and iron reducing pathways. *Environmental Science & Technology* 49: 2140–2146. https://doi.org/10.1021/es504618x.
Tian, J., Ni, L., Song, T. et al. (2018). An overview of operating parameters and conditions in hydrocyclones for enhanced separations. *Separation and Purification Technology* 206: 268–285. https://doi.org/10.1016/j.seppur.2018.06.015.
Usman, K., Al Jabri, H., Abu-dieyeh, M.H., and Alsafran, M.H.S.A. (2020). Comparative assessment of toxic metals bioaccumulation and the mechanisms of chromium (Cr) tolerance and uptake in *Calotropis procera*. *Frontiers in Plant Science* 11: 1–11. https://doi.org/10.3389/fpls.2020.00883.
Vakilchap, F., Mousavi, S.M., Baniasadi, M., and Farnaud, S. (2020). Development and evolution of biocyanidation in metal recovery from solid waste: a review. *Reviews in Environmental Science and Biotechnology* 19: 509–530. https://doi.org/10.1007/s11157-020-09544-y.
de Vargas, I., Macaskie, L.E., and Guibal, E. (2004). Biosorption of palladium and platinum by sulfate-reducing bacteria. *Journal of Chemical Technology & Biotechnology* 79: 49–56. https://doi.org/10.1002/jctb.928.
Velásquez, L. and Dussan, J. (2009). Biosorption and bioaccumulation of heavy metals on dead and living biomass of Bacillus sphaericus. *Journal of Hazardous Materials* 167 (1–3): 713–716.
Vera, M., Schippers, A., and Sand, W. (2013). Progress in bioleaching: fundamentals and mechanisms of bacterial metal sulfide oxidation—part A. *Applied Microbiology and Biotechnology* 97: 7529–7541. https://doi.org/10.1007/s00253-013-4954-2.
Verma, S.K., Trivedi, P., Gupta, A.K., and Verma, R.K. (2017). Aromatic plant–microbe associations: a sustainable approach for remediation of polluted soils. In: *Agro-Environmental Sustainability: Volume 2: Managing Environmental Pollution* (ed. J.S. Singh and G. Seneviratne), 85–103. Cham: Springer International Publishing. https://doi.org/10.1007/978-3-319-49727-3_5.
Vijayaraghavan, K. and Yun, Y.-S. (2008). Bacterial biosorbents and biosorption. *Biotechnology Advances* 26: 266–291. https://doi.org/10.1016/j.biotechadv.2008.02.002.
Vullo, D.L., Ceretti, H.M., Daniel, M.A. et al. (2008). Cadmium, zinc and copper biosorption mediated by *Pseudomonas veronii* 2E. *Bioresource Technology* 99: 5574–5581. https://doi.org/10.1016/j.biortech.2007.10.060.

Wang, J., Bai, J., Xu, J., and Liang, B. (2009). Bioleaching of metals from printed wire boards by *Acidithiobacillus ferrooxidans* and *Acidithiobacillus thiooxidans* and their mixture. *Journal of Hazardous Materials* 172: 1100–1105. https://doi.org/10.1016/j.jhazmat.2009.07.102.

Wasi, S., Tabrez, S., and Ahmad, M. (2011). Suitability of immobilized pseudomonas fluorescens SM1 strain for remediation of phenols, heavy metals, and pesticides from water. *Water, Air, & Soil Pollution* 220: 89–99. https://doi.org/10.1007/s11270-010-0737-x.

Whiting, S.N., de Souza, M.P., and Terry, N. (2001). Rhizosphere bacteria mobilize Zn for hyperaccumulation by *Thlaspi caerulescens*. *Environmental Science & Technology* 35: 3144–3150. https://doi.org/10.1021/es001938v.

Widmer, R., Oswald-Krapf, H., Sinha-Khetriwal, D. et al. (2005). Global perspectives on e-waste. *Environmental Impact Assessment Review* 25: 436–458. https://doi.org/10.1016/j.eiar.2005.04.001.

Williamson, A.J., Folens, K., Matthijs, S. et al. (2021). Selective metal extraction by biologically produced siderophores during bioleaching from low-grade primary and secondary mineral resources. *Minerals Engineering* 163: 106774. https://doi.org/10.1016/j.mineng.2021.106774.

Willner, J. and Fornalczyk, A. (2013). Extraction of metals from electronic waste. *Environment Protection Engineering* 39: 1–6. https://doi.org/10.5277/EPE130115.

Wood, J.L., Liu, W., Tang, C., and Franks, A.E. (2016). Microorganisms in heavy metal bioremediation: strategies for applying microbial-community engineering to remediate soils. *AIMS Bioengineering* 3: 211–229. https://doi.org/10.3934/bioeng.2016.2.211.

Wu, X., Jin, C.Z., Jin, F.J. et al. (2020). *Lacisediminimonas profundi* gen. nov., sp. nov., a member of the family *Oxalobacteraceae* isolated from freshwater sediment. *Antonie van Leeuwenhoek* 113: 253–264.

Xiang, Y., Wu, P., Zhu, N. et al. (2010). Bioleaching of copper from waste printed circuit boards by bacterial consortium enriched from acid mine drainage. *Journal of Hazardous Materials* 184: 812–818.

Yazdi, M.R.S., Abdollahy, M., and Mousavi, S.M. (2014). Adaptation of acidophilic thermophile *Acidianus brierleyi* with chalcopyrite concentrate for copper recovery via bioleaching. The 8th International Chemical Engineering Congress & Exhibition (IChEC 2014) Kish, Iran, 24–27 February, 2014.

Yi, S., Li, F., Wu, C. et al. (2022). Synergistic leaching of heavy metal-polycyclic aromatic hydrocarbon in co-contaminatedac soil by hydroxamate siderophore: role of cation-π and chelation. *Journal of Hazardous Materials* 424: 127514. https://doi.org/10.1016/j.jhazmat.2021.127514.

Yin, Y., Hu, Y., and Xiong, F. (2013). Biosorption properties of Cd(II), Pb(II), and Cu(II) of extracellular polymeric substances (EPS) extracted from *Aspergillus fumigatus* and determined by polarographic method. *Environmental Monitoring and Assessment* 185: 6713–6718. https://doi.org/10.1007/s10661-013-3059-9.

Yue, Z.-B., Li, Q., Li, C. et al. (2015). Component analysis and heavy metal adsorption ability of extracellular polymeric substances (EPS) from sulfate reducing bacteria. *Bioresource Technology* 194: 399–402. https://doi.org/10.1016/j.biortech.2015.07.042.

Zeldes, B.M., Loder, A.J., Counts, J.A. et al. (2019). Determinants of sulphur chemolithoautotrophy in the extremely thermoacidophilic *Sulfolobales*. *Environmental Microbiology* 21: 3696–3710. https://doi.org/10.1111/1462-2920.14712.

Zhu, J., Li, Q., Jiao, W. et al. (2012). Adhesion forces between cells of *Acidithiobacillus ferrooxidans*, *Acidithiobacillus thiooxidans* or *Leptospirillum ferrooxidans* and chalcopyrite. *Colloids and Surfaces B: Biointerfaces* 94: 95–100. https://doi.org/10.1016/j.colsurfb.2012.01.022.

13

Advances in Bioleaching of Rare Earth Elements from Electronic Wastes

Xu Zhang[1], *Ningjie Tan*[1], *Seyed Omid Rastegar*[2], *and Tingyue Gu*[3]

[1] *State Key Laboratory of Bioreactor Engineering, East China University of Science and Technology, Shanghai, China*
[2] *Department of Chemical Engineering, Faculty of Engineering, University of Kurdistan, Sanandaj, Iran*
[3] *Department of Chemical and Biomolecular Engineering, Institute for Sustainable Energy and the Environment, Ohio University, Athens, United States*

Abstract

Bioleaching rare earth elements (REEs) from low-grade resources, including wastes, is a promising green technology. The microorganisms used in bioleaching mainly include bacteria and fungi. Bioleaching REEs involves different interaction mechanisms between microorganisms and REEs. Bioleaching REEs from urban mines such as various electronic wastes has become a hotspot in research. Bioleaching of copper cores has been practiced at huge commercial scales. Thus, bioleaching of REEs has the potential to be commercialized. It will become an economically viable method soon. This work mainly reviews the technology of bioleaching REEs from electronic wastes, including the present situations and applications of REEs resources, various REEs recovery technologies, recovering REEs from WEEE (waste electrical and electronic equipment) using bioleaching, types of microbes and leaching mechanisms, and post-leaching/bioleaching process. In order to make bioleaching of REEs practical, more research is needed, especially in the development of new microbial strains and bioreactor technology.

Keywords *rare earth element; bioleaching; microorganism; waste electrical and electronic equipment; post-leaching recovery*

13.1 Introduction

At present, large volumes of various types of solid waste are generated by different industries. Their examples include wastes from the mining industry, coal ash from coal-firing power plants, spent catalysts from automotives, spent lithium-ion

Management of Electronic Waste: Resource Recovery, Technology and Regulation, First Edition.
Edited by Anshu Priya.

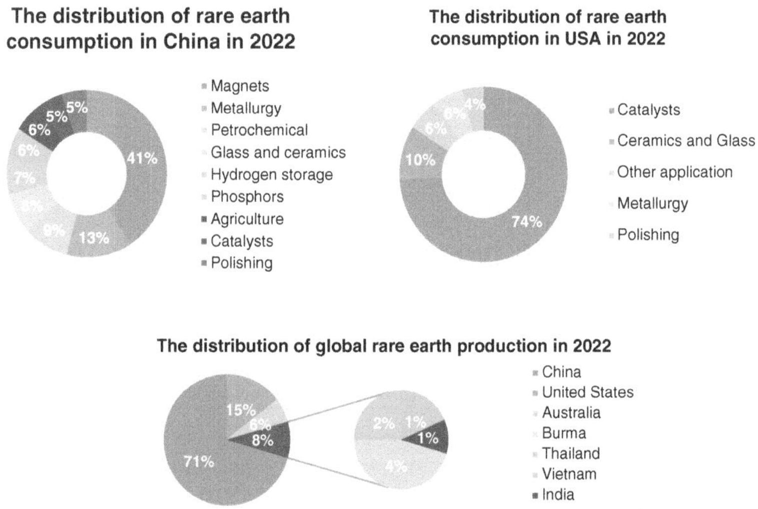

Figure 13.1 The distribution of rare earth consumption in China in 2022 and in United States of America in 2022, and the distribution of global rare earth production in 2022. Source: Data from Intelligence Research Group (2023), USGS (2022), and USGS (2022).

batteries (LIBs) from electronic devices, wastewater from electroplating, and sludge from sewage treatment plants. These wastes contain different metals including precious metals and rare earth elements (REEs) (Jadhav and Hocheng 2012; Konate et al. 2022). If untreated, their release into the environment will result in the pollution of soil, water, and air. Thus, the recovery of metals from solid wastes is a necessity for both environmental protection and economic opportunities (Lee et al. 2022).

REEs have useful magnetic properties, electrical conductivity, phosphorescence, catalysis, electrochemistry, and luminescence properties, which enable them to be used in many fields. The data for the production and applications of REEs are shown in Figure 13.1. Globally, most REEs are used in catalysts and high-performance magnets. For example, cerium and lanthanum are important catalysts for petroleum refining and diesel air pollution reduction. Neodymium is important in the production of magnets for conventional and low-carbon technologies. Electric motors for different types of electric vehicles, power-generating wind turbines, computer hard disks, consumer electronics, etc. The above three REEs are also important for the manufacture of alloys, fuel cells, and nickel-metal hydride batteries. Cerium, gallium, and neodymium are important in the manufacture of electronic products and are often used in liquid crystal displays (LCDs) and plasma displays, optical fibers, laser devices, medical imaging, etc.

Most REEs exist in the form of mineral deposits, and so far, more than 250 mineral types have been identified. They include carbonates, silicates, halides, phosphates, basic compounds, and oxides (Dushyantha et al. 2020). Monazite, bastnaesite, as well as xenotime, are commonly used ores to produce REEs (Zhou et al. 2017; Haque et al. 2014). Bastnaesite contains 70% rare earth element oxides (REOs). It is often found in soil and rocks such as carbonate–silicate rocks, pegmatites, and contact metamorphic zones. REOs are widely distributed across the globe, but they are most abundant in China, Brazil, and Vietnam. REO deposits are mainly in hydrothermal fluids and metamorphic rocks, and they are formed by weathering processes (Zhou et al. 2017). According to the U.S. Geological Survey (USGS) Inventory of Mineral Products (2016) report, about 34 countries have REO reserves. Table 13.1 displays the worldwide distribution (as of 2018) and production capacity of REEs (Dushyantha et al. 2020; Zhou et al. 2017). Global REE reserves are 120 Mt, of which 44 Mt come from China (36.6%), 22 Mt come from Brazil (18.3%), and Vietnam (18.3%).

Currently, the annual world-wide demand for REOs is approximately 60% of the growth rate due to the continuously growing demand for green and clean technologies for electronic and electrical equipment. It was about 118 kton in 2014 and is expected to reach 200 kton in 2025 (Van Gosen et al. 2017). Table 13.1 shows that over 90% of the global demand for REOs in 2018 came from China. According to the calculations by Kingsnorth (2016), the annual demand growth rate of REEs will reach 7–8%, and the price will also reach a new height. By 2025, the price of REEs per ton will reach 107 000–150 000 USD (Alonso et al. 2012; Dutta et al. 2016; Mertzman 2019). Entertainment, health, and transportation industries are increasingly using technologies across the globe, generating increased amounts of waste. These wastes are called waste electrical and electronic equipment (WEEE) (Ilyas and Lee 2014). The requirements for REEs have drawn attention to recycling REEs from WEEE for a sustainable society (Talebi et al. 2018; Ramprasad et al. 2022).

The extensive applications of REEs have also led to increased interest in the recovery of REEs. Revolutionary advances in green technologies, such as separate magnetic field generators, photovoltaic panels, strong magnets, and electric vehicles, have pushed the international trade of REEs. However, as a consequence, solid wastes containing REEs cause many health risks and ecological damages, such as pollution of fresh water, reduction of marine fish, and impacts on soil quality. Therefore, the largest REE supplier in the world, China, has formulated protective measures for REE exports, which to some extent increase the price of REEs around the world. The rising cost of raw materials has stimulated innovative methods for extracting REEs from both primary and secondary resources.

Research in this field has aroused great interest and made it a hot topic among many scientists, who want to discover innovative methods to obtain REEs from primary and secondary materials to meet the worldwide demand for REEs

Table 13.1 Distribution and production of rare earth oxides worldwide.

Location or country	Deposit	Type of ore	Main REEs present	REE reserve in million ton	REE production in kton
China	Bayan Obo	Carbonatite/hydrothermal	La, Ce, Nd	44	105
	Maoniu ping	Carbonatite	LREEs		
	Chinese ion adsorption deposits	Soils	La, Nd, HREEs		
Brazil	Catalao	Carbonatite	LREEs	22	2.0
	Araxa	Carbonatite	La, Ce, LREEs		
Vietnam	Dong Pao	Carbonatite	LREEs	22	0.1
Russia	Khibina and Lovenzero	Peralkaline igneous	LREE+Y, minor HREE	12	3.0
India	Chatra pur	Carbonatite	LREEs	6.9	1.5
Australia	Mount Weld	Laterite/carbonatite	LREE	3.3	20.0
	Nolans Bore	Vein	La, Ce, Nd		
United States	Mountain Pass	Carbonatite	LREEs	1.4	
Canada and Greenland	Holiday Lake	Vein	La, Ce, Pr, Nd	1.5	0.3
	Thor Lake	Alkaline igneous	La, Ce, Pr, Nd		
	Lake Strange and Lake Misery	Alkaline igneous/hydrothermal	La, Ce, Nd, HREEs		
South Africa, Malawi, and other African countries	Pilanesberg	Peralkaline igneous	La, Ce	1.68	0.6
	Steen kamps-Kraal	Vein	La, Ce, Nd		
	Kagankunde	Carbonatite	LREEs, Nd		
	Songwe	Carbonatite	LREEs and HREEs		
Other countries				5.22	1.0
Worldwide Total				120	134

LREE, light rare earth element; HREE, heavy rare earth element.
Reproduced from Dushyantha et al. (2020), Zhou et al. (2017), Voncken (2016), USGS (2022), Haque et al. (2014), Kumari et al. (2015).

(Maes et al. 2017; Gupta and Krishnamurthy 2005). Technologies to recover metals from industrial wastes are pyrometallurgy, hydrometallurgy, and electrochemical technology (Jadhav and Hocheng 2012; Rahimi et al. 2020; Habashi 1997), among which biohydrometallurgy technology is preferred because it is an effective, economical technology to obtain REEs from ores (Habashi 1997). Biohydrometallurgy employs microbes, mainly bacteria and fungi, to transform metal-containing chemicals into water-soluble ions via bioleaching. This is called bioleaching (Fuchs et al. 1995) or biooxidation (Andrès et al. 2003). It is a low-cost process, trading space (large reactors) with its lower efficiency, making it especially attractive for low-grade feeds (Hopfe et al. 2018).

Therefore, to meet the worldwide demand for REEs, researchers are working on bioleaching to recover REEs from waste products, retired products, and industrial wastes to help minimize environmental pollution from conventional mining and further processing of REEs. Bioleaching uses mild temperature and pressure. It only needs a simple reactor supplied with water, air, a few inexpensive nutrients, and microorganisms. The method can be economical and effective (Ilyas et al. 2007; Ilyas and Lee 2014). In fact, the world's largest bioreactors are bioleaching reactors for copper recovery (Gu et al. 2018). After REEs have been biosolubilized into leachate, they need to be recovered in a post-leaching process using precipitation, distributed crystallization, chromatography, solvent extraction, and ion exchange.

REEs in the WEEE need to be properly processed or recycled. Compared with conventional pyrometallurgical and hydrometallurgical techniques, bioleaching has the advantages of environmental friendliness, low energy consumption, simple operation, etc. Therefore, the recovery of REEs from WEEE by bioleaching is consistent with the requirements of sustainable development in a circular economy. It has both environmental and economic advantages. By utilizing WEEE, it can also alleviate the crisis of the shortage of earth's mineral resources.

13.2 REEs Recovery Technology

13.2.1 Classification and Characteristics of REEs Recovery and Treatment Technologies

In view of the recovery methods of REEs from various solid wastes, it is necessary to consider related factors such as the type of raw materials, the chemical content of REEs from WEEE, the total cost, and the environmental impact (Palaniandy et al. 2022). The operating cost of the whole recycling process should be low enough to be economically feasible. In order to be a green technology, the recovery of REEs should meet the low energy consumption target as well.

13.2.1.1 Pyrometallurgy

Due to the leapfrog growth in the demand for REEs in the world every year, REE recovery from WEEE has been put on the schedule (Akcil et al. 2018; Swain et al. 2020; Ujaczki et al. 2018). Pyrometallurgy is a traditional process to recover nonferrous and precious metals from ores and solid wastes. It uses a high-temperature range of 900 to 3000 °C (Priya and Hait 2017). Although pyrometallurgy is a relatively simple technology with a high recovery rate and tolerance of a wide range concentration of REEs in the raw materials, it is not an economical process for recovering REEs in WEEE, which can be compared to low-grade slag (Swain et al. 2020). Pyrometallurgy is known for its high energy consumption due to high operating temperatures, as well as causing the emission of harmful gases and the generation of dust particles (Jadhav and Hocheng 2012; Gu et al. 2018). These drawbacks limit its adoption under increasingly stringent environmental protection regulations all over the world (Priya and Hait 2017).

13.2.1.2 Hydrometallurgy

Hydrometallurgy involves the extraction of metals from complex matrices of materials. Chemicals such as acids, alkalis, and leaching agents are used (Binnemans et al. 2020). Moreover, the leachate requires further processing using unit operations including electrowinning, adsorption, or solvent extraction (Priya and Hait 2017; Moriwaki et al. 2012). Hydrometallurgical methods operate at mild temperatures with low energy consumption, and they are more selective compared with pyrometallurgical techniques (Binnemans et al. 2020). Recovering REEs from e-wastes by improving the efficiencies of pretreatment and acid-leaching processes has been investigated. The results can support the development of a sustainable, circular economy (Yuksekdag and Zeytuncu-Gokoglu 2022). Although hydrometallurgy requires a relatively small investment, making it the most popular leaching method on industrial scale, its daily operating expenses should not be overlooked (Borra et al. 2016a; Priya and Hait 2017; Qi 2018). At the same time, the reuse of leaching residues is relatively complicated. Workers risk exposure to dangerous acid solutions and smoke (Borra et al. 2016a). The strong acid and strong alkali used also need proper treatment before discharge for environmental protection (Syed 2006; Granados-Fernández et al. 2021).

13.2.1.3 Bioleaching

Bioleaching has been used to recover metals from different sources, such as ores, fly ashes, LIBs, PCBs (printed circuit boards), sludge, and sediments (Hubau and Bryan 2023; Gu et al. 2018; Stephen and Liang 2019). It has been practiced at an industrial scale (Gu et al. 2018). The most popular microbes for biosolubilization in bioleaching processes are *Acidithiobacillus thiooxidans*, *Acidithiobacillus ferrooxidans*, *Aspergillus niger*, and *Penicillium simplicissimum* (Gu et al. 2018; Castro et al. 2021).

13.2.1.3.1 Recovering REEs from WEEE using Bioleaching

Bioleaching is an attractive alternative to conventional hydrometallurgy and pyrometallurgy. It uses microorganisms to convert metals from ores or other materials to dissolved metal ions in the leachate (Panda 2020; Abdollahi et al. 2019; Panda et al. 2013; Esther et al. 2013). Bioleaching has several attractive features, such as mild operating conditions, environment friendliness, low operating expenses, and small energy expense (Chauhan et al. 2018; Priya and Hait 2017). It processes low-grade sulfide ores to extract metals from them (Johnson et al., 2013; Watling et al., 2016) and has been maturely industrialized. For example, Cu, Au, and U have been produced using bioleaching. Therefore, it meets the public's expectations of economical and sustainable production and shows a great potential for recovering REEs from urban mines (Priya and Hait 2017; Rasoulnia et al. 2021).

Like hydrometallurgy, bioleaching also requires subsequent steps to recover dissolved metal ions from leachates. For example, commercial bioleaching of copper from low-grade ores requires post-leaching precipitation (Riekkola-Vanhanen 2013) or electrowinning (Soto et al. 2013). The post-bioleaching unit operations include biosorption, bioprecipitation, and bioaccumulation, which can handle low-concentration feeds. The precipitation mechanism for dissolved metal ions in leachate is very complex, involving the interactions of a large amount of biomass and its metabolites with dissolved metal ions (Chauhan et al. 2018; Moriwaki et al. 2012; Omar et al. 1997; Ozaki et al. 2006). The leftover solution can be recirculated into the bioleaching reactor to minimize environmental pollution, such as wastewater, which was demonstrated in Talvivaara, Finland, and Escondida, Chile, mining operations (Soto et al. 2013; Riekkola-Vanhanen 2013).

Bioleaching efficiencies are impacted by redox potential, aeration, pH, temperature, slurry density, and metal ion toxicity of REEs. Compared with pyrometallurgy and chemical methods, bioleaching has a higher specificity and better economics at lower REE concentrations. Moreover, they can be used in the recycling process of low-grade electronic waste. Compared with other high-energy-consumption processes, bioleaching is more cost-effective and environmentally friendly. At present, the main limiting factors for bioleaching are its slow kinetics, large area requirement for industrial applications, and low recovery rate.

13.2.1.3.2 Type of Bioleaching Microbes and Mechanisms

The recovery rate of REEs can be improved through a green process using microorganisms to dissolve them (Oliveira et al. 2021; Omodara et al. 2019). Bioleaching microbes mainly include *A. ferrooxidans* (Oliveira et al. 2021), *Komagataeibacter hansenii, Zygosaccharomyces lentus* (Dev et al. 2020), and *Gluconobacter oxydans* (Dev et al. 2020; Zhang and Xu 2016; Ippolito et al. 2021). Organic acids are often produced during microbial metabolism. These organic acids are then released into the surrounding environment to accelerate biosolubilization, complexing and binding REEs and separating them from the solid wastes into the leachate (Dev et al. 2020; Ippolito et al. 2021; Bahaloo-Horeh and Mousavi 2022). These microbes need to be fed with nutrients such as acetate, butyrate, propionate, chitin, cellulose, and lignin as electron donors and carbon sources. A broad review of the biological processing of REEs was published recently (Macaskie et al. 2017). The different types of microbes utilized in bioleaching are shown in Table 13.2.

Using acidolysis, REEs can be separated from a solid phase. It is followed by complexation decomposition and a two-step oxidation–reduction process. In oxidation–reduction decomposition, electrons generated by ferrous ion oxidation to ferric ions are directly transferred from electronic wastes to microorganisms, and finally, the oxidation leaching of REEs from electronic wastes is realized. Microbes that can participate in this process include *A. thiooxidans*, *A. ferrooxidans*, and *Leptospirillum ferrooxidans*. Complexation decomposition is used to generate complexing agents and chelating agents through microbial metabolites such as iron carriers and organic acids to transfer target elements. With the advances in modern biotechnology, omics techniques are becoming more and more popular in the study of the mechanisms of bioleaching e-wastes, which can predict the results of microflora and maintain the activities of bioleaching microbes (Onwosi et al. 2023).

According to the principle of bioleaching to recover REEs from WEEE, the main functional microorganisms are divided into three categories: acidophilic bacteria, fungi, and cyanogenic microorganisms.

13.2.1.3.3 Acidophilic Bacteria

Acidophilic bacteria can grow under extremely acidic conditions and are often used for leaching common metals. Among them, *A. ferrooxidans*, *A. thiooxidans*, and *L. ferrooxidans* are the most widely used. They use CO_2 and inorganic nitrides as carbon and nitrogen sources, and obtain energy for growth by oxidizing metal elements, elemental sulfur, and reduced compounds (Fu et al. 2016). Sometimes, bacteria form biofilms on solid waste surfaces with the help of extracellular polymer substances (EPS) providing oxidation reaction space (Yang 2015) with a high volumetric sessile cell density. The leaching mechanisms of acidophilic bacteria include direct contact leaching as well as indirect contact leaching. In the former mechanism, bacteria-formed biofilms will attach to the waste's surface and leach metals by enzymolysis. In the indirect contact leaching mechanism, bacteria will produce Fe^{3+}, sulfuric acid, and other leachants and use oxidation to leach metals (Priya and Hait 2017). Both *A. ferrooxidans* and *L. ferrooxidans* can leach metal ions by oxidizing Fe^{2+} and continuously leach metal ions through two valence cycles (Figure 13.2). At the same time, *A. ferrooxidans* can also oxidize the low-valence sulfur compounds into sulfate radicals and use the generated sulfuric

Table 13.2 Types of microorganisms applied to rare earth element extraction.

Microorganism	REE	Recovery	Reference
Komagataeibacter xylinus, Lactobacillus casei, and *Yarrowia lipolytica.*	La, Eu, Tb, Y	up to 12.6%	Hopfe et al. (2018)
Gluconobacter oxydans	Sm, Y, Yb, Gd, Er	56.1%	Rasoulnia et al. (2021)
A. niger	Ce, La	25.9% Ce, 23.9% La	Hosseinzadeh et al. (2021)
	Ce	1.37 mg/L	Castro et al. (2021)
	Ce, Th	0.7 mg/L	Kishor et al., (2017)
A. ferrooxidans and *L. ferrooxidans*	Dy, Pr	up to 100%	Auerbach et al. (2019)
A. ferrooxidans	Ce, La	83%	Muddann and Baral (2021)
G. oxydans	Ce, La	up to 49%	Reed et al. (2016)
Penicillium expansum	La, Tb	390 ppm La, 1.52×10^3 ppm Tb	Piazza et al. (2016)
Aspergillus sp. or *A. ferrooxidans*	Nd, Gd	26.2% Nd, 62.8% Gd	Barnett et al. (2020)
Candida bombicola	Yb, Er	67.7% Yb, 64.6% Er	Park and Liang (2019)
Penicillum sp.	Ce, La, Nd, Pr	12.3 mg/L	Corbett et al. (2017)
Kombucha-consortium (Zygosaccharomyces lentus and *Komagataeibacter hansenii)*	Y, Eu	7.9%	Hopfe et al. (2017)
A. ferrooxidans and *Enterobacter aerogenes*	Ce, La, Nd, Pr, Y	5.84 mg/L total	Fathollahzadeh et al. (2018)
Micrococcus sp., *Micromonospora* sp., and *Streptomyces* sp.	Ce, La, Y	up to 548 μg/L total	Zhang et al. (2018)
Pseudomonas sp.	La	950 mg/g biomass dry wt	Kazy et al. (2006)
Agrobacterium sp. HN1	La, Ce	15 mg/L La,10 mg/L Ce	Xu et al. (2011)

Table 13.2 (Continued)

Microorganism	REE	Recovery	Reference
Monoraphidium sp. and *Penicillium* sp.	Nd	1.51 g/g dry cell	Palmieri et al. (2000)
Bacillus lichenifomis	Sm	333 mg/g dry cells	Tsuruta (2007)
Bacillus. subtilis	Sm	279–295 mg/g dry cells	
Brevibacterium helovolum	Sm	291 mg/g dry cells	
Rhodococcus elythmpolis	Sm	299 mg/g dry cells	
Penidiella sp. T9	Dy	47 mg/L	Horiike and Yamashita (2015)
Arthrobacter nicotianae	Sm	320 μmol/g dry cells	Tsuruta (2006)

Reproduced from Ramprasad et al. 2022and Liu et al. (2020).

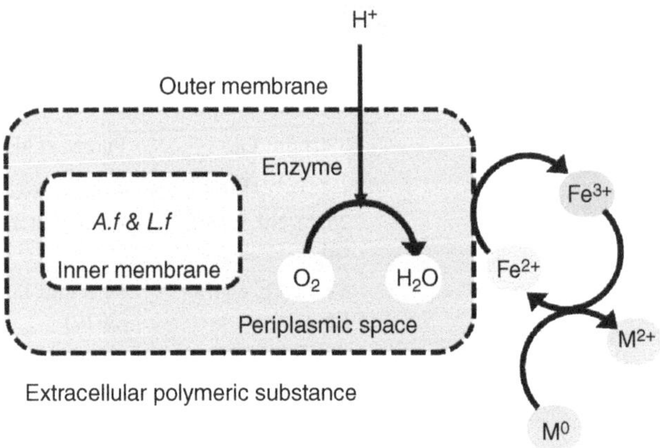

Figure 13.2 Schematic diagram of metal (M) leaching mechanism of *A. ferrooxidans* (*A.f*) and *L. ferrooxidans (L.f)*. Source: Adapted from Tan et al. (2020).

acid to leach metals, as shown in Reaction (1) and Reaction (3) with M representing metal. *A. thiooxidans* can produce acid by oxidizing sulfur elements for biosolubilization of metals as shown in Reaction (2) and Reaction (3).

$$MS_2 + 3.5O_2 + H_2O \xrightarrow{A.f} M^{2+} + 2SO_4^{2-} + 2H^+ \tag{13.1}$$

$$S^0 + 1.5O_2 + H_2O \xrightarrow{A.t} 2H^+ + SO_4^{2-} \tag{13.2}$$

$$M^0 + H_2SO_4 + 0.5O_2 \longrightarrow M^{2+} + SO_4^{2-} + H_2O \tag{13.3}$$

Some studies have shown that acidophilic bacteria can be used to leach REEs. *A. thiooxidans* was used to leach REEs from WEEE. At 10 g/L waste particle concentration, the leaching rates of Ce, Eu, and Nd reached over 99%, and the leaching rates of La and Y reached 80% within 8 days. Mineral phase analysis shows that REEs have good solubility and oxide morphology in H_2SO_4 solution, which promotes the leaching of REEs (Marra et al. 2018).

13.2.1.3.4 Fungi

A. niger, *Penicillium simplex*, *Penicillium chrysogenum*, and other fungi can produce organic acids such as citric acid, oxalic acid, lactic acid, and gluconic acid during their growth and metabolism (Yi et al. 2016; Horiike and Yamashita 2015; Kishor et al. 2017). These acids can leach precious metals such as Au and Ag and key metals such as Li, Co, and Ti from WEEE. Some marine fungi, such as *Dipodascaceae*, *Gymnoascaceae*, *Microascaceae*, and *Trichocomaceae*, were investigated for bioleaching and bioaccumulation of metals in e-wastes (Maleke et al. 2019). Platinum was found to be highly concentrated in fungal biomass (31–62%) (Galasso et al. 2022). Fungi can grow within a wide pH range, so their bioleaching of alkaline substances is more effective and can shorten the lag period of metal leaching from electronic wastes (Yi et al. 2013).

The bioleaching mechanism of fungi is mainly acid hydrolysis, complexation, and redox of metals, and these mechanisms can occur simultaneously (Arda et al. 2019). In the process of acidolysis, the biological acids produced by microorganisms protonated the oxygen atom covering the surface of the metal compound and simultaneously leached the metal (Baniasadi et al. 2019). Complexation refers to the dissolution of metals by the formation of ligands by biological chelating agents. In addition, the metal ions dissolved in the acidolysis process are also stabilized by complexation (Bahaloo-Horeh et al. 2018). The oxidation–reduction action is to dissolve the metals through the redox reactions. The energy necessary for fungal growth can be obtained through the acid-production process and respiration (Islam et al. 2020). For example, the reaction for *A. niger* to produce citric acid is shown in Reaction (4). Some studies using fungi are summarized In Table 13.2.

$$C_6H_{12}O_6 + 1.5O_2 \xrightarrow{A.niger} C_6H_8O_7 + 2H_2O + ATP + 3NAD(P)H_2 \qquad (13.4)$$

13.2.1.3.5 Cyanogenic Microorganisms

Cyanogenic microorganisms are microorganisms that can transform C and N elements in nutrients and generate secondary metabolite CN^-. They are widely used in the bioleaching of precious metals such as Ag, Au, Pt, Rh, Pd, and Ru. Biological cyanide is a secondary metabolite (Arda et al. 2019) formed by oxidative decarboxylation of glycine. The reaction is shown in Reaction (5). In the presence of CN^-, metals can form stable chelates with it and dissolve in water (Liang 2015), as shown in Reaction (6). Due to the different cyanide production capacities of microorganisms, their metal leaching efficiencies are different. Table 13.3 shows the leaching efficiencies of common cyanide-producing microorganisms for bioleaching precious metals from electronic wastes.

Table 13.3 Efficiency of bioleaching precious metals using cyanogen-producing microorganisms.

E-waste type	Chlorine-producing microorganism	Reactant concentration (g/L)	Leaching efficiency	Reference
Cell-phone circuit boards	*Pseudomonas chlororaphis*	16.0	8.2% Au, 12.1% Ag	Ruan et al. (2014)
	Chromobacterium violaceum	15.0	11.3% Au	Chi et al. (2011)
	Bacillus megaterium	8.13	50% Au	Arshadi et al. (2016)
Computer circuit boards	*Pseudomonas aeruginosa*	0.66	40.0% Au, 50.9% Ag	An (2019)
	Pseudomonas fluorescens	5.0	66.6% Au	Ge (2017)

$$NH_2CH_2COOH \xrightarrow{\text{Bacteria}} HCN + CO_2 + 2H_2 \tag{13.5}$$

$$4Au + 8CN^- + O_2 + 2H_2O \longrightarrow 4Au(CN)_2^- + 4OH^- \tag{13.6}$$

At present, the cyanogen-producing microorganisms used in the literature are mainly *C. violaceum*, *Bacillus megaterium* and *Pseudomonas* spp. *C. violaceum* has the strongest ability to generate CN^-, but it mainly lives in tropical and subtropical regions. Its particular living conditions limit its industrial application in recycling precious metals from electronic wastes (Ge 2017). Therefore, it is of great significance to select cyanogen-producing microorganisms with wider adaptability. Some studies using fungi are summarized in Table 13.3.

Literature reports have clearly demonstrated that REEs can be recycled from WEEE (Babu et al. 2007; Lee et al. 2007; Schluep and l'environnement 2009). Bioleaching is used to recover metals from the particles generated from the crushing of WEEE. *Thiobacillus ferrooxidans* and *Pseudomonas fetida* can reduce the pH of the leaching solution from 3.5 to 1.0 within 8 days. During this course, the yields of Ce and Eu exceeded 99%, and the yields of La and Y exceeded 80% (Marra et al. 2018).

13.2.1.4 Electrochemical Technology

Since the conventional REEs posttreatment process has the characteristics of large consumption of chemicals and energy, an efficient, highly selective, and sustainable leaching technology is desired (Kim et al. 2021a). Electrochemical methods, including electrocoagulation, electrodeposition, electrodialysis, and electrosorption, have the characteristics of high selectivity, multifunction, renewability, and scalability, so REEs can be selectively recovered from the leachate using the electrochemical method in absence of large amounts of chemicals, heat, and extreme pH conditions (Kim et al. 2021a; Irina et al. 2020; Kumari et al. 2015).

There are two different approaches to carry out molten salt electrolysis. One is to separate REEs from WEEE by the reaction between $AlCl_3$, FeF_3, AlF_3, $MgCl_2$, or ZnF_3 and REEs, and then separate REEs from molten metal halides using electrolysis (Yang et al. 2020). Albeit this approach is relatively effective, it suffers interference from Al, Fe, Zn, and Mg in REEs alloys, which have higher reduction potentials than REEs. Therefore, the REEs need to be further separated from the alloy. Additionally, WEEE can serve as anode materials that can be converted to halides. REEs obtained from WEEE electrodes are oxidized selectively and eluted as REE ions, which are then deposited on the cathode surface (Yang et al. 2020). For example, spent electrodes of permanent magnets could be incorporated into water tanks through the oxidation of $Nd_2Fe_{14}B$ and Nd–Pr alloys. Pr and Nd were oxidized selectively into REEs ions (Yang et al. 2020). Cathodic deposition of 13.2–14.7% of Nd was achieved with 0.1 M H_2SO_4, and 10.6–27.4% of Nd was recovered with 0.2 N H_2SO_4 and 0.1 N $H_2C_2O_4$ (Irina et al. 2020). As shown in Reactions (7) and (8). The electrochemical process based on weak ferric oxidants can reduce the consumption of chemical substances, is easy to control, and has lower energy consumption compared with hydrometallurgy and pyrometallurgy (Li et al. 2019). Moreover, the electrochemical process has faster reaction kinetics, high chemical stability, high conductivity, and a wider temperature range (Alvarado-Hernandez et al. 2019). However, its main drawback is that it is relatively costly. A recent report discussed electrochemical methods and techniques in detail (Kim et al. 2021a).

$$Nd_2Fe_{14}B + 17H_2SO_4 \longrightarrow 2Nd^{3+} + 14Fe^{2+} + 17SO_4^{2-} + 17H_2 + B \tag{13.7}$$

$$2B + 3H_2SO_4 \longrightarrow B_2O_3 + 3SO_2 + 3H_2O \tag{13.8}$$

13.2.1.5 Leaching Using Cell-Free Supernatant

In addition to the direct biological extraction of living bacterial cells, the metabolites produced by microbes can also promote the dissolution of REEs, but the leaching rate is relatively low when the cells are no longer present in a culture medium's supernatant. Circumstances such as no phosphate consumption, precipitation, and formation of secondary phosphate minerals lead to reduced total REE dissolution. As there is no microorganism in the cell-free supernatant, organic acid will not be continuously produced, resulting in a low REE leaching rate (Fathollahzadeh et al. 2018). The abiotic leaching rate of REEs in the synthetic organic acids (74 mmol L^{-1} citric acid and 14 mmol L^{-1} oxalic acid) of Egyptian monazite was at 58.8%, while the biological leaching rate was at 75.4% (Hassanoen et al. 2014). It was also found that three fungal strains could extract REEs from monazite, using monazite as a phosphate source and releasing REE cations into the solution. The dissolved concentrations of REE ions were significantly lower than those in microbial cell culture (*Aspergillus terroris* ML3-1 and *Paecilomyces* sp. WE3-F). The unknown metabolites secreted by the cells are more effective than the identified organic acids in dissolving REEs (Brisson et al. 2016). To a certain extent, synthetic organic acids can replace conventional leachants. However, using a live microbial culture, the leaching efficiencies of REEs improved (Fathollahzadeh et al. 2018; Corbett et al. 2017), which confirms the importance of the contact leaching mechanism in the biological leaching of monazite REEs. The use of biological methods to mine, recover, and recycle REE ores is an alternative and supplement to the current physical and chemical mining technologies. With different kinds of microorganisms, a bioleaching efficiency of 80–90% can be achieved (Brandl et al. 2016).

13.2.2 Recovery of REEs from WEEE

13.2.2.1.1 Permanent Magnet

Second-hand resources are a major source for REEs recycling. REEs such as Nd and Sm are mostly recovered from permanent magnets. Permanent magnet (NdFeB) is a major Nd resource, mainly including $Nd_2Fe_{14}B$. It contains REEs such as Nd, Pr, Ga, Dy, Tb, V, and Zr. Two kinds of magnets possess excellent conductivity and thermal properties. One is the Nd-Fe-B magnet, and the other is Sm–Co magnet. Nd-Fe-B magnets are known to have the advantage of superior performance at a low cost compared to Sm–Co magnets. The recovery of REEs from magnets in the form of direct reuse or powder has attracted attention (Binnemans et al. 2013; Ilka 2018). NdFeB alloy also exists in hard disk drives, speakers, smartphones, and LED lights. Each year, about 6–12 kt of NdFeB alloy is consumed to produce hard disks globally, according to published data in the literature (Gutfleisch et al. 2011; Jha et al. 2016). The hard disk crushing process can recover 60 tons of REEs every year, including the classical magnetic separation treatment. Hard disk drives contain permanent magnets, which can be separated by hydrogen under normal pressure. Under high pressure, hydrogen can demagnetize a NdFeB permanent magnet alloy powder. The powder from grinding hard disk drives is processed into sintered magnets, thus retaining the original magnetic properties (Yang et al. 2017).

There are many techniques for treating permanent magnets. Pyrometallurgy was used to extract REEs from Nd-Fe-B permanent magnets. The process included two steps: The first step was to oxidize the magnetic particles, and the second step was to reduce the generated oxides. The recovery rate of REEs was higher at higher temperatures (Bian et al. 2022). Sodium hydroxide was used for recovering REEs from NdFeB magnets at 200 °C for 1.5 hours. By adding NaOH and Na_2CO_3, borax led to crystallization. After leaching with tertiary carbonic acid, REEs loaded with tertiary carbonic acid and their residues were obtained. REEs were separated from NdFeB magnets by precipitation (Onal et al. 2017). In recent years, some research papers have been published on this subject. It has been found that the bioleaching efficiency of *A. thiooxidans* was only 0.02–0.3% after a week with 1% sulfur. Therefore, much needs to be done in this research area (Ilka 2018).

13.2.2.1.2 Fluorescent Materials

Waste fluorescent lamps are a source of REEs. Fluorescent lamps contain elements such as La, Y, and Eu, which can be recycled separately through chemical treatment. It was found that electronic wastes contain more REEs than some minerals in nature (Long et al. 2012). Phosphors contain one or several REEs. Fluorescent lamps rely on red, green, and blue phosphors that can emit white light. Y is a major REE in lamps. Red and blue phosphors containing Eu and Tb elements are present in green phosphors. Dilute HCl, H_2SO_4, and NaOH are utilized as leachants to recover REEs oxides from solid wastes containing $Y_2O_3:Eu^{3+}$ (YOX), $LaPO_4:Ce^{3+}$, Tb^{3+}(LAP), $BaMgAl_{10}O_{17}:Eu^{2+}$ (BAM), and (Ce, Tb)$MgAl_{11}O_{19}$ (CAT) tailings (Tan et al. 2015).

In the first step, dilute hydrochloric acid is used at 30 °C. In the second step, sulfuric acid is used at 90 °C. The third step is to use sodium hydroxide at 150 °C. REEs are extracted in the form of oxides (Binnemans et al. 2013). Fluorescent phosphors (FPs) have been found to contain 10% REE oxides, also known as triband dyes, which have become an important second-hand material for REE recycling. REEs can be recovered at a relatively high rate by co-culture of yeast and acetic-producing bacteria (Stefanie et al. 2017). *A. ferrooxidans* in indirect stepwise bioleaching was used to recover REEs from waste light-emitting diodes (WLEDs). At 20 g/L pulp density, the Ga recovery was 84%. The leaching period was also shortened from 30 days to 15 days (Pourhossein and Mousavi 2019). *A. ferrooxidans* in indirect stepwise bioleaching of Ga from light-emitting diodes (LEDs) has been studied. The results show that domesticated *A. ferrooxidans* is a suitable strain for bioleaching of LEDs, and its leaching efficiency of Ga can reach 60% (Pourhossein and Mousavi 2018).

13.2.2.1.3 Nickel Metal Hydride Batteries

Scientists have developed effective methods to extract REEs from nickel-hydrogen batteries (Jha et al. 2016; Kaindl et al. 2012). NiMH batteries are a major type of rechargeable battery that contains a significant amount of REEs. Hydrogen-absorbing alloys are used as anode and nickel hydroxide as cathode. NiMH batteries have energy densities 2–3 times larger than those of LIB. To minimize cost, the mixture of light REE metals is replaced by mixed metals in NiMH batteries. A NiMH battery contains 10% mixed metals, 4% cobalt, 36–42% nickel, and REEs such as Ce, Pr, Nd, and La (Müller and Friedrich 2006).

NiMH batteries have more REEs than other battery wastes. The REEs are separated from the solution by sulfuric acid leaching and extractant D2EHPA. Sulfuric acid is the most commonly used leachant for spent NiMH batteries. The positive and negative electrodes are mixed. Second-hand resources can be dissolved in 6 N sulfuric acid at 95 °C. At a high temperature, 94% of REEs can be leached out. The remaining REEs can be separated by solvent extraction with extractant P_2O_4. The REEs in the nickel-hydrogen battery can dissolve 80% of the REEs through 4 N sulfuric acid at 20 °C (Zhang et al. 1999).

Hydrochloric acid can also be used as the dissolution medium. By using 4 N HCl at 95 °C for 3 hours, 99% of REEs can be extracted. Sulfuric acid at 4 N concentration and 3 N hydrochloric acid are used to recover REEs, nickel, cobalt, etc. The extractant D2EHPA is further used for separation. Oxalic acid can also be used to treat the precipitate to separate REEs. The effects of hydrochloric acid and sodium hydroxide have also been studied. As a leaching agent, hydrochloric acid is superior to sulfuric acid, and sodium hydroxide is superior to sodium bicarbonate (Zhang et al. 2020). In the alkali leaching process, the maximum recoveries of Y, Tb, Ce, and Eu can reach 99%, 98%, 98.2%, and 97%, respectively. The bioleaching of spent NiMH batteries has been reported. Yeast extract was used instead of K_2HPO_4 to extract 19.5% of total REEs using *G. oxydans* within 24 hours (Rasoulnia et al. 2022).

13.3 Post-Leaching/Bioleaching Process

Solvent extraction is a common method for recovering REEs. The purity obtained from this method is better than from some other technologies (Dittrich et al. 2016). Scientists use effective and modern methods to extract REEs from both primary and secondary resources, but the biosolubilized REEs must be recovered using a post-leaching process.

A bioleaching process can turn REEs into soluble ions. And then, purification and separation of the metals from leachate are required. There are different methods that have been developed to recover metal ions including adsorption, precipitation, solvent extraction, and electrowinning, or combination of these methods. There are also other new and advanced technologies like microbial fuel cells (MFCs) and deep eutectic solvents (DESs) for the post-leaching recovery of metals from a leachate solution. The selection of these methods depends on the leaching system, content of target metals, and impurities in the leachate (Kumari and Samadder 2022; Ambaye et al. 2020; Dutta et al. 2022). Figure 13.3 represents several different methods.

13.3.1 Chemical Methods for Post-Leaching Recovery of Metals

13.3.1.1 Precipitation

Precipitation is another commonly used method for the recovery of different metals from leachate because it is simple and easily adoptable industrially (Chen

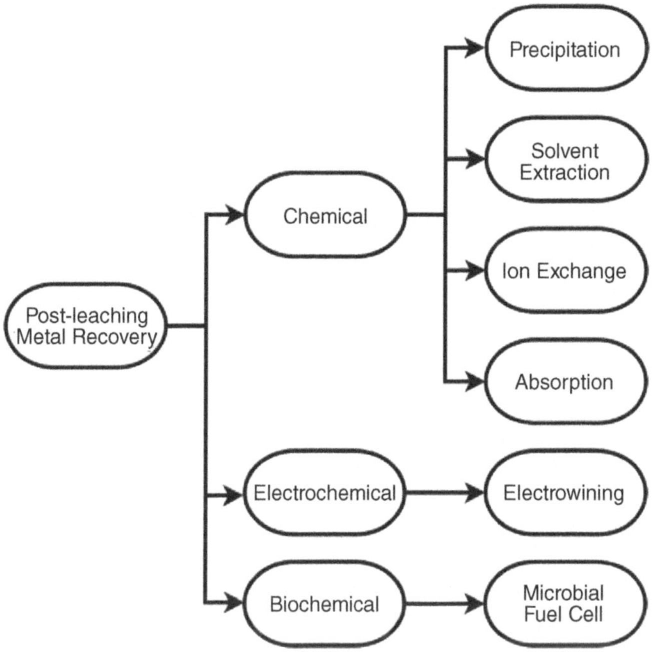

Figure 13.3 Various post-leaching metal recovery methods. Source: Kumari and Samadder (2022)/with permission of Elsevier.

et al. 2021). Precipitation involves adding chemical compounds to a leaching solution for selective separation of metals. Phosphate, carbonate, oxalate, and ammonium are the most popular salts used to precipitate metal ions in the leachant with adjusted pH and Eh values. However, it leads to the generation of secondary sludge at the bottom of the container. Different researchers used this method to recover metals, as reported in many literatures (León-Venegas et al. 2023; Yang et al. 2023; Chen et al. 2021). As one example, precipitation was used to recover vanadium after bioleaching. To precipitate V compounds selectively, NH_4Cl was added to the solution. After that, NH_4VO_3 was crystallized. Figure 13.4 shows a process flow chart V recovery from oil ash.

13.3.1.2 Solvent Extraction

This unit operation is also known as liquid–liquid extraction when partitioning organic solvent and water phases provides different solubilities as a basis for separation. Solvent extraction has been considered one of the most popular methods due to its high recovery efficiency, good flexibility, and easy operation. An organic agent interacts with the target component in the aqueous phase and transfers it from aqueous phase to organic phase. The extraction capacity is related to the distribution coefficient of different metal ions. A variety of solvents that are

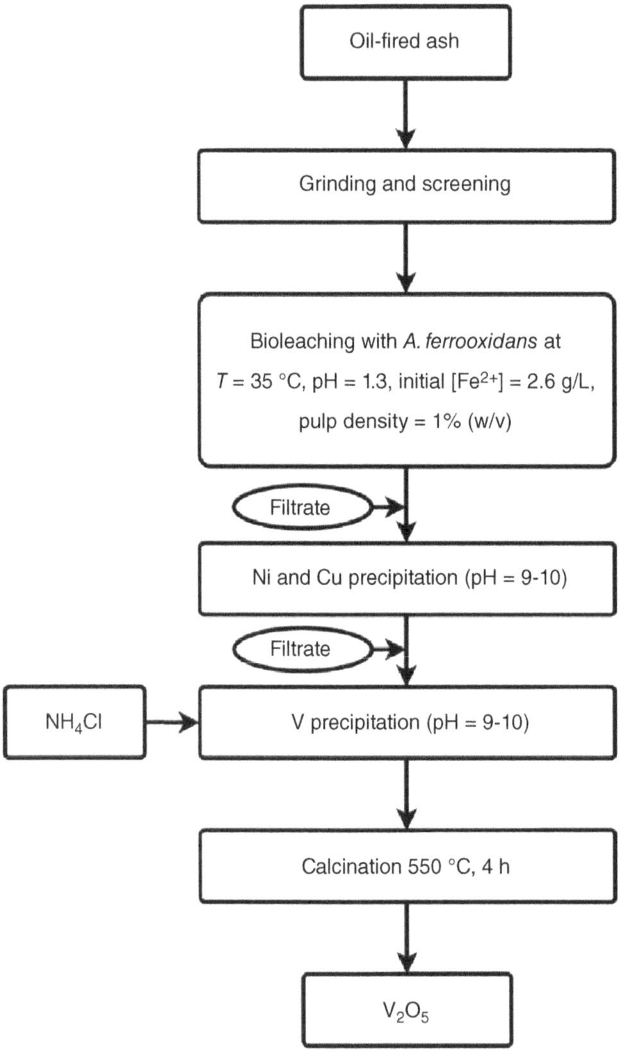

Figure 13.4 An integrated process for vanadium purification from burnt oil ash. Source: Adapted from Rastegar et al. (2015).

highly effective have been used, including D2EHPA, bis-2, 4, 4-trimethylpentyl dithiophosphinic acid (Cyanex 301), and LIX 84(Lei et al. 2022; Jiang et al. 2023).

Another form of solvent extraction is the extraction of metal ions from the leaching liquor (water content <50% by volume) with a DES as leaching agent. A DES is obtained when a hydrogen bond donor is mixed with a hydrogen bond acceptor, resulting in a melting point that is lower than the melting points of the two compounds. Common DES combinations include choline chloride

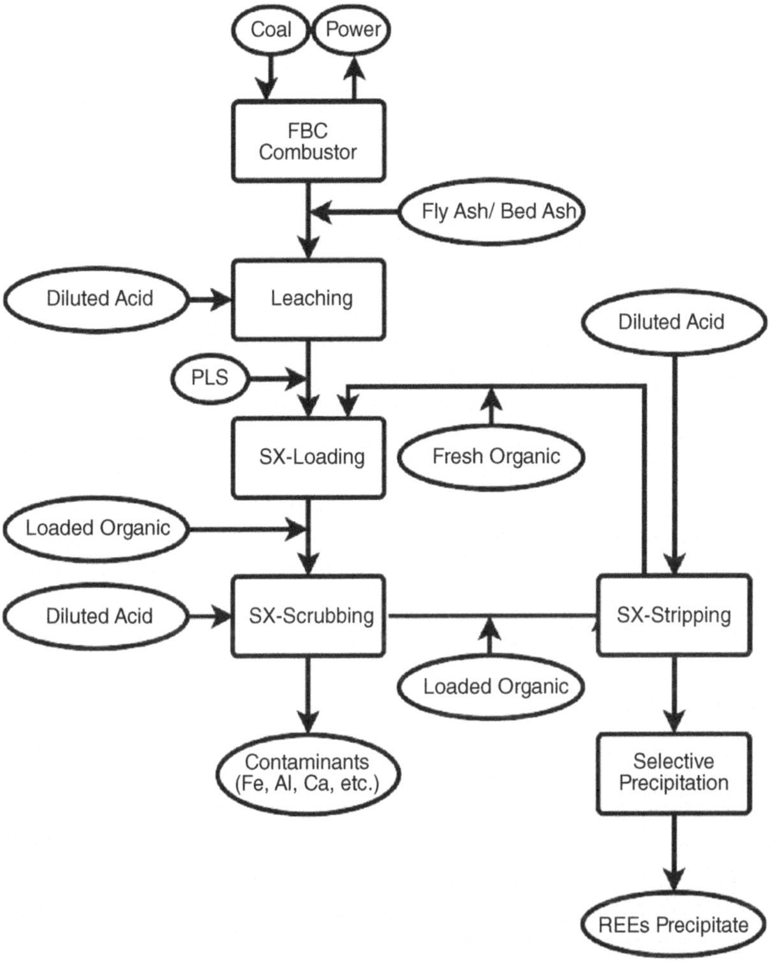

Figure 13.5 A process using solvent extraction (SX) for recovering REEs from coal combustion ash after acid leaching. Source: Adapted from Zhang et al. 2020.

(ChCl)/urea, ChCl/phenol, and ChCl/ethylene glycol (Yuan et al. 2022). A flow chart for recovering REEs using the solvent extraction process is shown in Figure 13.5. REEs in the leachate of coal fly ash are extracted using solvent extraction. In the end, a concentrated solution with high-purity REEs is obtained by selective precipitation using oxalic acid (Lei et al. 2022).

13.3.1.3 Ion Exchange

Ion exchange has been investigated to recover noble metals for quite some time due to its advantages, such as high separation resolution, relative ease of

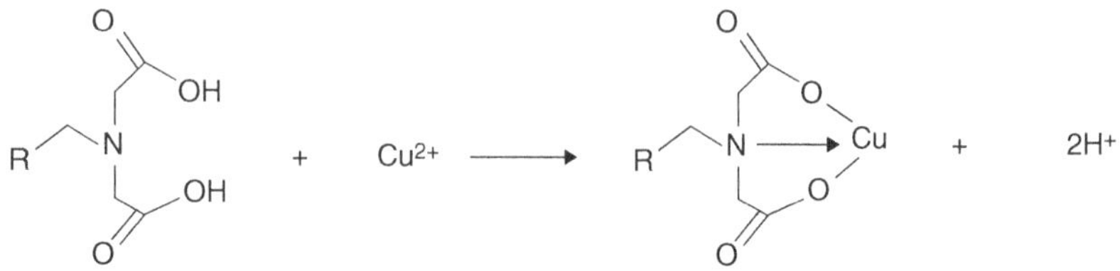

Figure 13.6 An example of ion exchange with a copper ion. Source: Junior et al. (2019)/with permission of Elsevier.

operation, minimal reagent use, no phase partitions, feasibility for a feed with low-concentration metal ions, and environmental safety. Polymer beads (resins) bind with metal ions and are then displaced by other strong ions or ions with higher concentrations in the mobile phase during elution for selective separation of desired metal ions (Goc et al. 2021; Wang et al. 2023). Figure 13.6 shows an example of iminodiacetate ligand on a resin forming a strong chelating complex with a copper ion (Junior et al. 2019).

Figure 13.7 shows a downstream process using ion exchange for post-leaching recovery of several REEs and Al (Zhang et al. 2020). The upstream leaching uses Na_2CO_3 roasting followed by water leaching.

13.3.1.4 Adsorption

Adsorption is a popular method for separating metal ions in post-leaching recovery of metals. It has many advantages because it is efficient, versatile, simple, cost-effective, practical, and environmentally friendly. The attachment of metal ions from an aqueous phase to the adsorbent through different interactions, including electrostatic attraction, surface coordination, complexation, and intermolecular van der Waals forces. Different adsorbents have been used to separate metals from aqueous solutions, such as leachates. They include carbonaceous adsorbents (activated carbon, graphene oxide, carbon nanotubes, etc.), magnetic adsorbents, ion-imprinted adsorbents, and different polymer-based adsorbents (Abidli et al. 2021; Vo et al. 2020; Hosomomi et al. 2013). Hydrous ferric oxyhydroxide was used to recover approximately 90% of W and 100% adsorption of V and Co from leachate of spent catalysts utilizing Mn and ferric compounds (Srivastava et al. 2012).

Another adsorbent is based on bio-based adsorbents. Thus, this process is termed biosorption. Several biosorbents, including chitosan, cellulose, and starch, have also been tested to remove different types of metals from leachates. Recently, several investigations have been conducted for the removal of REEs from leachates using bacteria, fungi, and algae as biosorption materials (Karavaiko et al. 1996; Philip et al. 2000; Takahashi et al. 2005; Vlachou et al. 2009; Wang 2008). Different biosorbents, including *Pseudomonas* sp., *Sargassum*, and *Agrobacterium* sp. HN1,

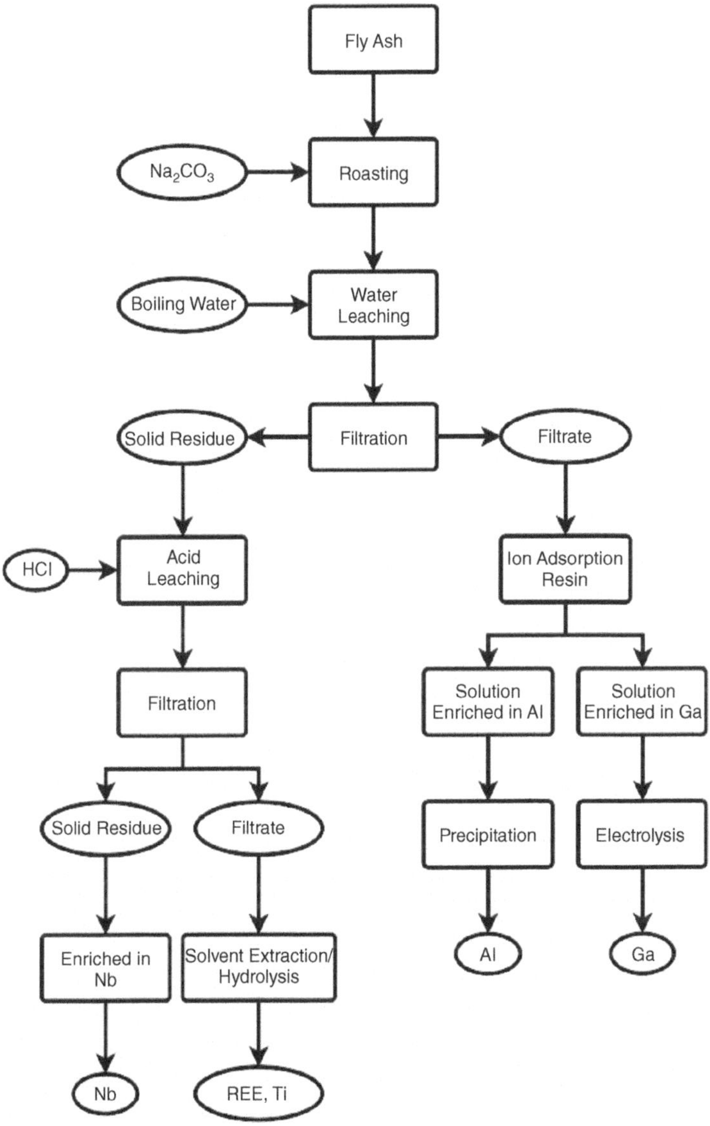

Figure 13.7 A flowsheet for recovering REEs, Ga, and Al from fly ash using alkali roasting. Source: Adapted from Zhang et al. (2020).

Saccharomyces cerevisiae, *Penicillium* sp., *Candida colliculosa*, *Monoraphidium* sp., *Kluyveromyces marxiamus,* and *Debaryomyces hansenii* biomass materials, were studied for recovering neodymium and lanthanum from leaching solutions. Recovery of Ce using a powder from *Platanus orientalis* leaves and the biomass of

Agrobacterium sp. HN1 were also carried out successfully in lab studies (Ambaye et al. 2020; Brown et al. 2023).

13.3.1.5 Electrochemical Method

Electrowinning is a popular technique to recover metals from metal ion solutions. The electrodeposition or electroplating method involves the oxidation of organic materials at the anode, and metal ions are reduced and electroplated or deposited at the cathode surface with the help of an external electric field. During the process, metal ions transfer to the cathode surface, where they are reduced and form a thin elemental metal layer on the surface of the cathode. This extracts nonferrous metals such as copper, tin, and nickel. The morphology of the deposited film is tuned by changing different parameters, including electrolyte, electrode distance, temperature, and applied current density. Electrowinning has the advantages of good economic feasibility, less secondary pollution, high-purity products, and environmental friendliness if electrolytes are properly treated before discharging (Kim et al. 2021b; Khan et al. 2022).

One research group used electrodeposition with an 86–96% max recovery rate for Cu, Eu, Nd, Ga, and Sc from unaligned carbon nanotubes that were encapsulated in polyvinyl alcohol filters (O'Connor et al. 2018). Another group deposited MnO_2 on the two sides of a steel electrode. The deposition was done in the galvanostatic mode, utilizing a current density of 1 mA cm^{-2}. A combination of electrodepositing and precipitation was used to recover Nd, Dy, and Pr with high purity of 99.4% from a leachate (Yousefi et al. 2012).

13.3.1.6 Bioelectrochemical Method

The recent development of bioelectrochemical systems, including MFCs, provides a new alternative for environmentally friendly recovery of metals. An MFC is an electrochemical method with a distinction that it uses a bioanode to harvest electrons from wastewater to produce bioelectricity. In the anode chamber, bacteria are used to decompose organic materials to produce electrons and protons. In the cathode chamber, noble metal (e.g. Cu) ions are reduced in the role of electron acceptors on the cathode surface, and this deposits high-purity elemental metals on the cathode (Noori et al. 2021; Mathuriya and Yakhmi 2014). If a metal is more noble (i.e. its ion has a higher redox potential), its ion is a more potent electron acceptor, which increases the MFC's voltage output. A schematic of MFC for recovering metals (using Cu as an example) is shown in Figure 13.8. MFCs can recover metals even at very low concentrations, without the need for another process. However, the method cannot be used for recovering active metals because they have a high thermodynamic tendency to be oxidized. Many studies using MFCs to recover different metals have been reported and they are listed in Table 13.4. By combining metal recovery with wastewater treatment,

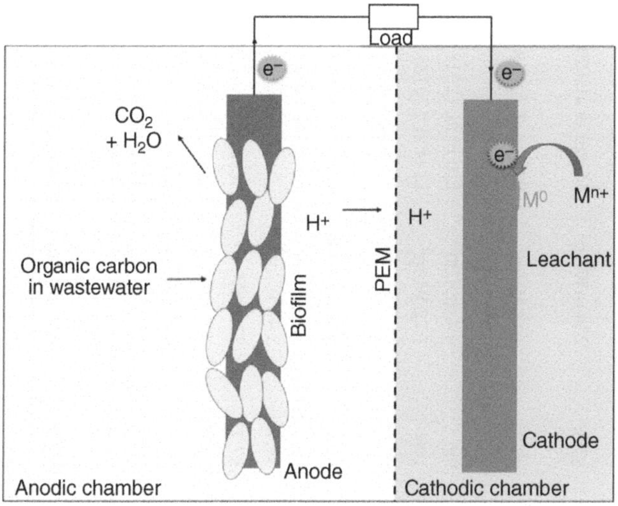

Figure 13.8 Illustration of MFC for wastewater treatment in the anode chamber and metal (e.g. Cu) recovery in the cathode chamber from leachate.

the hybrid MFC system has a much better chance for practical deployment than conventional MFCs using an oxygen cathode.

13.4 Conclusion and Outlook

With the rapidly escalating need for rare earth metals, it is necessary to develop new sustainable technologies to mine primary and secondary REEs. Bioleaching is among one of the practically promising technologies to recover REEs. Benefiting from the characteristics of low pollution and mild reaction conditions, bioleaching can extract REEs from low-grade ores and urban electronic wastes. However, compared with conventional REEs extraction methods, bioleaching requires a longer time and some nutrients for the growth of microorganisms, especially in heterotrophic microorganisms. Due to the complex compositions of e-wastes, further research is needed on the impact of leaching environment and metal ion toxicity on microorganisms, which can limit the waste loading. The research on the metabolic mechanisms of metallurgical microorganisms, and selections of more robust and effective microorganisms in the recovery of REEs from electronic wastes, including the technical feasibility of the industrialization of REEs bioleaching processes. In future research, improving leaching efficiency and developing bioleaching strains with strong stress resistance will become key issues to be solved. It is also necessary to develop efficient and green post-leaching separation processes to recover REEs.

Table 13.4 A summary of studies using MFC for recovery of metals from solutions.

Metal	Substrate	Anode	Cathode	Membrane	Removal efficiency	Power density	Reference
Cu(II)	Acetate	Carbon felt	Copper plate	AEM	99	5.5 W/m^2	Rodenas Motos et al. (2015)
V(V)	Glucose	Carbon fiber felt	Carbon fiber felt	PEM	76.8	419 mW/m^2	Zhang et al. (2015)
Cr(VI)	Acetate	Thin graphite sheets	Thin graphite sheets	salt bridge	95	89 mW/m^2	Sophia and Saikant (2016)
Cu(II)	Sodium lactate	Carbon felt	Carbon cloth	PEM	≤98	224 mW/m^2	Miran et al. (2017)
Cu(II)	Acetate	Carbon brush	Platinum-coated carbon cloth	–	98.3	10.2 W/m^3	Wu et al. (2018)
Cu(II)	Glucose	Graphite	Graphite	CEM	99	4.4 mA/m^2	Liu et al. (2018)
Cr(VI)	Acetate	Carbon felt	Carbon cloth	PEM	100	1.22 W/m^2	Li et al. (2018)
Co	Acetate	Granular activated carbon	Granular activated carbon	AEM	98.55	11.3 W/m^3	Huang et al. (2019)
Pt	Acetate	Graphite plate	Graphite plate	CEM	≤90	844 mW/m^2	Liu et al. (2019)
Pd, Pt, Rh	Acetate	Carbon brush	Carbon cloth	CEM	99.2, 99.5, 98.7	4.22, 5.04 2.44 W/m^2	Hu et al. (2019)
Cu(II)	Acetate	Carbon felt	Stainless steel wire mesh	PEM	87.7	370 mV	Wang et al. (2020)
Cr(VI)	Acetate	Carbon felt	Carbon cloth (PPY/SSC)	PEM	100	761 mW/m^2	Li et al. (2021)
Cu(II)	Acetate	Carbon felt	Stainless steel plate	CEM	100	65.8 mW/m^2	Zhang et al. (2021)
Cr(VI)	Sodium Formate	Granular activated carbon (Bio-PdNPs)	Carbon cloth	PEM	100	1.97 W/m^3	Matsena et al. (2021)
Cu(II)	Acetate	Carbon cloth	Carbon cloth	AEM	86.2	420 mW/m^2	Yang et al. (2021)

AEM, anion exchange membrane; PEM, proton exchange membrane; CEM, cation exchange membrane.
From Sobhani et al., 2022, courtesy of Kurdistan Univ Press.

References

Abdollahi, H., Noaparast, M., Shafaei, S.Z. et al. (2019). Prediction and optimization studies for bioleaching of molybdenite concentrate using artificial neural networks and genetic algorithm. *Minerals Engineering* 130: 24–35. https://doi.org/10.1016/j.mineng.2018.10.008.

Abidli, A., Huang, Y., Rejeb, Z.B. et al. (2021). Sustainable and efficient technologies for removal and recovery of toxic and valuable metals from wastewater: recent progress, challenges, and future perspectives. *Chemosphere* 133102: https://doi.org/10.1016/j.chemosphere.2021.133102.

Akcil, A., Akhmadiyeva, N., Abdulvaliyev, R. et al. (2018). Overview on extraction and separation of REEs from red mud: focus on scandium. *Mineral Processing & Extractive Metallurgy Review* 39: 145–151. https://doi.org/10.1080/08827508.2017.1288116.

Alonso, E., Sherman, A.M., Wallington, T.J. et al. (2012). Evaluating rare earth element availability: a case with revolutionary demand from clean technologies. *Environmental Science & Technology* 46 (6): 3406–3414. https://doi.org/10.1021/es203518d.

Alvarado-Hernandez, L., Lapidus, G.T., and Gonza´ lez, F. (2019). Recovery of rare earths from waste cathode ray tube (CRT) phosphor powder with organic and inorganic ligands. *Waste Manage* 95: 53–58. https://doi.org/10.1016/j.wasman.2019.05.057.

Ambaye, T.G., Vaccari, M., Castro, F.D. et al. (2020). Emerging technologies for the recovery of rare earth elements (REEs) from the end-of-life electronic wastes: a review on progress, challenges, and perspectives. *Environmental Science and Pollution Research* 27 (29): 36052–36074. https://doi.org/10.1007/s11356-020-09630-2.

An, N. (2019). *Study on gold leaching from the waste printed circuit boards by Pseudomonas aeruginosa (D).* Qingdao, China: Qingdao University of Science & Technology.

Andrès, Y., Texier, A.C., and Le Cloirec, P. (2003). REEs removal by microbial biosorption: a review. *Environmental Technology* 24 (11): 1367–1375. https://doi.org/10.1080/09593330309385681.

Arda, I., Hullebusch, E., Lenz, M. et al. (2019). Biotechnological strategies for the recovery of valuable and critical raw materials from waste electrical and electronic equipment (WEEE) A review. *Journal of Hazardous Materials* 362: 467–481. https://doi.org/10.1016/j.jhazmat.2018.08.050.

Arshadi, M., Mousavi, S.M., and Rasoulnia, P. (2016). Ehancement of simultaneous gold and copper recovery from discarded mobile phone PCBs using *Bacillus metagerium*: RSM absed optimization of effective factors and evaluation of thier interactions. *Waste Management* 57: 158–167. https://doi.org/10.1016/j.wasman.2016.05. 012.

Auerbach, R., Bokelmann, K., Stauber, R. et al. (2019). Critical raw materials advanced recycling technologies and processes: recycling of rare earth metals out of end of life magnets by bioleaching with various bacteria as an example of an intelligent recycling strategy. *Minerals Engineering* 134: 104–117. https://doi.org/10.1016/j.mineng.2018.12.022.

Babu, B.R., Parande, A.K., and Basha, C.A. (2007). Electrical and electronic waste: a global environmental problem. *Waste Management & Research* 25 (4): 307–318. https://doi.org/10.1177/0734242XO7076941.

Bahaloo-Horeh, N. and Mousavi, S.M. (2022). A novel green strategy for biorecovery of valuable elements along with enrichment of rare earth elements from activated spent automotive catalysts using fungal metabolites. *Journal of Hazardous Materials* 430: 128509. https://doi.org/10.1016/j.jhazmat.2022.128509.

Bahaloo-Horeh, N., Mousavi, S.M., and Baniasadi, M. (2018). Use of adapted metal tolerant *Aspergillus niger* to enhance bioleaching efficiency of valuable metals from spent lithium-ion mobile phone batteries. *Journal of Cleaner Production* 197: 1546–1557. https://doi.org/10.1016/j.jclepro.2018.06.299.

Baniasadi, M., Vakilchap, F., Bahaloo-Horeh, N. et al. (2019). Advances in bioleaching as a sustainable method for metal recovery from e-waste: a review. *Journal of Industrial and Engineering Chemistry* 76: 75–90. https://doi.org/10.1016/j.jiec.2019.03.047.

Barnett, M.J., Palumbo-Roe, B., Deady, E.A., and Gregory, S.P. (2020). Comparison of three approaches for bioleaching of rare earth elements from bauxite. *Minerals* 10 (8): 649. https://doi.org/10.3390/min10080649.

Bian, Y., Guo, S., Xu, Y. et al. (2022). Recovery of REEs from permanent magnet scraps by pyrometallurgical process. *Rare Metals* 41 (5): 1697–1702. https://doi.org/10.1007/s12598-015-0554-x.

Binnemans, K., Jones, P.T., Blanpain, B. et al. (2013). Recycling of rare earths: a critical review. *Journal of Cleaner Production* 51: 1–22. https://doi.org/10.1016/j.jclepro.2012.12.037.

Binnemans, K., Jones, P.T., Manjon, F., and A. and Masaguer, Torres, V. (2020). Hydrometallurgical processes for the recovery of metals from steel industry by-products: a critical review. *Journal of Sustainable Metallurgy* 6 (4): 505–540. https://doi.org/10.1007/s40831-020-00306-2.

Borra, C.R., Blanpain, B., Pontikes, Y. et al. (2016a). Recovery of rare earths and other valuable metals from bauxite residue (red mud): a review. *Journal of Sustainable Metallurgy* 2: 365–386. https://doi.org/10.1007/s40831-016-0068-2.

Brandl, H., Barmettler, F., Castelberg, C. et al. (2016). Microbial mobilization of REEs (REE) from mineral solids—a mini review. *AIMS Microbiology* 3 (2): 190–204. https://doi.org/10.3934/microbiol.2016.2.190.

Brisson, V.L., Zhuang, W.Q., and Alvarez-Cohen, L. (2016). Bioleaching of REEs from monazite sand. *Biotechnology and Bioengineering* 113 (2): 339–348. https://doi.org/10.1002/bit.25823.

Brown, R.M., Mirkouei, A., Reed, D., and Thompson, V. (2023). Current nature-based biological practices for rare earth elements extraction and recovery: bioleaching and biosorption. *Renewable and Sustainable Energy Reviews* 173: 113099. https://doi.org/10.1016/j.rser.2022.113099.

Castro, L., Blázquez, M.L., González, F., and Muñoz, J.Á. (2021). Biohydrometallurgy for rare earth elements recovery from industrial wastes. *Molecules* 26 (20): 6200. https://doi.org/10.3390/molecules26206200.

Chauhan, G., Jadhao, P.R., Pant, K.K., and Nigam, K.D.P. (2018). Novel technologies and conventional processes for recovery of metals from waste electrical and electronic equipment: challenges & opportunities – A review. *Journal of Environmental Chemical Engineering* 6 (1): 1288–1304. https://doi.org/10.1016/j.jece.2018.01.032.

Chen, J., Zhang, H., Zeng, Z. et al. (2021). Separation of lithium and transition metals from the leachate of spent lithium-ion battery by extraction precipitation with p-tert-butylphenoxy acetic acid. *Hydrometallurgy* 206: 105768. https://doi.org/10.1016/j.hydromet.2021.105768.

Chi, T.D., Lee, J., Pandey, B.D. et al. (2011). Bioleaching of gold and copper from waste mobile phone PCBs by using a cyanogenic bacterium. *Minerals Engineering* 24 (11): 1219–1222. https://doi.org/10.1016/j.mineng.2011.05.009.

Corbett, M.K., Eksteen, J.J., Niu, X.Z. et al. (2017). Interactions of phosphate solubilising microorganisms with natural rare-earth phosphate minerals: a study utilizing Western Australian monazite. *Bioprocess and Biosystems Engineering* 40 (6): 929–942. https://doi.org/10.1007/s00449-017-1757-3.

Dev, S., Sachan, A., Dehghani, F. et al. (2020). Mechanisms of biological recovery of rare-earth elements from industrial and electronic wastes: a review. *Chemical Engineering* 397: 124596. https://doi.org/10.1016/j.cej.2020.124596.

Dittrich, C., Kaya, S., Diekmann, F. (2016). Recovery of rare earth elements and scandium from European deposits by solvent extraction. ALTA Uranium-Ree Conference.

Dushyantha, N., Batapola, N., Ilankoon, I.M.S.K. et al. (2020). The story of REEs (REEs): occurrences, global distribution, genesis, geology, mineralogy and global production. *Ore Geology Reviews* 122: 103521. https://doi.org/10.1016/j.oregeorev.2020.103521.

Dutta, T., Kim, K.H., Uchimiya, M. et al. (2016). Global demand for rare earth resources and strategies for green mining. *Environmental Research* 150: 182–190. https://doi.org/10.1016/j.envres.2016.05.052.

Dutta, D., Rautela, R., Gujjala, L.K.S. et al. (2022). A review on recovery processes of metals from E-waste: a green perspective. *Science of the Total Environment* 859 (2): 160391. https://doi.org/10.1016/j.scitotenv.2022.160391.

Esther, J., Panda, S., Behera, S.K. et al. (2013). Effect of dissimilatory Fe (III) reducers on bio-reduction and nickel–cobalt recovery from Sukinda chromite-overburden.

Bioresource Technology 146: 762–766. https://doi.org/10.1016/j.biortech.2013.07.103.

Fathollahzadeh, H., Becke, T., Eksteen, J.J. et al. (2018). Microbial contact enhances bioleaching of rare earth elements. *Bioresource Technology Reports* 3: 102–108. https://doi.org/10.1016/j.biteb.2018.07.004.

Fu, K., Chen, H., Chen, S. et al. (2016). Application research status of bacteria in recycling waste printed circuit boards. *Metal Mine* 1: 176–181. CNKI:SUN:JSKS.0.2016-01-040.

Fuchs, T., Huber, H., Teiner, K. et al. (1995). *Metallosphaera prunae* sp. nov., a novel metal-mobilizing, thermoacidophilic archaeum, isolated from a uranium mine in Germany. *Systematic and Applied Microbiology* 18 (4): 560–566. https://doi.org/10.1016/S0723-2020(11)80416-9.

Galasso, C., Lekube, X., Cancio, I. et al. (2022). Marine fungi as potential eco-sustainable resource for precious metals recovery from electronic waste. *Waste and Biomass Valorization* 13 (2): 967–976. https://doi.org/10.1007/s12649-021-01587-8.

Ge, Z. (2017). *Study on Gold Leaching from the Waste Printed Circuit Boards by Pseudomonas fluorescens*. Qingdao: Qingdao University of Science&Technology.

Goc, K., Kluczka, J., Benke, G. et al. (2021). Application of ion exchange for recovery of noble metals. *Minerals* 11 (11): 1188. https://doi.org/10.3390/min11111188.

Granados-Fernández, R., Montiel, M.A., Díaz-Abad, S. et al. (2021). Platinum recovery techniques for a circular economy. *Catalysts* 11 (8): 937. https://doi.org/10.3390/catal11080937.

Gu, T., Rastegar, S.O., Mousavi, S.M. et al. (2018). Advances in bioleaching for recovery of metals and bioremediation of fuel ash and sewage sludge. *Bioresource Technology* 261: 428–440. https://doi.org/10.1016/j.biortech.2018.04.033.

Gupta, C. and Krishnamurthy, N. (2005). Extractive Metallurgy of Rare Earths. CRC Press*Boca Raton, Florida* 65: 70–75. https://vector.umd.edu/images/links/Extractive_Metallurgy_of_Rare_Earths_Gupta.pdf.

Gutfleisch, O., Willard, M.A., Brück, E. et al. (2011). Magnetic materials and devices for the 21st century: stronger, lighter, and more energy efficient. *Advanced Materials* 23 (7): 821–842. https://doi.org/10.1002/adma.201002180.

Habashi, F. (1997). *Handbook of Extractive Metallurgy, Volume*. Wiley-Vch: Federal republic of Germany.

Haque, N., Hughes, A., Lim, S., and Vernon, C. (2014). REEs: overview of mining, mineralogy, uses, sustainability and environmental impact. *Resources* 3 (4): 614–635. https://doi.org/10.3390/resources3040614.

Hassanoen, W.A.G., Desouky, O.A.N., and Hussien, S.S.E. (2014). Bioleaching of some REEs from Egyptian monazite using *Aspergillus ficuum* and *Pseudomonas aeruginosa*. *Walailak Journal Engineering and Physical Sciences* 11 (9): 809. https://doi.org/10.2004/WJST.V11I6.481.

Hopfe, S., Flemming, K., Lehmann, F. et al. (2017). Leaching of rare earth elements from fluorescent powder using the tea *fungus kombucha*. *Waste Management* 62 (APR.): 211–221. https://doi.org/10.1016/j.wasman.2017.02.005.

Hopfe, S., Konsulke, S., Barthen, R. et al. (2018). Screening and selection of technologically applicable microorganisms for recovery of rare earth elements from fluorescent powder. *Waste Management* 79: 554–563. https://doi.org/10.1016/j.wasman.2018.08.030.

Horiike, T. and Yamashita, M. (2015). A new fungal isolate, *Penidiella* sp. strain T9, accumulates the rare earth element dysprosium. *Applied & Environmental Microbiology* 81 (9): 3062. https://doi.org/10.1128/AEM.00300-15.

Hosomomi, Y., Baba, Y., Kubota, F. et al. (2013). Biosorption of REEs by *Escherichia coli*. *Journal of Chemical Engineering of Japan* 46 (7): 450–454. https://doi.org/10.1252/jcej.13we031.

Hosseinzadeh, F., Rastegar, S.O., and Ashengroph, M. (2021). Bioleaching of rare earth elements from spent automobile catalyst as pretreatment method to improve Pt and Pd recovery: process optimization and kinetic study. *Process Biochemistry* 105: 1–7. https://doi.org/10.1016/j.procbio.2021.03.022.

Hu, N., Cui, Y., and Choi, C. (2019). Recovery of platinum-group metals using a microbial fuel cell. *Trends in Diabetes and Metabolism* 2: 1–9. https://doi.org/10.15761/TDM.1000115.

Huang, T., Song, D., Liu, L., and Zhang, S. (2019). Cobalt recovery from the stripping solution of spent lithium-ion battery by a three-dimensional microbial fuel cell. *Separation and Purification Technology* 215: 51–61. https://doi.org/10.1016/j.seppur.2019.01.002.

Hubau, A. and Bryan, C.G. (2023). Metal recovery from E-wastes. *Biomining Technologies* 239–259: https://doi.org/10.1007/978-3-031-05382-5_14.

Ilka, A. (2018). *Bioleaching of Rare Earth Elements from Permanent Magnets (D).* Delft, the Netherlands: Host University.

Ilyas, S. and Lee, J.C. (2014). Biometallurgical recovery of metals from waste electrical and electronic equipment: a review. *ChembioEng Reviews* 1 (4): 148–169. https://doi.org/10.1002/cben.201400001.

Ilyas, S., Anwar, M.A., Niazi, S.B., and Ghauri, M.A. (2007). Bioleaching of metals from electronic scrap by moderately thermophilic acidophilic bacteria. *Hydrometallurgy* 88 (1–4): 180–188. https://doi.org/10.1016/j.hydromet.2007.04.007.

Intelligence Research Group. (2023) Research report on the operating situation and development trend of China's rare earth industry. https://www.chyxx.com/ku/1137249.html.

Ippolito, N.M., Ferella, F., Innocenzi, V. et al. (2021). Effect of mechanical activation on terbium dissolution from waste fluorescent powders. *Minerals Engineering* 167: 106906. https://doi.org/10.1016/j.mineng.2021.106906.

Irina, M., Ekaterina, S., Maria, O. et al. (2020). Electrochemical leaching of rare-earth elements from spent NdFeB magnets. *Hydrometallurgy* 192: 105264. https://doi.org/10.1016/j.hydromet.2020.105264.

Islam, A., Ahmed, T., Awual, M.R. et al. (2020). Advances in sustainable approaches to recover metals from e-waste: a review. *Journal of Cleaner Production* 244: 118815. https://doi.org/10.1016/j.jclepro.2019.118815.

Jadhav, U.U. and Hocheng, H. (2012). A review of recovery of metals from industrial waste. *Journal of Achievements in Materials and Manufacturing Engineering* 54 (2): 159–167. http://jamme.acmsse.h2.pl/papers_vol54_2/5422.pdf.

Jha, M.K., Kumari, A., Panda, R. et al. (2016). Review on hydrometallurgical recovery of rare earth metals. *Hydrometallurgy* 165: 2–26. https://doi.org/10.1016/j.hydromet.2016.01.035.

Jiang, T., Wang, P., Zhang, T. et al. (2023). A novel solvent extraction system to recover germanium from H_2SO_4 leaching liquor of secondary zinc oxide: extraction behavior and mechanism. *Journal of Cleaner Production* 383: 135399. https://doi.org/10.1016/j.jclepro.2022.135399.

Johnson, D.B., Grail, B.M., and Hallberg, K.B.A. (2013). New direction for biomining: extraction of metals by reductive dissolution of oxidized ores. *Minerals* 3 (1): 49–58. https://doi.org/10.3390/min3010049.

Junior, A.B.B., de Albuquerque Vicente, A., Espinosa, D.C.R., and Tenório, J.A.S. (2019). Recovery of metals by ion exchange process using chelating resin and sodium dithionite. *Journal of Materials Research and Technology* 8 (5): 4464–4469. https://doi.org/10.1016/j.jmrt.2019.07.059.

Kaindl, M., Luidold, S., and Poscher, A. (2012). Recycling von Seltene Erden aus Nickel-Metallhydrid-Akkumulatoren unter besonderer Berücksichtigung von Säurerückgewinnung. BHM Berg-und Hüttenm ¨annische. *Monatshefte* 157 (1): 20–26. https://doi.org/10.1007/s00501-012-0053-4.

Karavaiko, G.I., Kareva, A.S., Avakian, Z.A. et al. (1996). Biosorption of scandium and yttrium from solutions. *Biotechnology Letters* 18 (11): 1291–1296. https://doi.org/10.1007/BF00129957.

Kazy, S.K., Das, S.K., and Sar, P. (2006). Lanthanum biosorption by a *Pseudomonas* sp.: equilibrium studies and chemical characterization. *Journal of Industrial Microbiology & Biotechnology* 33 (9): 773–783. https://doi.org/10.1007/s10295-006-0108-1.

Khan, W.S., Asmatulu, E., Uddin, M.N. and Asmatulu, R. (2022). Recycling and Reusing of Engineering Materials: Recycling for Sustainable Developments. https://www.elsevier.com/books/recycling-and-reusing-of-engineering-materials/khan/978-0-12-822461-8.

Kim, S., Bang, J., Yoo, J. et al. (2021a). A comprehensive review on the pretreatment process in lithium-ion battery recycling. *Journal of Cleaner Production* 294: 126329. https://doi.org/10.1016/j.jclepro.2021.126329.

Kim, K., Candeago, R., Rim, G. et al. (2021b). Electrochemical approaches for selective recovery of critical elements in hydrometallurgical processes of complex feedstocks. *iScience* 24 (5): 102374. https://doi.org/10.1016/j.isci.2021.102374.

Kingsnorth, D. (2016). Rare earths—the China conundrum. In the 12th international rare earths conference Hong Kong 8th–10th.

Kishor, K., Keekan, J., and Chhaganbhai, J. (2017). Extraction of Ce and Th from monazite using REE tolerant *Aspergillus niger*. *Mineral Processing and Extractive Metallurgy Review* 38 (5): 312–320. https://doi.org/10.1080/08827508.2017.1350956.

Konate, F.O., Ancia, P., Soma, F. et al. (2022). Waste electrical and electronic equipments as urban mines in burkina faso: characterization and release of metal particles. *Waste Management* 139: 17–24. https://doi.org/10.1016/j.wasman.2021.12.014.

Kumari, R. and Samadder, S.R. (2022). A critical review of the pre-processing and metals recovery methods from e-wastes. *Journal of Environmental Management* 320: 115887. https://doi.org/10.1016/j.jenvman.2022.115887.

Kumari, A., Panda, R., Jha, M.K. et al. (2015). Process development to recover rare earth metals from monazite mineral: a review. *Minerals Engineering* 79: 102–115. https://doi.org/10.1016/j.mineng. 2015.05.003.

Lee, J.C., Song, H.T., and Yoo, J.M. (2007). Present status of the recycling of waste electrical and electronic equipment in Korea. *Resources Conservation & Recycling* 50 (4): 380–397. https://doi.org/10.1016/j.resconrec.2007.01.010.

Lee, L.C., Zhang, L., Chen, X. et al. (2022). An overview study on management and implementation of WEEE in China. *Environment, Development and Sustainability*. https://doi.org/10.1007/s10668-022-02489-y.

Lei, S., Sun, W., and Yang, Y. (2022). Solvent extraction for recycling of spent lithium-ion batteries. *Journal of Hazardous Materials* 424: 127654. https://doi.org/10.1016/j.jhazmat.2021.127654.

León-Venegas, E., Vilches-Arenas, L.F., Fernández-Baco, C., and Arroyo-Torralvo, F. (2023). Potential for water and metal recovery from acid mine drainage by combining hybrid membrane processes with selective metal precipitation. *Resources, Conservation and Recycling* 188: 106629. https://doi.org/10.1016/j.resconrec.2022.106629.

Li, M., Zhou, S., Xu, Y. et al. (2018). Simultaneous Cr (VI) reduction and bioelectricity generation in a dual chamber microbial fuel cell. *Chemical Engineering Journal* 334: 1621–1629. https://doi.org/10.1016/j.cej.2017.11.144.

Li, Z., Diaz, L.A., Yang, Z. et al. (2019). Comparative life cycle analysis for value recovery of precious metals and REEs from electronic waste. *Resources, Conservation and Recycling* 149: 20–30. https://doi.org/10.1016/j.resconrec.2019.05.025.

Li, M., Zhou, J., Bi, Y.G. et al. (2021). Polypyrrole/sewage sludge carbon as low-cost and high-effective catalyst for enhancing hexavalent chromium reduction and bio-power generation in dual chamber microbial fuel cells. *Separation and Purification Technology* 256: 117805. https://doi.org/10.1016/j.seppur.2020.117805.

Liang, C. (2015). Research progress of gold bio-leaching technology from waste printed circuit board. *Journal of Hanshan Normal University* 36 (3): 96–102. https://doi.org/10.3969/j.issn.1007-6883.2015.03.018.

Liu, W.P., Zha, M.Z., and Yin, X.F. (2018). Copper-containing wastewater treatment and copper recovery by using a continuous flow microbial fuel cell. In: *IOP Conference Series: Earth and Environmental Science*, vol. 191 (1), 012074. IOP Publishing.

Liu, Y., Song, P., Gai, R. et al. (2019). Recovering platinum from wastewater by charring biofilm of microbial fuel cells (MFCs). *Journal of Saudi Chemical Society* 23 (3): 338–345. https://doi.org/10.1016/j.jscs.2018.08.003.

Liu, X.L., Zao, Z.X., Gui, Z.Y., and Gong, A.J. (2020). Overview of microbial technology in the utilization of rare earth resources. *Chinese Journal of Engineering* 42 (1): 60–69. https://doi.org/10.13374/j.issn2095-9389.2019.09.12.003.

Long, K.R., Van Gosen, B.S., Foley, N.K., and Cordier, D. (2012). The principal REEs deposits of the United States: a summary of domestic deposits and a global perspective. In Non-Renewable Resource Issues. 131–155: https://doi.org/10.1007/978-90-481-8679-2_7.

Macaskie, L.E., Moriyama, S., Mikheenko, I. et al. (2017). Biotechnology processes for scalable, selective rare earth element recovery. *Rare Earth Element* 1: 1–40. https://doi.org/10.5772/intechopen.68429.

Maes, S., Zhuang, W., Rabaey, K. et al. (2017). Concomitant leaching and electrochemical extraction of REEs from monazite. *Environmental Science & Technology* 51 (3): 1654–1661. https://doi.org/10.1021/acs.est.6b03675.

Maleke, M., Valverde, A., Vermeulen, J.G. et al. (2019). Biomineralization and bioaccumulation of europium by a thermophilic metal resistant bacterium. *Frontiers in Microbiology* 10: 81. https://doi.org/10.3389/fmicb.2019.00081.

Marra, A., Cesaro, A., Rene, E. et al. (2018). Bioleaching of metals from WEEE shredding dust. *Journal of Environmental Management* 210: 180–190. https://doi.org/10.1016/j.jenvman.2017.12.066.

Mathuriya, A.S. and Yakhmi, J.V. (2014). Microbial fuel cells to recover heavy metals. *Environmental Chemistry Letters* 12 (4): 483–494. https://doi.org/10.1007/s10311-014-0474-2.

Matsena, M.T., Tichapondwa, S.M., and Chirwa, E.M. (2021). Improved chromium (VI) reduction performance by bacteria in a biogenic palladium nanoparticle enhanced microbial fuel cell. *Electrochimica Acta* 368: 137640. https://doi.org/10.1016/j.electacta.2020.137640.

Mertzman, S. (2019). What are rare earths, crucial elements in modern technology? 4 questions answered. In: *The Conversation*. http://theconversation.com/what-are-rareearths-crucial-elements-in-modern-technology-4-questions-answered-101364

Miran, W., Jang, J., Nawaz, M. et al. (2017). Mixed sulfate-reducing bacteria-enriched microbial fuel cells for the treatment of wastewater containing copper. *Chemosphere* 189: 134–142. https://doi.org/10.1016/j.chemosphere.2017.09.048.

Moriwaki, H., Koide, R., Yoshikawa, R. et al. (2012). Adsorption of rare earth ions onto the cell walls of wild-type and lipoteichoic acid-defective strains of *Bacillus subtilis*. *Applied Microbiology & Biotechnology* 97 (8): 3721–3728. https://doi.org/10.1007/s00253-012-4200-3.

Muddann, M.H. and Baral, S.S. (2021). Bioleaching of rare earth elements from spent fluid catalytic cracking catalyst using *Acidiothibacillus ferrooxidans*. *Journal of Environmental Chemical Engineering* 9: 104848. https://doi.org/10.1016/j.jece.2020.104848.

Müller, T. and Friedrich, B. (2006). Development of a recycling process for nickel-metal hydride batteries. *J. Power Sources* 158 (2): 1498–1509. https://doi.org/10.1016/j.jpowsour.2005.10.046.

Noori, M.T., Thatikayala, D., Pant, D., and Min, B. (2021). A critical review on microbe-electrode interactions towards heavy metal ion detection using microbial fuel cell technology. *Bioresource Technology* 347: 126589. https://doi.org/10.1016/j.biortech.2021.126589.

O'Connor, M.P., Coulthard, R.M., and Plata, D.L. (2018). Electrochemical deposition for the separation and recovery of metals using carbon nanotube-enabled filters. *Environmental Science: Water Research & Technology* 4 (1): 58–66. https://doi.org/10.1039/C7EW00187H.

Oliveira, R.P., Benvenuti, J., and Espinosa, D.C.R. (2021). A review of the current progress in recycling technologies for gallium and REEs from light-emitting diodes. *Renewable and Sustainable Energy Reviews* 145: 111090. https://doi.org/10.1016/j.rser.2021.111090.

Omar, N.B., Merroun, M.L., Peñalver, J.M.A. et al. (1997). Comparative heavy metal biosorption study of brewery yeast and *Myxococcus xanthus* biomass. *Chemosphere* 35 (10): 2277–2283. https://doi.org/10.1016/S0045-6535(97)00306-8.

Omodara, L., Pitk, aaho, S., Turpeinen, E. et al. (2019). Recycling and substitution of light REEs, cerium, lanthanum, neodymium, and praseodymium from end-of-life applications—a review. *Journal of Cleaner Production* 236: 117573. https://doi.org/10.1016/j.jclepro.2019.07.048.

Onal, M., Borra, C.R., Guo, M. et al. (2017). Hydrometallurgical recycling of NdFeB magnets: complete leaching, iron removal and electrolysis. *Journal of Rare Earths* 35: 574–584. https://doi.org/10.1016/S1002-0721(17)60950-5.

Onwosi, C.O., Nnaemeka, V.C., Onyishi, C. et al. (2023). Sustainable bioprospecting of electronic waste via omics-aided biometallurgy. *Conversion of Electronic Waste in to Sustainable Products* 39–77: https://doi.org/10.1007/978-981-19-6541-8-2.

Ozaki, T., Suzuki, Y., Nankawa, T. et al. (2006). Interactions of REEs with bacteria and organic ligands. *Journal of Alloys and Compounds* 408: 1334–1338. https://doi.org/10.1016/j.jallcom.2005.04.142.

Palaniandy, P., Yusoff, M.S., Wang, L.K., and Wang, M.H.S. (2022). Electronic and electrical equipment waste disposal. *Solid Waste Engineering and Management* 115–161: https://doi.org/10.1007/978-3-030-96989-9_3.

Palmieri, M.C., Garcia, O. Jr.,, and Melnikov, P. (2000). Neodymium biosorption from acidic solutions in batch system. *Process Biochemistry* 36 (5): 441–444. https://doi.org/10.1016/S0032-9592(00)00236-3.

Panda, S. (2020). Magnetic separation of ferrous fractions linked to improved bioleaching of metals from waste-to-energy incinerator bottom ash (IBA): a green approach. *Environmental Science and Pollution Research* 27: 9475–9489. https://doi.org/10.1007/s11356-020-07615-9.

Panda, S., Pradhan, N., and Mohapatra, U. (2013). Bioleaching of copper from pre and post thermally activated low grade chalcopyrite contained ball mill spillage. *Frontier Environmental Science & Engineering* 7: 281–293. https://doi.org/10.1007/s11783-013-0484-5.

Philip, L., Iyengar, L., and Venkobachar, C. (2000). Biosorption of U, La, Pr, Nd, Eu and Dy by *Pseudomonas aeruginosa*. *Journal of Industrial Microbiology & Biotechnology* 25 (1): 1–7. https://doi.org/10.1038/sj.jim.7000026.

Piazza, S.D., Cecchi, G., Cardinale, A.M. et al. (2016). *Penicillium expansum* link strain for a biometallurgical method to recover REEs from WEEE. *Waste Management* 60 (Feb.): 596–600. https://doi.org/10.1016/j.wasman.2016.07.029.

Pourhossein, F. and Mousavi, S.M. (2018). Enhancement of copper, nickel, and gallium recovery from LED waste by adaptation of *Acidithiobacillus ferrooxidans*. *Waste Management* 79: 98–108. https://doi.org/10.1016/j.wasman.2018.07.010.

Pourhossein, F. and Mousavi, S.M. (2019). A novel step-wise indirect bioleaching using biogenic ferric agent for enhancement recovery of valuable metals from waste light emitting diode (WLED). *Journal of Hazardous Materials* 378: 120648. https://doi.org/10.1016/j.jhazmat.2019.05.041.

Priya, A. and Hait, S. (2017). Comparative assessment of metallurgical recovery of metals from electronic waste with special emphasis on bioleaching. *Environmental Science & Pollution Research* 24: 6989–7008. https://doi.org/10.1007/s11356-016-8313-6.

Qi, D. (2018). Treatment of Wastewater, Off-Gas, and Waste Solid, Hydrometallurgy of Rare Earths https://doi.org/10.1016/b978-0-12-813920-2.00008-8.

Rahimi, G., Rastegar, S.O., Rahmani, F., and Gu, T. (2020). Correction: ultrasound-assisted leaching of vanadium from fly ash using lemon juice organic acids. *RSC Advances* 10 (63): 38378–38378. https://doi.org/10.1039/D0RA90109A.

Ramprasad, C., Gwenzi, W., and Chaukura, N. (2022). Strategies and options for the sustainable recovery of rare earth elements from electrical and electronic waste. *Chemical Engineering Journal* 442: 1. https://doi.org/10.1016/j.cej.2022.135992.

Rasoulnia, P., Barthen, R., Puhakka, J.A., and Lakaniemi, A.M. (2021). Leaching of rare earth elements and base metals from spent NiMH batteries using gluconate and its potential bio-oxidation products. *Journal of Hazardous Materials* 414: 125564. https://doi.org/10.1016/j.jhazmat.2021.125564.

Rasoulnia, P., Barthen, R., and Lakaniemi, A.M. (2022). Low residual dissolved phosphate in spent medium bioleaching enables rapid and enhanced solubilization of rare earth elements from end-of-life NiMH batteries. *Minerals Engineering* 176: 107361. https://doi.org/10.1016/j.mineng.2021.107361.

Rastegar, S.O., Mousavi, S.M., and Shojaosadati, S.A. (2015). Bioleaching of an oil-fired residual: process optimization and nanostructure NaV_6O_{15} synthesis from the bioleachate. *RSC Advances* 5 (51): 41088–41097. https://doi.org/10.1039/C5RA00128E.

Reed, D.W., Fujita, Y., Daubaras, D.L. et al. (2016). Bioleaching of rare earth elements from waste phosphors and cracking catalysts. *Hydrometallurgy* 166: 34–40. https://doi.org/10.1016/j.hydromet.2016.08.006.

Riekkola-Vanhanen, M. (2013). Talvivaara mining company - from a project to a mine. *Mineral Engineering* 48: 2–9. https://doi.org/10.1016/j.mineng.2013.04.018.

Rodenas Motos, P., Ter Heijne, A., van der Weijden, R. et al. (2015). High rate copper and energy recovery in microbial fuel cells. *Frontiers in microbiology* 6: 527. https://doi.org/10.3389/fmicb.2015.00527.

Ruan, J., Zhu, X., Qian, Y. et al. (2014). A new strain for recovering precious metals from waste printed circuit boards. *Waste Management* 34 (5): 901–907. https://doi.org/10.1016/j.wasman.2014.02.014.

Schluep, M. and l'environnement P.d.N.U. (2009), Recycling, from E-waste to resources. UNEP DTIE; StEP Initiative.

Sobhani, D., Rastegar, S.O., and Khamforoush, M. (2022). *Using Microbial fuel cell to separate valuable metals from wastewater of industrial solid waste leaching (D).* Kurdistan, Iran: University of Kurdistan.

Sophia, A.C. and Saikant, S. (2016). Reduction of chromium (VI) with energy recovery using microbial fuel cell technology. *Journal of Water Process Engineering* 11: 39–45. https://doi.org/10.1016/j.jwpe.2016.03.006.

Soto, P.E., Galleguillos, P.A., Seron, M.A. et al. (2013). Parameters influencing the microbial oxidation activity in the industrial bioleaching heap at Escondida mine, Chile. *Hydrometallurgy* 133: 51–57. https://doi.org/10.1016/j.hydromet.2012.11.011.

Srivastava, R.R., Mitta, N.K., Padh, B., and Reddy, B.R. (2012). Removal of tungsten and other impurities from spent HDS catalyst leach liquor by an adsorption route. *Hydrometallurgy* 127: 77–83. https://doi.org/10.1016/j.hydromet.2012.07.004.

Stephen, P. and Liang, Y.N. (2019). Bioleaching of trace elements and rare earth elements from coal fly ash. *International Journal of Coal Science & Technology* 6 (1): 74–83. https://doi.org/10.1007/s40789-019-0238-5.

Swain, B., Akcil, A., and Lee, J. (2020). Red mud valorization an industrial waste circular economy challenge; review over processes and their chemistry. *Critical Reviews in Environmental Science and Technology* 52 (4): 520–570. https://doi.org/10.1080/10643389.2020.1829898.

Syed, S. (2006). A green technology for recovery of gold from non-metallic secondary sources. *Hydrometallurgy* 82 (1–2): 48–53. https://doi.org/10.1016/j.hydromet.2006.01.004.

Takahashi, Y., Chatellier, X., Hattori, K.H. et al. (2005). Adsorption of REEs onto bacterial cell walls and its implication for REE sorption onto natural microbial mats. *Chemical Geology* 219 (1–4): 53–67. https://doi.org/10.1016/j.chemgeo.2005.02.009.

Talebi, A., Marra, A., and Cesaro, A. (2018). The recovery of rare earth metals from WEEE leaching solution via liquid-liquid extraction. *Global Nest Journal* 20 (4): 719–724. https://doi.org/10.30955/gnj.002646.

Tan, Q., Li, J., and Zeng, X. (2015). REEs recovery from waste fluorescent lamps: a review. *Critical Reviews in Environmental Science and Technology* 45 (7): 749–776. https://doi.org/10.1080/10643389.2014.900240.

Tan, X., Wang, Z., Ren, B. et al. (2020). Advances in bioleaching critical metals from electronic waste. *Metal Mine* 11: 553. https://doi.org/10.19614/j.cnki.jsks.202011011.

Tsuruta, T. (2006). Selective accumulation of light or heavy REEs using gram-positive bacteria. *Colloids Surf B* 52 (2): 117–122. https://doi.org/10.1016/j.colsurfb.2006.04.014.

Tsuruta, T. (2007). Accumulation of rare earth elements in various microorganisms. *Journal of Rare Earths* 25 (5): 526–532. https://doi.org/10.1016/S1002-0721(07)60556-0.

Ujaczki, E., Feigl, V., and Molnar, M. (2018). Re-using bauxite residues: benefits beyond (critical raw) material recovery. *Journal of Chemical Technology and Biotechnology* 93: 2498–2510. https://doi.org/10.1002/jctb.5687.

USGS. (2022). Mineral Commodity Summaries (Rare Earth). US, Geological Survey.

Van Gosen, B.S., Verplanck, P.L. and Seal II,, R.R. (2017). Rare-earth elements (no. 1802-O). US Geological Survey.

Vlachou, A., Symeopoulos, B.D., and Koutinas, A.A. (2009). A comparative study of neodymium sorption by yeast cells. *Radiochim Acta* 97 (8): 437–441. https://doi.org/10.1524/ract.2009.1632.

Vo, T.S., Hossain, M.M., Jeong, H.M., and Kim, K. (2020). Heavy metal removal applications using adsorptive membranes. *Nano convergence* 7 (1): 1–26. https://doi.org/10.1186/s40580-020-00245-4.

Voncken, J. and H.L. (2016). The ore minerals and major ore deposits of the rare earths. *Springer International Publishing* 2: 15–52. https://doi.org/10.1007/978-3-319-26809-5.

Wang, H.Q. (2008). *Studies on Adsorption of the Rare Earth Ions by Phanerochaete chrysosporium*. Zhengzhou: Henan Agricultural University.

Wang, H., Long, X., Zhang, J. et al. (2020). Relationship between bioelectrochemical copper migration, reduction and electricity in a three-chamber microbial fuel cell. *Chemosphere* 241: 125097. https://doi.org/10.1016/j.chemosphere.2019.125097.

Wang, H., Chen, S., Dong, Y. et al. (2023). Highly selective and rapid removal of metal ions from ionic liquids solutions through size-sieving and ion-exchange strategy. *Journal of Water Process Engineering* 52: 103540. https://doi.org/10.1016/j.jwpe.2023.103540.

Watling, H.R., Johnson, J.J., Shiers, D.W. et al. (2016). Effect of temperature and inoculation strategy on cu recovery and microbial activity in column bioleaching. *Hydrometallurgy* 189–201. https://doi.org/10.1016/j.hydromet. 2016. 05. 017.

Wu, Y., Zhao, X., Jin, M. et al. (2018). Copper removal and microbial community analysis in single-chamber microbial fuel cell. *Bioresource Technology* 253: 372–377. https://doi.org/10.1016/j.biortech.2018.01.046.

Xu, S.X., Zhang, S.M., Chen, K. et al. (2011). Biosorption of La^{3+} and Ce^{3+} by *Agrobacterium* sp. HN1. *Journal of Rare Earths* 29 (3): 265–270. https://doi.org/10.1016/S1002-0721(10)60443-7.

Yang, C. (2015). *Role and Mechanism of Extracellular Polymeric Substances in Bioleaching Waste Printed Circuit Boards by Acidophilic Bacteria*. Guangzhou: South China University of Technology.

Yang, Y., Walton, A., and Sheridan, R. (2017). REE recovery from end-of-life NdFeB permanent magnet scrap: a critical review. *Journal of Sustainable Metallurgy* 3 (1): 122–149. https://doi.org/10.1007/s40831-016-0090-4.

Yang, X., Liu, L., Tan, W. et al. (2020). Remediation of heavy metal contaminated soils by organic acid extraction and electrochemical adsorption. *Environmental Pollution* 264: 114745. https://doi.org/10.1016/j.envpol.2020.114745.

Yang, Z., Li, J., Chen, F. et al. (2021). Bioelectrochemical process for simultaneous removal of copper, ammonium and organic matter using an algae-assisted triple-chamber microbial fuel cell. *Science of The Total Environment* 798: 149327. https://doi.org/10.1016/j.scitotenv.2021.149327.

Yang, X., Liu, L., Wang, Y. et al. (2023). Sustainable and reagent-free cathodic precipitation for high-efficiency removal of heavy metals from soil leachate. *Environmental Pollution* 121002: https://doi.org/10.1016/j.envpol.2023.121002.

Yi, W.C., Santos, R.M., Monballiu, A. et al. (2013). Effects of bioleaching on the chemical, mineralogical and morphological properties of natural and waste-derived alkaline materials. *Minerals Engineering* 48: 116–125. https://doi.org/10.1016/j.mineng.2012.09.004.

Yi, X., Yang, K., Zhang, P. et al. (2016). Research progress in the recovery of precious metals from electronic waste. *Resource Recycling* 3: 62–64. https://doi.org/10.3969/j.issn.1007-1865.2016.03.030.

Yousefi, T., Golikand, A.N., Mashhadizadeh, M.H., and Aghazadeh, M. (2012). Template-free synthesis of MnO_2 nanowires with secondary flower like structure: characterization and supercapacitor behavior studies. *Current Applied Physics* 12 (1): 193–198. https://doi.org/10.1016/j.cap.2011.05.038.

Yuan, Z., Liu, H., Yong, W.F. et al. (2022). Status and advances of deep eutectic solvents for metal separation and recovery. *Green Chemistry* 24 (5): 1895–1929. https://doi.org/10.1039/D1GC03851F.

Yuksekdag, K. and Zeytuncu-Gokoglu, B. (2022). Process optimization for acidic leaching of rare earth elements (REE) from waste electrical and electronic equipment (WEEE). *Environmental Science and Pollution Research* 29 (5): 7772–7781. https://doi.org/10.1007/s11356-021-16207-0.

Zhang, L. and Xu, Z. (2016). A review of current progress of recycling technologies of metals from waste electrical and electronic equipment. *Journal of Cleaner Production* 127: 10–36. https://doi.org/10.1016/j.jclepro.2016.04.004.

Zhang, P., Yokoyama, T., Itabashi, O. et al. (1999). Recovery of metal values from spent nickel–metal hydride rechargeable batteries. *Power Sources* 77 (2): 116–122. https://doi.org/10.1016/s0378-7753(98)00182-7.

Zhang, B., Tian, C., Liu, Y. et al. (2015). Simultaneous microbial and electrochemical reductions of vanadium (V) with bioelectricity generation in microbial fuel cells. *Bioresource Technology* 179: 91–97. https://doi.org/10.1016/j.biortech.2014.12.010.

Zhang, L., Dong, H., Liu, Y. et al. (2018). Bioleaching of rare earth elements from bastnaesite-bearing rock by actinobacteria. *Chemical Geology* S0009254118301359. https://doi.org/10.1016/j.chemgeo.2018.03.023.

Zhang, W., Noble, A., Yang, X., and Honaker, R. (2020). A comprehensive review of rare earth elements recovery from coal-related materials. *Minerals* 10 (5): 451. https://doi.org/10.3390/min10050451.

Zhang, J., Wang, H., Zhou, X. et al. (2021). Simultaneous copper migration and removal from soil and water using a three-chamber microbial fuel cell. *Environmental Technology* 42 (28): 4519–4527. https://doi.org/10.1080/09593330.2020.1769743.

Zhou, B., Li, Z., and Chen, C. (2017). Global potential of rare earth resources and rare earth demand from clean technologies. *Minerals* 7 (11): 203. https://doi.org/10.3390/min7110203.

14

Bioprocessing of E-waste for Metal Recovery

Tannaz Naseri[1], *Ashkan Namdar*[2], *and Seyyed Mohammad Mousavi*[1,3]

[1] *Biotechnology Group, Chemical Engineering Department, Tarbiat Modares University, Tehran, Iran*
[2] *Faculty of Materials Science and Engineering, Khajeh Nasir Toosi University of Technology, Tehran, Iran*
[3] *Modares Environmental Research Institute, Tarbiat Modares University, Tehran, Iran*

Abstract

Nowadays, electronic waste (e-waste) is the fastest-growing solid waste stream globally. Approximately 44.7 million tons of e-waste were produced worldwide in 2016. However, by 2021, this rate had increased by 17%, totaling 52.2 million tons. E-waste contains toxic materials like heavy metals that present serious health risks to humans and the environment. This emphasizes the need for the recycling of e-waste. The recovery of metals from this type of waste has been widely studied using conventional methods such as high-temperature pyrometallurgical and hydrometallurgical processing. Bioprocessing approaches have the edge over other techniques due to their environmentally friendly attributes. Microbial technologies promise heavy metal removal since they are cost-effective and do not require complex equipment and toxic chemicals, thereby avoiding the generation of hazardous byproducts. This chapter discusses bioprocessing approaches for e-waste treatment to recover toxic and precious metals. Additionally, biologically-driven metal extraction technologies emphasizing bacterial and fungal leaching are reviewed.

Keywords *electronic waste; heavy metals; recycling; pyrometallurgical; hydrometallurgical; bioprocessing; secondary source; bacterial leaching; fungal leaching*

14.1 Introduction

Electronic waste, e-waste, or waste electrical and electronic equipment (WEEE) are discarded devices that are no longer of use to consumers and cannot be reused. E-waste stream has become one of the fastest-growing solid waste streams today (Naseri et al. 2023a). It consists of metals, ceramics, plastics, and other

Management of Electronic Waste: Resource Recovery, Technology and Regulation, First Edition.
Edited by Anshu Priya.

unwanted constituents. It was reported in 2020 that, there would be 74.7 million tons of e-waste in the world by 2030. According to reports, only 17.4% of this was collected correctly and recycled (Naseri et al. 2023c). Heavy metals found in e-waste are important secondary sources. Consumer electrical and electronic equipment (EEE) products have become increasingly complex owing to continuous innovation in their material properties and shapes (Chancerel and Rotter 2009). In addition to their variable and complex compositions, discarded devices contain a variety of elements (Bloodworth 2014). Modern smartphones have up to 58 elements with varying concentrations and chemical compositions. Therefore, innovative recycling strategies for complex materials should be guided by the principle of selectivity in metal recovery. Developing novel recycling strategies focused on resource recovery could reduce the environmental risks associated with improper WEEE management (Mahmoud et al. 2017).

Over the past two decades, most e-waste has been recycled using technologies such as mechanical processes and chemical leaching techniques (Awasthi and Li 2017). There are obvious limitations to these recycling approaches. For example, mechanical processes require low capital investment; hydrometallurgical processes consume chemicals that may pollute the environment, while pyrometallurgical processes incorporate high energy inputs and are more expensive. Biotechnological processes that are environmentally friendly have been the subject of considerable research in recent years. Biotechnology-based processes offer a promising means of recovering metallic fractions from e-waste. This approach extracts valuable metals from e-waste using naturally occurring microorganisms and their metabolites (Rozas et al. 2019; Naseri et al. 2019; Naseri et al. 2019).

14.2 Bioprocessing of E-waste for Metal Recovery

Bioprocessing of wastes is gaining popularity due to the fact that it is environmentally friendly and can reduce operating costs and energy requirements (Naseri et al. 2023b). Researchers have been working on the mechanism by which microorganisms use metal oxides and sulfides for biosolubilization since the 1940s. Biohydrometallurgy or bioleaching refers to a field of hydrometallurgy that uses microbes to produce soluble metal ions by interacting with insoluble metal oxides. This extraction might be because of metabolite activity directly on metals (direct bioleaching) or the effect of metabolic compounds made during their growth (indirect bioleaching) (Pollmann et al. 2018). After dissolution, recovery of metals from leaching media is what should be done (Hoque and Philip 2011). A group of bacteria known as *Acidithiobacillus ferrooxidans* and *Acidithiobacillus thiooxidans* is the most commonly used for bioleaching applications since they can facilitate metal dissolution through bioleaching reactions

(Gu et al. 2018). One major drawback to leachability of different sources is the toxicity of elements that might cause death of microorganisms. This limitation can be inhibited by adaptation (Gu et al. 2018). The bioleaching process may take place by abiotic dissolution, wherein the protons carry out the bio-oxidation and leaching reactions to carry out metal solubilization. However, passivation and other secondary reactions can affect metal solubilization, resulting in poor metal yield (Global NEST Journal 2018). The bioleaching process consists of three steps: one step, two step, and spent medium. The first two methods are based on microbial biomass being exposed to solid waste. In the spent medium, microbes are inoculated without solid powder until they reach the end of their logarithmic phase. Afterward, the microbial biomass is filtered and centrifuged, and then the bioleaching process is triggered by adding solid waste powder (Naseri et al. 2022). Numerous factors and parameters influence the process of bioleaching. Some are pulp density of biomass, toxicity of metals and the tolerance of microorganisms against them, metabolic behavior of organisms, pH, temperature, powder particle size, surface features, and so on (Gu et al. 2018). Bioprocessing of waste for metal recovery has been gaining academic and commercial interest due to: (1) potential for a better environmental profile; (2) an easier operational process; (3) economic advantage; and (4) developing future potential. In contrast, waste materials are very different from natural ores in that most metals in WEEE are found in their zero-valence elemental states, often alloyed with other metals (Naseri et al. 2022).

14.2.1 Autotrophic Bioleaching

Autotrophic bacteria are some of the most important biological factors in the bioleaching process, as they use carbon dioxide (CO_2) and inorganic sources for their nutrition. Inorganic compounds are used as energy generation sources. By consuming carbon, oxidation of relative elements takes place. Autotrophic bacteria are mainly separated into three different categories based on their growth temperature: mesophiles, moderate thermophiles, and extreme thermophiles. These organisms are capable of tolerating high levels of metal ions at low pH values. *Acidithiobacillus*, *Leptosirillium*, *Sulfolobus*, and *Acidicanus* are all acidophilic bacteria that can oxidize Fe^{2+} to Fe^{3+} and S^0 to H_2SO_4. Recently, Pourhossein et al. (2022) investigated bioleaching of Cu, Ni, and Ga from end-of-life light-emitting diode (LED) lamps using *Acidithiobacillus thiooxidans* (*A. thiooxidans*). The authors studied a direct multi-step regulation strategy for high extraction of heavy metals from 40 g/L pulp density of solid waste. The results showed that through a step-waste feeding method (10 g/day) the dissolution rates of Cu, Ni, and Ga increased from 67%, 92%, and 39% to 100%, 100%, and 75%. Besides, leaching yields and concentrations of biological sulfuric acid were higher when a smaller amount of solid power was used in various steps (Pourhossein et al. 2022).

The other important bacteria belonging to *Acidithiobacillus* species is *Acidithiobacillus ferrooxidans* (*A. ferrooxidans*), and Naseri et al. (2019) used this strain for bioleaching of critical metals from spent coin cells (SCCs). Through a two-step strategy, the authors succeeded in extracting 100% Li, 88% Co, and 20% Mn at a high pulp density of 40 g/L. The reduction of high-valence metals such as Co^{3+} and Mn^{4+} to soluble forms (Co^{2+} and Mn^{2+}) took place with the help of the Fe^{2+}/Fe^{3+} cycle and the formation of H^+ (Naseri et al. 2019).

14.2.2 Heterotrophic Bioleaching

The recovery of metals from solid wastes is also carried out exclusively by acidophilic autotrophs, but also by heterotrophic bacteria and fungi. Acidophiles are unable to tolerate a wide pH range, but heterotrophs can, and they are used to treat moderately alkaline wastes (Naseri and Mousavi 2022).

Chrombobacterium violaceum, *Pseudomonas aeruginosa*, and *Pseudomonas flourescens* are heterotrophic cyanogenic bacteria that utilize organic compounds as carbon sources and simple chemicals as electron and energy sources. During the late stationary phase of microbial growth, cyanide compounds are produced as a metabolic by-product from the decarboxylation of glycine, which can be released in solution and act as a ligand for metal dissolution, such as Au (Desmarais et al. 2020). Other heterotrophic microorganisms generate organic acids, amino acids, and other metabolites via acidolysis, comploxolysis, and redoxolysis for metal dissolution from e-waste. Figure 14.1 schematically illustrates the various bioleaching mechanisms. Microorganisms that produce cyanide and organic acids have been studied for their role in heterotrophic bioleaching of critical metals from e-waste (Desmarais et al. 2020). The most commonly used heterotrophic bacteria are *Bacillus* spp. and *Pseudomonas* spp. Among *Pseudomonas* spp., *P. aeruginosa*, *P. fluorescens*, and *Pseudomonas putida*, are used in the bioleaching of valuable metals from E-waste (Kaksonen et al. 2014). Besides, *Escherichia coli* and *Bacillus megaterium* have been successful at recovering Au and Cu from e-waste (mobile phones, printed circuit boards [PCBs]), with oxygen acting as an oxidant. Au, Ag, Pt, Zn, and Cu are recovered by a variety of *Pseudomonas* species from primary ores, metallurgical slags, and e-waste. *Chromobacterium violaceum* strains with genetically engineered cyanide production increased Au recovery from e-waste (Valix 2017).

Chi et al. (2011) investigated the removal of Au from mobile phone PCBs by using *C. violaceum*. They achieved an Au removal rate of 10.8% after increasing the medium pH (Chi et al. 2011). There are several strategies available for increasing biogenic cyanide production, such as the addition of nutrients, the modification of medium, and genetic engineering. Pourhossein et al. (2021) studied Au extraction from spent LEDs using *B. megaterium*. In their study, spent LED lamps were first pre-treated using biogenic ferric produced by *A. ferrooxidans* and valuable metals,

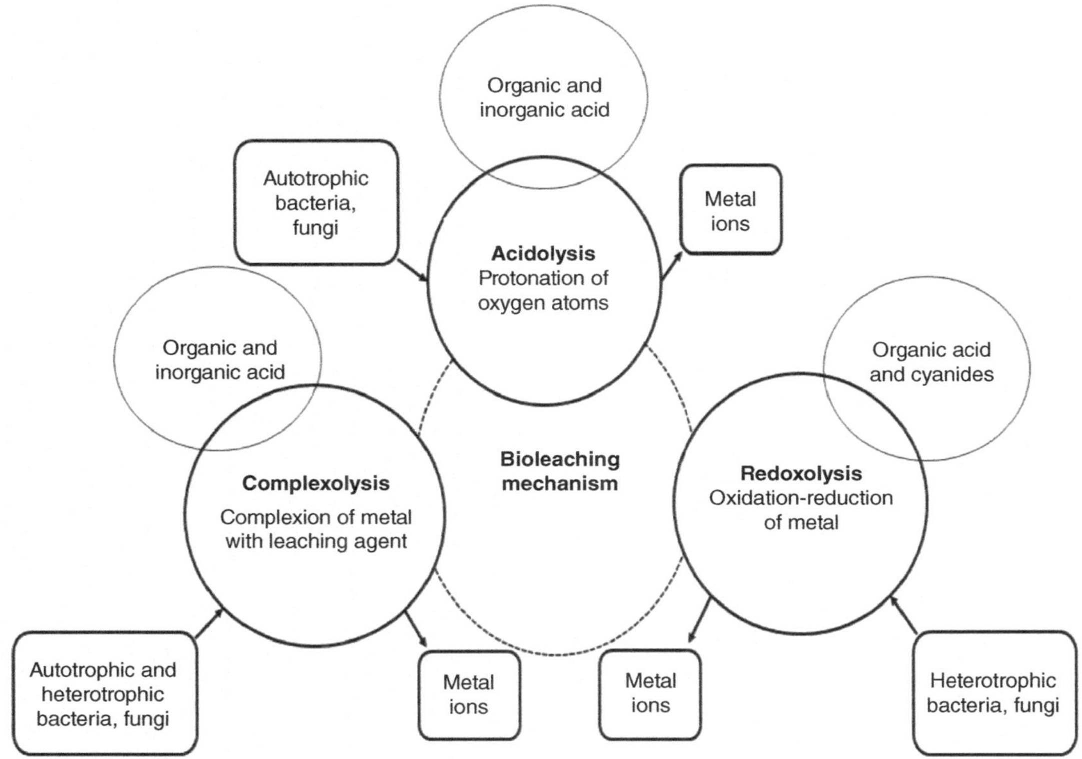

Figure 14.1 Bioleaching mechanism

such as 3% Sn, 94% Ni, 80% Cu, 68% Al, 60% Ga, 51% Pb, 46% Cr, and 35% Fe, were extracted. Then the treated residue was bioleached using *B. megaterium*. According to their results, 15 mg/L Cyanide was produced (in 14 h) by the bacteria under optimum conditions (methionine 10 g/L, glycine 2.5 g/L, and pH 7). At the pulp density of 10 g/L during direct bioleaching, 47% Au was extracted after 36 h (Pourhossein et al. 2021).

As we mentioned before, the other metabolite that is produced from heterotrophic bacteria is organic acid. In this case, Golzar-Ahmadi and Mousavi (2021) used *Bacillus foraminis* as an alkali-tolerant strain for removal of critical metals such as Ag, Mo, and Cu from active-matrix organic light-emitting diode (AMOLED) screens of discarded smartphones using organic acids. The authors examined three different factors, including pulp density, inoculation size, and initial pH of bioleaching process. The organic acids that affected the extraction of the heavy metals in their study were 12.1 mM tartaric acid, 21.5 mM acetic acid, 49.8 mm formic acid, 78.5 mM lactic acid, 2.7 mM citric acid, and 69.6 mM propionic acid. During 12 days of bioleaching experiment, the maximum extraction of 100% Ag, 56.8% Mo, and 41.4% Cu was observed (Golzar-Ahmadi and Mousavi 2021).

14.2.3 Fungal Bioleaching

Bioleaching by fungi generally facilitates metal dissolution by producing organic acids (acidolysis and complexolysis) and controlling the oxidation potential in leaching medium (redoxolysis), or a combination of these methods. In addition, other metabolites such as extracellular polymeric substances (EPS), proteins, amino acids, lipids, and exopolysaccharides are produced by fungi. It is advantageous to use fungi for bioleaching as they have a rapid growth rate, produce higher concentrations of organic acids compared to bacteria, and have a high pH growth capacity (Naseri and Mousavi 2022). The major fungi species used to treat e-waste are *Aspergillus niger* and *Penicillium simplicissimum*. It is worth saying that several fungi have high resistance to heavy metals due to the characteristics of their cell wall membrane. As a result, these fungi are ideal candidates for metal dissolution. Additionally, there are many reports of fungal bioleaching, but with less capacity than their bacterial counterparts (Dusengemungu et al. 2021).

The tolerance of fungi to metals at lab scale has been documented, but no evidence of their use at the industrial scale exists. As we mentioned before, fungi as a heterotrophic microorganism use organic carbon as a carbon source (Nili et al. 2022). *Aspergillus niger* was reported to play a role in the bioleaching of Al, Cd, Cu, Co, Fe, Li, Mn, Ni, Pb, and Zn from incinerator ash (Pethkar et al. 2001), and six *Aspergillus* species were studied in the bioleaching of Cd, Co, Mn, Ni, and Zn from spent Zn–Mn and Ni–Cd batteries (Kim et al. 2016). In this investigation, the authors mentioned that metal dissolution differed when various carbons were used, and this was due to the conversion of those to organic acids produced by the fungi. Sucrose is the best carbon source compared to glucose, lactose, and fructose. Industrializing this biological method requires finding an appropriate alternative to sucrose in the medium. It is one of the main aims to replace sucrose with something more economical, such as molasses, carob pod extract, glycerol, corn starch, and other substrates (Arshadi et al. 2019). Hence, in 2022, Nili et al. utilized molasses as a carbon source for fungal medium to extract Cu and Ni from mobile phone PCB waste. In this investigation, three important factors (pulp density, aeration, and molasses concentration) were optimized. According to their results, 96.94% of Cu and 71.51% of Ni were extracted. The optimized conditions for Cu were 8.8% (v/v) of molasses, 0.29 L/min aeration rate, and pulp density of 10 g/L, and for Ni was 1.9% (v/v) of molasses, 0.37 l/min aeration rate, and pulp density of 10 g/L (Nili et al. 2022).

Alavi et al. (2021) studied the bioleaching of Cu, Ni, and Al from computer PCBs using *Penicillium simplicissimum*. In this study, the authors adapted the fungi to PCB powder so that the fungi would be more resistant to the toxicity of waste powder and produce more organic acid. The adapted fungi were used for the bioleaching process. Under optimum condition of 60 g/L sucrose concentration, initial pH of 6, and pulp density of 16 g/L, 100% Cu, 70% Ni, and 98% Al were removed from the waste (Alavi et al. 2021).

14.2.4 The Bioleaching Reaction: Biochemical Mechanisms

It has long been debated how biological metals can be extracted from waste material. Due to its extraordinarily broad metabolic capabilities, the bacterial strain *A. ferrooxidans* has remained a focus of research attention. It can grow aerobically by reducing inorganic sulfur compounds to elemental sulfur or by oxidizing Fe^{2+}. However, anaerobic growth is likely to take place through oxidation of sulfur or hydrogen in combination with Fe^{3+} reduction (Ilyas et al. 2010). Eq. (14.1) shows that microorganisms are responsible for oxidizing S^0 to H_2SO_4. Typically, S^0 is not found in many e-wastes and needs to be added externally to the growth medium. In addition, Fe^{2+} is externally added to the growth medium and is also oxidized by microorganisms into Fe^{3+} (Eq. (14.2)) (Oguchi et al. 2012).

$$S^0 + 1.5O_2 + H_2O \rightarrow SO_4^{2-} + 2H^+ \tag{14.1}$$

$$4Fe^{2+} + O_2 + 2H^+ \rightarrow 4Fe^{3+} + 2OH^- \tag{14.2}$$

Xiang et al. (2010) and Wang et al. (2009) discussed several possible reactions as explanations for Cu dissolution from PCBs. Based on Eq. (14.3), the bio-oxidation of Fe^{2+} results in the production of Fe^{3+}, which is responsible for the conversion of Cu^0 into soluble form of Cu^{2+}. Also, for dissolving Ni^0 and Ga^0, the possible reactions (Eqs. (14.4) and (14.5)) will occur in the presence of biogenic Fe^{3+} (Pourhossein and Mousavi 2019).

$$2Fe^{3+} + Cu^0 \rightarrow 2Fe^{2+} + Cu^{2+} \quad (14.3)$$

$$2Fe^{3+} + Ni^0 \rightarrow 2Fe^{2+} + Ni^{2+} \quad (14.4)$$

$$2Fe^{3+} + Ga^0 \rightarrow 2Fe^{2+} + Ga^{2+} \quad (14.5)$$

Interestingly, Cu extraction was observed despite no iron, and S^0 was used as an energy source, pointing to the protons as responsible for solubilizing of Cu^0, even though O_2 is involved (Wang et al. 2009; Xiang et al. 2010).

$$2Cu^0 + 4H^+ + O_2 \rightarrow 2Cu^{2+} + 2H_2O \quad (14.6)$$

As we mentioned earlier, chemicals containing cyano-groups with the formula CN^- are referred to as cyanides. In the case of Au leaching by cyanide (tetra-cyanochelate), the reaction can be written as given by Eqs. (14.7) and (14.8).

$$NH_2CH_2OOH \xrightarrow{\text{Microorganisms}} HCN + CO_2 + 2H_2 \quad (14.7)$$

$$4Au + 8CN^- + O_2 + 2H_2O \rightarrow 4Au(CN)_2^- + 4HO^- \quad (14.8)$$

This mechanism occurs when metals interact chemically with cyanide formed by biological processes, causing metals (such as Au and other noble metals) to become soluble and water-soluble complexes to form. One of the most important factors in microbial cyanidation is pH, since it can greatly affect both microbial growth rate and the presence of cyanide ions. Also, there are some challenges involved in achieving the proper leaching conditions, such as the coexistence of other heavy metals. For instance, in the process of complexation with cyanide anions, Cu in the waste matrix can compete with Au (Işıldar et al. 2019).

In addition to bacteria, fungi dissolve metals by different mechanisms, which include the production of organic acids as a key component of metal dissolution. As a result of metal complexation, high organic acid concentration plays a critical role in metal dissolution. It is worth noting that some other metal oxides, such as MnO_2, which exist in lithium manganese oxides, need to be reduced to be in soluble form. The reduction of these kinds of oxides by organic acid occurs in four steps: (1) organic acid binds to metal oxide surfaces; (2) an organic acid transfers electrons to metal oxides; (3) the breakdown of metal oxide matrix and metal ions released into the medium; (4) the complexolysis reaction stabilizes metals (Naseri et al. 2022). Table 14.1 provides an example of some investigations on metal dissolution from e-waste using bacteria and fungi.

14.2.5 Industrial Scales of Bioleaching

Large-scale bioleaching is proven to be effective in the case of primary metal extraction. Almost 15% of Cu, 5% of Au, and some amounts of Ni and Zn are being

Table 14.1 Recent investigations on bioprocessing of metal recovery from e-waste

Microorganism	E-waste	Reaction conditions	Recovery (%)	Reference
Fungi				
Penicillium citrinum	SCCs	Direct bioleaching, Temp.: 30 °C, pulp density: 20 g/L, Time: 20 d	53% Mn, 70% Li	Naseri and Mousavi (2022)
Penicillium simplicissimum	PCBs	Direct bioleaching, sucrose Conc.: 60 g/L, pulp density: 16 g/L, initial pH: 6	100% Cu, 70% Ni, 98% Al.	Esmaeili et al. (2022)
Aspergillus niger	PCBs	Direct bioleaching, Temp.: 30 °C, pulp density: 25 g/L, Time: 7 d	60% Cu, 40% Zn	Becci et al. (2020)
Aspergillus niger	PCBs	Indirect bioleaching, time: 21 d, 120 rpm	100% Zn, 80.39% Ni, 85.88% Cu	Faraji et al. (2018)
Aspergillus niger KUC5254 *Aspergillus tubingensis KUC5037*	Ni–Cd batteries	Spent medium, Temp:27 °C, pulp density: 1 g/L, Time: 4d	Ni, Cd, and Zn: more than 90%	Kim et al. (2016)
Bacteria				
Acidithiobacillus ferrooxidans	PCBs	Bubble column bioreactor, aeration rate: 1.5 vvm, Fe^{2+} Conc.: 40 g/L, pulp density: 20 g/L, Temp: 30 °C, Time: 9 d	54% Cu, 75% Ni, 55% Fe	Arshadi et al. (2021)
Acidithiobacillus thiooxidans	PCBs	Indirect bioleaching, Temp: 30 °C, 1.6 g/L biochar, pulp density: 16 g/L, Time: 3d	98% Cu, 82% Ni	Kadivar et al. (2021)
Acidithiobacillus ferrooxidans	LIBs	Two-step bioleaching, Temp: 30 °C, pulp density: 100 g/L, Time: 72 h	90% Ni, 92% Mn, 82% Co, 89% Li	Jegan Roy et al. (2021)
Leptospirillum ferrooxidans *Sulfobacillus thermosulfidooxidans*	LIBs	Two-step bioleaching, Temp.: 42 °C, pulp density: 50 g/L, Time: 48 h	98.1% Li and 96.3% Co	Liu et al. (2020)
Acidithiobacillus ferrooxidans	LED	Step-wise indirect bioleaching, Temp.: 30 °C, pulp density: 20 g/L, Time: 15 d	83% Cu, 97% Ni, 84% Ga	Pourhossein and Mousavi (2019)
Acidithiobacillus thiooxidans	SCCs	Two-step bioleaching, Temp: 30 °C, pulp density: 30 g/L, Time: 16 d	99% Li, 60% Co, 20% Mn	Naseri et al. (2019)
Acidithiobacillus ferrooxidans	LED	Direct bioleaching, Temp.: 30 °C, pulp density: 20 g/L, Time: 30 d	86% Cu, 96% Ni, 60% Ga	Pourhossein and Mousavi (2018)
Acidithiobacillus thiooxidans *Leptospirillum ferriphilum*	Zn–Mn batteries	Two-step bioleaching, Temp: 35 °C, pulp density: 100 g/L	62.4% Mn 62.5% Zn	Niu et al. (2015)

extracted worldwide through biological leaching. Low-grade ores and wastes from different industries are one major potentiality of bioleaching processes (Pradhan et al. 2008). Most major industrial bioleaching processes are (1) dump bioleaching, (2) heap bioleaching, (3) *In situ* bioleaching, and (4) tank (bioreactor) bioleaching (Gu et al. 2018). Commercial and industrial interest in bioleaching can be due to its operative behavior (i.e. selectivity), cost-effectiveness, and possibility for development. The selectivity advantage comes in handy when a process is being conducted on e-waste, as it usually consists of different alloys (Işıldar et al. 2019). One other important factor in bioleaching on industrial scales is pulp density. Pulp density actually determines some other important parameters, for instance, size of bioreactor, medium, and cost. Higher pulp densities improve large-scale bioleaching of e-waste by lowering capital costs (Naseri et al. 2019). First attempts to recover valuable metals from e-waste on large scale and by bioleaching were conducted on Ni–Cd batteries, where sulfuric acid-producing *Acidithiobacilli* in a bioreactor resulted in 100% Cd and 66.1% Ni recovery after 50 days (Zhu et al. 2003). In another research, it has been provided that in the case of short bioleaching in plants with high loading rate, iron-oxidizing media are more efficient than sulfur-oxidizing media. On the other hand, for long bioleaching times, efficiency of sulfur-oxidizing media is higher due to excess acid generation (Zhao et al. 2008). Industrial bioleaching of PCBs has been under extensive studies recently. Mäkinen et al. (2015) attempted to recover metals from poor metal fractions of PCBs through a 3-step, continuous CSTR experiment. Using a perforated rotating drum reactor, Rodrigues et al. tried to bioleach sheets of PCB in a biological medium (Mäkinen et al. 2015).

14.3 Biosorption and Bioaccumulation of Metals

Absorbents such as clay or chitosan can recover/remove metallic ions, cost- and time-efficiently. Microorganisms, living or dead, active or inactive, can also act as absorbents (Bindschedler et al. 2017). Biosorption is a passive physio-chemical, and metabolically-independent process with the aim of removing metallic contaminants by use of a microorganism with possible mechanisms of absorption, adsorption, ion exchange, surface complexation, and precipitation (Işıldar et al. 2019). Biosorption can be defined as binding of metals and the biomass or their metabolites. Valuable metals' uptake by biosorption is possible, under different terms from what bioleaching provides (Awasthi et al. 2019) The complex structure and behavior of microorganisms indicate a variety of parameters affecting biosorption. Cell wall and cell surface structure of microorganisms, functional groups like carboxylate, hydroxyl, sulfate, phosphate, and amino acids; temperature and pH are some of the most important constraints (Işıldar et al. 2019). Bioelectrochemical sorption has also shown great potential in the removal of metallic ions from an e-waste medium (Awasthi et al. 2019). Surfaces and walls of biomass can be chemically or genetically enhanced, which can extensively improve sorption capacity and selectivity. *Rhodopseudomonas palustris* had elevated metal capacity and selectivity after acid pretreatment. *P. chrysogenum* and *P. spinulosum* have shown efficient selective uptake of Au (Kwak et al. 2013).

The first actual difference between biosorption and bioaccumulation is the use of only living organisms for accumulating metals. The two-stage process of bioaccumulation starts with binding of metallic ions to the extracellular, reactive surfaces of organisms. Subsequently, in a metabolism- and energy-dependent step, metal transfers into the cytoplasm of microorganisms. If the toxicity and respiration of metal ions are higher than microorganisms' tolerance, accumulation can be stopped (Hoque and Philip 2011). Yeasts pose higher capacities for metal uptake, whereas bacteria show speedier rates of reaction. Studies indicate the accumulation of Pb, Cd, Zn, and Hg by fruity mushrooms and macrofungi. Depending on the species, fungi showed selective bioaccumulation of Au, Sb, and Ag (Bindschedler et al. 2017).

14.4 Perspective and Future Aspects

In the next few years, there will be an increase in the generation of WEEE, especially in developing countries. This is one of the most important secondary sources of critical raw materials. Some of these metals are particularly crucial in helping society move toward a greener, more sustainable future. Since their secure supply is under threat, alternative sources of supply, such as waste material, should be considered. However, the economic implications associated with the recovery of valuable material should not be the only reason for the development of these technologies. There is an environmental and human health risk associated with improper disposal of WEEE and other secondary sources containing heavy metals and toxic elements. There is a growing interest in the bioprocessing of secondary wastes for metal recovery due to its potential to reduce pollution and recover resources. The cation of biotechnology will enhance conventional technology for recovering and reusing valuable metals from secondary sources (OECD 2009; Dominguez-Benetton et al. 2018; Lee and Pandey 2012).

Today, the bioleaching approach to metal extraction from secondary wastes is not only a promise of the future but also a viable and efficient economic and environmental alternative. The use of biological sulfuric acid and ferric ions for the leaching of valuable metals from e-waste has been extensively studied. Studies have also used cyanide and organic acid produced by heterotrophic microorganisms for extracting critical metals. It is worth mentioning that, historically, bioleaching took place using autotrophic microorganisms capable of conserving the energy from metal oxides and sulfide ores. Consequently, it is necessary to provide the microbes with additional energy sources. This challenge needs the development of new strategies for recovering metals from e-waste. However, bioleaching from e-waste is not currently being used on a large scale, and the technology is only available at laboratory scale. The microbial leaching on an industry scale is expected to expand in the near future, which requires more specialized research to develop commercial prospects. As a part of biotechnological strategies for metal recovery from e-waste, scale-up studies, and techno-economic analysis will be considered, along with environmental sustainability.

Acknowledgments

This study was financially supported by Tarbiat Modares University under grant number IG-39701.

References

Alavi, N., Partovi, K., Majlessi, M. et al. (2021). Bioleaching of metals from cellphones batteries by a co-fungus medium in presence of carbon materials. *Bioresource Technology Reports* 15: 100768. https://doi.org/10.1016/j.biteb.2021.100768.

Arshadi, M., Nili, S., and Yaghmaei, S. (2019). Ni and Cu recovery by bioleaching from the printed circuit boards of mobile phones in non-conventional medium. *Journal of Environmental Management* 250: 109502. https://doi.org/10.1016/j.jenvman.2019.109502.

Arshadi, M., Pourhossein, F., Mousavi, S.M., and Yaghmaei, S. (2021). Green recovery of Cu-Ni-Fe from a mixture of spent PCBs using adapted *A. ferrooxidans* in a bubble column bioreactor. *Separation and Purification Technology* 272: 118701. https://doi.org/10.1016/j.seppur.2021.118701.

Awasthi, A.K. and Li, J. (2017). Management of electrical and electronic waste: a comparative evaluation of China and India. *Renewable and Sustainable Energy Reviews* 76: 434–447. https://doi.org/10.1016/j.rser.2017.02.067.

Awasthi, A.K., Hasan, M., Mishra, Y.K. et al. (2019). Environmentally sound system for E-waste: biotechnological perspectives. *Current Research in Biotechnology* 1: 58–64. https://doi.org/10.1016/j.crbiot.2019.10.002.

Becci, A., Karaj, D., Merli, G., and Beolchini, F. (2020). Biotechnology for metal recovery from end-of-life printed circuit boards with *Aspergillus niger*. *Sustainability* 1–10.

Bindschedler, S., Vu Bouquet, T.Q.T., Job, D. et al. (2017). Fungal biorecovery of gold from e-waste. *Advances in Applied Microbiology* 53–81. https://doi.org/10.1016/bs.aambs.2017.02.002.

Bloodworth, A. (2014). Resources: track flows to manage technology-metal supply. *Nature* 505: 19–20. https://doi.org/10.1038/505019a.

Bosecker, K. (1997). Bioleaching: metal solubilization by microorganisms. *FEMS Microbiology Reviews* 20: 591–604. https://doi.org/10.1111/j.1574-6976.1997.tb00340.x.

Bosecker, K. (2001). Microbial leaching in environmental clean-up programmes. *Hydrometallurgy* 59: 245–248. https://doi.org/10.1016/S0304-386X(00)00163-8.

Chancerel, P. and Rotter, S. (2009). Recycling-oriented characterization of small waste electrical and electronic equipment. *Waste Management* 29: 2336–2352. https://doi.org/10.1016/j.wasman.2009.04.003.

Chi, T.D., Lee, J., Pandey, B.D. et al. (2011). Bioleaching of gold and copper from waste mobile phone PCBs by using a cyanogenic bacterium. *Minerals Engineering* 24: 1219–1222. https://doi.org/10.1016/j.mineng.2011.05.009.

Desmarais, M., Pirade, F., Zhang, J., and Rene, E.R. (2020). Biohydrometallurgical processes for the recovery of precious and base metals from waste electrical and electronic equipments: current trends and perspectives. *Bioresource Technology Reports* 11: 100526. https://doi.org/10.1016/j.biteb.2020.100526.

Dominguez-Benetton, X., Varia, J.C., Pozo, G. et al. (2018). Metal recovery by microbial electro-metallurgy. *Progress in Materials Science* 94: 435–461. https://doi.org/10.1016/j.pmatsci.2018.01.007.

Dusengemungu, L., Kasali, G., Gwanama, C., and Mubemba, B. (2021). Overview of fungal bioleaching of metals. *Environmental Advances* 5: 100083. https://doi.org/10.1016/j.envadv.2021.100083.

Esmaeili, A., Arshadi, M., and Engineering, P. (2022). Simultaneous leaching of Cu, Al , and Ni from computer printed circuit boards using *Penicillium simplicissimum*. *Resources, Conservation and Recycling* 177: 105976. https://doi.org/10.1016/j.resconrec.2021.105976.

Faraji, F., Golmohammadzadeh, R., Rashchi, F., and Alimardani, N. (2018). Fungal bioleaching of WPCBs using *Aspergillus niger*: observation, optimization and kinetics. *Journal of Environmental Management* 217: 775–787.

(2018). WEEE management in a circular economy perspective: an overview. *Global NEST Journal* 20: 743–750. https://doi.org/10.30955/gnj.002623.

Golzar-Ahmadi, M. and Mousavi, S.M. (2021). Extraction of valuable metals from discarded AMOLED displays in smartphones using *Bacillus foraminis* as an alkali-tolerant strain. *Waste Management* 131: 226–236. https://doi.org/10.1016/j.wasman.2021.06.006.

Gu, T., Rastegar, S.O., Mousavi, S.M. et al. (2018). Advances in bioleaching for recovery of metals and bioremediation of fuel ash and sewage sludge. *Bioresource Technology* 261: 428–440. https://doi.org/10.1016/j.biortech.2018.04.033.

Hoque, M.E. and Philip, O.J. (2011). Biotechnological recovery of heavy metals from secondary sources—an overview. *Materials Science and Engineering: C* 31: 57–66. https://doi.org/10.1016/j.msec.2010.09.019.

Ilyas, S., Ruan, C., Bhatti, H.N. et al. (2010). Column bioleaching of metals from electronic scrap. *Hydrometallurgy* 101: 135–140. https://doi.org/10.1016/j.hydromet.2009.12.007.

Işıldar, A., van Hullebusch, E.D., Lenz, M. et al. (2019). Biotechnological strategies for the recovery of valuable and critical raw materials from waste electrical and electronic equipment (WEEE) – a review. *Journal of Hazardous Materials* 362: 467–481. https://doi.org/10.1016/j.jhazmat.2018.08.050.

Jegan Roy, J., Srinivasan, M., and Cao, B. (2021). Bioleaching as an eco-friendly approach for metal recovery from spent NMC-based lithium-ion batteries at a high pulp density. *ACS Sustainable Chemistry & Engineering* 9: 3060–3069. https://doi.org/10.1021/acssuschemeng.0c06573.

Kadivar, S., Pourhossein, F., and Mousavi, S.M. (2021). Recovery of valuable metals from spent mobile phone printed circuit boards using biochar in indirect bioleaching. *Journal of Environmental Management* 280: 111642.

Kaksonen, A.H., Mudunuru, B.M., and Hackl, R. (2014). The role of microorganisms in gold processing and recovery—a review. *Hydrometallurgy* 142: 70–83. https://doi.org/10.1016/j.hydromet.2013.11.008.

Kim, M.-J., Seo, J.-Y., Choi, Y.-S., and Kim, G.-H. (2016). Bioleaching of spent Zn–Mn or Ni–Cd batteries by Aspergillus species. *Waste Management* 51. https://doi.org/10.1016/j.wasman.2015.11.001.

Kwak, I.S., Won, S.W., Chung, Y.S., and Yun, Y.-S. (2013). Ruthenium recovery from acetic acid waste water through sorption with bacterial biosorbent fibers. *Bioresource Technology* 128: 30–35. https://doi.org/10.1016/j.biortech.2012.10.146.

Lee, J.-C. and Pandey, B.D. (2012). Bio-processing of solid wastes and secondary resources for metal extraction – a review. *Waste Management* 32: 3–18. https://doi.org/10.1016/j.wasman.2011.08.010.

Liu, X., Liu, H., Wu, W. et al. (2020). Oxidative stress induced by metal ions in bioleaching of $LiCoO_2$ by an acidophilic microbial consortium. *Frontiers in Microbiology* 10: 1–17. https://doi.org/10.3389/fmicb.2019.03058.

Mahmoud, A., Cézac, P., Hoadley, A.F.A. et al. (2017). A review of sulfide minerals microbially assisted leaching in stirred tank reactors. *International Biodeterioration & Biodegradation* 119: 118–146. https://doi.org/10.1016/j.ibiod.2016.09.015.

Mäkinen, J., Bachér, J., Kaartinen, T. et al. (2015). The effect of flotation and parameters for bioleaching of printed circuit boards. *Minerals Engineering* 75: 26–31. https://doi.org/10.1016/j.mineng.2015.01.009.

Naseri, T. and Mousavi, S.M. (2022). Insights into the polysaccharides and proteins production from *Penicillium citrinum* during bioleaching of spent coin cells. *International Journal of Biological Macromolecules* https://doi.org/10.1016/j.ijbiomac.2022.04.042.

Naseri, T., Bahaloo-Horeh, N., and Mousavi, S.M. (2019). Environmentally friendly recovery of valuable metals from spent coin cells through two-step bioleaching using *Acidithiobacillus thiooxidans*. *Journal of Environmental Management* 235. https://doi.org/10.1016/j.jenvman.2019.01.086.

Naseri, T., Bahaloo-Horeh, N., and Mousavi, S.M. (2019). Bacterial leaching as a green approach for typical metals recovery from end-of-life coin cells batteries. *Journal of Cleaner Production* 220. https://doi.org/10.1016/j.jclepro.2019.02.177.

Naseri, T., Beigi, V., Namdar, A. et al. (2022). Biohydrometallurgical recycling approaches for returning valuable metals to the battery production cycle. In: *Nanotechnology in Battery Recycling, Remanufacturing, and Reuse*, 217–246. Elsevier https://doi.org/10.1016/B978-0-323-91134-4.00005-4.

Naseri, T., Beiki, V., Mousavi, S.M. and Farnaud, S. (2023a). A comprehensive review of bioleaching optimization by statistical approaches: recycling mechanisms, factors affecting, challenges, and sustainability. *RSC Advances* 13: 23570–23589. https://doi.org/10.1039/D3RA03498D.

Naseri, T., Mousavi, S.M., and Kuchta, K. (2023b). Environmentally sustainable and cost-effective recycling of Mn-rich Li-ion cells waste: Effect of carbon sources on the leaching efficiency of metals using fungal metabolites. *Waste Management* 157: 47–59. https://doi.org/10.1016/j.wasman.2022.11.043.

Naseri, T., Mousavi, S.M., Liese, A. et al. (2023c). Bioleaching of valuable metals from spent LIBs followed by selective recovery of manganese using the precipitation method: Metabolite maximization and process optimization. *Journal of Environmental Management* 343: 118197. https://doi.org/10.1016/j.jenvman.2023.118197.

Naseri, T., Pourhossein, F., Mousavi, S.M. et al. (2022). Manganese bioleaching: an emerging approach for manganese recovery from spent batteries. *Reviews in Environmental Science and Biotechnology* 21: 447–468. https://doi.org/10.1007/s11157-022-09620-5.

Nili, S., Arshadi, M., and Yaghmaei, S. (2022). Fungal bioleaching of e-waste utilizing molasses as the carbon source in a bubble column bioreactor. *Journal of Environmental Management* 307: 114524. https://doi.org/10.1016/j.jenvman.2022.114524.

Niu, Z., Huang, Q., Wang, J. et al. (2015). Metallic ions catalysis for improving bioleaching yield of Zn and Mn from spent Zn-Mn batteries at high pulp density of 10%. *Journal of Hazardous Materials* 298. https://doi.org/10.1016/j.jhazmat.2015.05.038.

The bioeconomy to 2030. *OECD*; 2009. https://doi.org/10.1787/9789264056886-en.

Oguchi, M., Sakanakura, H., Terazono, A., and Takigami, H. (2012). Fate of metals contained in waste electrical and electronic equipment in a municipal waste treatment process. *Waste Management* 32: 96–103. https://doi.org/10.1016/j.wasman.2011.09.012.

Pethkar, A., Kulkarni, S., and Paknikar, K. (2001). Comparative studies on metal biosorption by two strains of *Cladosporium cladosporioides*. *Bioresource Technology* 80: 211–215. https://doi.org/10.1016/S0960-8524(01)00080-3.

Pollmann, K., Kutschke, S., Matys, S. et al. (2018). Bio-recycling of metals: recycling of technical products using biological applications. *Biotechnology Advances* 36. https://doi.org/10.1016/j.biotechadv.2018.03.006.

Pourhossein, F. and Mousavi, S.M. (2018). Enhancement of copper, nickel, and gallium recovery from LED waste by adaptation of *Acidithiobacillus ferrooxidans*. *Waste Management* 79: 98–108.

Pourhossein, F. and Mousavi, S.M. (2019). A novel step-wise indirect bioleaching using biogenic ferric agent for enhancement recovery of valuable metals from waste light emitting diode (WLED). *Journal of Hazardous Materials* 378: 120648.

Pourhossein, F., Mousavi, S.M., Beolchini, F., and Lo, M.M. (2021). Novel green hybrid acidic-cyanide bioleaching applied for high recovery of precious and critical metals from spent light emitting diode lamps. *Journal of Cleaner Production* 298. https://doi.org/10.1016/j.jclepro.2021.126714.

Pourhossein, F., Mousavi, S.M., and Beolchini, F. (2022). Innovative bio-acid leaching method for high recovery of critical metals from end-of-life light emitting diodes. *Resources, Conservation and Recycling* 182: 106306. https://doi.org/10.1016/j.resconrec.2022.106306.

Pradhan, N., Nathsarma, K.C., Srinivasa Rao, K. et al. (2008). Heap bioleaching of chalcopyrite: a review. *Minerals Engineering* 21: 355–365. https://doi.org/10.1016/j.mineng.2007.10.018.

Rozas, E.E., Mendes, M.A., Custódio, M.R. et al. (2019). Self-assembly of supramolecular structure based on copper-lipopeptides isolated from e-waste bioleaching liquor. *Journal of Hazardous Materials* 368: 63–71. https://doi.org/10.1016/j.jhazmat.2019.01.038.

Valix, M. (2017). Bioleaching of electronic waste. *Current Developments in Biotechnology and Bioengineering* 407–442. https://doi.org/10.1016/B978-0-444-63664-5.00018-6.

Wang, J., Bai, J., Xu, J., and Liang, B. (2009). Bioleaching of metals from printed wire boards by *Acidithiobacillus ferrooxidans* and *Acidithiobacillus thiooxidans* and their mixture. *Journal of Hazardous Materials* 172: 1100–1105. https://doi.org/10.1016/j.jhazmat.2009.07.102.

Xiang, Y., Wu, P., Zhu, N. et al. (2010). Bioleaching of copper from waste printed circuit boards by bacterial consortium enriched from acid mine drainage. *Journal of Hazardous Materials* 184: 812–818. https://doi.org/10.1016/j.jhazmat.2010.08.113.

Zhao, L., Zhu, N.-W., and Wang, X.-H. (2008). Comparison of bio-dissolution of spent Ni–Cd batteries by sewage sludge using ferrous ions and elemental sulfur as substrate. *Chemosphere* 70: 974–981. https://doi.org/10.1016/j.chemosphere.2007.08.011.

Zhu, N., Zhang, L., Li, C., and Cai, C. (2003). Recycling of spent nickel–cadmium batteries based on bioleaching process. *Waste Management* 23: 703–708. https://doi.org/10.1016/S0956-053X(03)00068-0.

15

State-of-the-Art Biotechnological Recycling Processes

Mital Chakankar, Franziska Lederer, Rohan Jain, Sabine Matys, Sabine Kutschke, and Katrin Pollmann

Department of Biotechnology, Helmholtz Institute Freiberg for Resource Technology, Helmholtz-Zentrum Dresden-Rossendorf, Dresden, Germany

Abstract

Electronic waste is a very heterogeneous yet attractive secondary source of many base and critical metals. Various bioprocesses are being studied extensively as an environmentally friendly route not only for resource recovery from wastes but also for pollution mitigation. Consequently, the application of biotechnological processes for metal recovery from electronic waste is evolving as a green and sustainable method for resource recovery. This chapter explores various state-of-the-art bioprocesses for recycling electronic wastes, including bioleaching, biosorption, and bioflotation. The chapter draws attention to the use of microorganisms and various biomolecules in such processes and facilitates a better understanding of fundamental mechanisms. In addition, it provides future directions for the use of novel biotechnologies and interdisciplinary approaches for such recycling processes.

Keywords *biotechnological approach; bioleaching; biosorption; bioreduction; bioflotation; metals; recovery*

15.1 Introduction

A wide range of electrical and electronic equipment are discarded as waste at the end of their life and the exponential growth of this e-waste is a crucial challenge. E-waste is the highest-growing municipal waste category globally, with a growth rate of 3–5% every year (Afroz et al. 2013). Nearly 53.6 million tons of e-waste was generated in 2019 and is predicted to increase to 74.7 million tons by the year 2030 (Barragan et al. 2021; Forti et al. 2020). E-waste is very heterogeneous and is composed of different metallic and nonmetallic components coming

Management of Electronic Waste: Resource Recovery, Technology and Regulation, First Edition. Edited by Anshu Priya.

from printed circuit boards (PCBs), batteries, lamps, magnets, monitors or screens, mobile phones, computers, etc.

Heavy metals such as As, Be, Cd, Cr, Hg, Pb, and Sb as well as plastics with chlorinated and brominated flame retardants in this waste are hazardous to health as well as the environment (Julander et al. 2014). Whereas, metals such as Al, Ba, Bi, Cu, Fe, Ni, Au, Ag, and various rare earth elements (REEs) are critical as well as precious. The high concentration of such valuable metals in the e-waste as compared to the primary ores makes them an attractive secondary resource. However, landfilling of the e-waste without any treatment results in the release of toxic metals into the aquatic, terrestrial, and atmospheric environments and at the same time results in the loss of valuable metals. Recycling this waste is imperative to reduce the environmental impacts of waste disposal as well as the mining of new metals. Recycling and recovery of metals from this secondary source will also help in closing the gap between the supply and demand of the necessary metals. However, the recovery of these metals is still at an early stage and has not been adopted for the commercial scale (Akcil et al. 2021). In addition, emerging technological developments mean that the composition of incoming e-waste keeps on changing continuously, which makes it difficult to have a flow sheet that will work for all types of e-waste and that also distinguishes it from other types of waste.

Current procedures for metal recovery from WEEE mostly rely on pyrometallurgical processes. Such processes are used on an industrial scale and include methodologies such as incineration, sintering, roasting and melting at high temperatures, and smelting in a blast or plasma arc furnace (Lee et al. 2007; Hoffmann 1992). However, all these processes consume high amounts of energy, limiting their economic benefits (Priya and Hait 2017). Further, thermal treatment of organic compounds such as flame retardants, PVC, and other polymer compounds may release hazardous gases such as dioxin into the atmosphere, thus requiring additional treatment procedures (Cui and Zhang 2008). In addition, the presence of alloys with high stability and complex compositions complicates the efficient recovery of pure metals. Furthermore, low-concentrated elements are not recovered and are lost.

These limitations make alternative processes such as hydrometallurgical processes generally attractive (Thakur and Kumar 2020). The WEEE is dismantled and crushed into small particles. After sorting, the metal-containing particles are treated with different chemicals, and the metals are mostly transformed into an ionic state. Subsequently, the dissolved metals are recovered by various techniques such as electrolysis, precipitation, crystallization, and chemical reduction (Rimaszeki et al. 2012). For these processes, different kinds of leaching reagents are used. For the leaching of precious metals, the best-known, oldest, and still-used method is cyanide leaching (Tuncuk et al. 2012). At alkaline conditions, cyanide exists in the form CN- that is dissolving the precious metals from WEEE via complexations. However, cyanide is highly toxic and harmful to

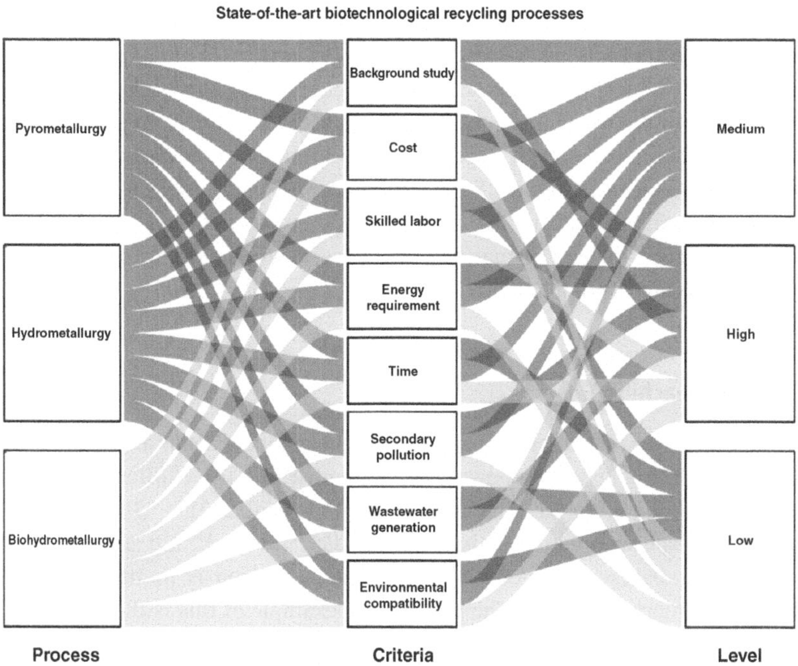

Figure 15.1 Comparison of different processes for metal recovery.

the environment, causing a lot of environmental hazards in the past (Quinet et al. 2005). Therefore, non-cyanide methods have received attention as more environmentally friendly alternatives (Abdelbasir et al. 2018). Leaching with thiourea is carried out under very acidic conditions at $pH < 1.5$. Thiourea complexes with gold, which is released again by the oxidation of Fe^{3+}, which is added to the solution at the same time. In halide leaching, halogens such as bromide or iodine are added, which form complexes with gold. However, all these methods have disadvantages and require special precautions for environmental protection.

Biotechnological methods that employ microorganisms and their metabolic products are considered to be novel, environmentally friendly, energy-saving, and economic methods for metal extraction from waste (Li et al. 2021). Figure 15.1 shows the comparison of different methods for metal recovery.

In this chapter, we summarize existing biotechnological approaches that are being applied for metal recovery from e-waste, along with the challenges and opportunities associated with each approach. The scope of other technologies that are being used in bioremediation and could be combined or adapted for this application is also discussed. Further, we highlight the future potential of a few approaches as well as technical limitations that must be countered to develop such methods into a functional approach.

15.2 State-of-the-art Biotechnological Processes

State-of-the-art biotechnological or biohydrometallurgical approaches are being explored all over the world for sustainable metal recovery from primary as well as secondary resources. However, large-scale applications are hindered with respect to limitations, as seen in Figure 15.1. Thus, biotechnological applications require more attention and technical know-how. The different biotechnological processes that are employed include primarily bioleaching, bioreduction, or bioprecipitation, as well as biosorption and bioflotation.

15.2.1 Bioleaching

The bioleaching approach relies on diverse microbial mechanisms and metabolite formation. Principally, the processes can be divided into three different mechanisms: redoxolysis, acidolysis, and complexolysis (Li et al. 2021).

During redoxolysis, microorganisms convert Fe^{2+} into Fe^{3+} and subsequently, the Fe^{3+} ions oxidize the metals. The now reduced iron is in turn oxidized again by the microorganisms. The best-known example of this is the bacterium *Acidithiobacillus ferrooxidans*. During acidolysis, the microorganisms produce acids such as sulfuric acid (in the case of sulfur-oxidizing bacteria) that dissolve the metals. In complexolysis, lixiviants such as citric acid, oxalic acid, or siderophores are formed by the microorganisms, which interact with the metals and dissolve them as soluble metal complexes. Often, there are mixed forms in which several mechanisms play a role.

In the past few years, various microorganisms have been investigated for their bioleaching capabilities for various kinds of WEEE. Acidophilic chemolithoautrotrohpic bacteria such as *A. ferrooxidans*, *Leptospirillum ferriphilum*, and *Acidithiobacillus caldus*, which are widely used for bioleaching of low-grade sulfidic minerals, have been used for the bioleaching of electronic scrap, e.g. of printed circuit boards, either as single culture or as a mixed consortium (Anaya-Garzon et al. 2021; Xia et al. 2017; Yang et al. 2017). For example, Yang et al. (2017) reported the use of a new isolate *A. ferrooxidans* Z1 for the bioleaching of metal concentrates of PCBs, reaching leaching rates of >90% for copper and zinc and >80% for aluminum (Yang et al. 2017). Other studies used mixed cultures of iron-oxidizing bacteria, e.g. *L. ferriphilum* (Anaya-Garzon et al. 2021). All these leaching procedures required the addition of Fe^{2+} or sulfur, hampering subsequent recovery processes. Further, bacterial activity was inhibited by higher PCB concentrations due to the toxicity of dissolved metals. Prior adaption of bacteria to higher PCB concentrations reduced the toxic effect (Anaya-Garzon et al. 2021). To avoid these obstacles, other researchers pre-produced the biolixiviants, e.g. by biooxidation of pyrite, and used the bacteria-free supernatant as a bioleaching agent (Wu et al. 2018; Bryan et al. 2015), thus obtaining high leaching rates of, e.g. >90% of copper (Wu et al. 2018).

Other promising bacteria are heterotrophic cyanogenic bacteria such as *Chromobacterium violaceum*, *Pseudomonas fluorescens*, *Pseudomonas pseudoalcaligenes*, and *Bacillus megaterium* (Luque-Almagro et al. 2005; Campbell et al. 2001; Shin et al. 2015; Arshadi and Mousavi 2015; Li et al. 2020). These bacteria produce cyanide at alkaline conditions that is complexed with precious metals from electronic scrap. However, this methodology principally corresponds to the cyanide leaching of chemical leaching producing the same environmental risks, thus requiring the treatment of generated cyanide wastes. Moreover, high amounts of base metals like Cu may interfere with the recovery of precious metals like Au due to the formation of Cu complex with cyanide.

Other approaches used heterotrophic microorganisms such as *Aspergillus niger* and *Penicillium simplicissimum*, which produce different kinds of organic acids and complexing metabolites such as citric acid, oxalic acid, malic acid, and gluconic acid (Faraji et al. 2018; Horeh et al. 2016; Rasoulnia and Mousavi 2016; Esmaeili et al. 2022). For example, Jadhav et al. (2016) incubated NaOH-pretreated PCBs with *A. niger* supernatant and H_2O_2, thus solubilizing 100% of all metals (Jadhav et al. 2016). Other researchers used metal-tolerant *Streptomyces* strains (Saravanan et al. 2022) or magnetotactic bacteria (*Magnetospirillum*) (Sannigrahi and Suthindhiran 2019).

Karwowska et al. (2014), studied the use of biosurfactant-producing strains in consortium with sulfide-oxidizing bacteria for bioleaching of metals from PCBs at neutral pH, which may be useful due to the alkalinity of the waste (Karwowska et al. 2014). Another interesting approach is the usage of enzymatic processes. Jadhav and Hocheng (2014) used an enzymatic bio-Fenton process for the bioleaching of metals from PCBs (Jadhav and Hocheng 2014). The glucose oxidase enzyme was used for the in situ production of hydrogen peroxide (H_2O_2) that was further used in Fenton's reaction for the conversion of Fe(II) into Fe(III) as a leaching agent. However, this approach required the initial addition of ferrous sulfate. To reduce the time required for bioleaching, some authors also experimented with the use of different catalysts like metal ions (Zeng et al. 2012), activated carbon (Pathak et al. 2017), biochar (Baniasadi et al. 2016), graphene (Gu et al. 2017), nitrogen-doped carbon nanotubes (Bai et al. 2016), and surfactants (Liu et al. 2015) to improve the kinetics of bioleaching.

Recently, some studies have reported sequential bioleaching and hybrid bioleaching as an alternative to overcome some limitations of the classical bioleaching approach. In a sequential bioleaching approach, Isildar et al. (2016) applied the chemolithotrophic acidophilic bacteria in the first step to remove Cu from PCBs, followed by the application of cyanide produced by cyanogenic heterotrophic bacteria to leach Au from the Cu-leached residue. This is an interesting approach with respect to the removal of base metals in high concentration in the first step, which would otherwise interfere with the leaching of other metals that are present in low concentrations as well as selectively recovering the target metals in the process (Isildar et al. 2016). Another interesting approach of combining different methods was employed by Mäkinen et al. (2015) for the recovery of metals from PCBs. The group used the flotation method to separate the metal-rich concentrate from the metal-poor concentrate. The metal-rich concentrate was suitable for pyrometallurgical metal recovery, whereas the metal-poor concentrate, which still had significant metal content, was subjected to bioleaching using an acidophilic consortium. Despite promising results, such studies need further process optimization to improve recovery efficiency (Mäkinen et al. 2015).

Acidithiobacillus ferrooxidans was employed for the recovery of metals from PCBs along with the addition of citric acid as a lixiviant in a hybrid bioleaching and hydrometallurgy approach. The use of citric acid in this study not only improved the recovery of base metals but also improved the exopolymer production by *A. ferrooxidans,* which in turn decreased jarosite formation (Priya and Hait 2018). In another study, Erust and coworkers employed a two-stage hybrid-sequential approach where bioleaching of PCBs was followed by electrochemical treatment to obtain highly pure Cu. To avoid the negative effect of iron on electro-winning treatment, it was precipitated as ferric hydroxide and eliminated by solvent displacement crystallization, which resulted in an increased Cu concentration in the solutions. This step significantly improved the efficiency of electrowinning treatment. Such integration of approaches is necessary for further purification and concentration of the metals from bioleaching liquors (Erust et al. 2021).

Although high leaching rates were obtained with all these approaches there are still some challenges, these are generally low metal selectivity, low metal tolerance of the microbial strains, long bioleaching times, and costly and energy-consuming cultivation and metabolite production. These issues have to be addressed, e.g. by optimization of culture conditions and growth substrates, recycling of leaching agents, a combination of different strains and leaching agents or pretreatment of waste material, two-step approaches, etc. (Li et al. 2021).

Aiming to improve the metal recovery by biotechnological processes, a series of advancements/developments have been made not only for improving the existing methods but also for researching and developing the use of biolixiviants and/or molecules in these processes. As discussed earlier, various biolixiviants or metabolites like organic acids, siderophores, and biosurfactants play a role in metal recovery. Another interesting molecules in this regard are peptides. Peptides are small biomolecules consisting of amino acids that are very robust regarding temperature, ionic strength, pH, and toxic elements in the environment. Compared to complex microorganisms or even proteins, peptides are less structured and due to that reason less attackable by harmful treatments. Peptides in nature can differentiate their target from background competitors with exclusive selectivity and sensitivity (Tamerler et al. 2010). Organic as well as inorganic materials are possible targets for peptides. The interaction of peptides with inorganic target materials is nicely described by Schwaminger and co-workers, who divided these interactions into two different groups: physisorption as long-range interactions and chemisorption as short-range interactions. The binding energy of long-range interactions is in the range of 1–5 kcal/mol, occurs via electrostatic interactions of electrically charged functional groups, and depends strongly on the pH and the presence of buffer ions forming an electrical double layer. In contrast, the chemisorption in the form of covalent bonding results in stronger interaction energies per bond (>10 kcal/mol) due to the shared electron pair of two atoms. The major interactions of biomolecules with their inorganic targets include hydrophobic interactions, hydrogen interactions, and short-distance van-der-Waals interactions (Schwaminger et al. 2018).

In nature, many highly specific processes are based on the interaction of biomolecules with inorganic targets. The hemoglobin in mammalian red blood cells, for example, holds iron as a central ion, which enables oxygen transport through the body. Many newly developed industrial applications such as biosensors, drug delivery, and medical applications for implants are based on interactions between inorganics and biomolecules.

15.2.1.1 Biohydrometallurgy Based on Naturally Occurring Peptides

The specific molecular recognition feature of peptides coupled with their high affinity to inorganic surfaces offers the potential redefinition of metal extraction from ores or waste by applying ion- and solid-binding peptides in biohydrometallurgy (Donatan et al. 2012; Schönberger et al. 2021). The application of peptides in biotechnological recycling processes is a growing and promising research field where several groups have started to develop material-specific applications. Metallothioneins (MTs) and phytochelatins (PCs), which are cysteine-rich peptides that chelate metal ions via thiolate bonds, occur in a wide range of organisms (García-García et al. 2016; Capeness and Horsfall 2020). These biomolecules increase the resistance of the organisms to toxic metals such as cadmium, lead, and zinc by keeping them away from essential cellular functions (Cobbett 2000). Numerous studies investigated the role of MTs and PCs in heavy-metal detoxification processes and their potential applications in bio-recovery approaches (Cobbett 2000; Choi et al. 2018; Diep et al. 2018; Edmundson and Horsfall 2015). Urbina and co-workers developed a new method for metal recovery from e-waste. In their research, the team focused on copper-binding peptides as a proof-of-concept for a new bioremediation and biomining process. The researchers compared naturally occurring copper-binding peptides with those derived in silico by predicting a possible copper-binding motif. The copper-binding motifs were later combined with chitin-binding domains to functionalize a mycelium-based filter for the recovery of copper from metal-containing solutions (Urbina et al. 2019). Similar to that, Johnston and co-workers investigated the application of delftibactin, a siderophore-like non-ribosomal peptide from *Delfetia acidovorans* with the ability to accumulate gold, for the recovery and recycling of the valuable metal from low-concentrated primary and secondary resources (Johnston et al. 2013). The role of lanmodulin LanM as a naturally occurring protein in lanthanide-utilizing bacteria with remarkable affinity and selectivity for lanthanides was described earlier (Cotruvo et al. 2018). Gutenthaler and co-workers used four lanmodulin peptides – the so-called EF-hands, which are the specific metal-binding motifs of LanM. In their study, the affinity and selectivity of the four different EF hands were tested for their lanthanide binding characteristics. Although the affinity for lanthanides in the structure-giving LanM protein is significantly higher, the four EF hands are also characterized by an increased selectivity for lanthanides and actinides over calcium (Gutenthaler et al. 2022). Park and co-workers focused on the construction of bioengineered cells that express lanthanide-binding tags at the outer membrane protein OmpA. The researchers were able to test the modified *Escherichia coli* cells with leachates from metal-mine tailings and rare earth deposits. During their studies, they proved the enhanced extraction of rare earth elements through functionalized bacterial surfaces (Park et al. 2016, 2017). A smart peptide expression and application idea was investigated by Ishida and co-workers. The research team constructed a transgenic silkworm in which the gene for a rare-earth-recognizing peptide was fused to the fibrion L protein. The purified fusion protein from the transgenic silkworm was able to recognize the rare earth element dysprosium (Ishida et al. 2020).

15.2.2 Biosorption

15.2.2.1 Biomass and Siderophores

Biosorption is a phenomenon where the metals of interest are adsorbed on the surface of the biomass. It is an easy, relatively cost-effective technology for the separation and up-concentration of metal ions, especially from their low-concentrated leachates (Giese 2020). Over the years, extensive research has been carried out in using different kinds of biomass for the recovery of precious and rare-earth metals using biosorption from different electronic waste (Gupta et al. 2019; Tunali and Yenigun 2021; Kaur et al. 2022; Ortinero et al. 2022). Biosorption unlike bioaccumulation is a surface phenomenon. The interaction between biomass and metals is due to the presence of several surface functional groups – OH, COOH, NH_2, CONH, SO_3, and PO_4 (Gupta et al. 2019). The interactions are electrostatic, surface complexation, ion transfer, and precipitation in single metal or in a mixture (Giese 2020). The factors that affect the adsorption of the metals are initial metal concentration, presence of different metals, background electrolytes, pH of metal solutions, temperature, time of contact, biomass dosage, and biomass pore distribution (Giese 2020). The different biosorption rates on important metals from electronic waste are given in Table 15.1.

The pH of the adsorption reaction is the most important factor for not only the adsorption of the metal but also the selectivity of the metal. This is because the speciation of the metal changes with the pH as well as the protonation of the surface functional groups on the biomass. In general, a decrease in pH leads to a decrease in the adsorption capacity, while an increase in the pH leads to a decrease in selectivity (Furuhashi et al. 2019). Contact time is another important factor that determines the usefulness of the biosorption. As can be imagined, an increase in contact time increases the adsorption (Furuhashi et al. 2019). The volume that needs to be treated in the hydrometallurgy is very large, and thus a short contact time makes a biosorbent-based technology scalable. Typically, a 15-minute adsorption process is good. Biomass, due to the presence of several functional groups, can easily achieve it (Thirunavukkarasu et al. 2021). Biomass dosage and initial metal concentration work in different directions for mg of metal adsorbed per g of biomass (Qe) and removal percentage. While increasing the initial metal concentrations will increase the Qe but increasing the dosage will decrease the Qe. On the other hand, the percentage removal will increase when the dosage increases but will decrease when the initial metal concentration is increased. Pore distribution is another factor that increases the attractiveness of biomass (Akar et al. 2009). Higher pore distribution leads to a higher probability of contact between the functional groups and metal ions, leading to better adsorption (Akar et al. 2009).

In almost every scenario, the metal leachate will contain a large number of metal ions along, thus selectivity is a key in the adsorption process. However, the presence of large functional groups, which give biomass enhanced adsorption capacity, also diminishes their ability to be very selective. Nevertheless, the selectivity in the biosorption still exists and is largely due to different stability of the functional groups and metal complexes. The stability of the complexes is again due to the factors such as charge on the metal ions and electronegativity of the metal ions.

Microorganisms are also able to produce metallophores that can be highly selective toward certain metals due to cage-like structures satisfying the metal's need for its denticity and ionic radius (Hider and Kong 2010). For instance, siderophores that are known to bind very strongly to Fe^{3+} can also bind to Ga and recycle it. Desferrioxamine B and E have been demonstrated to complex Ga from electronic industry wastewater under relevant conditions. Methanobactin SB2, pyochelin, and pseudobactin are known to strongly complex Cu^{2+} while desferrioxamine B has shown some complexing ability with Pd^{2+} and Pt^{2+}; however, weaker than Ga and Fe (Hofmann et al. 2020). The complexing abilities of different siderophores with Ga, In, Pd, Pt, Cu, and rare earth elements are given in Table 15.2.

The application of biosorption in the industrial scale is typically carried out in the packed bed column (Thirunavukkarasu et al. 2021). In these columns, the biomass will be packed while the metal-containing solution at optimized

Table 15.1 Biosorption capacity of different biomasses for precious metals and rare earths.

Metal	Organisms	Qe (mg/g)	pH	Reference
Au^{3+}	*Streptomyces erythraeus*	5.9	4.0	Cui and Zhang (2008) and references therein
Au^{3+}	*Spirulina platensis*	5.1	4.0	
Au^{3+}	*Saccharomyces cerevisiae*	5.1	5.0	
Au^{3+}	*Cladosporium cladosporioides*	98.5	4.0	
Au in wastewater	*Cladosporium cladosporioides*	35.5	4.0	
Au^{3+}	*Cladosporium cladosporioides* strain 1	78.8	4.0	
Au^{3+}	*Cladosporium cladosporioides* strain 2	98.5	4.0	
Au^{3+}	*Chlorella vulgaris*	98.5	2.0	
Au^{3+}	*Sargassum natans*	413.6	2.5	
Au^{3+}	*Ascophyllum nodosum*	29.5	2.5	
Au^{3+}	Alginate cross-linked with $CaCl_2$	289.5	2.0	
Au^{3+}	Alginate cross-linked with $Ca(OH)_2$	67.0	2.0	
Au^{3+}	Dealginated seaweed waste	78.8	3.0	
Au^{3+}	Alfalfa	35.5	5.0	
Au^{3+}	Glutaraldehyde crosslinked chitosan	571.2	1.6	
Au^{3+}	Sulfur derivative of chitosan (RADC)s	630.3	3.2	
Au^{3+}	*Fucus vesiculosus* (Brown algae)	74.1	7.0	
Au^{3+}	*Turbinaria conoides* (Brown algae)	34.5	2.0	
Au^{3+}	*Aspergillus niger*	197.0	2.5	
Au^{3+}	*Rhizopus arrhizus*	157.6	2.5	
Au^{3+}	PVA-immobilized biomass (*Fomitopsis carnea*)	94.5	1 to 13	
$Au(CN)_2^-$	*Penicilium chrysogenum*	1.4	2.0	
$Au(CN)_2^-$	*Bacillus subtilis*	1.6	2.0	
$Au(CN)_2^-$	*Sargassum fluitans*	0.6	2.0	
$Au(CN)_2^-$	Hen eggshell membrane	132.0	3.0	

Table 15.1 (Continued)

Metal	Organisms	Qe (mg/g)	pH	Reference
$Au(CN)_2^-$	Acid-washed *Ucides cordatus* (waste crab shells)	33.5	3.4	
Ag^+	*Cladosporium cladosporioides* Strain 1	43.1	4.0	
Ag^+	*Cladosporium cladosporioides* Strain 2	12.9	4.0	
Ag^+	*Aspergillus niger*	97.1	5 to 7	
Ag^+	*Neurospora crassa*	64.7	5 to 7	
Ag^+	*Fusarium oxysporium*	53.9	5 to 7	
Ag^+	*Chlorella vulgaris*	53.9	6 to 7	
Ag^+	*Saccharomyces cerevisiae*	41.7	–	
Ag^+	Exopolysaccharides produced by marine bacteria	333.0	5.7	Tunali and Yenigun (2021)
$Ag(CN)_2^-$	Dairy waste sludge	0.5	6.0	Bhat et al. (2012) and references therein
$Ag(CN)_2^-$	*Parthenium* sp.	0.5	6.0	
$Ag(CN)_2^-$	Yeast biomass	0.3	5.0	
$Ag(CN)_2^-$	Penicillium waste	0.4	6.0	
$Ag(CN)_2^-$	Streptomyces waste	0.3	6.0	
Cu^{2+}	*Ulva fasciata* (Green algae)	73.5	5.0	
Cu^{2+}, Zn^{2+}, Cd^{2+}, PB^{2+}	*Ulva fasciata* (Green algae)	Cu = 32.80, Zn = 30, Cd = 49, Pb = 54.3	5.0	
Cu^{2+}, Ni^{2+}, Cr^{6+}	*Chlorella vulgaris* (Green algae)	Cu = 46.8, Ni = 55.6, Cr = 33.8	5.0	
Pt^{4+}	*Desulfovibrio desulfuricans*	62.4	2.0	Cui and Zhang (2008) and references therein
Pt^{4+}	*Desulfovibrio fructosivorans*	33.2	2.0	
Pt^{4+}	*Desulfovibrio vulgaris*	33.2	2.0	
Pt^{4+}	Bayberry tannin immobilized collagen fiber (BTICF) membrane	44.9	3.0	
Pt^{4+}	Thiourea derivative of chitosan	390.1	2.0	
Pt^{4+}	Chitosan derivatives	624.2	2.0	

(*Continued*)

Table 15.1 (Continued)

Metal	Organisms	Qe (mg/g)	pH	Reference
Pd^{2+}	*Racomitriumlanuginosum* biomass	37.2	5.0	
Pd^{2+}	*Desulfovibrio desulfuricans*	127.7	2.0	
Pd^{2+}	*Desulfovibrio fructosivorans*	106.4	2.0	
Pd^{2+}	Bayberry tannin immobilized collagen fiber (BTICF) membrane	34.1	4.0	
Pd^{2+}	Chitosan derivatives	372.5	2.0	
Pd^{2+}	Thiourea derivative of chitosan	270.3	2.0	
Ce^{3+}	*Turbinaria conoides*	152.7	4.9	Giese (2020) and references there in
Eu^{3+}	*Chlorella vulgaris*	12.2	4.0	
Eu^{3+}	*Sargassum polycustum*	147.4	5.0	
Eu^{3+}	*Sargassum* sp.	95.7	5.0	
Eu^{3+}	*Bacillus thuringiensis*	160.0	4.0	
Gd^{3+}	*Sargassum* sp.	105.4	5.0	
La^{3+}	*Chlorella vulgaris*	154.2	3.0	
La^{3+}	*Sargassum fluitans*	73.6	5.0	
La^{3+}	*Sargassum polycustum*	138.9	5.0	
La^{3+}	*Ulva pertusa*	129.2	4.0	
La^{3+}	*Schizymenia dubyi*	90.3	4.0	
La^{3+}	*Chloroidium saccharophilum*	129.2	6.0	
La^{3+}	*Stichococcus bacillaris*	50.0	6.0	
La^{3+}	*Desmodesmus multivariabilis*	100.0	6.0	
La^{3+}	*Saccharomyces cerevisiae*	70.0	4.0	
Nd^{3+}	*Ankistrodesmus gracilis*	142.8	1.5	
Nd^{3+}	*Ankistrodesmus densus*	106.7	1.5	
Nd^{3+}	*Monoraphidium* sp.	137.0	1.5	
Nd^{3+}	*Chlorella minutissima*	102.4	1.5	
Nd^{3+}	*Chlorella vulgaris*	157.2	5.0	
Nd^{3+}	*Kluyveromyces marxianus*	10 to 12	1.5	
Pr^{3+}	*Sargassum* sp.	40.9	5.0	
Pr^{3+}	*Turbinaria conoides*	131.0	5.0	
Pr^{3+}	*Sargassum wightii*	109.9	5.0	

Table 15.1 (Continued)

Metal	Organisms	Qe (mg/g)	pH	Reference
Pr^{3+}	*Turbinaria conoides*	146.5	5.0	
Pr^{3+}	*Sargassum wightii*	118.4	5.0	
Yb^{3+}	*Sargassum polycustum*	155.7	5.0	
Yb^{3+}	*Turbinaria conoides*	133.2	4.0	
La^{3+}, Eu^{3+}, Yb^{3+}	*Pseudomonas aeruginosa*	La = 55.14, Eu = 44.06, Yb = 56.4	5.0	
La^{3+}, Sm^{3+}	*Bacillus subtilis*-alginate	La = 81.10; Sm = 83.5	3.0	
La^{3+}, Ce^{3+}	*Agrobacterium* sp. HN1	La = 23.26; Ce = 31.15	6.8	
La^{3+}, Nd^{3+}	*Sargassum* sp.	La = 31.9, Nd = 40.3	5.0	
Sm^{3+}, Pr^{3+}	*Sargassum* sp.	Sm = 58.6, Pr = 46.5	5.0	

Table 15.2 Stability constant of metal and siderophore complexes.

Metal	Siderophores	Log β	References
Ga^{3+}	Desferrioxamine B	28.17	Hofmann et al. (2020) and references therein
In^{3+}	Desferrioxamine B	20.6	
Cu^{2+}	Methanobactin-Ob3b	≥18.8	
Cu^{2+}	Methanobactin SB2	27	
Cu^{2+}	Desferrioxamine B	11.2–14.12	
Cu^{2+}	Desferrioxamine E	13.69	
Cu^{2+}	Pyochelin	25.0	
Cu^{2+}	Pseudobactin	22.3	
Pt^{2+}	Desferrioxamine B	16.2–17.2	
Pd^{2+}	Desferrioxamine B	19.3–23.2	
Er^{3+}	Desferrioxamine B	14.66	
Eu^{3+}	Desferrioxamine B	13.67–15	
La^{3+}	Desferrioxamine B	10.09–12	
Nd^{3+}	Desferrioxamine B	12.33	
Yb^{3+}	Desferrioxamine B	15–17	

pH determined in laboratory experiments is carried out. The metal solutions are injected into a column at a particular flow rate, and metal concentration is measured when the leachate exits the column. In the beginning, the concentration of the metal at the exit of the column will be close to zero and will continue to increase. At a certain time, the concentration will be approximately 5% of the inlet concentration. At this concentration, breakthrough of the column is achieved. The 5% is generally accepted value, and it can be of any value depending on the economics of the process. Upon reaching the breakthrough, elution in a smaller volume is carried out to recover the metal(s) and up-concentrate. The biosorbent is also regenerated with elution. It is ideal to operate the columns at the highest superficial velocity so that the column can treat the maximum amount of leachate in the smallest amount of time. This will reduce the capital and operational expenditures of the process.

15.2.2.2 Artificial Metal-Binding Peptides

Material-specific peptides identified by genetics, in vivo techniques, and addressed as genetically engineered peptides for inorganics (GEPI) (Donatan et al. 2012; Hnilova et al. 2008) do not suffer from barriers of other biological systems since there is no need for substrate or specific operational conditions (Braun et al. 2018b). Moreover, their mitigated biological/chemical waste and carbon footprint (Setyawati et al. 2014) turned them into the recent focus of many studies. They have been broadly applied in the realms of biomedicine, biosensors, crystal quality control, and bio-nanotechnology as bio-linkers for assembling functional nanostructures and the synthesis of nanoparticles (Iori et al. 2009; Vreuls et al. 2011).

One major identification technique for GEPI is the so-called Phage Surface Display (PSD). The central tool in this technique is bacterial viruses. Bacteriophages, the most abundant virus particles that only infect bacteria, have exhibited outstanding sorption capacity for inorganics such as gold. They owe this biosorption/bioreduction ability to their phage protein coat and its covalent or ionic interactions with different surfaces (Setyawati et al. 2014). F-specific filamentous phage (Ff group) made up of a single-stranded DNA (ssDNA) covered by major coat protein p8 (VIII) α-helical subunits, and minor coat proteins p7, p9 and p3 (III), p6 on its head and tail (Wang et al. 2006). By PSD, this protein coat can be manipulated. The technique uses a combination of phage genotype and phenotype. A DNA sequence attributed to a short peptide motif is added to a part of the ssDNA encoding the coat protein. Consequently, the phage displays the peptides on its surface at N-terminus of pIII or pVIII (Braun et al. 2018a; Lederer et al. 2017). The large population of random peptides (109) in libraries is employed in identifying the target-specific peptides (Brown 1997). Short peptides (7–14 amino acids) that bind to metals such as gold, copper, nickel, and cobalt are now available (Hnilova et al. 2008; Vodnik et al. 2011). Recently, the novel application of PSD in resource technology and biomining has been introduced (Lederer et al. 2019). The discovery of a peptide capable of distinctive separation of complex components in mineralogy by Curtis et al. was a breakthrough in biohydrometallurgy. In recent years, target-specific phages or peptides have been successfully used as new bio-reagents for the selective separation of sphalerite, chalcopyrite, enargite, and pyrite minerals (Pollmann et al. 2018; Curtis et al. 2009; Curtis et al. 2011). The group of Lederer and co-workers has recently deployed an innovative method in the recovery of a rare earth element in the form of Lanthanum phosphate from e-waste (Lederer et al. 2017). However, safety rules, mutation risks, and being economically impractical deter the extensive application of bacteriophage in recycling processes. A more practical approach is to use peptides instead (Lederer et al. 2019). Therefore, this is of great importance that the contributing peptide retain its conformation and biological selectivity (binding properties) out of the protein and the phage displaying it (Causa et al. 2013).

15.2.2.3 Peptide-Based Biohybrid Tools for Resource Recovery

Vreuls and co-workers used ZnO-binding peptides immobilized on magnetic beads to separate ZnO particles from a suspension containing ZnO and Cu_2O and showed a selective separation of the target particles as well as their easy reclamation and reusability of peptide-functionalized magnetic beads (Vreuls et al. 2011). Cetinel and co-workers developed an alternative biomining approach for molybdenite (MoS_2) based on artificial peptides. In their study, the researchers identified MoS_2 binding peptides, immobilized them to magnetic nanoparticles, and proved the successful capture and pull-down of MoS_2 from aqueous solutions (Cetinel et al. 2018). Similarly, the BioKollekt research group develops a biogenic approach based on REE binding peptides, which are immobilized to magnetic beads. The so-called biocollectors are applied in the sequential selective separation of the different REE particles from fluorescent phosphor powder that is part of compact fluorescent light bulbs (Boelens et al. 2021, 2022).

Ionic metal resources were focused on in numerous phage display approaches. Nian and co-workers analyzed the application of phage-display derived Pb^{2+}-ion binding peptides for their recovery (Nian et al. 2010). The identification and application of cobalt and nickel-binding peptides were focused on in the project Ecometals for the development of a new biomining approach (Braun et al. 2017, 2018b). Schoenberger and co-workers focused on the separation of gallium ions from low-concentrated industrial wastewater. The researchers were able to identify several selective gallium-binding peptides via a chromatopanning process. Using computational modeling, the interaction of these peptides with gallium ions was investigated. A gallium separation approach based on columns functionalized with gallium-binding peptides was developed for industrial

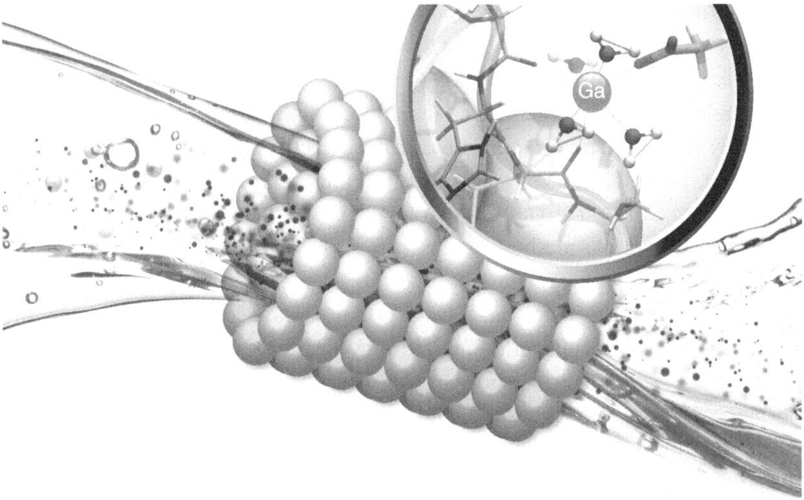

Figure 15.2 Illustration of a peptide-functionalized column for the selective recovery of gallium ions from industrial wastewater. Source: Schönberger et al. (2021)/with permission of Elsevier.

wastewaters (Schönberger et al. 2019a, b, 2021; Gupta and Shukla 2016) (Figure 15.2).

Although peptides and amphiphilic peptides are suitable biological alternatives to conventional flotation reagents, industrial use of peptide-based flotation processes relies extensively on finding a simple and cost-effective method for their production (Lederer et al. 2019). Successful production of recombinant peptides depends on the selection of a suitable cloning vector, expression elements, and host microorganisms, the generation of an appropriate recombinant clone, the development of an optimized fermentation process, and the identification of a simple but efficient purification strategy (Lederer et al. 2019; Gupta and Shukla 2016). Peptide-based resource recovery processes are still under development, however, industrial applications are imminent.

15.2.3 Bioreduction

Many microorganisms have been identified that can reduce Fe(III) to Fe(II) in the environment. Multiple mechanisms, either indirect or direct, have been identified that promote iron reduction. In indirect reactions, reductive metabolites such as oxalic acid or humic acids are produced by the microorganisms that react with the Fe^{3+} forming organo-metallic complexes or reducing Fe^{3+} to Fe^{2+} (Taxiarchou et al. 1997a, b). Humic acids can be reduced by microorganisms such as *Geobacter metallireducens* and *Shewanella algae*, the reduced humic acids themselves can reduce Fe^{2+} to Fe^{3+}, thus working as an electron shuttle

(Lovley et al. 1996). However, these indirect mechanisms can be regarded as a byproduct of microbial metabolism that is not beneficial for the organism itself. On the contrary, in direct mechanisms, the iron itself is targeted by the microbial cells. The microorganisms generate energy from the transfer of electrons from a donor to an acceptor. Instead of oxygen, which is used as an electron acceptor by aerobic microorganisms, some microorganisms can use Fe^{2+} as an acceptor in anaerobic environments. These microorganisms use iron in their respiration process. The prominent iron-reducing microorganism that was first described to use iron in respiration is *G. metallireducens* (Lovley et al. 1987). Meanwhile, it was demonstrated that this capability is widespread, and numerous iron-reducing microorganisms have been described (a comprehensive list is given in [Eisele and Gabby 2014]). Most of these microorganisms are obligate aerobes, but some are able to switch from aerobic iron-oxidizing metabolism to iron-reducing metabolism in anaerobic conditions.

Bioreduction has a high impact on the biogeochemical cycle (Eisele and Gabby 2014; Ullman et al. 1996). However, the microorganisms are also highly attractive and useful for many industrial applications. Iron-reducing bacteria have been used for bioleaching approaches as an alternative to the usage of acidophilic chemolithoautotrophic bacteria (Hallberg et al. 2011; He et al. 2011). In other approaches, iron-reducing bacteria are used for the beneficiation of kaolin clay, silica, or bauxite (Taxiarchou et al. 1997b; Hosseini et al. 2007; Papassiopi et al. 2010; Štyriaková and Lovás 2007; Styriaková and Styriak 2000; Ubaldini et al. 1996). Iron impurities are removed from these materials by bacterial activity, thus increasing their quality and making them usable for multiple purposes. As an example, the whiteness of kaolin clay is a key quality parameter for its application, e.g. as porcelain. For this a low content of iron is necessary. Beneficiation of these materials has been achieved by the removal of iron with oxalic acid produced by *Aspergillus niger* (Hosseini et al. 2007) or with iron reduction mediated by anaerobically grown *Bacillus cereus* (Styriaková and Styriak 2000). Other examples are the beneficiation of silica or bauxite (Papassiopi et al. 2010). Bioreducing bacteria have also been used for iron recovery from low-grade ores via bioleaching. Organisms collected from anaerobic samples were used to leach iron from a low-grade magnetite ore (Eisele and Gabby 2014). The acidophilic heterotrophic bacterium *Acidiphilium* SJH was shown to catalyze the reductive dissolution of a wide range of ferric iron-containing minerals such as akageneite, goethite, jarosite, natrojarosite, and of the mixed ferrous/ ferric mineral magnetite (Toni and Bridge 2000). However, the disadvantages of these approaches are the requirement of anaerobic conditions and the slow growth rate of the organisms, thus resulting in a long leaching time of several weeks to months (Eisele and Gabby 2014). Another more effective biomining approach is the biorecovery of dissolved metals from leachates by bioreduction of ferric ions coupled with bio-mediated sulfate reduction for H_2S reduction. This approach was used for the reduction of Cu(II) to Cu(I) leachates, resulting in the precipitation of copper sulfide (Hedrich et al. 2013).

However, besides iron, bioreducing organisms are generally also capable of reducing a variety of other metals such as V, U, Se, Cr, Au, Pa, or Pt. Therefore, applications that have been developed for iron reduction can be transferred to other metals. Se, U, and Cr bioreduction have been applied for the bioremediation of soils and mining waters. Hexavalent Cr(VI) is a highly toxic pollutant occurring in industrial wastewater. Bioreduction is one approach to remove Cr(VI) from polluted waters (Pradhan et al. 2017). Similarly, selenium, which is toxic in higher concentrations, has been removed from mining waters and industrial wastewater via bioreduction (Stefaniak et al. 2018). Kim et al. (2017) used a consortium of metal-reducing bacteria for the bioreduction of Cr(VI) and Se(V) that precipitated as C(III)-hydroxide and Se(0) nanoparticles (Kim et al. 2017).

All these approaches that have been originally developed and applied for mineral treatment or other mining approaches could be principally used also for processing secondary resources such as metal recovery from electronic waste materials or wastewater cleaning. Further, many microorganisms have been described that form inorganic nanoparticles such as Au, Ag, Pd, Pt, and others via bioreduction that are highly interesting for technical applications. Thus, the bioreductive pathway can be directly used for the synthesis of defined inorganic nanoparticles from metal-containing wastewater. This mechanism has been used for the biorecovery of precious metals from wastewater, e.g. from E-waste recycling processes, by several researchers. Especially the reduction of Pd(II) has been extensively investigated, and many applications have been developed. In 2009, Yong et al. used an *E. coli* mutant for the synthesis of Pd nanoparticles from wastewater (Yong et al. 2009). The generated nanoparticles were used as catalysts in a fuel cell for power generation. Martins et al. (2013) used an anaerobic bacterial consortium composed of *Clostridium*, *Bacteroides*, and *Citrobacter* species derived from a sludge sample from a municipal wastewater treatment plant for the removal of Pd(II) from model solutions and reached recovery rates of 60% with simultaneous Pd(0) formation (Martins et al. 2013). Other researchers used anaerobic methanogenic granular sludge in a reactor system for the removal of Pd(II) and recovery as bio-Pd(0) (Pat-Espadas et al. 2016). In all cases, the formed bio-Pd was catalytically active. Bio-Pd(0) was used for the degradation of pollutants (Humphries et al. 2007; Mabbett et al. 2006; Tuo et al. 2013), dehalogenation (e.g. PCBs, trichloroethylene, pharmaceuticals (Baxter-Plant et al. 2004; De Windt et al. 2005), or as a catalyst for diverse chemical syntheses, e.g. the hydrogenation of organic molecules or coupling reactions in synthetic organic chemistry (Creamer et al. 2007). Principally, these mechanisms can be transferred to other precious metals, thus enabling their recovery from waste stream leachates.

Another interesting approach to bioreduction for metal recycling is its application in microbial fuel cells. Huang et al. (2013) designed microbial fuel cells to effectively reduce Co(II) in lithium cobalt oxide that occurs in lithium-ion batteries with concomitant energy generation (Huang et al. 2013). Such a process could enable the recovery of cobalt and the recycling of spent lithium-ion batteries with concomitant energy generation. The anodic biofilm was mainly composed of *Geobacter sulfurreduscens*, *Bacteriodetes* sp., *Pseudomonas* sp., and others. Microbial fuel cells were also used for the reduction and recovery of Cu(II) (Qian et al. 2019). Cu(II) reduction was accompanied by acetate production from inorganic carbon and hydrogen evolution, thus contributing to CO_2 sequestration and production of "green" hydrogen. A biofilm of *Serratia marcescens* Q1 was used in this system. Other researchers applied biocathodes with power generation for Cr(VI) removal as $Cr(OH)_3$ (Tandukar et al. 2009; Huang et al. 2011), for cobalt recovery with simultaneous production of methane and acetate (Huang et al. 2014), for selenium recovery (Catal et al. 2009), for zinc recovery from industrial effluents (Lim et al. 2021), or silver with simultaneous antibiotic removal (Wu et al. 2019). However, to our knowledge, although being quite promising, none of these technologies have been applied at an industrial scale yet, probably due to difficulties in upscale, efficiency, and costs.

15.2.4 Bioflotation

Flotation is a simple separation technology originally applied in the mineral industry to extract valuable minerals from ores (Zhang et al. 2021), which has been proven to be efficient in water and wastewater treatments. Many studies have investigated the use of flotation for the separation of dissociation products of waste PCBs on the basis of the difference in the hydrophobicity of the particles. Li-ion batteries pretreated with ultrasound-assisted pyrolysis were subjected to flotation in an effort to separate lithium cobalt oxide and graphite. They used N-dodecane and methyl iso-butyl methonal as collectors and frothers, respectively (Zhang et al. 2018). Dodecylamine and 2-octanol were reportedly used as collectors and frothers for the recovery of indium tin oxide from e-waste (Wang et al. 2018).

Some studies investigated the reverse flotation technology for separation as well as recovery of metals from electronic waste. Flores-Campos et al. (2017) utilized methyl isobutyl carbinol as a reagent to separate the nonmetallic portion from metals (Flores-Campos et al. 2017). Whereas, kerosene and turpentine oil were used as collectors and frothers for the recovery of metal components from waste PCBs (He and Duan 2017). These studies show the feasibility of metal recovery from e-waste by flotation. However, there is a need to explore new flotation reagents that are environmentally benign (Zhu et al. 2019). Zhu et al. (2019) studied the preparation of renewable collectors from waste oil and tested their applicability for the flotation recovery of metals from waste printed circuit boards (WPCBs). Waste oil was saponified for the preparation of a renewable collector composed of n-hexadecanoic acid, oleic acid, and octadecanoic acid. Analysis of collector dosage and pH for flotation was analyzed along with the feasibility of pretreatment of multiple sorting test processes, which resulted in improved Cu recovery (Zhu et al. 2019, 2020).

In addition, ion flotation is another separation technology for the recovery of metal ions from leach solutions, dilute effluents, and other waste solutions (Doyle 2003). It involves the removal of non-surface active ions by the addition of a surfactant or collector that forms the complex with ions that are collected by passing gas bubbles through the solution. However, new environmentally friendly flotation reagents need more attention to make these separation technologies more sustainable. Microbial biosurfactants composed of hydrophilic and hydrophobic groups are known to complex with various metal ions and thus can act as flotation reagents in both froth flotation and ion flotation methods. Moreover, they are termed green reagents considering their production using microorganisms, biodegradability, and low toxicity (Abdel-Mawgoud et al. 2010). Such an ion flotation process could be useful in the processing of the solutions (leach liquors) obtained after bioleaching, which are usually a multi-metal mixture.

Few reports have investigated the use of rhamnolipid biosurfactants for metal recovery from solutions. A mixture of mono and dirhamnolipid efficiently recovered Cd from mixed solutions (Bodagh et al. 2013). Mono-rhamnolipid was used as an ion collector for the recovery of Cs^{+}, Cd^{2+}, and La^{3+} from water (Hogan et al. 2018) and U from contaminated groundwater (Hogan et al. 2022). Shetty et al. (2020) compared the use of SDS and mono-rhamnolipid as collectors in ion flotation for REE recovery. They observed the complete recovery of REE using rhamnolipid in the presence of gangues, except for Al^{3+} and Zn^{2+} (Shetty et al. 2020).

Very few studies have investigated the use of biosurfactants for metal recovery by ion flotation. These studies suggest that various factors like pH, target metal, collector ratio, solution constituents, etc. affect the process efficiency. Further studies are required to overcome these limitations as well as to know the process's mechanism. Moreover, the low-cost, large-scale production of these biosurfactants is another challenge to overcome in order to apply such techniques at an industrial level. Such production can be obtained by employing genetically modified strains.

15.3 Conclusion and Future Perspectives

The biotechnological recycling processes are a viable and eco-friendly alternative to the classical recycling processes. These processes are developing rapidly at the lab scale, but very few studies at the pilot scale or above are conducted. So, there is a need to develop these technologies at high technology readiness levels and prove their feasibility at the commercial level. Further, having an in-depth knowledge of how bacterial behavior changes physiologically as well as at the molecular level, in terms of changes in the environment and the presence of toxic metals at various concentrations would help in improving the metal recovery efficiencies. In this regard, the OMICS approach, wherein various innovative platforms like genomics, proteomics, and metabolomics in combination with bioinformatics tools could provide new insights into the process fundamentals, development of conceptual and functional metabolic pathways, analysis of microbe–metal interaction, as well as advance the knowledge on the use of various microorganisms and different biomolecules and improve their potential to ensure high recovery efficiencies. It will also help to take the biotechnological approaches to the next level in the applications, with regards to the use of modified strains with enhanced bioleaching capacities, decrease the time as well as increase production of biomolecules for industrial-scale applications.

Another way to take the biotechnological approaches ahead of traditional methods in resource recovery is by using them to separate the target metals from the metal mixture. The use of various biomolecules like peptides, siderophores, and biosurfactants in the approaches of biosorption, bioaccumulation, and bioflotation could be very interesting and efficient in such instances. Moreover, the hybrid strategy using the combination of biotechnological as well as hydrometallurgical processes also shows great potential for a recycling application. As seen from the literature, many steps are already being taken with respect to the technological development of the process, however; on the other hand, the research should also focus on the equipment development for such processes.

Many of these studies, as said earlier, could be employed for the separation and recovery of metals post-bioleaching. At the same time, one should also make sure that the residue left behind after these treatments is also assessed for toxicity so as to avoid further pollution after land disposal. Another way could be an assessment of the utilization of such residues/their use as additives in cement manufacturing, bricks, and other construction materials. These approaches will help in reducing the carbon footprint associated with e-waste, improve resource recovery, and close the loop in a more sustainable and economically viable way.

References

Abdelbasir, S.M., El-Sheltawy, C.T., and Abdo, D.M. (2018). Green processes for electronic waste recycling: a review. *Journal of Sustainable Metallurgy* 4: 295–311.

Abdel-Mawgoud, A.M., Lépine, F., and Déziel, E. (2010). Rhamnolipids: diversity of structures, microbial origins and roles. *Applied Microbiology and Biotechnology* 86: 1323–1336.

Afroz, R., Masud, M.M., Akhtar, R., and Duasa, J.B. (2013). Survey and analysis of public knowledge, awareness and willingness to pay in Kuala Lumpur, Malaysia – a case study on household WEEE management. *Journal of Cleaner Production* 52: 185–193.

Akar, T., Kaynak, Z., Ulusoy, S. et al. (2009). Enhanced biosorption of nickel(II) ions by silica-gel-immobilized waste biomass: biosorption characteristics in batch and dynamic flow mode. *Journal of Hazardous Materials* 163: 1134–1141.

Akcil, A., Ibrahim, Y.A., Meshram, P., and Panda, S. (2021). Hydrometallurgical recycling strategies for recovery of rare earth elements from consumer electronic scraps: a review. *Journal of Chemical Technology & Biotechnology* 96: 1785–1797.

Anaya-Garzon, J., Hubau, A., Joulian, C., and Guezennec, A.G. (2021). Bioleaching of e-waste: influence of printed circuit boards on the activity of acidophilic iron-oxidizing bacteria. *Frontiers in Microbiology* 12: 669738.

Arshadi, M. and Mousavi, S.M. (2015). Enhancement of simultaneous gold and copper extraction from computer printed circuit boards using *Bacillus megaterium*. *Bioresource Technology* 175: 315–324.

Bai, J., Gu, W., Dai, J. et al. (2016). The catalytic role of nitrogen-doped carbon nanotubes in bioleaching copper from waste printed circuit boards. *Polish Journal of Environmental Studies* 25: 951–957.

Baniasadi, M., Tugnoli, A., Conti, R. et al. (2016). Waste to energy valorization of poultry litter by slow pyrolysis. *Renewable Energy* 90: 458–468.

Barragan, J.A., Alemán Castro, J.R., Peregrina-Lucano, A.A. et al. (2021). Leaching of metals from e-waste: from its thermodynamic analysis and design to its implementation and optimization. *ACS Omega* 6: 12063–12071.

Baxter-Plant, V.S., Mikheenko, I.P., Robson, M. et al. (2004). Dehalogenation of chlorinated aromatic compounds using a hybrid bioinorganic catalyst on cells of *Desulfovibrio desulfuricans*. *Biotechnology Letters* 26: 1885–1890.

Bhat, V., Rao, P., and Patil, Y. (2012). Development of an integrated model to recover precious metals from electronic scrap – a novel strategy for e-waste management. *Procedia – Social and Behavioral Sciences* 37: 397–406.

Bodagh, A., Khoshdast, H., Sharafi, H. et al. (2013). Removal of cadmium(II) from aqueous solution by ion flotation using rhamnolipid biosurfactant as an ion collector. *Industrial & Engineering Chemistry Research* 52: 3910–3917.

Boelens, P., Lei, Z., Drobot, B. et al. (2021). High-gradient magnetic separation of compact fluorescent lamp phosphors: elucidation of the removal dynamics in a rotary permanent magnet separator. *Minerals* 11: 1116.

Boelens, P., Bobeth, C., Hinman, N. et al. (2022). Peptide functionalized Dynabeads for the magnetic carrier separation of rare–earth fluorescent lamp phosphors. *Journal of Magnetism and Magnetic Materials* 169956.

Braun, R., Matys, S., Schönberger, N. et al. (2017). Simplified expression and production of small metal binding peptides. *Solid State Phenomena* 262: 447–451.

Braun, R., Bachmann, S., Schonberger, N. et al. (2018a). Peptides as biosorbents - promising tools for resource recovery. *Research in Microbiology* 169: 649–658.

Braun, R., Bachmann, S., Schönberger, N. et al. (2018b). Peptides as biosorbents – promising tools for resource recovery. *Research in Microbiology* 169: 649–658.

Brown, S. (1997). Metal-recognition by repeating polypeptides. *Nature Biotechnology* 15: 269–272.

Bryan, C.G., Watkin, E.L., Mccredden, T.J. et al. (2015). The use of pyrite as a source of lixiviant in the bioleaching of electronic waste. *Hydrometallurgy* 152: 33–43.

Campbell, S.C., Olson, G.J., Clark, T.R., and Mcfeters, G. (2001). Biogenic production of cyanide and its application to gold recovery. *Journal of Industrial Microbiology and Biotechnology* 26: 134–139.

Capeness, M.J. and Horsfall, L.E. (2020). Synthetic biology approaches towards the recycling of metals from the environment. *Biochemical Society Transactions* 48: 1367–1378.

Catal, T., Bermek, H., and Liu, H. (2009). Removal of selenite from wastewater using microbial fuel cells. *Biotechnology Letters* 31: 1211–1216.

Causa, F., Della Moglie, R., Iaccino, E. et al. (2013). Evolutionary screening and adsorption behavior of engineered M13 bacteriophage and derived dodecapeptide for selective decoration of gold interfaces. *Journal of Colloid and Interface Science* 389: 220–229.

Cetinel, S., Shen, W.-Z., Aminpour, M. et al. (2018). Biomining of MoS2 with peptide-based smart biomaterials. *Scientific Reports* 8: 3374.

Choi, Y., Park, T.J., Lee, D.C., and Lee, S.Y. (2018). Recombinant *Escherichia coli* as a biofactory for various single- and multi-element nanomaterials. *Applied Biological Sciences* 115: 5944–5949.

Cobbett, C.S. (2000). Phytochelatin biosynthesis and function in heavy-metal detoxification. *Current Opinion in Plant Biology* 3: 211–216.

Cotruvo, J.A., Featherston, E.R., Mattocks, J.A. et al. (2018). Lanmodulin: a highly selective lanthanide-binding protein from a lanthanide-utilizing bacterium. *Journal of the American Chemical Society* 140: 15056–15061.

Creamer, N.J., Mikheenko, I.P., Yong, P. et al. (2007). Novel supported Pd hydrogenation bionanocatalyst for hybrid homogeneous/heterogeneous catalysis. *Catalysis Today* 128: 80–87.

Cui, J. and Zhang, L. (2008). Metallurgical recovery of metals from electronic waste: a review. *Journal of Hazardous Materials* 158: 228–256.

Curtis, S.B., Hewitt, J., Macgillivray, R.T., and Dunbar, W.S. (2009). Biomining with bacteriophage: selectivity of displayed peptides for naturally occurring sphalerite and chalcopyrite. *Biotechnol Bioeng* 102: 644–650.

Curtis, S.B., Macgillivray, R.T., and Dunbar, W.S. (2011). Effects of bacteriophage on the surface properties of chalcopyrite ($CuFeS_2$), and phage-induced flocculation of chalcopyrite, glacial till, and oil sands tailings. *Biotechnology and Bioenginering* 108: 1579–1590.

De Windt, W., Aelterman, P., and Verstraete, W. (2005). Bioreductive deposition of palladium (0) nanoparticles on *Shewanella oneidensis* with catalytic activity towards reductive dechlorination of polychlorinated biphenyls. *Environmental Microbiology* 7: 314–325.

Diep, P., Mahadevan, R., and Yakunin, A.F. (2018). Heavy metal removal by bioaccumulation using genetically engineered microorganisms. *Frontiers in Bioengineering and Biotechnology* 6.

Donatan, S., Sarikaya, M., Tamerler, C., and Urgen, M. (2012). Effect of solid surface charge on the binding behaviour of a metal-binding peptide. *Journal of the Royal Society Interface* 9: 2688–2695.

Doyle, F.M. (2003). Ion flotation—its potential for hydrometallurgical operations. *International Journal of Mineral Processing* 72: 387–399.

Edmundson, M.C. and Horsfall, L. (2015). Construction of a modular arsenic-resistance operon in *E. coli* and the production of arsenic nanoparticles. *Frontiers in Bioengineering and Biotechnology* 3: 160.

Eisele, T.C. and Gabby, K.L. (2014). Review of reductive leaching of iron by anaerobic bacteria. *Mineral Processing and Extractive Metallurgy Review* 35: 75–105.

Erust, C., Akcil, A., Tuncuk, A. et al. (2021). A novel approach based on solvent displacement crystallisation for iron removal and copper recovery from solutions of semi-pilot scale bioleaching of WPCBs. *Journal of Cleaner Production* 294.

Esmaeili, A., Arshadi, M., and Professor of Chemical and Petroleum Engineering Department Soheila Yaghmaei (2022). Simultaneous leaching of Cu, Al, and Ni from computer printed circuit boards using *Penicillium simplicissimum*. *Resources, Conservation and Recycling* 177: 105976.

Faraji, F., Golmohammadzadeh, R., Rashchi, F., and Alimardani, N. (2018). Fungal bioleaching of WPCBs using *Aspergillus niger*: observation, optimization and kinetics. *Journal of Environmental Management* 217: 775–787.

Flores-Campos, R., Estrada-Ruiz, R.H., and Velarde-Sánchez, E.J. (2017). Study of the physicochemical effects on the separation of the non-metallic fraction from printed circuit boards by inverse flotation. *Waste Management* 69: 400–406.

Forti, V., Balde, C.P., Kuehr, R., and Bel, G. (2020). *The Global E-Waste Monitor 2020: Quantities, Flows and the Circular Economy Potential*. Bonn, Geneva and Rotterdam: United Nations University/United Nations Institute for Training and Research, International Telecommunication Union, and International Solid Waste Association.

Furuhashi, Y., Honda, R., Noguchi, M. et al. (2019). Optimum conditions of pH, temperature and preculture for biosorption of europium by microalgae *Acutodesmus acuminatus*. *Biochemical Engineering Journal* 143: 58–64.

García-García, J.D., Sánchez-Thomas, R., and Moreno-Sánchez, R. (2016). Bio-recovery of non-essential heavy metals by intra- and extracellular mechanisms in free-living microorganisms. *Biotechnology Advances* 34: 859–873.

Giese, E.C. (2020). Biosorption as green technology for the recovery and separation of rare earth elements. *World Journal of Microbiology and Biotechnology* 36: 52.

Gu, W., Bai, J., Dong, B. et al. (2017). Catalytic effect of graphene in bioleaching copper from waste printed circuit boards by *Acidithiobacillus ferrooxidans*. *Hydrometallurgy* 171: 172–178.

Gupta, S.K. and Shukla, P. (2016). Advanced technologies for improved expression of recombinant proteins in bacteria: perspectives and applications. *Critical Reviews in Biotechnology* 36: 1089–1098.

Gupta, N.K., Gupta, A., Ramteke, P. et al. (2019). Biosorption-a green method for the preconcentration of rare earth elements (REEs) from waste solutions: a review. *Journal of Molecular Liquids* 274: 148–164.

Gutenthaler, S.M., Tsushima, S., Steudtner, R. et al. (2022). Lanmodulin peptides – unravelling the binding of the EF-hand loop sequences stripped from the structural corset. *Inorganic Chemistry Frontiers* 9: 4009–4021.

Hallberg, K.B., Grail, B.M., Plessis, C., and a. D. & Johnson, D. B. (2011). Reductive dissolution of ferric iron minerals: a new approach for bio-processing nickel laterites. *Minerals Engineering* 24: 620–624.

He, J. and Duan, C. (2017). Recovery of metallic concentrations from waste printed circuit boards via reverse floatation. *Waste Management* 60: 618–628.

He, Q.-X., Huang, X., Chen, Z.J., and a. C. S. (2011). Influence of organic acids, complexing agents and heavy metals on the bioleaching of iron from kaolin using Fe(III)-reducing bacteria. *Applied Clay Science* 51: 478–483.

Hedrich, S., Du Plessis, C., Mora, N., and Johnson, D.B. (2013). Reduction and complexation of copper in a novel bioreduction system developed to recover base metals from mine process waters. *Advanced Materials Research* 825: 483–486.

Hider, R.C. and Kong, X. (2010). Chemistry and biology of siderophores. *Natural Product Reports* 27: 637–657.

Hnilova, M., Oren, E.E., Seker, U.O.S. et al. (2008). Effect of molecular conformations on the adsorption behavior of gold-binding peptides. *Langmuir* 24: 12440–12445.

Hoffmann, J.E. (1992). Recovering precious metals from electronic scrap. *JOM* 44: 43–48.

Hofmann, M., Retamal-Morales, G., and Tischler, D. (2020). Metal binding ability of microbial natural metal chelators and potential applications. *Natural Product Reports* 37: 1262–1283.

Hogan, D.E., Curry, J.E., and Maier, R.M. (2018). Ion flotation of La^{3+}, Cd^{2+}, and Cs^{+} using monorhamnolipid collector. *Colloids and Interfaces* 2.

Hogan, D.E., Stolley, R.M., Boxley, C. et al. (2022). Removal of uranium from contaminated groundwater using monorhamnolipids and ion flotation. *Journal of Environmental Management* 301: 113835.

Horeh, N.B., Mousavi, S.M., and Shojaosadati, S.A. (2016). Bioleaching of valuable metals from spent lithium-ion mobile phone batteries using *Aspergillus niger*. *Journal of Power Sources* 320: 257–266.

Hosseini, M.R., Pazouki, M., Ranjbar, M., and Habibian, M. (2007). Bioleaching of iron from highly contaminated kaolin clay by *Aspergillus niger*. *Applied Clay Science* 37: 251–257.

Huang, L., Chai, X., Chen, G., and Logan, B.E. (2011). Effect of set potential on hexavalent chromium reduction and electricity generation from biocathode microbial fuel cells. *Environmental Science & Technology* 45: 5025–5031.

Huang, L., Li, T., Liu, C. et al. (2013). Synergetic interactions improve cobalt leaching from lithium cobalt oxide in microbial fuel cells. *Bioresource Technology* 128: 539–546.

Huang, L., Jiang, L., Wang, Q. et al. (2014). Cobalt recovery with simultaneous methane and acetate production in biocathode microbial electrolysis cells. *Chemical Engineering Journal* 253: 281–290.

Humphries, A.C., Penfold, D.W., and Macaskie, L.E. (2007). Cr(VI) reduction by bio and bioinorganic catalysis via use of bio-H2: a sustainable approach for remediation of wastes. *Journal of Chemical Technology and Biotechnology* 82: 182–189.

Iori, F., Di Felice, R., Molinari, E., and Corni, S. (2009). GolP: an atomistic force-field to describe the interaction of proteins with Au(111) surfaces in water. *Journal of Computational Chemistry* 30: 1465–1476.

Ishida, N., Hatanaka, T., Hosokawa, Y. et al. (2020). Direct recovery of the rare earth elements using a silk displaying a metal-recognizing peptide. *Molecules* 25.

Isildar, A., Van De Vossenberg, J., Rene, E.R. et al. (2016). Two-step bioleaching of copper and gold from discarded printed circuit boards (PCB). *Waste Management* 57: 149–157.

Jadhav, U. and Hocheng, H. (2014). Enzymatic bioleaching of metals from printed circuit board. *Clean Technologies and Environmental Policy* 17: 947–956.

Jadhav, U., Su, C., and Hocheng, H. (2016). Leaching of metals from printed circuit board powder by an *Aspergillus niger* culture supernatant and hydrogen peroxide. *RSC Advances* 6: 43442–43452.

Johnston, C.W., Wyatt, M.A., Li, X. et al. (2013). Gold biomineralization by a metallophore from a gold-associated microbe. *Nature Chemical Biology* 9: 241–243.

Julander, A., Lundgren, L., Skare, L. et al. (2014). Formal recycling of e-waste leads to increased exposure to toxic metals: an occupational exposure study from Sweden. *Environment International* 73: 243–251.

Karwowska, E., Andrzejewska-Morzuch, D., Lebkowska, M. et al. (2014). Bioleaching of metals from printed circuit boards supported with surfactant-producing bacteria. *Journal of Hazardous Materials* 264: 203–210.

Kaur, P., Sharma, S., Albarakaty, F.M. et al. (2022). Biosorption and bioleaching of heavy metals from electronic waste varied with microbial genera. *Sustainability* 14: 935.

Kim, Y., Oh, J.M., and Roh, Y. (2017). Microbial precipitation of Cr(III)-hydroxide and Se(0) nanoparticles during anoxic bioreduction of Cr(VI)- and Se(VI)-contaminated water. *Journal of Nanoscience and Nanotechnology* 17: 2302–2304.

Lederer, F.L., Curtis, S.B., Bachmann, S. et al. (2017). Identification of lanthanum-specific peptides for future recycling of rare earth elements from compact fluorescent lamps. *Biotechnology and Bioengineering* 114: 1016–1024.

Lederer, F.L., Braun, R., Schöne, L.M., and Pollmann, K. (2019). Identification of peptides as alternative recycling tools via phage surface display – how biology supports geosciences. *Minerals Engineering* 132: 245–250.

Lee, J.-C., Song, H.T., and Yoo, J.-M. (2007). Present status of the recycling of waste electrical and electronic equipment in Korea. *Resources, Conservation and Recycling* 50: 380–397.

Li, J., Wen, J., Guo, Y. et al. (2020). Bioleaching of gold from waste printed circuit boards by alkali-tolerant *Pseudomonas fluorescens*. *Hydrometallurgy* 194: 105260.

Li, J., Xu, T., Liu, J. et al. (2021). Bioleaching metals from waste electrical and electronic equipment (WEEE) by *Aspergillus niger*: a review. *Environmental Science and Pollution Research* 28: 44622–44637.

Lim, S.S., Fontmorin, J.-M., Pham, H.T. et al. (2021). Zinc removal and recovery from industrial wastewater with a microbial fuel cell: experimental investigation and theoretical prediction. *Science of The Total Environment* 776: 145934.

Liu, W., Yang, H.-Y., Song, Y., and Tong, L.-L. (2015). Catalytic effects of activated carbon and surfactants on bioleaching of cobalt ore. *Hydrometallurgy* 152: 69–75.

Lovley, D.R., Stolz, J.F., Nord, G.L., and Phillips, E.J.P. (1987). Anaerobic production of magnetite by a dissimilatory iron-reducing microorganism. *Nature* 330: 252–254.

Lovley, D.R., Coates, J.D., Blunt-Harris, E.L. et al. (1996). Humic substances as electron acceptors for microbial respiration. *Nature* 382: 445–448.

Luque-Almagro, V.M., Blasco, R., Huertas, M.J. et al. (2005). Alkaline cyanide biodegradation by *Pseudomonas pseudoalcaligenes* CECT5344. *Biochemical Society Transactions* 33: 168–169.

Mabbett, A.N., Sanyahumbi, D., Yong, P., and Macaskie, L.E. (2006). Biorecovered precious metals from industrial wastes: single-step conversion of a mixed metal liquid waste to a bioinorganic catalyst with environmental application. *Environmental Science & Technology* 40: 1015–1021.

Mäkinen, J., Bachér, J., Kaartinen, T. et al. (2015). The effect of flotation and parameters for bioleaching of printed circuit boards. *Minerals Engineering* 75: 26–31.

Martins, M., Assunção, A., Martins, H. et al. (2013). Palladium recovery as nanoparticles by an anaerobic bacterial community. *Journal of Chemical Technology and Biotechnology* 88: 2039–2045.

Nian, R., Kim, D.S., Nguyen, T. et al. (2010). Chromatographic biopanning for the selection of peptides with high specificity to Pb2+ from phage displayed peptide library. *Journal of Chromatography A* 1217: 5940–5949.

Ortinero, C.V., Meim, H.N., and Paragas, D.S. (2022). Biosorbents from *Mangifera indica* L. peel for the recovery of gold from electronic waste. *Journal of Ecological Engineering* 23: 102–108.

Papassiopi, N., Vaxevanidou, K., and Paspaliaris, I. (2010). Effectiveness of iron reducing bacteria for the removal of iron from bauxite ores. *Minerals Engineering* 23: 25–31.

Park, D.M., Reed, D.W., Yung, M.C. et al. (2016). Bioadsorption of rare earth elements through cell surface display of lanthanide binding tags. *Environmental Science & Technology* 50: 2735–2742.

Park, D.M., Brewer, A., Reed, D.W. et al. (2017). Recovery of rare earth elements from low-grade feedstock leachates using engineered bacteria. *Environmental Science & Technology* 51: 13471–13480.

Pat-Espadas, A.M., Field, J.A., Razo-Flores, E. et al. (2016). Continuous removal and recovery of palladium in an upflow anaerobic granular sludge bed (UASB) reactor. *Journal of Chemical Technology and Biotechnology* 91: 1183–1189.

Pathak, A., Morrison, L., and Healy, M.G. (2017). Catalytic potential of selected metal ions for bioleaching, and potential techno-economic and environmental issues: a critical review. *Bioresource Technology* 229: 211–221.

Pollmann, K., Kutschke, S., Matys, S. et al. (2018). Bio-recycling of metals: recycling of technical products using biological applications. *Biotechnology Advances* 36: 1048–1062.

Pradhan, D., Sukla, L.B., Sawyer, M., and Rahman, P.K.S.M. (2017). Recent bioreduction of hexavalent chromium in wastewater treatment: a review. *Journal of Industrial and Engineering Chemistry* 55: 1–20.

Priya, A. and Hait, S. (2017). Comparative assessment of metallurgical recovery of metals from electronic waste with special emphasis on bioleaching. *Environmental Science and Pollution Research* 24: 6989–7008.

Priya, A. and Hait, S. (2018). Extraction of metals from high grade waste printed circuit board by conventional and hybrid bioleaching using *Acidithiobacillus ferrooxidans*. *Hydrometallurgy* 177: 132–139.

Qian, Y., Huang, L., Zhou, P. et al. (2019). Reduction of Cu(II) and simultaneous production of acetate from inorganic carbon by *Serratia marcescens* biofilms and plankton cells in microbial electrosynthesis systems. *Science of The Total Environment* 666: 114–125.

Quinet, P., Proost, J., and Van Lierde, A. (2005). Recovery of precious metals from electronic scrap by hydrometallurgical processing routes. *Mining, Metallurgy & Exploration* 22: 17–22.

Rasoulnia, P. and Mousavi, S.M. (2016). Maximization of organic acids production by *Aspergillus niger* in a bubble column bioreactor for V and Ni recovery enhancement from power plant residual ash in spent-medium bioleaching experiments. *Bioresource Technology* 216: 729–736.

Rimaszeki, G., Kulcsar, T., and Kekesi, T. (2012). Application of HCl solutions for recovering the high purity metal from tin scrap by electrorefining. *Hydrometallurgy* 125–126: 55–63.

Sannigrahi, S. and Suthindhiran, K. (2019). Metal recovery from printed circuit boards by magnetotactic bacteria. *Hydrometallurgy* 187: 113–124.

Saravanan, T.G., Loganathan, A., Rajendran, G. et al. (2022). Bioleaching of heavy metals by *Streptomyces avermitilis* – BBA4 isolate from coal mine soil of Tamil Nadu. *Current Microbiology* 79: 303.

Schönberger, N., Braun, R., Matys, S. et al. (2019a). Chromatopanning for the identification of gallium binding peptides. *Journal of Chromatography A* 1600: 158–166.

Schönberger, N., Zeitler, C., Braun, R. et al. (2019b). Directed evolution and engineering of gallium-binding phage clones – a preliminary study. *Biomimetics (Basel)* 4.

Schönberger, N., Taylor, C., Schrader, M. et al. (2021). Gallium-binding peptides as a tool for the sustainable treatment of industrial waste streams. *Journal of Hazardous Materials* 414: 125366.

Schwaminger, S., Blank-Shim, S.A., Borkowska-Panek, M. et al. (2018). Experimental characterization and simulation of amino acid and peptide interactions with inorganic materials. *Engineering in Life Sciences* 18: 84–100.

Setyawati, M.I., Xie, J., and Leong, D.T. (2014). Phage based green chemistry for gold ion reduction and gold retrieval. *ACS Applied Materials & Interfaces* 6: 910–917.

Shetty, S., Chernyshova, I.V., and Ponnurangam, S. (2020). Foam flotation of rare earth elements by conventional and green surfactants. *Minerals Engineering* 158.

Shin, D., Park, J., Jeong, J., and Kim, B.-S. (2015). A biological cyanide production and accumulation system and the recovery of platinum-group metals from spent automotive catalysts by biogenic cyanide. *Hydrometallurgy* 158: 10–18.

Stefaniak, J., Dutta, A., Verbinnen, B. et al. (2018). Selenium removal from mining and process wastewater: a systematic review of available technologies. *Journal of Water Supply: Research and Technology-Aqua* 67: 903–918.

Štyriaková, I. and Lovás, M. (2007). Biological purification of silicate minerals. *Advanced Materials Research* 20–21: 126–129.

Styriaková, I. and Styriak, I. (2000). Iron removal from kaolins by bacterial leaching. *Ceramics - Silikaty* 44: 135–141.

Tamerler, C., Khatayevich, D., Gungormus, M. et al. (2010). Molecular biomimetics: GEPI-based biological routes to technology. *Biopolymers* 94: 78–94.

Tandukar, M., Huber, S.J., Onodera, T., and Pavlostathis, S.G. (2009). Biological chromium(VI) reduction in the cathode of a microbial fuel cell. *Environmental Science & Technology* 43: 8159–8165.

Taxiarchou, M., Panias, D., Douni, I. et al. (1997a). Dissolution of hematite in acidic oxalate solutions. *Hydrometallurgy* 44: 287–299.

Taxiarchou, M., Panias, D., Douni, I. et al. (1997b). Removal of iron from silica sand by leaching with oxalic acid. *Hydrometallurgy* 46: 215–227.

Thakur, P. and Kumar, S. (2020). Metallurgical processes unveil the unexplored "sleeping mines" e-waste: a review. *Environmental Science and Pollution Research* 27: 32359–32370.

Thirunavukkarasu, A., Nithya, R., and Sivashankar, R. (2021). Continuous fixed-bed biosorption process: a review. *Chemical Engineering Journal Advances* 8: 100188.

Toni, A.M. and Bridge, D.B.J. (2000). Reductive dissolution of ferric iron minerals by *Acidiphilium* SJH. *Geomicrobiology Journal* 17: 193–206.

Tunali, M. and Yenigun, O. (2021). Biosorption of Ag+ and Nd3+ from single- and multi-metal solutions (Ag+, Nd3+, and Au3+) by using living and dried microalgae. *Journal of Material Cycles and Waste Management* 23: 764–777.

Tuncuk, A., Stazi, V., Akcil, A. et al. (2012). Aqueous metal recovery techniques from e-scrap: hydrometallurgy in recycling. *Minerals Engineering* 25: 28–37.

Tuo, Y., Liu, G., Zhou, J. et al. (2013). Microbial formation of palladium nanoparticles by *Geobacter sulfurreducens* for chromate reduction. *Bioresource Technology* 133: 606–611.

Ubaldini, S., Piga, L., Fornari, P., and Massidda, R. (1996). Removal of iron from quartz sands: a study by column leaching using a complete factorial design. *Hydrometallurgy* 40: 369–379.

Ullman, W.J., Kirchman, D.L., Welch, S.A., and Vandevivere, P. (1996). Laboratory evidence for microbially mediated silicate mineral dissolution in nature. *Chemical Geology* 132: 11–17.

Urbina, J., Patil, A., Fujishima, K. et al. (2019). A new approach to biomining: bioengineering surfaces for metal recovery from aqueous solutions. *Scientific Reports* 9: 16422.

Vodnik, M., Zager, U., Strukelj, B., and Lunder, M. (2011). Phage display: selecting straws instead of a needle from a haystack. *Molecules* 16: 790–817.

Vreuls, C., Genin, A., Zocchi, G. et al. (2011). Genetically engineered polypeptides as a new tool for inorganic nano-particles separation in water based media. *Journal of Materials Chemistry* 21: 13841–13846.

Wang, Y.A., Yu, X., Overman, S. et al. (2006). The structure of a filamentous bacteriophage. *Journal of Molecular Biology* 361: 209–215.

Wang, S., He, Y., Yang, J., and Feng, Y. (2018). Enrichment of indium tin oxide from colour filter glass in waste liquid crystal display panels through flotation. *Journal of Cleaner Production* 189: 464–471.

Wu, W., Liu, X., Zhang, X. et al. (2018). Bioleaching of copper from waste printed circuit boards by bacteria-free cultural supernatant of iron–sulfur-oxidizing bacteria. *Bioresources and Bioprocessing* 5: 10.

Wu, D., Lu, D., Sun, F., and Zhou, Y. (2019). Process optimization for simultaneous antibiotic removal and precious metal recovery in an energy neutral process. *Science of The Total Environment* 695: 133914.

Xia, M.C., Wang, Y.P., Peng, T.J. et al. (2017). Recycling of metals from pretreated waste printed circuit boards effectively in stirred tank reactor by a moderately thermophilic culture. *Journal of Bioscience and Bioengineering* 123: 714–721.

Yang, C., Zhu, N., Shen, W. et al. (2017). Bioleaching of copper from metal concentrates of waste printed circuit boards by a newly isolated *Acidithiobacillus ferrooxidans* strain Z1. *Journal of Material Cycles and Waste Management* 19: 247–255.

Yong, P., Mikheenko, I.P., Deplanche, K. et al. (2009). Biorecovery of precious metals from wastes and conversion into fuel cell catalyst for electricity production. *Advanced Materials Research End* 71–73: 729–732.

Zeng, G., Deng, X., Luo, S. et al. (2012). A copper-catalyzed bioleaching process for enhancement of cobalt dissolution from spent lithium-ion batteries. *Journal of Hazardous Materials* 199–200: 164–169.

Zhang, G., He, Y., Feng, Y. et al. (2018). Pyrolysis-ultrasonic-assisted flotation technology for recovering graphite and LiCoO2 from spent lithium-ion batteries. *ACS Sustainable Chemistry & Engineering* 6: 10896–10904.

Zhang, Y., Jiang, H., Bian, K. et al. (2021). Is froth flotation a potential scheme for microplastics removal? Analysis on flotation kinetics and surface characteristics. *Science of The Total Environment* 792: 148345.

Zhu, X.N., Nie, C.C., Zhang, H. et al. (2019). Recovery of metals in waste printed circuit boards by flotation technology with soap collector prepared by waste oil through saponification. *Waste Management* 89: 21–26.

Zhu, X.-N., Cui, T., Li, B. et al. (2020). Metal recovery from waste printed circuit boards by flotation technology with non-ionic renewable collector. *Journal of Cleaner Production* 255.

16

Biorecovery of Critical and Precious Metals

Shivangi Mathur[1], Nirmaladevi Saravanan[2], Soumya V. Menon[3], and Biswaranjan Paital[4]

[1] *Department of Biotechnology, President Science College, Gujarat University, Ahmedabad, Gujarat, India*
[2] *Department of Biochemistry and Biotechnology, Avinashilingam University for Women, Coimbatore, Tamil Nadu, India*
[3] *Department of Chemistry and Biochemistry, School of Sciences, Jain (Deemed to be) University, Bengaluru, Karnataka, India*
[4] *Redox Regulation Laboratory, Department of Zoology, Odisha University of Agriculture and Technology, Bhubaneswar, Odisha, India*

Abstract

Disposal of electronic waste (e-waste) has increased many folds post-Covid 2019 (TBRC 2022). E-waste can be an alternate ore to leach out precious metals as compared to conventional techniques, but only a few regions like Europe (42% recycling rate) and Asia (12% recycling rate) are accountable for effective recycling at the global scale; others majorly follow unsafe strategies to recycle e-waste (Cui and Zhang 2008), releasing hazardous contents into the environment and leading to many complications (Loganathan and Masunaga 2015). This chapter gives insights on various methods used for leaching metals from e-waste, their limitations, and biorecovery using microorganisms, as an alternative strategy to extract valuable metals such as gold, copper, palladium, silver, iron, and platinum. Mechanisms of microbial leaching with examples of bacteria used are also discussed. Interventions of biotechnological approaches, like developing new strains with high microbial leaching efficiency, are also suggested for sustainable development.

Keywords *e-waste; recycling; critical metals; precious metals; biorecovery*

16.1 Introduction to Critical and Precious Metals for Recovery

Precious metals have universal use in the production of electronic items due to their highly conductive nature and resistance to corrosion (Steffens et al. 1989;

Management of Electronic Waste: Resource Recovery, Technology and Regulation, First Edition. Edited by Anshu Priya.

Richards 2006). Globally, resources of metals from the environment are meager; hence, their recovery is very much necessary. One of the rapidly growing segments of waste is electronic waste, also termed identically as WEEE (waste electrical and electronic equipment) and/or e-waste. The worldwide e-waste generation was about 54 million metric tons in 2019, and the increased economy, accessibility, and ease of use are a few reasons due to which e-waste has been accumulated in the past few years, has increased many folds, marking it as swiftly developing waste stream globally. Moreover, it is speculated that by the end of 2030, yearly e-waste accumulation at a global scale might be enhanced by approximately 30 percent. The WEEE is a very dangerous waste type, and it includes relatively high concentrations of metals. Due to its inappropriate use, it may affect the health of every organism (Grant et al. 2013). Valuable metals such as silver, palladium, gold, copper, iron, silver, and platinum comprise around $57b global market but are mostly dumped instead of collected/recycled from e-waste. The value of precious metals in e-waste is worth $14bn but only $4bn is currently being recovered.

Electronic equipment dumped consists of various materials such as: (a) Exquisite metals such as silver, gold, palladium, platinum, and ruthenium in lesser quantities, (b) Nonmetals and metalloids such as selenium and tellurium, (c) Nonferrous metals such as nickel, iron, aluminum, tin, cobalt, zinc, copper, indium, and gallium, (d) Toxic elements such as cadmium, beryllium, arsenic, and mercury, and (e) Halogens such as bromine, chlorine, and fluorine and other substances such as plastic, glass, ceramic, and organic substances (Corty and Hagelüken, 2010).

Dimitrijević et al. (2013) studied the composition of printed circuit board (PCB) parts in terms of mass and found that 42.03% comprises phenol formaldehyde (PF) resins, "black plastic" weighing 103.4 g. The second major component (34.06%) was FR-4 glass epoxy board with printed connections with 83.8 g weight. Metal contacts comprised 17.32% weighing 45.10 g, metal solder comprised 2.44% (5.72 g), other minor components comprised 2.44% (6.0 g), and technological loss comprised 1.70% (4.2 g). This study enhanced the purview of the recapturing of valuable metals from e-waste.

16.2 Precious Metal E-waste Recovery in the International Market

The international market for precious metal recovery is increasing in the post-pandemic Covid-2019 time period due to the increase in the accumulation of computer-assisted tools for enforcement of restrictive containment measures and the need to maintain social distance. This recapturing of valuable metals business is predicted to increase from the year 2021 ($9.22 billion) to 2022 ($9.61 billion) at a CAGR (compound annual growth rate) of 4.28% and will grow as high as $11.51 billion with a CAGR of 4.61% in 2026 (TBRC, 2022).

In recycling rate comparison, Europe ranks first with 42% of the recycling rate, second is Asia with 12%, but other regions have much smaller recycling rates as compared to developed regions. Recycling of e-waste is done in some low-income countries too, but the recycling process involves unsafe methods like burning circuit boards to collect copper, causing the discharge of highly toxic metals, e.g. lead, cadmium, etc. leaving detrimental effects on the health of the population near e-waste activities.

16.2.1 Expected Fastest-Growing E-waste Recovery: Copper

Due to high electrical conductivity, good machinability, high ductility, good heat conductivity, and corrosion resistance, copper is one of the fastest-growing metal recovery target components from e-waste, projected to increase rapidly from 2020 to 2025 (TBRC 2022). With the use of copper in electronic manufacturing, production of industrial machinery, wiring in heating and cooling systems, and as an integral part of telecommunication systems, its demand has increased many folds.

16.2.2 Expected Thriving Local Segment for Valuable Metals Electronic Waste Recapturing: Europe and the Asia Pacific

After Europe, Asia is the second-largest region to recycle and recover precious metals from e-waste. Asian countries such as South Korea, Japan, China, Taiwan, and Malaysia are the key countries that have made major contributions to the recovery of metals. The availability of cheap raw materials and labor has made these countries a preferable manufacturing region for electronics and automobiles (Terazono et al. 2004).

16.3 E-waste Sources and Progression

Sources of e-waste are numerous, such as electronic appliances, e.g. personal computers, laptops, digital versatile disc (DVD), television, mobile phones, mp3 players, etc., used in households and a large number of electronic appliances on a commercial scale. Some of the electric appliances such as washing machines, vacuum cleaners, drying machines, refrigerators, irons, air conditioners, and coffee machines also contribute to the accumulation of e-waste after they are discarded or dumped at dumpsites. The WEEE category also involves other materials such as sewing machines, electronic toys, sodium lamps, lawnmowers, coin slot machines, fluorescent tubes, and surveillance equipment. All such items constitute a major part of waste at dumpsites, which could be target sites for the mining of precious metals. Other sources of metal-rich waste are mine waste, sludges, slags, contaminated soils, fly ashes, spent catalysts, etc. (Potysz et al. 2016).

16.4 Conventional E-waste Metal Recovery Methods and Their Limitations

The different conventional methods followed for metal recovery are chemical leaching, photocatalysis, hydrometallurgic methods, etc. Microbial processing and recovery of metals are recognized technologies mainly for their low hazardous effects on the environment and cost-effectiveness.

16.4.1 Chemical Leaching

Leaching is one of the solid-phase extraction methods used for the separation and extraction of critical and precious metals. A huge solid mass comprising solvable and unsolvable components is disintegrated and separated by one of the most widely accepted macroscopic techniques known as leaching. Under the same lineage, even chemical processes were discovered for e-waste such as PCBs, liquid crystal displays (LCDs), and alkaline batteries (see Figure 16.1). Moreover, many researchers have gained a keen interest in refining the process of Leaching with the usage of several chemical and biological methods (Akcil et al. 2021; Panda et al. 2021). From recent studies, it is reported that the concentration of base (nonferrous) metals is higher in e-waste (even higher than that present in the ores), and so, holding this thought in mind, the ideology revolves around the fact that even precious metals can be extracted. Various studies showed that chemical leaching could recover about 84–100% of valuable metals like gold, silver, and copper.

For recapturing gold (Au), one of the chemical leaching methods employed is thiourea leaching. Although gold extraction from e-waste by undergoing the general leaching step is quite a toiled job as the final recovery stages are an incomprehensible task that seeks a lot of refinement (Örgül and Atalay, 2000; Celik, 2004). Specifically, while undergoing the process of leaching by making use of thiourea, several of the contending components can hinder gold solubility, and the solubility of thiourea along with an oxidant can generate some conflicting results by solubilizing gold.

In the previous investigation, the leaching of several precious metals from ICs (Integrated Circuits) present in discarded circuit boards using thiourea was discovered by Lee et al. (2011). Studies reported dosage of Thiourea along with the leaching time and its efficacy towards recovery rate. Thiourea dosage content was detected from 2.5 g/L to 100 g/L with an optimal concentration of 70g/L.

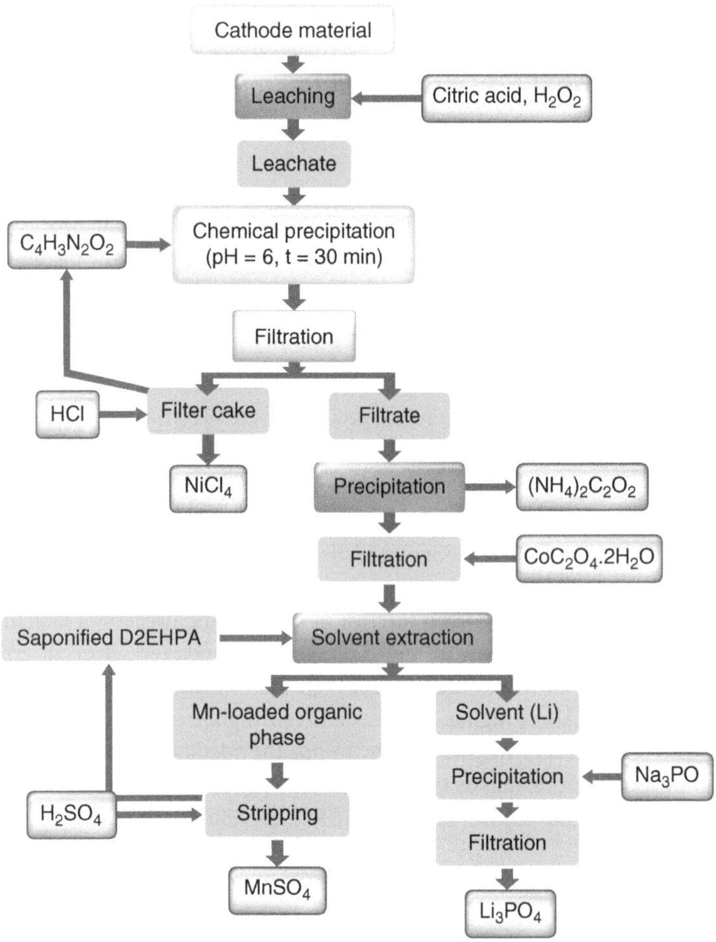

Figure 16.1 Recapturing of metals using chemical method.

Additionally, the system was also comprised of 13 g/L iron (III) with 1.8M H2SO4 at optimum temperature, giving 100% recovery of gold after 7 hours of leaching time. Another study was done by Gurung et al. (2012) with chemical modifications by adding persimmon tannin gel along with an acidic leaching method with a view to recovering valuable metals such as Cu, Ag, Au, and platinum (Pt) from mobile phones and PCBs. This study, also stated that a novel gel adsorbent

'BTU-PT gel' had the potential to selectively adsorb Au and Pt rather than using base metals in the acidic leach liquor (Gurung et al. 2012).

16.4.1.1 Pretreatment of E-waste

Before undergoing the leaching process, pretreatment is required, which is undertaken by using physical and chemical treatment methods. During this process of chemical treatment, base metals such as Pb and Sn are present in the filtrate, whereas precious metals form the residue, which is further isolated and purified. Some of the frequently used chemicals in chemical pretreatment methods are: hydrogen peroxide, sulfuric acid, and nitric acid (Zhang et al. 2012). However, Zhu and Gu (2002) concluded from their studies that at very high or low acid conditions, recovery of gold and copper was found to be very low, but when other conditions like 1:2 solid: liquid ratio, 20 ml of sulfuric acid, and hydrogen peroxide (3:1) were used, from 10 g of waste material, all of the copper and gold were almost recaptured. The desquamation rate of gold and the recovery rate of copper reach 98.75% and 99.43%, respectively.

Besides these leaching methods, there are also other leaching methods such as thiosulfate leaching, halide leaching, and cyanide leaching that have been employed for the recovery of precious metals from e-waste. The major advantage of opting for the chemical method is recovering precious metals at a very high rate at a low cost, but the chemicals or reagents used are toxic and unstable in nature and also found to be hazardous to the environment.

16.4.2 Physical Methods (Grinding and Pulverizing)

Treatment of e-waste and valuable metal recovery can be attained by physical methods, for example, crushing, gravitation, separation using magnets, electrostatic separation, eddy current separation, etc. There are different techniques involved in recycling e-waste, especially PCB waste.

The e-waste physical recycling process is divided into three main phases (see Figure 16.2).

16.4.2.1 Disassembly

The elimination of toxic chemicals such as AI radiators, capacitors, and batteries and the recovery of recyclable materials needs a variety of tools. Disassembly is an orderly method that involves the separation of contents such as subassemblies from a product or partial disassembly, the detachment, of contents into all their

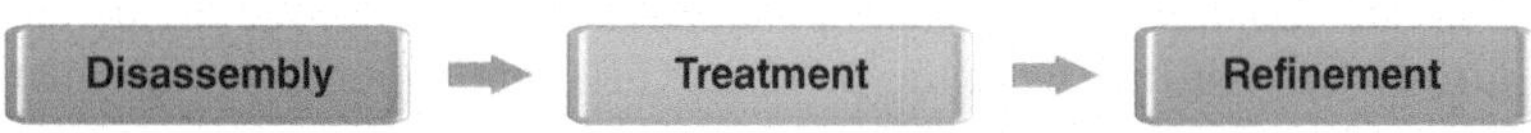

Figure 16.2 Recycling methods.

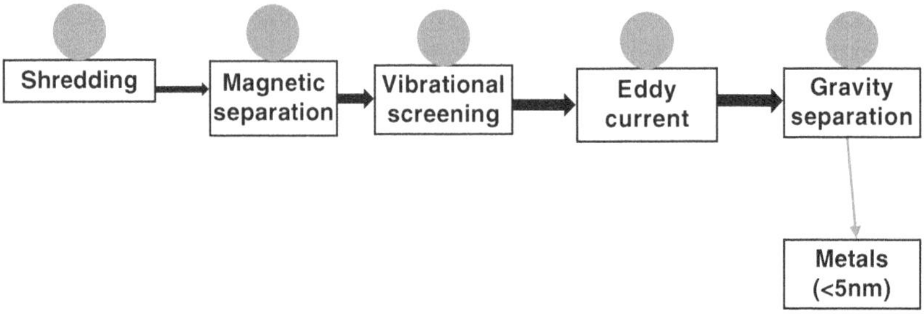

Figure 16.3 Treatment process – steps.

parts such as full disassembly (Gungor and Gupta 1998). There are different steps of the disassembly process in a system plan, i.e.

1. Input/Output: hazardous components and reusable material selection
2. Arrangement of assembly components
3. Analysis of defective parts/ components
4. Destructive/nondestructive disassembly

16.4.2.2 Treatment

Different treatments are involved in e-waste recycling such as shredding, sieving, grinding, and magnetic separation. Shredding is reducing the size (3–5 mm) of the particle for processing samples further. A variety of techniques and instruments, including metal shredders, hammers, and knife mills, are used to crush and grind the e-waste (Kumar et al. 2017; Hubau et al. 2019). Additionally, the removal of toxic material is achieved by washing for better metal recovery (Ruan et al. 2018). Several components can be separated from such types of waste materials after crushing and separation using various methods (see Figure 16.3), such as density separation and froth flotation, to separate plastics from metals (Moyo et al. 2020).

All the steps separate the metals into ferrous and nonferrous metals. Among these, Cu, which is a precious metal, can be separated and recovered mainly by magnetic and eddy current separation techniques. But generally, density gradients and eddy current separation are not effective for non-ferrous precious metals (Vermesan et al. 2020). Once the disassembly and various treatment methods are performed, the precious metal separated needs refining and purification.

16.4.2.3 Refinement: Porphyrin Polymers

Porphyrin, derived from the basic organic molecule porphine, consists of four pyrrole units establishing connecting links with the element pyrrole nuclei. Porphyrins linked in a polymer possess pores, making it easy to trap the metal, which further helps in the recycling process. The coordination polymers that consist of porphyrin are applicable to optical and electronic devices. One major advantage of using porphyrin polymers is their high binding affinity (scavenging) and selectivity for precious metals (Dolphin, 1978). Porphyrin and its derivatives are identified for their activity as photosensitizers and photoredox catalysts (Rybicki-Jasinska et al. 2016). In one of the studies conducted by Hong et al. 2021, they reported economic and efficient porphyrin-based COPs (covalent organic polymers) for valuable metal recapture. These units were connected by a fast one-step FC (Friedel–Crafts) reaction via three separate linkers, i.e. CHCL3 (chloroform), DCE (1,2- dichloroethane), and DCM (dichloromethane) to produce three polymers of COP-210, COP-211, and COP-212, respectively. Porphyrin polymers, especially COP-212, were applied in recovering precious metals from e-waste. These polymers exhibited enhanced selective adsorption for valuable metals such as Au, Ag, and Pt in a metal test where acid was used for 30 different metals. In one of the similar studies, circuit boards from a junkyard were kept in an acid bath to recapture precious metals. After this, it was mixed with polymer and kept for stirring for a few days (Although other tests showed that 99 percent of gold can be scavenged in about 30 minutes.) This was filtered, and polymer and gold were separated. The addition of acid again caused precipitation of gold in the form of nuggets, comprising 94% of gold recaptured from wasteyard circuit boards.

Nguyen et al. (2020) synthesized a porphyrin-phenazine-based polymer, and these polymers had a rise in the selectivity toward gold and other precious metals. These polymers could recover gold 7-fold more than the theoretical estimate based on their photochemical adsorption reduction mechanism. The main advantage of this method is that the adsorbents are robust and recyclable.

16.4.3 Photocatalysis

It is a chemical process that happens as the acceleration of a photoreaction in the presence of a catalyst. Generally, Semi-conductors or transition-metal oxides, for example, TiO_2, ZnO, WO_3, etc., have no energy levels to induce the joining of photogenerated electrons (e-) and holes (h+) in their structure, for this reason, they are used as a photocatalyst. A higher photocatalytic removal of Au was performed by the combination of TiO_2/SiO_2 than bare TiO_2. The occurrence of methanol (hole scavenger) stimulates the accumulation of gold on the surface of the used photocatalyst (Kunthakudee et al. 2022). Another photocatalyst used was polyoxometalate (W 100324-) hybrid under the pH value of 3–6. Recapturing of gold using a photocatalyst can be achieved by optimizing the amount of sacrificial agent and loading of catalysts. Recapturing of gold was achieved by using commercial TiO_2 as photocatalyst (see Figure 16.4), with the loading of 4 g/L and pH of 6.5 due to its increased surface area. By this method, complete gold was recovered within 5–15 min in the presence and absence of mixed complexing agents (Kunthakudee et al. 2022).

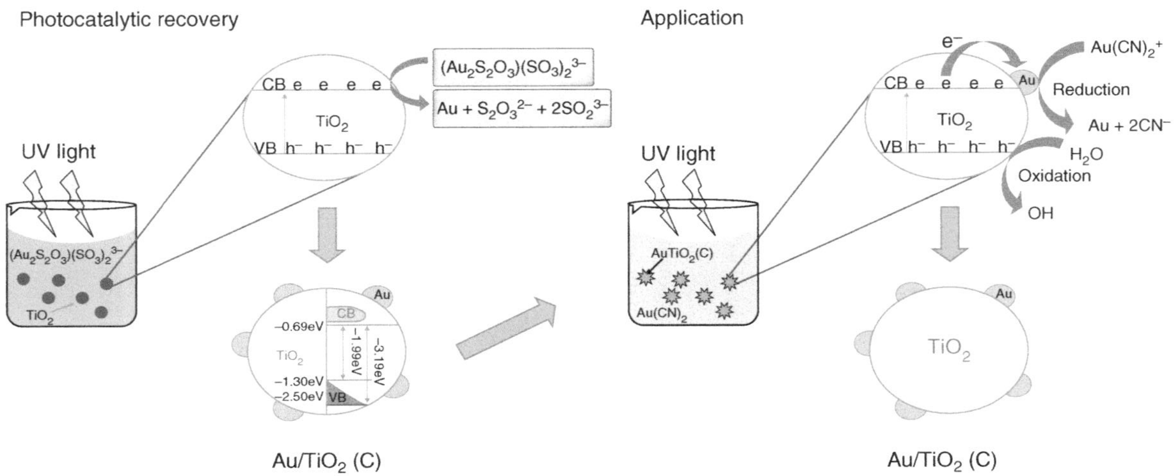

Figure 16.4 Recapturing of gold by the photocatalytic method based on TiO_2/SiO_2 combination.

There are different photocatalysts based on their function, which are classified into:

(a) TiO_2 photocatalyst
(b) TiO_2/SiO_2 photocatalyst
(c) Bi-based semiconductor photocatalyst
(d) Nonoxide photocatalysts: CuS, CdS, and ZnS catalysts

All the photocatalysts function differently in the valuable metal recovery from e-waste, but the suitable method employed is TiO_2 based on the enormous research conducted. However, there are challenges associated with this method such as less quantum efficiency, less recyclability, and less efficiency in utilization of sunlight. Henceforth, there is a need to develop a strong and efficient photocatalyst to improve this technology.

16.4.4 Pyrometallurgy

Pyrometallurgy is a traditional method to recapture valuable and nonferrous metals from e-waste. In this method, waste material is crushed and then liquefied in a furnace at high temperatures to remove plastics. The pyrometallurgy process not only comprises crushing but also sintering, smelting, incineration, and chemical reactions in the gas phase (see Figure 16.5). This technique is being used to recover gold from e-waste for many years now. Impurities are removed using heat (volatilized by chemical reaction) or may be converted to slag (comes at the surface of molten metal) or sludge (sinks at the bottom) (Hoffmann, 1992; Lee et al. 2007; Sum, 1991). Among all feasible techniques, pyrometallurgy is considered as a favorable approach for rapid and large-scale recovery due to its independency of nature and composition of material. As a conventional technology, pyrometallurgy has been used for the recovery of gold from old materials and by the chlorination process, mainly using calcium chloride. This method could recover Au close to 99% at a temperature of 873 K and at a reaction time of 90 min (Ojedaa et al. 2009).

An overview is given in Figure 16.5, about the pyro- and hydro-metallurgy processes involved in e-waste recycling in the recovery of valuable metals (Rao et al. 2020).

16.4.4.1 Process of Pyrometallurgy

The pyrometallurgical method comprises smelting in a plasma arc furnace, sintering, and melting waste at high temperatures. This technology includes mainly two steps: roasting and smelting. At a very high temperature below the melting point, e-waste is processed in roasting, while in smelting, the material is completely melted at a high temperature, which is further parted into two separate liquid layers. One layer may have the metals ready for the refining process

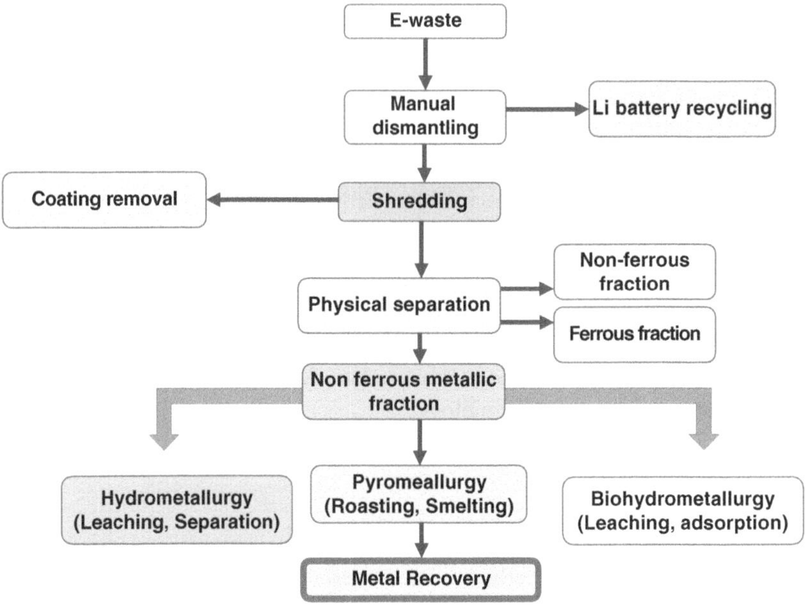

Figure 16.5 Summary of the process of pyrometallurgy in recovery from e-waste.

(Li and Zeng, 2012; Syed, 2012). In pyrometallurgy, pre-treatment of the e-waste is not much required other than disassembly and crushing of the components (Rao et al. 2020). Because of the high content of copper found in circuit boards, the outcome from the smelter for e-waste is considered as copper bullion. After parting from the waste and electrowinning, the copper is recaptured, which is further refined for other precious metals present on circuit boards.

16.4.4.2 Limitations and Drawbacks of Pyrometallurgy

Pyrometallurgical recycling processes for electronic-waste recycling, even though they are cost-effective, this technique occurs as a challenge for the environment due to toxic fumes released into the environment. This technology has less efficient selectivity for specific metals, and it requires many phases of purification and refining, which are time-consuming and needs extensive labor too. Some of the other issues associated with this method are loss of valuable metals, loss of precious metals, inconvenient recapturing of precious metals, and a large amount of slag generation. Another major issue is the less efficient and unsafe recycling of plastic parts of PCBs, mobile phones, etc. Moreover, during the process of smelting dioxin is released from the plastic waste, which is highly dangerous to the environment.

16.4.5 Hydrometallurgy

Hydrometallurgy is a process that occurs in a similar manner to the pyrometallurgical process. This is a very traditional method for extraction and purification of metals from ores, and it is another important technique employed in the recovery and refining of valuable metals from waste. The green metallurgical process is another method for recapturing precious metal from e-waste. For example, Yang et al. 2017 recovered tin via co-processing of waste circuit boards. For sustainability of environment, such green metallurgical processes should be developed at industrial scale.

The hydrometallurgical method comprises a sequence of steps involving the leaching of gold from gold-containing waste. Disassembly and purification methods, for example, solvent extraction, adsorption, precipitation of impurities, ion exchange, etc., to concentrate metals of interest from the waste are followed by electrowinning, cementation, crystallization, or chemical reduction (Sadegh et al. 2007).

Wang (1981), applied the hydrometallurgy process to extract precious metals, for example, Au, Ag, Te, Se, etc., from copper anode sludge. The hydrometallurgy process was used to leach precious metals like Au, Ag, Se, and tellurium (Te) from copper anode sludge. By this process, anode sludge without copper is extracted as Ag, Se, and Au. First, the copper-less leaching process is initiated in nitric acid at a temperature of 40–115°C. Ag, Se, and Te are melted at this stage; the waste leached into nitric acid is treated at a high temperature of 40°C to 110°C with aqua regia to dissolve Au. At this step, precious metals along with other metals are leached. Impurities and residues are removed by oxalic acid, and gold is extracted from an organic extractor. Finally, the retrieved gold is then washed with alcohol such as ethanol and distilled water to extract pure sponge. Up to 99% of gold was recovered using this method.

The recapturing of gold by e-waste was almost 99.8% in the acid-leaching process (Ghasem and Khoramnejadian, 2015). Another method was coupling the hydrometallurgical process with polyaniline powder, films, cotton fibers, etc., and the recovery of the precious metals was around 90% (Wu et al. 2017).

The benefit of the hydrometallurgy method is that it requires less capital, is more environment-friendly as compared to pyrometallurgy, offers a high potential for selective metal recovery, and results in a significant increase in high-purity metals. In the conventional mining process, hydrometallurgy is more suitable for the recovery of metals from mixed-metal, lower-grade ores as compared to pyrometallurgical routes. Nevertheless, challenges are also there such as strong acids required for leaching, organic solvent loss, and chemical reagent loss during the separation process.

16.5 Biorecovery of Valuable Metals from Electronic Waste

Though electronic and electrical equipments are devices that have a short life span but generate a large fraction of waste called e-waste or WEEE. Among all extraction processes like physical, chemical, and biological methods, biological extraction methods are mostly preferred because they are economically and ecologically efficient processes for the recovery of metals.

16.5.1 Microbial Mobilization

Microbial mobilization exploits microorganisms and their metabolites for the solubilization of metals into the aqueous phase from mineral ores, metal concentrates, and e-waste. Microbial-mediated methods are easy to regulate because microorganisms can be maintained at ambient temperature and pressure under mild conditions.

16.5.1.1 Extraction Through Biologically Mediated Reactions

Bioleaching or biomining is a fundamental process in biohydrometallurgy, where the metals are converted by low-grade ores to soluble metal concentrations by utilizing microorganisms. Native microorganisms in the mine environments are helpful in metal extraction from ores (Brierley and Brierley 2013). Kennecott Mining Company got its first patent in 1958 for the bioleaching process using *Acidithiobacillus ferrooxidans* (reclassified from *Thiobacillus ferrooxidans*) for the recapturing of Cu metal from low-grade ore. The leaching solutions used in the above process are ferric sulfate ($FeSO_4$) and sulfuric acid (H_2SO_4). Rio Tinto mines in the Iberian Peninsula were one among many companies that used microorganisms on a large scale for metal extraction1961. Lo Aguirre mine, Chile, initiated the commercial applications of biohydrometallurgy in 1980, which operated for 16 years (1980–1996).

Nowadays, bioleaching plays a prominent role in the large-scale production of base metals such as copper, lead, cobalt, nickel, molybdenum, and zinc, as well as metalloids, including gallium, arsenic, and antimony, and platinum group metals like rhodium, palladium, iridium, rubidium, and osmium (Watling, 2015).

16.5.1.2 Principles and Mechanism of Microbial Leaching

(i) *Bioleaching of metals*:
Bioleaching is known as a microbially catalyzed metal leaching process, whereas bio-oxidation is referred to as mineral oxidation by microorganisms. Acidophilic microorganisms survive in low pH environments like natural waters, corrosive sewer pipes, caves forming snottites (slimy mats

of bacteria), hydrothermal vents, and geysers. Acidophiles obtain their energy through iron and sulfur oxidation. On the other hand, heterotrophic cyanide-generating microorganisms have a higher affinity for binding transition metals. Examples: *Chromobacterium violaceum*, *Pseudomonas fluorescens*, *Pseudomonas aeruginosa*, *Pleurotus ostreatus*, and *Chlorella Vulgaris* are used in precious metal extraction (Işíldar et al. 2017). Three methods involved in bioleaching processes are illustrated in Figure 16.6.

(ii) *Thiosulfate and polysulfide mechanisms for dissolution of metal sulfides*:
All elemental metals dissolve differently due to their different chemical compositions. Dissolution is generally achieved by combining acidic leaching (proton attack) and oxidation processes (Vera et al. 2013).
The thiosulfate mechanism is for acid-insoluble sulfides, such as pyrites (FeS_2), depending on the ferric ion oxidation in the solution (see Figure 16.7). Similarly, acid-soluble sulfides follow the polysulfide mechanism, e.g. chalcopyrite ($CuFeS_2$). The conversion of Fe^{3+} to Fe^{2+} ions is usually catalyzed by iron-oxidizing bacteria, where the mineral leaches metal cations and transient sulfur compounds. An additional proton attack is performed in the thiosulfate mechanism, where sulfur-oxidizing bacteria oxidize the sulfur compounds. The main reaction products of thiosulfate and polysulfide mechanisms are elemental sulfur and sulfuric acid, respectively (Işíldar et al. 2017).

(iii) *Physical contact mechanisms*:
Mineral dissolution of metals takes place directly by the cell's metabolism or subsidiarily from the cell's metabolic products. Direct or contact mechanisms make close contact with the mineral, and the bacteria get attached to the surface and gain electrons straight from the mineral by enzymatic reactions. The direct mechanism is also evident by the release of synthetic sulfides free of iron, which is possible when bacteria come in contact with the surface of waste material. In the indirect or noncontact mechanism, mineral oxidation catalyzed by the microorganisms produces a leaching agent. In cooperative leaching, direct and indirect methods are included where oxidizing chemicals and microorganisms adhered to the surface leach out the metals.
The metal oxidation evolved from the metabolic processes of iron- and sulfur-oxidizing and acidophilic chemolithotrophic microorganisms (*Acidithiobacillus thiooxidans*, *Acidithiobacillus ferrooxidans*, *Acidithiobacillus caldus*, and *Sulfolobus acisocaldarius*). During bioleaching, microorganisms metabolize lixiviants through direct or indirect electron transfer, creating a reaction center between the mineral surface and the cell wall. The EPS (extracellular polymeric substances) released from microorganisms also may build a reaction space, where many advanced bioleaching chemical reactions occur inside this EPS layer as compared to the solution (Işíldar et al. 2017), as depicted in Figure 16.8.

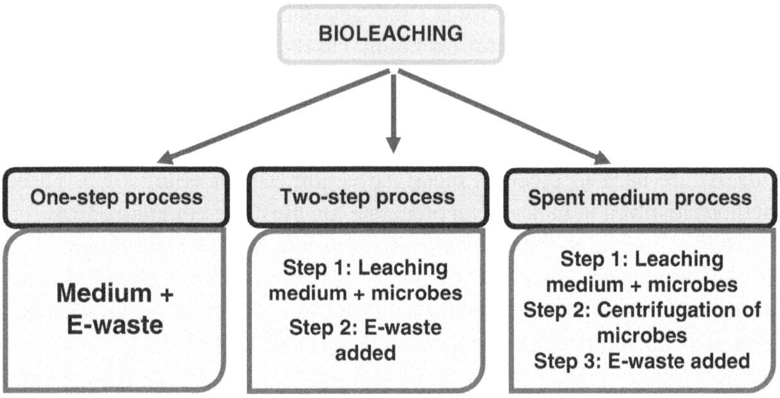

Figure 16.6 Types of bioleaching processes.

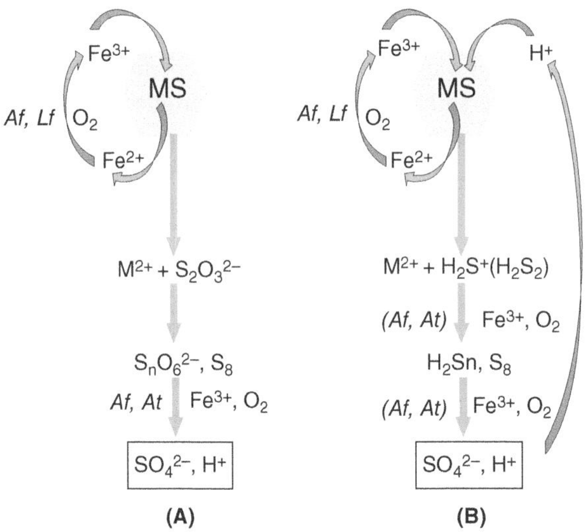

Figure 16.7 Metal sulfides dissolution using (A) thiosulfate and (B) polysulfide mechanisms.

16.5.2 Metal Mobilization Mechanism

Metal mobilization relies upon microbially catalyzed processes for metal extraction from different raw materials. The three primary mechanisms of metal mobilization are acidolysis, radiolysis, and complexolysis (see Figure 16.9).

(i) *Acidolysis*:

Acidolysis is carried out by many autotrophic sulfur oxidizers (*Acidithiobacillus ferrooxidans*), bacteria, and heterotrophic fungi. Microorganisms produce metabolic acids such as acetic, formic, pyruvic, citric, and gluconic acids, which help dissolve metals. The above acids weaken the metal ion bonds by donating protons to the oxygen atoms present on the exterior of the

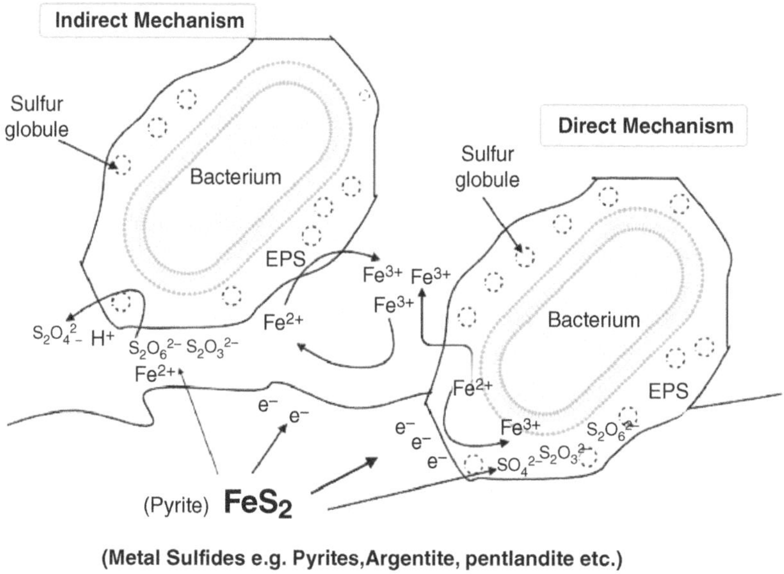

Figure 16.8 Direct and Indirect mechanisms of bioleaching.

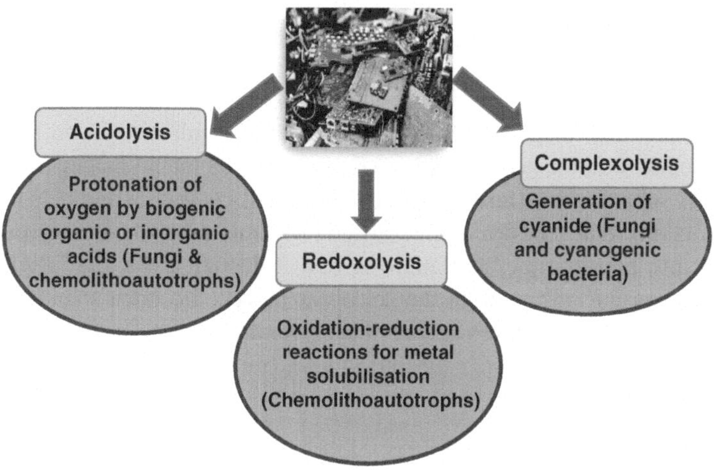

Figure 16.9 Types of metal mobilization mechanisms. Source: KPixMining/Adobe Stock.

metal. After the oxygen protonation, the metal ions interact with water and get converted into soluble metals. H_2SO_4, an inorganic acid formed by microorganisms from the external addition of ferric and ferrous sulfates, can also protonate the oxygen atom similarly to the metabolic acids.

$$Fe_2O_3 + 2H^+Fe2^+ + H_2O$$

(ii) *Redoxolysis*:

This is an oxidation–reduction method where microorganisms can reduce ferric iron (Fe^{3+}) to ferrous iron (Fe^{2+}) under anaerobic conditions. In this process, Fe^{3+} acts as an oxidizing agent, leading to metal dissolution. It is a cyclic process in which iron oxidizers produce ferric iron (Fe^{3+}), which in turn is converted to Fe^{2+} (ferrous iron) during the bioleaching process, then re-oxidized to Fe^{3+} by iron oxidizers. The Fe^{3+} is recycled constantly by the oxidation–reduction reaction so that Fe^{3+} can be reused for further metal solubilization.

The Redoxolysis reactions of Cu, Zn, and Ni under normal conditions are given in the following equations:

$$Cu^0 + 2Fe^{3+}Cu^{2+} + 2Fe^{2+}(\Delta G^0 = -347.1\ kJ)$$

$$Ni^0 + 2Fe^{3+}Ni^{2+} + 2Fe^{2+}(\Delta G^0 = -822.6\ kJ)$$

$$Zn^0 + 2Fe^{3+}Zn^{2+} + 2Fe^{2+}(\Delta G^0 = -1235.9\ kJ)$$

(iii) *Complexolysis*:

Complexolysis is ligand-induced solubilization; here, the formation of chelates and metal-ligand formations lead to the leaching of metal ions in the solution. Ligands or complexing agents involved in this process include siderophores, cyanides, and organic acids, which many bacteria and fungi secrete. Siderophores are one of the strongest soluble irons (Fe^{3+}) binding agents, and they also bind and solubilize other metals. The best examples of ligand and metal complexes are tartaric acid with Fe, Mg, and Al, citric acid with magnesium (Mg), and oxalic acid with Al, Fe, and Mg. Heterotrophic cyanide-generating bacteria help recover noble metals from secondary sources by the formation of metal-cyanide complexes and help in metal solubilization. The cyanogenic bacteria will automatically decompose cyanide into nontoxic β-cyan alanine so that it doesn't cause any health hazards. Some metal ions from the acidolysis process are often stabilized through a complexolysis mechanism (Işíldar et al. 2017).

The dissolution of Au is given in the following equations. This reaction is also known as Elsner's equation:

$$4Au + 8CN\ 4Au(CN)^2 + 4e^-$$

$$O_2 + 2H_2O\ 4e^- + 4OH^-$$

$$4Au + 8CN + O_2 + 2H_2O\ 4Au(CN)^2 + 4OH$$

16.5.3 Microorganisms Involved in Bioleaching

A diverse group of microorganisms that use inorganic or organic compounds for energy production perform the bioleaching process (Bharadwaj and Ting, 2012).

Some bacteria present naturally in ores or a metal-rich environment flourish on the surface of the metal-containing materials. The main groups of microorganisms involved in the bioleaching process are (Ahonen and Tuovinen, 1989) Archaea and bacteria, which are chemolithotrophs, and (Akcil et al. 2021) heterotrophs, including bacteria and fungi (Johnson and Du Plessis, 2015).

16.5.3.1 Chemolithoautotrophs

In both processes of acidolysis and Redoxolysis, chemolithoautotrophs are primarily included through inorganic acids such as H_2SO_4 production and the cycling of Fe^{2+} and Fe^{3+}respectively. These are usually acidophilic autotrophic prokaryotes using inorganic carbon as a carbon source and hydrogen or sulfur electrons from inorganic compounds as a source of energy. Examples of chemolithoautotrophs are *Acidithiobacillus ferrooxidans*, *Metallosphaera sedula*, *Leptospirillum ferrooxidans*, *Acidithiobacillus thiooxidans*, *Sulfobacillus thermosulfidooxidans*, and *Acidianus brierleyi*, usually grow at 45–75 °C temperatures (Paul Chen and Lim, 2005).

The following table illustrates the list of microorganisms involved in the leaching process, and the conditions required to carry out the process are also given (Table 16.1).

16.5.3.2 Heterotrophs

Heterotrophs endure a broad pH range and are used in the treatment of abstemiously alkaline wastes (Natarajan et al. 2015). Cyanogenic heterotrophs (*Pseudomonas flourescens, Pseudomonas aeruginosa, and Chromobacterium violaceum*) act as ligands in the complexolysis process and utilize organic components as carbon sources and simple molecules of water as energy and electrons sources (Gu et al. 2015). These heterotrophs have the power to convert Fe^{2+} to Fe^{3+} through the redoxoloysis mechanism (Hassanshahiana et al. 2019).

Heterotrophic fungi use organic acids such as gluconic acid, oxalic acid, and citric acid as bioleaching agents. These organic acids stimulate metals to leach out from ores and waste materials through acidolysis and redoxolysis mechanisms. The most common fungi used in the bioleaching process are *Aspergillus niger* and *Penicillium simplicissimum*. Fungal bioleaching occurs at a higher pH of 9.0–10.5.

16.5.4 Bioreactors used for Bioleaching

Bioprocessing metals from crucial ores has evolved into a profitable method with the help of bioreactors (Brierley and Brierley, 2013). Bioreactors used for metal extraction should be selected based on the system's physical, chemical, and biological characteristics. One of the key criteria for the reactor should be the autocatalytic nature of microbial growth. According to this, the first choice would be a continuous stirred tank reactor (CSTR). CSTR is a batch reactor with an impeller

Table 16.1 Microorganisms used for bioleaching of precious metals.

Mechanism	Microorganism	pH
Archaea		
Redoxolysis (Fe^{3+}) and Acidolysis (H_2SO_4),	*Acidianus Brierleyi* (70 °C)	1.5–2
	Ferroplasma acidarmanus (42 °C)	0.5–1.2
	Metallosphaera sedula (75 °C)	1.0–4.0
	Metallosphaera yellowstonensis (65 °C)	1.0–4.5
	Sulfolobus acisocaldarius (55–85 °C)	1.0–4.5
	Sulfolobus solffataricus (55–87 °C)	2.5–3.5
	Sulfolobus metallicus (50–75 °C)	2.0–3.0
	Thermoplasma acidophilum (60 °C)	1.0–2.0
Acidolysis (H_2SO_4)	*Metallosphaera hakonenisis* (55–80 °C)	1.0–4.0
	Metallospharea Prunae (55–80 °C)	1.0–4.5
	Sulfolobus yangmeingensis (80 °C)	4.0
Redoxolysis (Fe^{3+})	*Ferroplasma cupricumulans* (54–63 °C)	1.0–1.2
	Ferroplasma acidiphilum (15–45 °C)	1.3–2.2
Bacteria		
Acidolysis (H_2SO_4) and Redoxolysis (Fe^{3+})	*Acidithiobacillus ferrooxidans* (28–35 °C)	2.5
	Sulfobacillus sibiricus (45–55 °C)	1.7–2.0
	Sulfobacillus thermosulfioxidans (40–60 °C)	1.7–2.0
Complexolysis (CN) and Acidolysis (citrate),	*Pseudomonas putida* (25–35 °C)	7.0–8.2
Acidolysis (H_2SO_4)	*Acidithiobacillus caldus* (45–50 °C)	2.0–2.5
Redoxolysis (Fe^{3+})	*Acidimicrobium ferrooxidans* (45–50 °C)	2.0
	Ferrimicrobium acidiphilum (35 °C)	1.4–2.0
	Ferrithrix thermotolerans (43–50 °C)	1.6–1.8
	Ferrovum myxofaciens (20–30 °C)	1.0–2.0
	Leptospirillum ferriphilum (42 °C)	1.2–1.6
	Leptospirillum ferrooxidans (28–35 °C)	1.8
	Leptothrix discophora (15–40 °C)	5.8–7.8
Complexolysis (CN)	*Chromobacterium violaceum* (25–37 °C)	7.0–7.5
	Pseudomonas aeruginosa (25–35 °C)	7.0–7.8
	Pseudomonas fluorescens (25–35 °C)	7.0–7.8
Fungi		
Acidolysis (citrate) and complexolysis (CN)	*Aspergillus awamori* (28 °C)	6.8–7.2
	Aspergillus fumigatus (30 °C)	6.6–7.2
Acidolysis (organic acids)	*Aspergillus flavus* (30 °C)	6.5
	Aspergillus niger (30 °C)	6.4–7.3
	Caldosporium oxysporium (30 °C)	6.5
	Penicillium chrysogenum (30 °C)	6.7
	Penicillium simplicissimum (30 °C)	6.0–7.0

Note: Mesophilic (15–40 °C); moderately thermophilic (41–60 °C); thermophilic (61–75 °C); extremely thermophilic (76–88 °C).

and other mixing devices that can be operated at a low cost. Other bioreactors are tubular plug flow reactors, percolation columns, air-lift columns, Pachuca tanks, and rotary reactors.

Harvie and de Mello (2021) studied the selective extraction of copper ions from aqueous solutions using a flow reactor with a 98% yield. This process completes in two steps with the help of liquid-liquid separators and passive mixers. The first stage is an organic phase containing the extractant molecules to which selected target metal ions are transferred. In the latter step, recirculation of metal ions takes place due to the presence of acid, which strips the metal ions from the organic phase and regenerates extractant molecules, which results in recirculation through the reactor.

16.5.5 Biosorption of Precious Metals

Biosorption is an innovative and cost-effective technique of metal accumulation by microbial biomass. Certain types of inactive or dead microbial biomass bind and concentrate metals from aqueous solutions and industrial effluents. Biosorption occurs within the cell wall or through many metabolites (polysaccharides, metal-binding peptides, and EPS) . The reactive groups present in the cell wall, such as hydroxyl groups, phosphoryl groups, and carboxylamine, play a major role in biosorption (Ilyas and Lee, 2014).

Algae, bacteria, yeasts, and fungi act as biosorbents for precious metals. *Fucus vesiculosus*, the dead biomass of brown algae, was an efficient biosorbent for recovering metallic gold nanoparticles. Silver biosorption was also exhibited by *Pseudomonas mendocinu* bacterial strain, isolated from a copper mine. Biopolymers also serve as potential biosorbents for metal recovery. The deacetylated derivative compound of chitin, i.e. Chitosan, is the second most commonly found biopolymer in the world. GCC (Glutaraldehyde-crosslinked chitosan), sulfur-grafted chitosan known as RADC, and crosslinked chitosan resin chemically modified with l-lysine are used as biosorbents for Au (III) from aqueous solutions. Other substances used as biosorbents for Au^{3+} recovery are rice husk carbon, bio thiourea derivative of resins, condensed-tannin gel, alfalfa, hen eggshell membrane (ESM), and de-alginate seaweed waste.

16.5.6 Biomineralization

Biomineralization is a process in which microorganisms form minerals, which help in metal solubilization, providing a means of detoxification and biorecovery (Gadd and Pan, 2016). *Cupriavidus metallidurans*, a spore-forming bacteria, usually isolated from gold-copper mines, stores inert gold (Au) nanoparticles inside its protoplasm to prevent themselves from gold (Au) metal solubilization. *Delftia acidovorans*, a gold-resistant bacterium, formerly known as

Comamonas acidovorans or *Pseudomonas acidovorans*. It secretes a metabolite called delftibactin against toxicity, enabling the biomineralization of elemental Au^0 (Johnston et al. 2013).

A novel mechanism of Pd(II) biomineralization was catalyzed chiefly by a molybdenum-containing enzyme in the presence of oxygen by cultures of *Escherichia coli* (Foulkes et al. 2016). Wild-type strains could biomineralize the metal within 30 minutes, whereas bacterial cultures without molybdoenzymes were able to reduce the palladium in 7 hours. *Phanerochaete chrysosporium*, a white-rot fungus, helps biomineralize palladium nanoparticles from e-wastes (Tarver et al. 2019). This method is a feasible option for the selective recapturing of precious metals from mixed e-wastes. Additionally, bacterial cultures are unique to distinct metals, which advances this technique as compared to other techniques.

16.6 Factors Affecting Biorecovery of Precious Metals

Though there is increasing demand for electronic goods in the global market post-pandemic, there are several factors due to which the recovery of precious metals doesn't correlate with its potential. According to Mijke Hertoghs (UN International Telecommunication Union), a major hurdle in establishing an efficient recovery system is that there is no proper collection of e-waste. Extraction of precious metals from tailing gold mine sites, e-waste, and other modern sources is often difficult and unfeasible due to the low level of gold in gold-entrapped sulfides (Gericke and Pinches, 2006; Natarajan et al. 2001, Reith et al. 2007; Marsden and House, 2006). There are several other factors mentioned below that affect the bioleaching process.

16.6.1 Oxygen Supply

An adequate concentration of oxygen is needed for the optimum growth of bacterial cultures used for the recovery of metals if aerobic cultures are used. On a laboratory scale, this can be achieved by shaking and stirring, but on a large scale or in the case of dump or heap leaching this is difficult to achieve.

16.6.2 pH

Microbial cultures used for bioleaching generally prefer a low pH value so for optimum growth pH generally ranges from 2.0 to 2.5. However, bacterial populations are adaptive in nature so some of the bacteria, e.g. *T. ferrooxidans*, tolerate pH changes to some extent (Tuovinen and Kelly, 1973).

16.6.3 Mineral Substrate

Constituents present in e-waste material directly affect the growth of microorganism, especially carbonates present in plastics (e.g. Blends of Polycarbonates, Francisco, and Sigbritt, 2008). E-waste releases carbonates, which increase pH and retard the growth of bacteria. The addition of acid from outside can solve this purpose, but this further causes an increase in the cost of leaching and precipitate formation.

16.6.4 Nutrients

Generally, chemolithoautotrophic bacteria for metal extraction of sulfide materials are used for the bioleaching process, where they need only inorganic substances for growth. Generally, the minerals are obtained from waste material to be leached but for optimum growth of bacteria used for bioleaching, some supplements, for example, phosphate, ammonium, and magnesium salts can be added.

16.6.5 Temperature

Bacteria have a diverse range of optimum temperatures. Bactria used in iron and sulfide oxidation, e.g. *T. ferooxidans* grows between 28 and 30 °C. However, metal solubilization of cobalt, copper, zinc, and nickel has been observed by a few researchers at as low as 4 °C (Ahonen and Tuovinen, 1989).

16.6.6 Presence of Organic Surfactants and Extractants

The presence of organic surfactants and extractants at the surface of waste decreases the bioleaching rate as it causes a decrease in the mass transfer of oxygen and surface area (Torma et al. 1976; Torma and Itzkovitch, 1976; Bosecker, 1995). For this reason, solvent extraction is currently preferred, but it enhances the level of the aqueous phase in solvents, which should be removed before recirculating the solvents in the reactor.

16.6.7 Concentration of Heavy Metals

A high concentration of heavy metals may play an inhibitory role in bioleaching as microorganisms cannot survive at high concentrations of heavy metals. However, in some of the cultures especially, Thiobacilli have been reported to exhibit tolerance to heavy metals, and research has shown that few strains can even tolerate 55 g/l Cu, 50 g/l Ni, and 112 g/l Zn. Additionally, few strains adapt to a higher concentration of heavy metals if the concentration of metal or substrate is increased gradually (Bosecker, 1977).

Methods used for extraction are another contributing factor to the low recovery of the precious metal, as conventional extraction methods like pyrometallurgical and hydrometallurgical processes, which have many limitations, generate hazardous emissions (Khaliq et al. 2014: Kaya, 2016: Cui and Zhang, 2008) and cyanide used in hydrometallurgical processes is highly toxic (Cui and Zhang, 2008). Several other factors such as temperature, humidity, pH, chemical atmosphere, and mixed composition with other waste items at e-waste dumping sites, also play a role in metal recovery.

To optimize the recovery of precious metals such factors should be taken into consideration such as physical separation (Chehade et al. 2012) and chemical pretreatment (Sun et al. 2012) of e-waste could be employed before any extraction method for enhanced recovery.

16.7 Confirmatory Tests for Recovered Metals from E-waste

After collection of precious metals, these can be tested with many common methods such as Touchstone and acid test, Aqua regia, Fire assay and cupellation method, electronic gold detectors, and X-Ray fluorescence tests. These methods are simple yet effective in testing recovered metal. Other advanced techniques like Gamma Activation Analysis are also employed to detect precious metals from waste sources. This involves irradiating samples with X-rays varied from a few tens to MeV. In this technique, photons are generated with energies equal to electron beam energy by accelerating electrons onto a metal surface. In the target element, sufficient energy X Rays can induce various nuclear reactions. Radioisotopes produced from these nuclear reactions can be measured by counting decay products. The diversity of the range of radioactive material's half-lives offers flexibility to select the duration of irradiation and energy of incident beams giving the advantage of using this technique to find out different elements after recovery.

16.8 Biorecovery and Environment Sustainability

Processes employed to recover valuable metals from waste such as the Hydrometallurgical process involves extensive leaching time up to 1–2 weeks, apart from this, substances released are sometimes toxic to microorganisms. Process optimization and improvement in kinetics could enhance recovery efficiency and recovery period (Isildar et al. 2017) In addition to this, the adaptive nature of microorganisms provides a further scope of improvement as these can tolerate high concentrations of metals and toxic conditions in a bioreactor, in addition to it novel biotechnological tools can be implied to increase biorecovery of precious metals (Isildar et al. 2017). These biotechnological tools like genetic engineering, studying the mechanism of genes involved in bioleaching, and adapting microorganisms to harsh environments can also be implied to optimize and scale up biorecovery. Chalcopyrite is one such example where thermophilic microorganisms are used to scale up the bio-recovery (Gahan et al. 2012).

Biorecovery using microorganisms offers a promising future to collect precious metals from e-waste or other unconventional sources as biogenic leaching agents released from bacterial activities from metal hydrophilic compound formations. Operation costs, energy input, and skilled labor requirements are less in this as compared to other recovery methods. The biggest advantage of using the biorecovery method is that it has less detrimental effects on the environment due to efficient effluent detoxification. Further studies on optimization and using advanced tools of biotechnology will contribute to making the biorecovery of precious metals a more favorable and eco-friendly process for environmental sustainability.

16.9 Biorecovery and Socio-economic Sustainability

In the post-Covid-19 time period, there has been an increase in e-waste generation globally due to an increase in demand for IT-based applications and solutions. However, the popularization of biomining shows a positive commencement of sustainable biorecovery methods. The study of genes involved in the biorecovery process of precious metals, for example, platinum, silver, and gold, has opened up a new segment of metabolic engineering of these harvesters by altering their pathways and gene expression. Bioinformatics (e.g. high-throughput analytical technologies) has opened the possibility to explore diverse groups of bacteria that have a higher capacity for bioleaching, which could be a crucial field of research for the recovery of precious metals from e-waste. The study of changing microbial dynamics and interspecies crosstalk is the area of future study to make the recovery of precious metals not only an eco-friendly process but also a socio-economical process.

16.10 Conclusion

Diverging e-waste towards recovery is very important, as not only does it emit tons of carbon dioxide but also creates a nonbiodegradable, polluted landfill and generates toxic scrap. To produce one ton of gold, 10,000 tons of CO_2 are generated, which increases the carbon emission load. The best way to leach precious metals from e-waste is to make a strategy that can selectively leach out metals from waste. For example, Silica and DMAEMA (N, N-dimethyl amino ethyl methacrylate) with covalently bonded imidazole on polyethylene-coated skin core fiber have shown their potential in selective adsorption of precious metals from waste (Kang et al. 2004: Liu et al. 2017). In another approach, cyclodextrin was used to co-precipitate gold particles (Liu et al. 2013), and using microorganisms that can selectively extract precious metals from waste, e.g. *Trichoderma harzinum* was used to extract silver from waste dump rocks (Cecchi et al. 2017) and *Rhodopseudomonas palustris* strains (purple non-sulfur bacteria) was used to leach out Ruthenium specifically (Colica et al. 2012).

Study and exploration of new methods for collecting exquisite metals from e-waste are not only justified because of their environmental impact but also for the value and profit generated from the waste material and significant reduction in resource utilization and energy consumption. However, there are several factors in the biorecovery of precious metals from unconventional sources such as the availability of efficient bioleaching microorganisms and their survival rates in toxic environments of e-waste. The lack of characterization of microorganisms with high leaching capacity, and the lack of developed techniques for bioleaching such as the availability of genetically modified microorganisms that could efficiently and rapidly leach selective metals from the waste are a few more limitations associated with this method. Nevertheless, this is a preferred method due to its lower environmental impact and newer trends in microbiology like bioinformatics, which could lead to development of genetically engineered strains to perform fast and efficient bioleaching.

References

Ahonen, L. and Tuovinen, O.H. (1989). Effect of temperature on the microbiological leaching of sulfide ore material in percolators containing chalcopyrite, pentlandite, sphalerite, pyrrhotite as main minerals. *Biotechnology Letters* 11: 331–336.

Akcil, A., Ibrahim, Y.A., Meshram, P. et al. (2021). Hydrometallurgical recycling strategies for recovery of rare earth elements from consumer electronic scraps: a review. *Journal of Chemical Technology and Biotechnology* 96: 1785–1797. https://doi.org/10.1002/jctb.6739.

Bharadwaj, A. and Ting, Y.P. (2012). From biomining of mineral ores to bio urban mining of industrial waste. In: . Environmental Technology and Management Conference, 4th, EMTC, Indonesia.

Bosecker, K. (1977). Studies on the bacterial leaching of nickel ores. In: *Conference Bacterial Leaching* (ed. W. Schwartz), 139–144. Weinheim: Verlag Chemie.

Bosecker, K. (1995). Influence of solvent extraction reagents on the activity of *Thiobacillus ferrooxidans* and *Leptospirillum ferrooxidans*. In: *Biohydrometallurgical Processing*, vol. 1 (ed. T. Vargas, C.A. Jerez, J.V. Wiertz, and H. Toledo), 283–291. Santiago de Chile: University of Chile.

Brierley, C.L. and Brierley, J.A. (2013). Progress in bioleaching: Part B: applications of microbial processes by the minerals industries. *Applied Microbiology and Biotechnology* 97: 7543–7552.

Cecchi, G., Marescotti, P., Di Piazza, S., and Zotti, M. (2017). Native fungi as metal remediators: silver myco-accumulation from metal contaminated waste-rock dumps (Libiola Mine, Italy). *Journal of Environmental Science and Health, Part B: Pesticides, Food Contaminants, and Agricultural Wastes* 52: 191–195.

Celik, H. (2004). Extraction of gold and silver from a Turkish gold ore through thiourea leaching. *Mining, Metallurgy & Exploration* 21: 144–148.

Chehade, Y., Siddique, A., Alayan, H. et al. (2012). Recovery of gold, silver, palladium, and copper from waste printed circuit boards. In: *International conference on Chemical, Civil and Environmental Engineering (ICCEE 2012)* March 24–25, 2012. Dubai: Planetary Scientific Research center.

Colica, G., Caparrotta, S., and De Philippis, R. (2012). Selective biosorption and recovery of Ruthenium from industrial effluents with *Rhodopseudomonas palustris* strains. *Applied Microbiology and Biotechnology* 95: 381–387.

Corty, C.W. and Hagelüken, C. (2010). Recycling of gold from electronics: cost-effective use through design for recycling. *Gold Bulletin* 43 (3): 209.

Cui, J. and Zhang, L. (2008). Metallurgical recovery of metals from electronic waste: a review. *Journal of Hazardous Materials* 158: 228–256.

Dimitrijević, S., Trujić, V., Ivanović, A. et al. (2013). Recycling of Precious Metals from E-scrap. *Iranian Journal of Chemistry and Chemical Engineering* 32 (4): 17–23. https://doi.org/10.30492/ijcce.2013.6740.

Dolphin, D. (1978). *The Porphyrins*. New York: Academic.

Foulkes, J., Deplanche, K., Sargent, F. et al. (2016). A novel aerobic mechanism for reductive palladium biomineralization and recovery by *Escherichia coli*. *Geomicrobiology Journal* 33: 230–236. https://doi.org/10.1080/01490451.2015.1069911.

Francisco, V. and Sigbritt, K. (2008). *Macromolecular Materials and Engineering* 293: 274–297.

Gadd, G.M. and Pan, X. (2016). Biomineralisation, bioremediation, and biorecovery of toxic metals and radionuclides. *Geomicrobiology Journal* 33 (3–4): 175–178.

Gahan, C.S., Srichandan, H., Kim, D.-J., and Akcil, A. (2012). Biohydrometallurgy and biomineral processing technology: a review on its past. *Present and Future.* 1 (10): 85–99.

Gericke, M. and Pinches, A. (2006). Biological synthesis of metal nanoparticles. *Hydrometallurgy* 83: 132–140. https://doi.org/10.1016/j.hydromet.2006.03.019.

Ghasem, A.M.H. and Khoramnejadian, S. (2015). The extraction of gold from e-waste by hydrometallurgy. *Oriental Journal of Chemistry* 31: 113–120.

Grant, K., Goldizen, F.C., Sly, P.D. et al. (2013). Health consequences of exposure to e-waste: a systematic review. *Lancet Glob Health* 1 (6): e350–61. https://doi.org/10.1016/S2214-109X(13)70101-3. Epub 2013 Oct 30. PMID: 25104600.

Gu, T., Xu, D., Zhang, P. et al. (2015). Microbiologically influenced corrosion and its impact on metals and other materials. In: *Microbiology for Minerals, Metals, Materials and the Environment*, (1st ed. D. Pandey Bhilashb, and K.A. Natarajan), 383–408. CRC Press, Taylor and Francis Group.

Gungor, A. and Gupta, S.M. (1998). Disassembly sequence planning for products with defective parts in product recovery. *Computers and Industrial Engineering* 35: 161–164.

Gurung, M., Adhikari, B.B., Kawakita, H. et al. (2012). Selective recovery of precious metals from acidic leach liquor of circuit boards of spent mobile phones using chemically modified persimmon tannin gel. *Industrial & Engineering Chemistry Research* 51: 11901–11913. https://doi.org/10.1021/ie3009023.

Harvie, A.J. and de Mello, J.C. (2021). Efficient extraction of metal ions using a recirculating two-stage flow reactor. *Chemistry-Methods* 1 (12): 494–501.

Hassanshahiana, M., Toorani, S., Roghanian, R. et al. (2019). Characterization of heterotrophic iron and sulfur-oxidizing bacteria from acid mine drainage in Sarcheshmeh copper mine, Iran. *Desalination and Water Treatment* 147: 264–272. https://doi.org/10.5004/dwt.2019.23723.

Hoffmann, J.E. (1992). Recovering precious metals from electronic scrap. *The Journal of The Minerals, Metals & Materials Society (TMS)* 44: 43–48.

Hong, Y., Rozyyev, V., and Yavuz, C.T. (2021). Alkyl-linked porphyrin porous polymers for gas capture and precious metal adsorption. *Small Science* 1: 2000078.

Hubau, A., Chagnes, A., Minier, M. et al. (2019). Recycling-oriented methodology to sample and characterize the metal composition of waste Printed Circuit Boards. *Waste Manag.* 91: 62–71. https://doi.org/10.1016/j.wasman.2019.04.041. Epub 2019 May 3. PMID: 31203943.

Ilyas, S. and Lee, J.C. (2014). Biometallurgical recovery of metals from waste electrical and electronic equipment: a review. *ChemBioEng Reviews* 1: 148–169.

Isildar, A., Rene, E.R., Hullebusch, E.D.V., and Lens, P.N.L. (2017). Two-step leaching of valuable metals from discarded printed circuit boards, and process optimization using response surface methodology. *Advances in Recycling & Waste Management* 2: 132. https://doi.org/10.4172/2475-7675.1000132.

Işíldar, A., Vossenberg, J.V.D., Rene, E.R. et al. (2017). Biorecovery of metals from electronic waste. In: *Sustainable Heavy Metal Remediation*, 241–278. Cham: Springer. https://doi.org/10.1007/978-3-319-61146-4_8.

Johnson, D.B. and Du Plessis, C.A. (2015). Biomining in reverse gear: using bacteria to extract metals from oxidized ores. *Minerals Engineering* 75: 2–5.

Johnston, C.W., Wyatt, M.A., Li, X. et al. (2013). Gold biomineralization by a metallophone from a gold-associated microbe. *Nature Chemical Biology* 4: 241–243. https://doi.org/10.1038/nchembio.1179. Epub 2013 Feb 3. PMID: 23377039.

Kang, T., Park, Y., Choi, K. et al. (2004). Ordered mesoporous silica (SBA-15) derivatized with imidazole-containing functionalities as a selective adsorbent of precious metal ions. *J. Mater. Chem.* 14: 1043–1049.

Kaya, M. (2016). Recovery of metals and nonmetals from electronic waste by physical and chemical recycling processes. *Waste Management* 57: 64–90.

Khaliq, A., Rhamdhani, M.A., Brooks, G., and Masood, S. (2014). Metal extraction processes for electronic waste and existing industrial routes: a review and Australian perspective. *Resources* 3: 152–179.

Kumar, A., Holuszko, M., and Espinosa, D.C.R. (2017). E-waste: an overview on generation, collection, legislation and recycling practices. *Resources, Conservation & Recycling* 122: 32–42.

Kunthakudee, N., Puangpetch, T., Ramakul, P., and Hunsom, M. (2022). Photocatalytic recovery of gold from a non-Cyanide gold plating solution as Au nanoparticle-decorated semiconductors. *ACS Omega* 7: 7683–7695.

Lee, J.-C., Yoo, J.M., Jeong, J., Jha, M.K., 2007. Novel Pre-treatment Process for Liberation of Metals from Waste Printed Circuit Boards Using Organic Solution, International Patent Application Number WO/2009/064063, Dated: 28th July 2009.

Lee, C., Tang, L., and Popuri, S.R. (2011). A study on the recycling of scrap integrated circuits by leaching. *Waste Management & Research* 29: 677–685.

Li, J. and Zeng, X. (2012). *Recycling Printed Circuit Boards*. Woodhead Publishing Limited.

Liu, Z., Frasconi, M., Lei, J. et al. (2013). Selective isolation of gold facilitated by second-sphere coordination with α-cyclodextrin. *Nat Commun.* 4: 1855. https://doi.org/10.1038/ncomms2891. PMID: 23673640; PMCID: PMC3674257.

Liu, X., Ao, J., Yang, X. et al. (2017). Green and efficient synthesis of an adsorbent fiber by preirradiation-induced grafting of PDMAEMA and its Au(III) adsorption and reduction performance. *Journal of Applied Polymer Science* 134: 44955. https://doi.org/10.1002/app.44955.

Loganathan, B.G. and Masunaga, S. (2015). PCBs, dioxins and, furans: Human exposure and health effects. In: *Handbook of Toxicology of Chemical Warfare Agents*, seconde, 239–247.

Marsden, J. and House, I. (2006). *The Chemistry of Gold Extraction*. Littleton, CO: Society for Mining, Metallurgy and Exploration, Inc.

Moyo, T., Chirume, B.H., and Petersen, J. (2020). Assessing alternative pre-treatment methods to promote metal recovery in the leaching of printed circuit boards. *Resour. Conserv. Recycle.* 152: 104545.

Natarajan, K., Modak, J., and Raichur, A. (2001). Bioreactor engineering for treating refractory gold-bearing concentrates: an Indian experience. *Process Metallurgy* 11: 183–190.

Natarajan, G., Tay, S.B., Yew, W.S., and Ting, Y.P. (2015). Engineered strains enhance gold biorecovery from electronic scrap. *Minerals Engineering* 75: 32–37.

Nguyen, T.S., Hong, Y., Dogan, N.A., and Yavuz, C.T. (2020). Gold Recovery from E-Waste by Porous Porphyrin–Phenazine Network Polymers. *Chemistry of Materials* 32: 5343–5349.

Ojedaa, M.W., Perinob, E., Ruiz, M., and Del, C. (2009). Gold extraction by chlorination using a pyrometallurgical process. *Minerals Engineering* 22: 409–411.

Örgül, S. and Atalay, Ü. (2000). Gold Extraction from Kaymaz gold ore by thiourea leaching. *Developments in Mineral Processing* 13: C6–C28.

Panda, S., Costa, R.B., Shah, S.S. et al. (2021). Biotechnological trends and market impact on the recovery of rare earth elements from bauxite residue (red mud)—A review. *Resources Conservation and Recycling* 171: 105645. https://doi.org/10.1016/j.resconrec.2021.105645.

Paul Chen, J. and Lim, L.L. (2005). Recovery of precious metals by an electrochemical deposition method. *Chemosphere* 60 (10): 1384–1392. https://doi.org/10.1016/j.chemosphere.2005.02.001. PMID: 16054907.

Potysz, A., Lens, P.N.L., van de Vossenberg, J. et al. (2016). Comparison of Cu, Zn and Fe bioleaching from Cu-metallurgical slags in the presence of *Pseudomonas fluorescens* and *Acidithiobacillus thiooxidans*. *Applied Geochemistry* 68: 39–52. https://doi.org/10.1016/j.apgeochem.2016.03.006.

Rao, M.D., Singh, K.K., Morrison, C.A., and Love, J.B. (2020). Challenges and opportunities in the recovery of gold from electronic waste. *RSC Advances.* 10: 4300–4309.

Reith, F., Rogers, S.L., McPhail, D., and Brugger, J. (2007, 2006). Potential for the utilization of micro-organisms in gold processing In World Gold. In: *The Chemistry of Gold Extraction* (ed. J. Marsden and I. House), 1–8. Littleton, CO: Society for Mining, Metallurgy and Exploration, Inc.

Richards, J. (2006). "Precious" metals: the case for treating metals as irreplaceable. *Journal of Cleaner Production* 14 (3–4): 324–333. ISSN 0959-6526.

Ruan, J., Qin, B., and Huang, J. (2018). Controlling measures of micro-plastic and nano pollutants: a short review of disposing of waste toners. *Environment International* 118: 92–96.

Rybicki-Jasinska, K., Shan, W., Zawada, K. et al. (2016). Porphyrins as photoredox catalysts: experimental and theoretical studies. *J Am Chem Soc.* 138 (47): 15451–15458. https://doi.org/10.1021/jacs.6b09036. Epub 2016 Nov 17. PMID: 27933929.

Sadegh, M.S., Bafghi, M.S., Moradkhani, D., and Ojaghi-Llhchi, M. (2007). A review on hydrometallurgical extraction and recovery of cadmium from various resources. *Minerals Engineering* 20: 211–220.

Steffens, H.-D., Wielage, B., and Brandl, W. (1989). Corrosion testing of silver-brazed stainless-steel joints in tap water. In: , vol. 68, 151s–157s.

Sum, E. (1991). *The Journal of The Minerals, Metals & Materials Society (TMS)* 43.

Sun, L.X., Zhang, X., Tan, W.S., and Zhu, M.L. (2012). Effect of agitation intensity on the bio-oxidation process of refractory gold ores by *Acidithiobacillus ferrooxidans*. *Hydrometallurgy* 127: 99–103.

Syed, S. (2012). Recovery of gold from secondary sources—A review. *Hydrometallurgy* 115–116: 30–51.

Tarver, S., Gray, D., Loponov, K. et al. (2019). Biomineralisation of Pd nanoparticles using *Phanerochaete chrysosporium* as a sustainable approach to turn platinum group metals (PGMs) wastes into catalysts. *International Biodeterioration & Biodegradation* 143: 104724. https://doi.org/10.1016/j.ibiod.2019.104724.

TBRC (2022) (thebusinessresearchcompany.com) Metal and Mineral Market Size (2022) And Growth Analysis (Link: Metal And Mineral Market Size 2022 And Growth Analysis (thebusinessresearchcompany.com))

Terazono, A., Yoshida, A., Yang, Y. et al. (2004). Material cycles in Asia: especially the recycling loop between Japan and China. *Journal of Material Cycles and Waste Management* 6: 82–96.

Torma, A.E. and Itzkovitch, I.J. (1976). Influence of organic solvents on chalcopyrite oxidation ability of *Thiobacillus ferrooxidans. Applied and Environmental Microbiology* 32: 102–107.

Torma, A.E., Gabra, G.G., Guay, R., and Silver, M. (1976). Effects of surface active agents on the oxidation of chalcopyrite by *Thiobacillus ferrooxidans. Hydrometallurgy* 1: 301–309.

Tuovinen, O.H. and Kelly, D.P. (1973). Studies on the growth of *Thiobacillus ferrooxidans* I. Use of membrane filters and ferrous iron agar to determine viable numbers, and comparison with 14CO2-fixation and iron oxidation as measures of growth. *Archives of Microbiology* 88: 285–298.

Vera, M., Schippers, A., and Sand, W. (2013). Progress in bioleaching: fundamentals and mechanisms of bacterial metal sulfide oxidation—part A. *Applied Microbiology and Biotechnology* 97 (17): 7529–7541.

Vermesan, H., Tiuc, A.-E., and Purcar, M. (2020). Advanced recovery techniques for waste materials from IT and telecommunication equipment printed circuit boards. *Sustainability* 12: 74.

Wang, T.G. (1981). Hydrometallurgical process for recovering precious metals from anode slimes, US. Patent 4293332

Watling, H.R. (2015). Review of biohydrometallurgical metals Extraction from polymetallic mineral resources. *Minerals* 5: 1–60.

Wu, Y., Fanga, Q., Yia, X. et al. (2017). Recovery of gold from hydrometallurgical leaching solution of electronic waste via spontaneous reduction by polyaniline. *Progress in Natural Science: Materials International* 27: 514–519.

Yang, C., Tan, Q., Liu, L. et al. (2017). Recycling Tin from electronic waste: a problem that needs more attention. *ACS Sustainable Chemistry & Engineering* 5: 9586–9598.

Zhang, Y., Liu, S., Xie, H. et al. (2012). Current status on leaching precious metals from waste printed circuit boards. *Procedia Environmental Sciences* 16: 560–568. https://doi.org/10.1016/j.proenv.2012.10.077.

Zhu, P. and Gu, G.B. (2002). Recovery of gold and copper from waste printed circuits. *Chinese Journal of Rare Metals* 26 (3): 214–216.

17

Biohydrometallurgical Metal Recycling/Recovery from E-waste: Current Trend, Challenges, and Future Perspective

Shital C. Thacker[1,2], Devayani R. Tipre[1], and Shailesh R. Dave[3]

[1]*Department of Microbiology and Biotechnology, School of Sciences, Gujarat University, Ahmedabad 380009, India*
[2]*Department of Microbiology, Silver Oak Institute of Sciences, Silver Oak University, Ahmedabad 382481, India*
[3]*Xavier Research Foundation, Loyola Centre for Research and Development, St. Xavier's College Campus, Ahmedabad 380009, India*

Abstract

The volume of e-waste is growing rapidly and is becoming a secondary resource for metal recovery. Currently, the modern mining industry needs the recycling of secondary resources to meet the industrial and domestic demand for metals. Biohydrometallurgy is generally applied practically for the solubilization of metal(s) from secondary resources using microorganisms, which is the most promising process for recovery. There is a need to develop the process in a manner that is less energy-consuming and environmentally friendly. This chapter provides brief information about the conventional processes involved in the biohydrometallurgy approach and its significance, along with critical challenges for waste management practices and metal recovery. The development of bioengineering and technological engineering has improved current situations to some extent, but some future perspectives still need to be implemented. Finally, the emphasis on advanced techniques that could be used in the future is discussed, which will enable their sustainable implementation.

Keywords *bioleaching; biohydrometallurgy; e-waste; omics; bioinformatics; bioreactor; iron oxidation; cyanogenic*

17.1 Introduction

Electronic waste (e-waste) is pervasive globally and contains a wide variety of elements, some of which have been classified as treacherous if they are present at

Management of Electronic Waste: Resource Recovery, Technology and Regulation, First Edition. Edited by Anshu Priya.

higher than permitted levels. These elements include arsenic (As), selenium (Se), cadmium (Cd), lead (Pb), chromium (Cr^{6+}), and mercury (Hg). But if processed sustainably for a reliable supply of contained metals, they might constitute significant secondary resources. This may be achieved by resource recycling and pollution extenuation both using bacteria and fungi for metal extraction from e-waste. It is suggested that recovery through biological means enables the cycling of metals through a progression that is similar to ecological biogeochemical cycles, decreasing the need for resources like energy, landfill space, or ores. They are referred to as technology metals collectively. Metals are used for a wide range of purposes in technology, and because of their high demand and unequal geographic distribution, their supply is in jeopardy. For a sustainable and circular economy, there arises a necessity for reliable supply. Secondary sources for these metals are becoming highly popular. In this article, the current status of e-waste around the world and the most recent advancements in metal recovery and management from electronic garbage are discussed. Regarding their crucial metal content, attention is placed on printed circuit boards (PCBs), batteries, hard disk drives (HDDs), and displays.

The different forms of global communication, entertainment, transportation, education, and healthcare continue to undergo revolutionary change as a result of the global spread of technology. There are no indications that the revolution will slow down soon. This has caused waste to significantly increase whenever a product has reached the end of its useful life or no longer meets technological standards. E-waste is also described as waste electric and electronic equipment (WEEE). A large number of technical as well as nontechnical tools have been established to accomplish WEEE sustainably, with a universal approach that incorporates the entire WEEE management process comprising collection, pretreatment, recycling, and final disposal of waste (Kiddee et al. 2013). In contrast to primary mining, the majority of secondary raw materials, are complex metal mixtures, elemental alloys, and poly metals (Ongondo et al. 2015). As technology advanced and more components were required to produce electronic gadgets, the complexity of WEEE grew as well. Up to 60 elements in different metal combinations can be found in modern electronic equipment (Bloodworth 2014). Highly complicated alloys present a barrier to the development of effective metal recovery technologies for WEEE from the perspective of recyclers. A variety of components are present in e-waste, with iron and steel making up about half of it. As shown in Table 17.1, when elements like arsenic (As), cadmium (Cd), mercury (Hg), chromium (Cr^{6+}), lead (Pb), selenium (Se), and flame-retardants are present in more than the permissible amounts, e-waste is categorized as toxic (Onwughara et al. 2010; CPCB 2007; Scharnhorst et al. 2005; Bandyopadhyay 2008; Hageluken 2006; Hageluken and Art 2006; Cui and Zhang 2008; Kang and Schoenung 2005).

In general, organic molecules are simple to break down through various biological, biochemical, as well as physical methods (Bosecker 2001; Lewandowski and

Table 17.1 Metals content (% and ppm values) in different e-wastes.

Electronic waste	Metal present (%)					Metal concentration (ppm)			References
	Lead (Pb)	Nickel (Ni)	Aluminum (Al)	Iron (Fe)	Copper (Cu)	Palladium (Pd)	Gold (Au)	Silver (Ag)	
Calculator waste	0.1	0.5	5	4	3	5	50	260	Bandyopadhyay (2008), Hageluken (2006), Hageluken and Art (2006)
Computer PCB waste	2.2	1.1	2.8	4.5	14.3	124	566	639	Cui and Zhang (2008)
Computer waste	6.3	0.85	14.1	20.4	6.93	3	16	189	Cui and Zhang (2008), Kang and Schoenung (2005)
DVD player waste	0.3	0.05	2	62	5	4	15	115	Bandyopadhyay (2008), Hageluken (2006), Hageluken and Art (2006)
Mobile phone waste	0.3	0.1	1	5	13	210	350	1340	Bandyopadhyay (2008), Hageluken (2006), Hageluken and Art (2006)
PC board waste	1.5	1	5	7	20	110	250	1000	Bandyopadhyay (2008), Hageluken (2006), Hageluken and Art (2006), Cui and Zhang (2008)
PCB waste	3.5	2	7	12	16	–	110	280	CPCB (2007)
Portable audio waste	0.14	0.03	1	23	21	4	10	150	Bandyopadhyay (2008), Hageluken (2006), Hageluken and Art (2006)
Printed wiring board	2.6	29	29	29	22	–	0.04	–	Onwughara et al. (2010)
Printed wiring board	3	0.3	4.8	10.6	3.5	–	–	–	Scharnhorst et al. (2005)
Television board waste	1	0.3	10	28	10	10	17	280	Bandyopadhyay (2008), Hageluken (2006), Hageluken and Art (2006)
Television waste	0.2	0.04	1.2	–	3.8	27.1	27.1	27.1	Cui and Zhang (2008), Kang and Schoenung (2005)

De Fillipi 1998; Skippers and Turco 1995), but heavy metals containing inorganic materials would not be degraded. Due to their inherent solubility at different pH levels, redox traits, complex formation ability, and mobility of heavy metal in aqueous systems is the reason for concern in the e-waste treatment process. The e-waste contains a variety of toxic heavy metals, such as cadmium, cobalt, copper, chromium, lead, molybdenum, nickel, vanadium, and zinc, threatening the health of living beings and the environment. A waste must undergo a toxicity characteristic leaching procedure (TCLP) test, which determines the toxicity parameters in amounts that are equal to or greater than the statutory limits, making it a hazardous waste to identify if it is toxic (Salkin 2003). A viable strategy to reduce toxicity is to apply the appropriate method for extraction and recovery of the hazardous and precious metals present in the waste.

It is frequently advised to use biological processes for development because they are expected to have a big impact on the future of the chemical and metallurgical industries, as well as how waste and by-products containing metal are handled. In this direction, a strategy known as "process-integrated biotechnology" (CD, O.E. 2001) has gained widespread acceptance. This contains attributes like incorporating microorganisms into already-existing processes, swapping out industrial processes, and creating an entirely new approach (Brandl and Faramarzi 2006), but it must economically compete with the prevailing physicochemical processes.

Biological systems have been used in the reactions for the production and processing of chemicals, materials, and energy for the past few decades. Mining and biohydrometallurgical metal recovery from low-grade ores successfully employ microbiological processes. Through a process called microbiological leaching in which metals that are found in the trash in a solid matrix are converted to solubilized form. Generally, acidophilic microorganisms carrying out the biological activity of metal leaching play a major role in the biohydrometallurgical processes. In addition to the bioleaching of metals in an alkaline environment, cyanogenic bacteria are also involved. Chemolithotrophic acidophilic microbial consortia consisting of *A. ferrooxidans*, *L. ferrooxidans*, *A. thiooxidans*, and archaea, such as *Sulfolobus* sp., are the most frequently employed major families of bacteria. Additionally, certain fungi like *Aspergillus niger* and *Penicillium* sp. are also utilized in bioleaching during the solubilization of metal(s) from various wastes.

17.2 Overview of Biological Approach for Recycling of Metals

17.2.1 Bioleaching

The bioleaching procedure as compared to traditional methods is less expensive and simpler to carry out. The advantage of microorganisms is their adaptability;

they can live in a variety of challenging environments. With natural biogeochemical metal cycles, biohydrometallurgical processing of solid waste lowers the need for resources like ores, energy, and landfill space. This method is regarded as green because it produces less trash and is therefore environmentally benign (Brombacher et al. 1998; Mishra et al. 2008).

Bioleaching describes the biological oxidation and complexation processes that mobilize metal cations from frequently nearly insoluble materials. On the crust of the earth, microorganisms play a role in both the production and degradation of numerous inorganic and organic materials. The basis for bioleaching is the inherent capacity of microorganisms to convert solid metallic compounds into their soluble and leachable forms. The three main types of microbes engaged in the bioleaching of metals are autotrophic bacteria (such as *Thiobacilli* sp.), heterotrophic bacteria (such as *Pseudomonas* sp., *Bacillus* sp.), and heterotrophic fungi (such as *Aspergillus* sp., *Penicillium* sp.) (Schinner and Burgstaller 1989). The most significant mineral-decomposing microorganisms are chemolithoautotrophic iron- and sulfur-oxidizing microorganisms that fix atmospheric carbon dioxide as the sole carbon source for growth (Nagpal et al. 1993; Clark and Norris 1996). However, the action of leaching bacteria is exclusively used to solubilize impeding metals before cyanidation treatment to recover precious metals.

The process mechanism for metal extraction through bioleaching is briefly illustrated in Figure 17.1. *Acidithiobacillus ferrooxidans* is of prime importance for the bioleaching process, but other Fe- or S^0-oxidizing bacteria or archaea serve as potential microbial means for metal extraction at high temperatures or acidic pH environments (Pant et al. 2012).

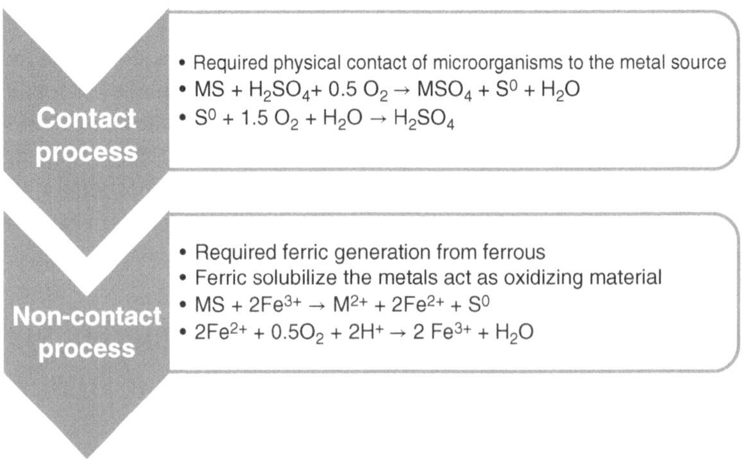

Figure 17.1 Contact and noncontact bioleaching mechanisms. Source: Adapted from Pant et al. (2012).

Chemolithotrophic acidophilic bacterial strains with moderate thermophilic nature were also employed to bioleach metals from e-waste. Other than that, heterotropic microorganisms were also applied for the bio-solubilization process. These include an undiscovered acidophilic heterotroph isolated from nearby habitats, as well as *Sulfobacillus thermosulfidooxidans*. These strains were also used to bioleach metals from e-waste in columns.

Parameters such as the type of microorganisms, amount of ferrous iron in the system, pH, composition of the pulp (waste), type of e-waste, the presence of the toxic compound, and particle size are a few variables that affect the metal bioleaching from e-waste. The response is significantly influenced by temperature and time as well (Kiddee et al. 2013; Brandl et al. 2001; Faramarzi et al. 2004; Yang et al. 2009; Ilyas et al. 2010). The bioleaching process may be hampered by the material heterogeneity of waste, particularly if there are alkaline particle components in the solid phase (Pant et al. 2012; Faramarzi et al. 2004). Here, in transferring metals from the solid state to the solution, factors such as pH, Fe^{3+} ion concentration, and the number of employed microorganisms are critical (Faramarzi et al. 2004). The ability of the system to extract copper more effectively from the waste was enhanced by adding a greater dose of *A. ferrooxidans* and iron ions Fe^{2+}, as well as by maintaining a pH between 1.5 and 2.0.

Metals may chemically precipitate during the bioleaching process (Kiddee et al. 2013; Lewandowski and De Fillipi 1998; Choi et al. 2004) and create sludge, as, reported for Pb and Sn, which precipitate most likely in the form of $PbSO_4$ and SnO, respectively. To optimize pH levels for bacterial growth, the process has to be carried out beneath carefully monitored environments with the proper concentration of input Fe ions and the addition of a complexing agent. An essential factor that ensures the action of bacteria is temperature. Temperatures between 28 and 35 °C are ideal for the genus *Acidithiobacillus* (Pacholewska and Farbiszewska 2003). Experiments with bioleaching on electronic trash were done at 30 °C (Lewandowski and De Fillipi 1998; Salkin 2003). Also, the efficacy of the method was investigated at 22–25 °C (Cwalina et al. 2009; Watling 2006).

The kinetics of bioleaching of the metals mainly depend on the particle size and the proportion of the solid phase to the liquid phase. So, it is critical to correctly prepare e-waste and determine its pretreatment process. According to the experimental findings (Salkin 2003), it was observed that the particle size of the e-waste sample was inversely proportional to the amount of Cu, Pb, and Zn solubilization. Smaller the particle size greater the metal(s) solubilization. *A. thiooxidans*, *A. ferrooxidans*, and a mixed culture of these bacteria dissolve copper in fractions of 0.5–1.0 mm in 99.0 percent, 74.9 percent, and 99.9 percent, respectively, while maintaining a solid phase density of 7.8 g/dm^3. According to the classification of the waste (Table 17.1), which can contain a variable amount of basic metals, valuable metals, and/or dangerous compounds, reports on the extraction of metals from those waste materials have been undertaken using a variety of microbes. E-waste is an intriguing type of waste since it is made up of a variety of metals, including Cu, Al, Fe, and Ni, as well as their alloys. The primary method of transporting precious metals in the garbage is printed electronic circuit boards (PCBs). For instance, roughly 28 percent of mobile phone components are made of metal, which includes 10–20 percent copper, 1–5 percent lead, 1–3 percent nickel, and about 0.3–0.4 percent of valuable metals like Ag, Au, and Pt. The remaining materials include glass and ceramics. The value of precious metals is typically the main economic driver for recycling e-waste, followed by the recovery of the highest-containing metals like Cu, while the value of the recovery of metals alike Al, Pb, and Fe is significantly lower.

In the last decade, microorganisms have been used to recover metal(s) from e-waste in addition to conventional ways of processing (hydrometallurgical and pyrometallurgical procedures). Table 17.2 provides an overview of microorganisms utilized for the extraction of different metals from e-waste.

The extraction of Al, Co, Ni, Pb, Sn, and Zn from e-waste has been the subject of numerous experimental studies using bacterial leaching. In the presence of cyanogenic bacteria, such as *Chromobacterium violaceum* and *Pseudomonas fluorescens*, bioleaching of valuable metals (Ag, Au) using crushed e-waste samples was also conducted (CPCB 2007; Hageluken 2006). The hydrocyanic acid (HCN) that these bacteria can dissolve gold. Table 17.3 provides details on the microbes and levels of different metal(s) leached from e-waste.

The pyrohydrometallurgical method is utilized commercially for the recovery of valuable metals, and current advancements in the hydrometallurgical techniques are evaluated by Cui and Zhang (Cui and Zhang 2008). However, as was already noted, the use of microorganisms for bioleaching is one of the most encouraging technologies in the metallurgical segment. The use of the *Thiobacilli* group of bacteria (Kiddee et al. 2013; Brandl et al. 2001; Choi et al. 2004; Yang et al. 2009) or fungi such as *A. niger* and *P. simplicissimum* (Brandl et al. 2001) is reported. In the presence of metals such as Fe, Au, Cu, Ag, Pd, and Ni in the metallic state and other metals in the form of inorganic complexes, these microorganisms are applied for the bioleaching process. A few researchers are also mentioned as employing this methodology to extract gold from the e-wastes utilizing cyanogenic bacteria such as *P. fluorescens*, *C. violaceum*, and *P. plecoglossicida* (Brandl and Faramarzi 2006; Norton et al. 2005; Brandl et al. 2008; Tran et al. 2011; Kita et al. 2006).

Modern research (Brandl and Faramarzi 2006; Choi et al. 2004) demonstrates how cyanogenic bacteria, in particular *Pseudomona fluorescens*, *C. violaceum*, and *Pseudomonas plecoglossicida*, work in the occurrence of various metal-bearing e-wastes, such as e-scrap having gold, jewelry waste comprising silver, and automobile catalyst converters containing platinum. The synthesis of more dicyanoaurate (68.5% in 168 h) at a lower pulp density of 5 g/L as compared to *Pseudomonas fluorescence* after a 3 d lag phase made *C. violaceum* the most efficient cyanogenic bacteria for mobilizing Au from the waste PCBs. The sorption

Table 17.2 Microorganisms and optimum conditions applied for the solubilization of various metals from e-waste.

Recovered metals	Microorganisms involved in the process of recovery	pH	Temperature (°C)	References
Silver (Ag)	*Acidithiobacillus* sp.	2–3	45–60	Norton et al. (2005), Olson (1994)
	Leptospirillum sp.	2–3	45	
	Ferromicrobium	2–3	45	
	Acidiphilium	2–3	45	
Aluminum (Al)	*B. circulans* and *B. mucilaginosus*	4–5	35	Brandl et al. (2001), Groudev (1987)
	A. ferrooxidans	2–3	45	
	A. thiooxidans	3–4	45	
	A. niger	2–6	30	
	Penicillium simplicissimum strain of F3-02	2–6	32	
Gold (Au)	*Chromobacterium violaceum*	3	30–37	Norton et al. (2005), Olson (1994), Mrvčić et al. (2012), Davis et al. (2003)
	Pseudomonas fluorescens	7–8	25	
	Desulfovibrio desulfuricans	2	35–43	
	Acidithiobacillus sp.	3–4	45	
	Leptosprillum ferrooxidans	2–3	45	
	Ferromicrobium	2–3	45	
	Acidiphilum	2–3	45	
Copper (Cu)	*Acidithiobacillus ferrooxidans*	2–3	45	Brandl et al. (2001), Wang et al. (2009), Suzuki (2001), Imai and Saitoh (2001), Wu et al. (2007), Johnson et al. (2008), Mulligan and Kamali (2003), Xiang et al. (2010), Wang et al. (2009)
	Acidithiobacillus thiooxidans	3–4	45	
	Aspergillus niger	2–6	30	
	Penicillium simplicissimum	2–6	32	
	Gallionella sp.	3	30	
	Leptospirillum sp.	2–3	45	
Nickel (Ni)	*Acidithiobacillus thiooxidans*	3–4	45	Brandl et al. (2001)
	A. ferrooxidans	2–3	45	
	Aspergillus niger	2–6	30	
Lead (Pb)	*Acidithiobacillus ferrooxidans*	2–3	45	Johnson et al. (2008), Groudev (1987), Pacholewska (2004)
	Acidithiobacillus thiooxidans	3–4	45	
	Aspergillus niger	2–6	30	
	Penicillium simplicissimum	2–6	32	
Zinc (Zn)	*Acidithiobacillus ferrooxidans*	2–3	45	Pacholewska (2004), Brandl et al. (2001), Johnson et al. (2008)
	Acidithiobacillus thiooxidans	3–4	45	
	Aspergillus niger	2–6	30	
	Penicillium simplicissimum	2–6	32	

Table 17.3 Organisms and percent metal leaching from e-waste.

Microorganisms involved in bioleaching of e-waste	Metal	Percent leached	Reference
Acidithiobacillus ferrooxidans + *Acidithiobacillus thiooxidans*	Copper Nickel, Aluminum, Zinc	>90 >90 >90 >90	Pant et al. (2012)
Acidithiobacillus ferrooxidans	Copper	99	Faramarzi et al. (2004)
Acidithiobacillus ferrooxidans	Copper	99	Wang et al. (2009)
Acidithiobacillus thiooxidans	Copper	74.9	
Acidithiobacillus ferrooxidans + *Acidithiobacillus thiooxidans*	Copper	99.9	
Aspergillus niger + *Penicillium simplicissimum*	Copper Aluminum Nickel Lead Zinc	65 >95 >95 >95 >95	Pant et al. (2012)
Chromobacterium violaceum	Gold	68.5	Brandl et al. (2008)
Sulfobacillus thermosulfidooxidans	Nickel Copper Aluminum Zinc	81 89 79 83	Kiddee et al. (2013)
Sulfobacillus thermosulfidooxidans + *Thermoplasma acidophilum*	Copper Zinc Aluminum Nickel	86 80 64 74	Ilyas et al. (2010)

process by the biomass or the metal-cyanide degradation was reported to make the gold complex non-stable in the presence of *Pseudomonas fluorescence* (Choi et al. 2004). A significant amount of Cu solubilization from the e-waste was also seen due to its high content and instant reactivity in the presence of cyanide in the system.

17.2.2 Biosorption

The term "biosorption" refers to the low-cost biological removal of metal or metal-like particles from a solution. All biological materials can serve as excellent biosorbents for the sequestration of metals, except for mobile alkali metal cations like Na^+ and K^+ ions, and this can be a substantial passive process in both live and dead animals. Numerous low-cost plant and animal-origin biosorbents are used for metal removal. Generally speaking, it was discovered that all types of biomaterials employed for biosorption had good biosorption capabilities toward most metal ions. Several studies examine the biosorption process for removing heavy metals, which uses either microorganisms cultured in a lab or biomass generated as waste in fermentation industries and wastewater treatment plants. The major purpose of biosorption is to remove metals from effluents, as they are not biodegradable. Even though their separation, screening, and harvesting on a larger scale may be challenging, it is still one of the most effective ways to remove toxins. To eliminate various metal ions, various bacterial strains are employed. Bacteria have developed a lot of effective detoxification methods for metal ions; they mostly do this for survival. The attachment of metal ions (sorbate) to the exterior of the biosorbent, which is made of biological matter, is referred to as biosorption, and it is a straightforward metabolically passive physicochemical process (Mrvčić et al. 2012). Utilizing microorganisms, materials generated from plants, agricultural waste, or industrial waste are all examples of biological sorbents. Instead of oxidation, it is a reversible, fast process that involves the attachment of ions onto the functional groups represented on the surface of the biosorbent in an aqueous solution over different interactions (Davis et al. 2003). The benefits of this process include its ease of use, lack of additional nutrient requirements, minimal sludge production, low costs, higher efficiency, biosorbent regeneration, and lack of an increase in COD values of the water, which is the main drawback of the conventional techniques (Chojnacka 2010). Due to their toxicity at ppb levels, biosorption has particular relevance for the elimination of heavy metals and can remove pollutants even at diluted concentrations. Live and dead microorganisms, as well as other industrial and agricultural waste, can be employed as biosorbents for the biosorption process (Chojnacka 2009).

With the release of significant quantities of metal-contaminated wastewater from industries, heavy metals like Zn, Hg, Cd, Pb, Cr, Ni, and Cu are major contaminants that harm freshwater reservoirs. They enter into the food chain, creating major health problems because of their poisonous and nonbiodegradable nature. The removal of these metals has been accomplished over the past few decades using a variety of conventional treatment techniques. Generally utilized techniques include chemical precipitation, electrowinning, ultrafiltration, reverse osmosis, ion exchange, and phytoremediation (Joshi 2017).

17.2.3 Bioaccumulation

Heavy metals bioaccumulate in microbes and through their entry into the food chain affect plants, animals, and humans as well as cause a variety of physiological diseases. Thus, heavy metal pollution is a serious issue (Ahemad and Malik 2011). Physical extraction and chemical procedures such as the use of tartaric acid in the past to remove heavy metals; however, both techniques are expensive and very inefficient. Scientists have recently adopted a novel strategy for removing heavy metals from the environment by employing microorganisms. Although in larger quantities, heavy metal ions can combine to form hazardous chemicals in cells, microbes perform a significant part in the transformation, breakdown, and bioaccumulation (Nies 1999). Additionally, poisonous soluble metals might be changed by microorganisms into insoluble materials that are easy to get rid of (Rodriguez-Montelongo et al. 1993). Microorganisms have developed several methods to resist the uptake of heavy metal ions and thrive in metal-stressed environments (Schmidt et al. 2003). These include the stockpile, complexation, and precipitation of metals inside the cell, the outflow of metal ions outside the cell, and the reduction into a less hazardous state. When compared to conventional remediation techniques, biological solutions can be more efficient and cost-effective and can remove pollutants more quickly. Bioaccumulation, which relies on intrinsic biochemical and structural features, physiological and genetic adaptability, environmental alteration of metal specification, availability, and toxicity, is the term most often used to describe the active method of metal accumulation in living cells. Bioaccumulation refers to the uptake of toxicants by living cells and their transport across the cell. It is dependent on the growth of cells, which is facilitated only by living biomass (Jaafar et al. 2015).

17.2.4 Bioprecipitation

According to Hussain et al. (Hussain et al. 2016), bioprecipitation is a technique to remove toxic metals from industrial waste streams conventionally using sulfate-reducing bacteria (SRB), which precipitates the soluble metal ions into insoluble sulfidic forms. Processes for the reduction of sulfate are employed in the mining and metal recycling industries. Additionally, bacteria that are metabolizing substances may release other substances (such as phosphates) that can be employed in processes of precipitation and/or effect a change in the aqueous environment, like a change in pH that enhances precipitation (Park et al. 2010). Janyasuthiwong et al. (Janyasuthiwong et al. 2016) effectively used SRB to precipitate Cu from PCB leachates and > 90% of the Cu was recovered in a continuous system.

In recent years, direct bioreduction of valuable metals to metallic nanoparticles has gained more attention. Microorganisms can be used to produce nanomaterials from waste materials, which is especially encouraging due to:

(1) nanomaterials produced can be utilized straightway in industrial and catalytic processes,
(2) microorganisms are supposed to become low-cost catalysts, and
(3) production can be done at normal atmospheric pressure and temperatures.

Palladium (Pd) bioreduction in the form of biomass-associated nanopalladium (Pd^0 or bio-Pd) has drawn the attention of numerous researchers (Wang et al. 2011). *In situ,* synthesis of reducing agents, like hydrogen, for Pd(II) reduction and the following catalytic events appear to be where the microbial activity looks to be most helpful (Pat-Espadas et al. 2016). For instance, Pd(II) was recovered from industrial automotive catalysts and the leachates of lead-containing materials through the microbial-mediated reduction to bio-Pd from Pd(II) (Deplanche et al. 2014; Creamer et al. 2006). Previous research on the formation of various biogenic metallic (nano) particles revealed the possibility of producing directly marketable products from waste streams. Deplanche et al. (Deplanche et al. 2007), for instance, concentrated on the biomanufacturing of catalytically active Au nanomaterials for glycerol oxidation and Au recovery from jewelry wastewaters using *Escherichia coli*. By mixing microorganisms and a surfactant (Ilyas et al. 2017), research work showed how to fabricate novel stable bio-Au nanocomposites and gold nanostructures with good optical characteristics. While De Gusseme et al. (Gusseme et al. 2010) created membranes containing biogenic Ag^0 precipitates with antifouling capabilities and also highlighted the possible application of biogenic Ag(0) and Ce particles for disinfection of viruses. Moreover, with the least amount of biogenic silver (i.e. 140 mg bio-Ag(0)), the 'bio-Ag-0/PES composite membranes' are still effective in the formation of biofilms. De Corte et al. (De Corte et al. 2011) have created biosupported Pd–Au bimetallic nanocatalysts that can be utilized for the dechlorination of environmental pollutants. Even though all these reports show great promise, several of these studies prepared the biogenic particles and demonstrated how they might be utilized to extract and recover precious metals from waste using synthetic media. Therefore, establishing experiments centered on the production of biogenic particles from actual wastes like WEEE and scaling up the techniques to pilot-scale is the need of the time.

17.2.5 Biomineralization

Biomineralization is the process by which minerals are produced by microorganisms. In these processes, toxic metal ions associated with anions or ligands get precipitated (Tabak et al. 2005). Biomineralization is generally used to eliminate harmful metals from polluted water and impound them in a recyclable form.

There are two classes of bacterial mineralization:

(1) Biological-induced mineralization (BIM) and
(2) Biological-controlled mineralization (BCM).

In BIM, the minerals are synthesized extracellularly because of the metabolic potential of organisms, while in BCM minerals are produced at a certain location inside or on the cell under precise circumstances. The formations of minerals give the microbial community to treat the wastewater and could also be used for the treatment of waste. Biomineralization can be studied by relating to phylogenetically different microorganisms isolated from untouched and contaminated ecosystems (Lens 2016). Microorganisms that have been studied for biomineralization from e-waste are *Bacillus fusiformis* (Pb), *Cupriavidus metallidurans* (Cd), *Desulfotomaculum auripigmentum* (As), *Cupriavidus metallidurans* (Au), *Aspergillus flavus* (Pb), and *Sporosarcina ginsengisoli* (As) (Patel and Kasture 2014).

17.2.6 Biomining

Using geochemistry and microbial capabilities, biomining, which developed from biohydrometallurgy, extracts metals from natural source materials. For instance, basalts are a good source of geopolitically significant trace metals such as titanium (Ti), aluminum (Al), and iron (Fe) that are not typically exploited. In a similar vein, it is not a routine practice to recover trace metals from waste electronic components for use in new components. Due to the significantly lower concentrations of usable metals present compared to the sulfide ores utilized in bioleaching procedures, former substrates are not a typical source of metals (Jain and Sharma 2004). Recent studies have concentrated on using bioextraction technologies on discarded electronic parts to recover the metals present for recycling. About 16 distinct types of precious rare-earth elements or metals (REE) and semiconductor metalloids can be found in the PCBs, solar cells, catalytic converters, and electronic components used in these devices. These materials are held together in a silicate (glass) matrix. Electronics waste can be a source of 150 kg of copper, 5 kg of gold, and 100 g of palladium from a tonne (1000 kg) of materials, such as electric shavers, toasters, hair dryers, and cell phones. The United States does not yet have the infrastructure or capability to recover these important metals. The majority of hydrometallurgical methods for recovering metals from electronic trash involve the leaching of valuable metals with cyanide and halides (Ilyas et al. 2017). But as a developing method for recovering primary and secondary sources, biotechnology exhibits great potential (Akcil and Deveci 2010). In a report, a consortium comprising thermophilic iron- and sulfur-oxidizing bacteria, *S. thermosulfidooxidans* and *Sulfobacillus acidophilus*, was used to solubilize Al, Cu, Ni, and Zn from pretreated e-scrap, and more than 80% of the metals were extracted successfully (Ilyas et al. 2017). An adapted culture of *Leptospirillum*-dominated culture was used for the extraction of various metal(s) by Shah et al. from computer PCBs and achieved extraction of Cu (99.99%), Zn (99.49%), and Ni (84.21%) (Shah et al. 2015); from television PCBs, Cu (90%), Zn (93%), and Ni (87%) by Sodha et al. (Ilyas et al. 2017; Sodha et al. 2019); from cell phone PCBs, Ni (91.5 %), Zn (95.2%), and Cu (84.2%) by Khatri et al. (Khatri et al. 2018); and from mobile PCBs, Cu (96.86%), Zn (90.69%), and 93.65% (Ni) by Thacker et al. (Thacker et al. 2022).

This review emphasizes the numerous advancements in the reclamation of resources from unusual source materials and the processes that go beyond biomining that are conventionally used to process sulfide minerals like copper (Cu) from chalcopyrite ($CuFeS_2$), iron (Fe) from pyrite (FeS), etc. The financial costs should be considered in the following development steps (Plessis et al. 2007).

17.3 Existing E-waste Management Challenges

E-waste disposal is a rising environmental and public health concern on a global scale since it is part of the official municipal trash stream that is increasing the fastest (Dahl 2002). Since most Indians do not know how to properly dispose of electronic garbage, the majority of it is kept in homes. The intricate nature of this ever-growing trash makes it a rich source of metals like Au, Ag, and Cu that may be recovered and recycled into the production process. So, e-waste trade and recycling associations employ many groups of people (Baud et al. 2001) in India. In Delhi alone, there are over 25,000 employees, including minors, employed in crude dismantling plants where 10,000–20,000 tonnes of e-waste are processed annually with bare hands. E-waste is dangerous to human health and our ecology when it is improperly disassembled and processed. So, it has become clear that efficient e-waste management is necessary (Forti et al. 2020). Reviewing the threats to public health as well as available defenses against this expanding threat is essential.

With the rise of e-waste in India, managing solid trash has become a challenging task. Generation of electronic and electrical equipment (WEEE/e-waste) was 53.6 million tons in 2019; just 17.4% of e-waste was gathered and reprocessed (Forti et al. 2020), and the remaining 82.6% of e-waste was uncertain. Only 3% of this made it to the premises of the authorized recyclers. This is due, in part, to the fact that many industrialized nations now dump their waste in India. According to research by the Basel Action Network (BAN), 50–80% of the e-waste that the United States collects is sold to countries including Africa, China, India, Pakistan, and Taiwan (CPCB 2008). India is presently among the fastest-expanding economies of the world, and the domestic demand for durable goods in the country has been soaring. Domestic appliance sales, including both large and small ones, increased globally from 1998 to 2002 by 53.1 percent. (Euromonitor 2022). According to another study, businesses and households in India throw away 1.38 million personal computers annually, which accelerates the country's annual pace of e-waste generation by almost 10% (Agarwal et al. 2003; ELCINA-DSIR 2021). This has an impact on environmental health indices. (Mehra 2004).

Conventional methods like acid leaching, hydraulic bed separation, and incineration used for recycling WEEE are currently implemented in many countries in informal and formal ways (Thiery et al. 2009). These processes produce secondary pollutants. The conventional hydrometallurgy process is acid leaching, which is used to recover metals. Acids like $HClO_4$/HCl/HNO_3 are utilized as leaching solvents to extract metals such as Pb, Cu, and Zn from WEEE. Moreover, valuable metals like Au and Ag are solubilized using a cyanide solution (Collard et al. 2008; Sheng and Etsell 2007). In this method, leaching agent, "cyanide," has a toxic quality and causes environmental toxicity. These old techniques have undesirable qualities that will ultimately affect the environment. Biological processes are done at ambient temperature and pressure and could produce a minimal amount of effluent that affects the environment, but this process has its restriction factors.

17.3.1 Biotic Factor Restrictions

The biological process for metal(s) solubilization could be inhibited by biotic factors like the microorganism tolerance of the metals and toxicity. Heavy metals present within the e-waste will inhibit the metabolic activity of the microorganisms and their growth. The toxicity due to the heavy metal concentration limits the activity of the microorganism by inhibiting the enzymes necessary for its growth. Other than that, toxicity also occurs due to epoxy plastic and brominated flame retardant (BFR) present in the e-waste (Valix 2017). Microorganisms adapting to the e-waste before the process can help to decrease the toxicity of the metals to microorganisms and increase their tolerance. The adaptation could be achieved by serial subculturing of the culture with e-waste. The organic content of e-waste does not degrade by the microorganisms which will affect bacterial growth. The BFR inhibits the bacterial enzymes involved in carbon dioxide fixation and decreases the iron oxidation rate. This affects the activity of solubilization of metals, so large-scale bioprocess could not be achieved (Işıldar et al. 2016). The microorganisms also produce limited metabolites that will solubilize the metals, which is another limiting factor for the large-scale process of bioleaching. Gene manipulation and r-DNA technology could be used to overcome this restriction at some level.

17.3.2 Abiotic Factor Restrictions

Many abiotic factors like pulp density, precipitation, coating formation on the exterior of the e-waste, and pH also limit the biological process for metal(s) recovery. Precipitation occurs due to the bioleaching process, which will form the jarosite formation and also decrease the metals leaching efficiency. The jarosite formation occurs due to the presence of precipitating agents in the solution. The heavy metals are trapped in the lattice of jarosite crystals (Nazari et al. 2014, Vestola et al. 2010). So, jarosite formation is an undesirable by-product that decreases the metal(s) solubilization during the biological solubilization process due to precipitation on the waste. Oxalate complexes like nickel oxalate and cobalt oxalate also precipitate during the fungal bioleaching process. This will inhibit the leaching efficiency of Cd, Co, Ni, and Zn.

17.4 Advance Technology for Recycling Metals

17.4.1 Biohydrometallurgical Engineering

Biohydrometallurgy comprises fields of hydrometallurgy and biology, where the latter has a major role in the natural ecosystem through a variety of biological reactions and their interactions with metals present in mineral and soluble forms. This results in the usage of microorganisms in a varied variety of extraction and recovery of metals and water treatment processes. Yet the invention of new biohydrometallurgy applications continues to this day. Copper recovery from mining solutions is from times memorial (Olson et al. 2003) but metal extraction due to bacterial oxidation was discovered in 1670 at the Rio Tinto mine in Spain (Brierley and Le Roux 1978). Biohydrometallurgy began with the isolation and identification of *Thiobacillus ferrooxidans* (now *Acidithiobacillus ferrooxidans*) from a coal mine by Colmer and Hinckle (Colmer et al. 1950) while solving the environmental issue of the coal mine drainage that was due to the activity of this bacteria (Colmer et al. 1950). Then onwards, *Acidithiobacillus ferrooxidans* and several other microorganisms are employed to speed up the leaching of metal sulfides coupled with copper and other metal(s) (Bryner et al. 1954; Bryner and Anderson 1957; Bryner and Jameson 1958; Razzell and Trussell 1963). The most commonly studied and exploited microorganisms are *Acidithiobacillus ferrooxidans* (formerly *Thiobacillus ferrooxidans*), *Leptospirillum ferrooxidans*, and *Acidithiobacillus thiooxidans* (formerly *Thiobacillus thiooxidans*). These bacteria are chemolithoautotrophic and utilize CO_2 for the growth of cells. Attempts are also being put on to use other microbes including archaea.

Commercially, biohydrometallurgy was exploited in the 1950s at Kennecott Copper Mine, Salt Lake City, USA (Brierley and Brierley 2001). Research on the activity of microorganisms in the leaching of copper from its ore dump suggested that the population of bacteria was in significant numbers in the ore dump within the upper 2.5 m of the surface because of the inaccessibility of oxygen at depth (Bhappu et al. 1969). Around 1980 in Chile, commercial copper bioextraction from engineered heaps demonstrated the significance of shallow stacking and that finer particles were responsible for enhanced conditions and extractions (Bustos et al. 1993). Further innovations in bioleaching technologies gave birth to commercial-scale bioleaching and biooxidation processes like BioCOP, BIOX, GEOCOAT, BioHeap, BioZinc, BioNIC, and BIOPRO for the recovery of various metals (Rawlings 2013; Brierley 2009). Currently, thermophilic microorganisms such as *Acidianus*, *Solfolobus*, and *Metallosphaera* archaea are exploited for mineral oxidation at high temperatures owing to their several advantages over conventional mesophiles (Rawlings 2013). Mainly, the large-scale bioleaching and biooxidation processes are applied for the recovery of gold and copper but now the processes are extended for the recovery of other metals like zinc, cobalt, and nickel, but still have a long way to go (Rawlings 2013; Brierley 2009; Rawlings 2013).

17.4.2 rDNA Technology Involved in Microorganism

E-waste recycling is becoming progressively important in the growing worldwide stream of e-waste, not only to reduce the risks to human health and the environment but also to recover valuable metals and other nonmetal wealth. Biologically aided treatment of e-waste presents a possible solution by leveraging the intrinsic ability of specific microbes to degrade plastics or interact with metals to create eco-friendlier choices than conventional techniques of e-waste recycling. The creation of new or improved skills is required to promote bioremediation and resource recovery. These could potentially be enhanced by developing genetic techniques with the aid of synthetic biology by altering substrate specificities, enhancing toxin tolerance, and enhancing enzyme catalytic abilities. However, because of its enormous volume, complicated component makeup, and related toxicity, managing e-waste poses severe hurdles. Before promising lab-scale advancements in bioremediation can be applied to practical commercial settings, they need to overcome several obstacles. The use of microbes for commercial applications in e-waste recycling has nevertheless been pioneered by vested parties, including both start-ups and established businesses. Gaps or deficiencies in our existing understanding or methodologies must be realized and addressed by focusing intensive research efforts to go past technological barriers to enhance the biological approach to managing e-waste. Even though it shows promise, the use of biologically modified microorganisms for the bioremediation of e-waste is still in its infancy, and much more research and development must be done before the technique can be used commercially. Improvements have been achieved in cell tolerance to e-waste toxicity, the necessity to recover numerous metals in one procedure, and the effectiveness of bioleaching recovery. However, the scientific community lacks research on scaling up these lab-scale results, which limits the commercial application of bioremediation of e-waste. Research groups and companies must collaborate to enable a soft switch from the lab-scale bioremediation process to the operational field scale. Additionally, the process of increasing their metal recovery efficiencies is being slowed down by the restricted supply of molecular toolkits for genetically modifying unusual bacteria that have tremendous potential for e-waste bioremediation. These crucial molecular toolkits can be created more quickly by implementing the 'design, construct, test, and learn' paradigm in automation and synthetic biology. The development toward the eventual commercialization of the bioremediation of e-waste must also solve major gaps in proper e-waste management and workflow planning. After these technical obstacles are defeated, it is crucial to consider whether the technology needs to be integrated with additional secondary processes for full e-waste disposal and whether the workflow is suitable from an ecological and economic standpoint and suitable for decentralized adoption.

17.5 Future Development Strategies for E-waste Management

17.5.1 Application of Omics Technology for Biohydrometallurgy

OMICS technologies include genomics, metagenomics, proteomics, and metaproteomics, which aid in giving an idea about the microorganisms involved in the bioleaching and mining processes and their metabolic pathways to solubilize metals, other than that their adaptive responses (Figure 17.2) are also studied (De Sousa et al. 2018). Omics studies have significantly transformed our understanding and concept of biomining. The genomic research gives a deeper understanding of the metabolic pathway and adaptation to a bioleaching environment. TIGR (The Institute for Genome Research) has sequenced the genome of *A. ferrooxidans* (Selkov et al. 2000). Currently, complete genome sequences of 36 archeal and 55 bacterial biomining microorganisms are available in the NCBI database, which is free to access (Martinez et al. 2015). Various molecular biology techniques like qPCR, FISH, and DGGE used in genome sequencing enabled us to understand the microbial diversity and delve into extreme leaching ecosystems for novel biomining microbes with superior leaching potential (Christel et al. 2018; Buetti-Dinh et al. 2020).

Along with omic technology, bioinformatics can be exploited as a tool to identify the traits of microorganisms involved in the biomining processes and improve their capacity.

Transmissible metabolic traits and gene regulation in free and biofilm-forming acidophiles in environmental conditions are key areas for advanced research. Gene manipulation and strain improvement are also possible to improvise the bioleaching and biomining processes to recover the metals.

17.5.2 Combined Multi-omic and Bioinformatics Technology

A thorough grasp of how bio-mining microorganisms react to harsh environments has been made possible by omics technology and bioinformatics. The development of metabolic models for specific biomining microbial communities is aided by the application of bioinformatics, genomics, transcriptomics, proteomics, and

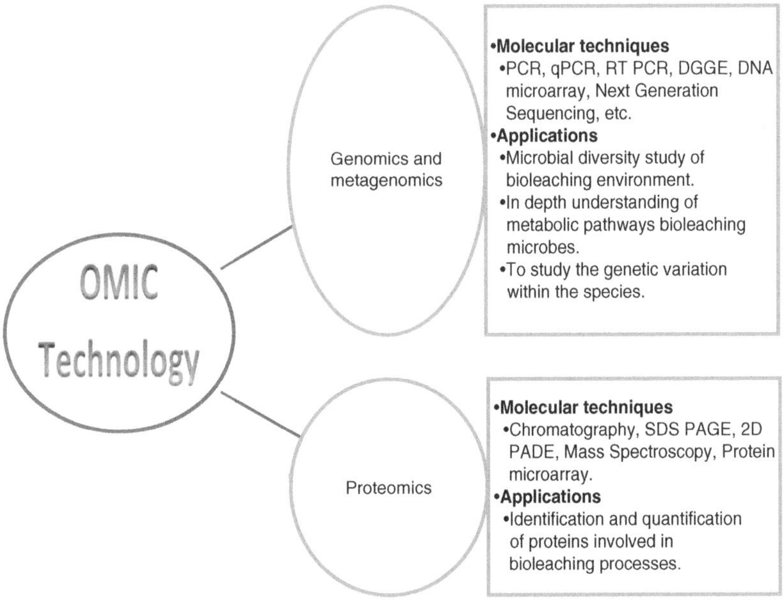

Figure 17.2 Application of bioinformatics for biohydrometallurgy.

metabolomics. They can also make assumptions about the evolution of strains, find novel chemicals, and research the relationships between proteins and genes, which can be characterized using metagenomics, metatranscriptomes, and metaproteomics. These methods can help identify underrepresented individuals and highlight their crucial roles. The limitation of metabolic capacity revealed by genomes, the poor reproducibility of transcriptomics and proteomics, and the challenge of identifying metabolism are just a few of the drawbacks of individual omics technologies. Multi-omics can offer more thorough and precise information, but there is still more work to be done in terms of examining the actual behaviors of communities in the always-shifting severe environment. Therefore, integrated multi-omics research will undoubtedly lead to a breakthrough in our understanding of bio-mining microbes. Last but not least, future development will focus on the investigation of resistance mechanisms and the use of omics data to create highly effective engineered strains.

Genetic modification of cyanogenic organisms used for bioleaching metals from PCBs was accomplished by Natarajan and Ting (2014). Nitroso-N-ethyl urea was used as a mutagen, and the culture of *C. violaceum* was exposed to different alkaline pHs of 9, 9.5, and 10 to obtain a high pH resistant mutant. In this reported study, elevated recovery of gold was achieved at pH 9.5 in comparison to the mutated strains obtained at pH 9 and 10 (Natarajan and Ting 2014).

The metabolic characteristics and adaptation mechanisms of complex microbial communities can be better understood in detail using a multi-omics approach rather than using single omics (Jansson and Baker 2016). The adaptation strategy of *L. ferriphilum* growing on chalcopyrite was disclosed by genomics, transcriptomics, and proteomics, which in particular uncovered a yet unknown nitrogenase cluster responsible for atmospheric nitrogen fixation (Ma et al. 2019). The potential strategies adapted to grow under multi-extreme conditions by *A. ferrivorans* YL15 were revealed by genomic and transcriptome analyses. (Peng et al. 2017). The first thorough understanding of aerobic and anaerobic hydrogen metabolism of *A. ferrooxidans* at low pH came from transcriptomic and proteomic research (Kucera et al. 2020).

Omics can simultaneously study the metabolic relationships of the microbial communities and disclose the taxonomic breakdown and functional division (Denef et al. 2010). The primary roles of seven main bacteria in the cycle of inorganic resources (arsenic, iron, and sulfur) and organic resources were determined by metagenomics and metaproteomics, and a metabolic interaction model was developed (Bertin et al. 2011). Comparably, metagenomic and metatranscriptomic research identified the distinctive gene transcription patterns of microbes that are associated with responses and environmental adaption (Chen et al. 2015). In a different study, metagenomics and metatranscriptomics highlighted the crucial role that uncommon group activities carried out by microorganisms, such as nitrogen fixation and sulfur oxidation, play in the general operation and community assembly of the AMD (Hua et al. 2015). *Leptospirillum* group IV UBA BS is a novel species that was uncovered using community genomes, transcriptomics, and proteomics. Its cooperative connection with *L. ferrodiazotrophum* in hydrogen metabolism was also proposed (Goltsman et al. 2013). However, the multi-omics technology faces limitations in detecting the microorganisms that are present in low numbers in a population and elucidating interactions among the microbial community members.

17.6 Conclusion and Recommendation

The recycling of electronic trash is a crucial topic from the perspective of both waste management and the recovery of precious metals. The valuable metals from discarded electronic equipment has traditionally been accomplished by biohydrometallurgy. Environmental concerns have, however, presented some difficulties. Modern smelters are hence heavily dependent on investments. The microbially mediated extraction of precious metals has gathered attention in recent years as a result of several large-scale bio-hydrometallurgical processes practiced in past decades for Cu, Au, and U by heap/dump and bioreactor leaching. Living or dead biomass, such as algae, bacteria, fungi, and yeast, can be used to retrieve valuable metals from e-waste. The search for and modification of biomass with a high capacity for absorption, including the use of advanced bioleaching processes, should continue. One of the most promising e-waste processing technologies is thought to be bioleaching. For many years, metals have been recovered from their ores via bioleaching. Recently, bioleaching from solutions has attracted a lot of importance and attention. A bioleaching-based approach has a variety of benefits over conventional processes, including reduced chemical volume and low operating costs. Considering the enormous amount of e-waste that needs to be managed and reduce the cost of chemicals wasted during treatment, it is possible to conclude that biological leaching is more practical than conventional leaching from the standpoints of the economy and the environment. However, the operational costs for biological leaching are significantly higher. These findings collectively imply that biological recovery may be a different method for treating e-waste to get rid of hazardous heavy metals.

References

Agarwal, R., Ranjan, R., and Sarkar, P. (2003). *Scrapping the Hi-tech Myth: Computer Waste in India*. New Delhi: Toxics Link.

Ahemad, M. and Malik, A. (2011). Bioaccumulation of heavy metals by zinc resistant bacteria isolated from agricultural soils irrigated with wastewater. *Bacteriology Journal* 2 (1): 12–21.

Akcil, A. and Deveci, H. (2010). *Mineral biotechnology of sulphides*, 101–137. Enfield, New Hampshire, USA: Geomicrobiology. Science Publishers.

Bandyopadhyay, A. (2008). A regulatory approach for e-waste management: a cross-national review of current practice and policy with an assessment and policy recommendation for the Indian perspective. *International Journal of Environment and Waste Management* 2 (1–2): 139–186.

Baud, I.S.A., Grafakos, S., Hordijk, M., and Post, J. (2001). Quality of life and alliances in solid waste management: contributions to urban sustainable development. *Cities* 18 (1): 3–12.

Bertin, P.N., Heinrich-Salmeron, A., Pelletier, E. et al. (2011). Metabolic diversity among main microorganisms inside an arsenic-rich ecosystem revealed by meta-and proteo-genomics. *The ISME Journal* 5 (11): 1735–1747.

Bhappu, R.B., Johnson, P.H., Brierley, J.A., and Reynolds, D.H. (1969). Theoretical and practical studies on dump leaching. *AIME Transactions* 244: 307–320.

Bloodworth, A. (2014). Resources: track flows to manage technology-metal supply. *Nature* 505 (7481): 19–20.

Bosecker, K. (2001). Microbial leaching in environmental clean-up programmes. *Hydrometallurgy* 59 (2–3): 245–248.

Brandl, H. and Faramarzi, M.A. (2006). Microbe-metal-interactions for the biotechnological treatment of metal-containing solid waste. *China Particuology* 4 (2): 93–97.

Brandl, H., Bosshard, R., and Wegmann, M. (2001). Computer-munching microbes: metal leaching from electronic scrap by bacteria and fungi. *Hydrometallurgy* 59 (2–3): 319–326.

Brandl, H., Lehmann, S., Faramarzi, M.A., and Martinelli, D. (2008). Biomobilization of silver, gold, and platinum from solid waste materials by HCN-forming microorganisms. *Hydrometallurgy* 94 (1–4): 14–17.

Brierley, C.L. (2009). Biohydrometallurgy: what is its future? In: *Advanced Materials Research*, vol. Vol. 71, 3–10. Trans Tech Publications Ltd.

Brierley, J.A. and Brierley, C.L. (2001). Present and future commercial applications of biohydrometallurgy. *Hydrometallurgy* 59 (2–3): 233–239.

Brierley, C.L. and Le Roux, N.W. (1978). Bacterial leaching. *CRC Critical Reviews in Microbiology* 6 (3): 207–262.

Brombacher, C., Bachofen, R., and Brandl, H. (1998). Development of a laboratory-scale leaching plant for metal extraction from fly ash by *Thiobacillus* strains. *Applied and Environmental Microbiology* 64 (4): 1237–1241.

Bryner, L.C. and Anderson, R. (1957). Microorganisms in leaching sulfide minerals. *Industrial and Engineering Chemistry* 49: 1721–1724.

Bryner, L.C. and Jameson, A.K. (1958). Microorganisms in leaching sulfide minerals. *Applied Microbiology* 6 (4): 281–287.

Bryner, L.C., Beck, J.V., Davis, D.B., and Wilson, D.G. (1954). Microorganisms in leaching sulfide minerals. *Industrial and Engineering Chemistry* 46 (12): 2587–2592.

Buetti-Dinh, A., Herold, M., Christel, S. et al. (2020). Reverse engineering directed gene regulatory networks from transcriptomics and proteomics data of biomining bacterial communities with approximate Bayesian computation and steady-state signalling simulations. *BMC Bioinformatics* 21 (1): 1–15.

Bustos, S., Castro, S., and Montealegre, R. (1993). The Sociedad Minera Pudahuel bacterial thin-layer leaching process at Lo Aguirre. *FEMS Microbiology Reviews* 11 (1–3): 231–235.

CD, O.E. (2001). *(Organisation for Economic Cooperation, Development), the application of biotechnology to industrial sustainability–Christian A.* Paris: OECD Publication Service, Hansen.

Chen, L.X., Hu, M., Huang, L.N. et al. (2015). Comparative metagenomic and metatranscriptomic analyses of microbial communities in acid mine drainage. *The ISME Journal* 9 (7): 1579–1592.

Choi, M.S., Cho, K.S., Kim, D.S., and Kim, D.J. (2004). Microbial recovery of copper from printed circuit boards of waste computer by *Acidithiobacillus ferrooxidans*. *Journal of Environmental Science and Health, Part A* 39 (11–12): 2973–2982.

Chojnacka, K. (2009). *Biosorption and Bioaccumulation in Practice*, 137. UK: Nova Science Publishers.

Chojnacka, K. (2010). Biosorption and bioaccumulation—the prospects for practical applications. *Environment International* 36 (3): 299–307.

Christel, S., Herold, M., Bellenberg, S. et al. (2018). Multi-omics reveals the lifestyle of the acidophilic, mineral-oxidizing model species *Leptospirillum ferriphilum* T. *Applied and Environmental Microbiology* 84 (3): e02091-17.

Clark, D.A. and Norris, P.R. (1996). Acidimicrobium ferrooxidans gen. nov., sp. nov.: mixed-culture ferrous iron oxidation with Sulfobacillus species. *Microbiology* 142 (4): 785–790.

Collard, J.M., Bertrand, S., Dierick, K. et al. (2008). Drastic decrease of Salmonella Enteritidis isolated from humans in Belgium in 2005, shift in phage types and influence on foodborne outbreaks. *Epidemiology and Infection* 136 (6): 771–781.

Colmer, A.R., Temple, K.L., and Hinkle, M.E. (1950). An iron-oxidizing bacterium from the acid drainage of some bituminous coal mines. *Journal of Bacteriology* 59 (3): 317–328.

CPCB, 2007. Draft guidelines for environmentally sound management of electronic waste, pp. 10–25 (http://ewasteguide.info/newsandevents/new-dr).

CPCB (2008). *Guidelines for environmentally sound management of e-waste (As approved vide MoEF letter No. 23-23/2007-HSMD)*. Delhi: Ministry of Environment and Forests, Central Pollution Control Board March 2008. Available from: http://www.cpcb.nic.in [last accessed on 2022 Mar 12].

Creamer, N.J., Baxter-Plant, V.S., Henderson, J. et al. (2006). Palladium and gold removal and recovery from precious metal solutions and electronic scrap leachates by Desulfovibrio desulfuricans. *Biotechnology Letters* 28 (18): 1475–1484.

Cui, J. and Zhang, L. (2008). Metallurgical recovery of metals from electronic waste: a review. *Journal of Hazardous Materials* 158 (2–3): 228–256.

Cwalina, B., Pacholewska, M., Sozańska, M., and Cabala, J. (2009). Microenvironments determining growth of acidophilic bacteria in Zn-Pb flotation tailings of neutral reaction. *Environment Protection Engineering* 35 (2): 113.

Dahl, R. (2002). Who pays for e-junk? *Environmental Health Perspectives* 110 (4): A196–A199.

Davis, T.A., Volesky, B., and Mucci, A. (2003). A review of the biochemistry of heavy metal biosorption by brown algae. *Water Research* 37 (18): 4311–4330.

De Corte, S., Hennebel, T., Fitts, J.P. et al. (2011). Biosupported bimetallic Pd–Au nanocatalysts for dechlorination of environmental contaminants. *Environmental Science & Technology* 45 (19): 8506–8513.

De Sousa, C.S., Hassan, S.S., Pinto, A.C. et al. (2018). Microbial omics: applications in biotechnology. In: *Omics Technologies and Bio-Engineering*, 3–20. Academic Press.

Denef, V.J., Kalnejais, L.H., Mueller, R.S. et al. (2010). Proteogenomic basis for ecological divergence of closely related bacteria in natural acidophilic microbial communities. *Proceedings of the National Academy of Sciences* 107 (6): 2383–2390.

Deplanche, K., Attard, G.A., and Macaskie, L.E. (2007). Biorecovery of gold from jewellery wastes by *Escherichia coli* and biomanufacture of active Au-nanomaterial. In: *Advanced Materials Research*, vol. Vol. 20, 647–650. Trans Tech Publications Ltd.

Deplanche, K., Bennett, J.A., Mikheenko, I.P. et al. (2014). Catalytic activity of biomass-supported Pd nanoparticles: influence of the biological component in catalytic efficacy and potential application in 'green'synthesis of fine chemicals and pharmaceuticals. *Applied Catalysis B: Environmental* 147: 651–665.

ELCINA-DSIR (2021). E-waste menace needs urgent technological and market interventions. In: Global SMT and Packaging India. Available from: http://www.global-smtindia.in/indexphp?option=com_contentandtask=viewandid=26548 anditemid=7 [last accessed on 2021 Dec].

Euromonitor, 2022 World market for domestic electrical appliances. US: Euromonitor; Euromonitor. Available from: http://www.nautilus.org [last accessed on 2022 Feb 28]

Faramarzi, M.A., Stagars, M., Pensini, E. et al. (2004). Metal solubilization from metal-containing solid materials by cyanogenic *Chromobacterium violaceum. Journal of Biotechnology* 113 (1–3): 321–326.

Forti, V., Balde, C. P., Kuehr, R., & Bel, G. (2020). *The Global E-waste Monitor 2020: Quantities, flows and the circular economy potential.*

Goltsman, D.S.A., Dasari, M., Thomas, B.C. et al. (2013). New group in the Leptospirillum clade: cultivation-independent community genomics, proteomics, and transcriptomics of the new species "Leptospirillum group IV UBA BS". *Applied and Environmental Microbiology* 79 (17): 5384–5393.

Groudev, S.N. (1987). Use of heterotrophic microorganisms in mineral biotechnology. *Acta Biotechnologica* 7 (4): 299–306.

Gusseme, B.D., Laing, G.D., Hennebel, T. et al. (2010). Virus removal by biogenic cerium. *Environmental Science & Technology* 44 (16): 6350–6356.

Hageluken, C. (2006, May). Improving metal returns and eco-efficiency in electronics recycling-a holistic approach for interface optimisation between pre-processing and integrated metals smelting and refining. In: *Proceedings of the 2006 IEEE International Symposium on Electronics and the Environment, 2006*, 218–223. IEEE.

Hageluken, C. and Art, S. (2006). Recycling of E-scrap in a global environment – chances and challenges: umicore precious metals refining. In: . Indo–European training workshop, 4–5 and 8–9 May 2006, Bangalore and New Delhi.

Hua, Z.S., Han, Y.J., Chen, L.X. et al. (2015). Ecological roles of dominant and rare prokaryotes in acid mine drainage revealed by metagenomics and metatranscriptomics. *The ISME Journal* 9 (6): 1280–1294.

Hussain, A., Hasan, A., Javid, A., and Qazi, J.I. (2016). Exploited application of sulfate-reducing bacteria for concomitant treatment of metallic and non-metallic wastes: a mini review. *3 Biotech* 6 (2): 1–10.

Ilyas, S., Ruan, C., Bhatti, H.N. et al. (2010). Column bioleaching of metals from electronic scrap. *Hydrometallurgy* 101 (3-4): 135–140.

Ilyas, S., Kim, M.S., Lee, J.C. et al. (2017). Bio-reclamation of strategic and energy critical metals from secondary resources. *Metals* 7 (6): 207.

Imai, T. and Saitoh, A. (2001). *U.S. Patent No. 6,168,766*. Washington, DC, U.S. Patent and Trademark Office.

Işıldar, A., van de Vossenberg, J., Rene, E.R. et al. (2016). Two-step bioleaching of copper and gold from discarded printed circuit boards (PCB). *Waste Management* 57: 149–157.

Jaafar, R., Al-Sulami, A., Al-Taee, A. et al. (2015). Biosorption and bioaccumulation of some heavy metals by Deinococcus radiodurans isolated from soil in Basra Governorate-Iraq. *Journal of Biotechnology and Biomaterial* 5 (2): 1–5.

Jain, N. and Sharma, D.K. (2004). Biohydrometallurgy for nonsulfidic minerals—a review. *Geomicrobiology Journal* 21 (3): 135–144.

Jansson, J.K. and Baker, E.S. (2016). A multi-omic future for microbiome studies. *Nature Microbiology* 1 (5): 1–3.

Janyasuthiwong, S., Ugas, R., Rene, E.R. et al. (2016). Effect of operational parameters on the leaching efficiency and recovery of heavy metals from computer printed circuit boards. *Journal of Chemical Technology & Biotechnology* 91 (7): 2038–2046.

Johnson, D.B., Okibe, N., Wakeman, K., and Yajie, L. (2008). Effect of temperature on the bioleaching of chalcopyrite concentrates containing different concentrations of silver. *Hydrometallurgy* 94 (1–4): 42–47.

Joshi, N.C. (2017). Heavy metals, conventional methods for heavy metal removal, biosorption and the development of low cost adsorbent. *European Journal of Pharmacy and Medical Research* 4: 388–393.

Kang, H.Y. and Schoenung, J.M. (2005). Electronic waste recycling: A review of US infrastructure and technology options. *Resources, Conservation and Recycling* 45 (4): 368–400.

Khatri, B.R., Sodha, A.B., Shah, M.B. et al. (2018). Chemical and microbial leaching of base metals from obsolete cell-phone printed circuit boards. *Sustainable Environment Research* 28 (6): 333–339.

Kiddee, P., Naidu, R., and Wong, M.H. (2013). Electronic waste management approaches: an overview. *Waste Management* 33 (5): 1237–1250.

Kita, Y., Nishikawa, H., and Takemoto, T. (2006). Effects of cyanide and dissolved oxygen concentration on biological Au recovery. *Journal of Biotechnology* 124 (3): 545–551.

Kucera, J., Lochman, J., Bouchal, P. et al. (2020). A model of aerobic and anaerobic metabolism of hydrogen in the extremophile *Acidithiobacillus ferrooxidans*. *Frontiers in Microbiology* 11: 610836.

Lens, P.N.L. (2016). Biotechnological applications for electronic waste water processing. *Linnaeus Eco-Tech* 103–103.

Lewandowski, G.A. and De Fillipi, L.J. (ed.) (1998). *Biological Treatment of Hazardous Wastes*. New York: Wiley.

Ma, L., Wang, H., Wu, J. et al. (2019). Metatranscriptomics reveals microbial adaptation and resistance to extreme environment coupling with bioleaching performance. *Bioresource Technology* 280: 9–17.

Martinez, P., Vera, M., and Bobadilla-Fazzini, R.A. (2015). Omics on bioleaching: current and future impacts. *Applied Microbiology and Biotechnology* 99 (20): 8337–8350.

Mehra, H. C. (2004). PC waste leaves toxic taste. The Tribune, 22nd March.

Mishra, D., Kim, D.J., Ralph, D.E. et al. (2008). Bioleaching of spent hydro-processing catalyst using acidophilic bacteria and its kinetics aspect. *Journal of Hazardous Materials* 152 (3): 1082–1091.

Mrvčić, J., Stanzer, D., Šolić, E., and Stehlik-Tomas, V. (2012). Interaction of lactic acid bacteria with metal ions: opportunities for improving food safety and quality. *World Journal of Microbiology and Biotechnology* 28 (9): 2771–2782.

Mulligan, C.N. and Kamali, M. (2003). Bioleaching of copper and other metals from low-grade oxidized mining ores by *Aspergillus niger*. *Journal of Chemical Technology & Biotechnology: International Research in Process, Environmental & Clean Technology* 78 (5): 497–503.

Nagpal, S., Dahlstrom, D., and Oolman, T. (1993). Effect of carbon dioxide concentration on the bioleaching of a pyrite–arsenopyrite ore concentrate. *Biotechnology and Bioengineering* 41 (4): 459–464.

Natarajan, G. and Ting, Y.P. (2014). Pretreatment of e-waste and mutation of alkali-tolerant cyanogenic bacteria promote gold biorecovery. *Bioresource Technology* 152: 80–85.

Nazari, B., Jorjani, E., Hani, H. et al. (2014). Formation of jarosite and its effect on important ions for Acidithiobacillus ferrooxidans bacteria. *Transactions of Nonferrous Metals Society of China* 24 (4): 1152–1160.

Nies, D.H. (1999). Microbial heavy-metal resistance. *Applied Microbiology and Biotechnology* 51 (6): 730–750.

Norton, A., de Klerk Batty, J., Dew, D.W., and Basson, P. (2005). *U.S. Patent No. 6,860,919*. Washington, DC: U.S. Patent and Trademark Office.

Olson, G.J. (1994). Microbial oxidation of gold ores and gold bioleaching. *FEMS Microbiology Letters* 119 (1–2): 1–6.

Olson, G.J., Brierley, J.A., and Brierley, C.L. (2003). Bioleaching review part B. *Applied Microbiology and Biotechnology* 63 (3): 249–257.

Ongondo, F.O., Williams, I.D., and Whitlock, G. (2015). Distinct urban mines: exploiting secondary resources in unique anthropogenic spaces. *Waste Management* 45: 4–9.

Onwughara, N.I., Nnorom, I.C., Kanno, O.C., and Chukwuma, R.C. (2010). Disposal methods and heavy metals released from certain electrical and electronic equipment wastes in Nigeria: adoption of environmental sound recycling system. *International Journal of Environmental Science and Development* 1 (4): 290.

Pacholewska, M. (2004). Bioleaching of galena flotation concentrate. *Fizykochemiczne Problemy Mineralurgii/Physicochemical Problems of Mineral Processing* 38: 281–290.

Pacholewska, M. and Farbiszewska, T. (2003). Chalcopyrite concentrate leaching using sulphur-and iron-oxidizing bacteria. *Polish Journal of Chemical Technology* 5 (4): 40–43.

Pant, D., Joshi, D., Upreti, M.K., and Kotnala, R.K. (2012). Chemical and biological extraction of metals present in E waste: a hybrid technology. *Waste Management* 32 (5): 979–990.

Park, D., Yun, Y.S., and Park, J.M. (2010). The past, present, and future trends of biosorption. *Biotechnology and Bioprocess Engineering* 15 (1): 86–102.

Patel, S. and Kasture, A. (2014). E (electronic) waste management using biological systems-overview. *International Journal of Current Microbiology and Applied Sciences* 3 (7): 495–504.

Pat-Espadas, A.M., Field, J.A., Razo-Flores, E. et al. (2016). Continuous removal and recovery of palladium in an upflow anaerobic granular sludge bed (UASB) reactor. *Journal of Chemical Technology & Biotechnology* 91 (4): 1183–1189.

Peng, T., Ma, L., Feng, X. et al. (2017). Genomic and transcriptomic analyses reveal adaptation mechanisms of an Acidithiobacillus ferrivorans strain YL15 to alpine acid mine drainage. *PLoS One* 12 (5): e0178008.

Plessis, C.A.D., Batty, J.D., and Dew, D.W. (2007). *Commercial Applications of Thermophile Bioleaching*, 57–80. Berlin, Heidelberg: Biomining Springer.

Rawlings, D.E. (ed.) (2013). *Biomining: Theory, Microbes and Industrial Processes*, 21–25. Springer Science & Business Media. C.L.

Rawlings, D.E. (ed.) (2013). *Biomining: Theory, Microbes and Industrial Processes*, 103–110. Springer Science & Business Media.C.L.

Razzell, W.E. and Trussell, P.C. (1963). Microbiological leaching of metallic sulfides. *Applied Microbiology* 11 (2): 105–110.

Rodriguez-Montelongo, L., Lilia, C., Farias, R.N., and Massa, E.M. (1993). Membrane-associated redox cycling of copper mediates hydroperoxide toxicity in *Escherichia coli*. *Biochimica et Biophysica Acta (BBA)-Bioenergetics* 1144 (1): 77–84.

Salkin, F.I. (2003). Conventional and alternative technologies for the treatment of infectious waste. *Journal of Material Cycles and Waste Management* 5: 9–12.

Scharnhorst, W., Althaus, H.J., Classen, M. et al. (2005). The end of life treatment of second generation mobile phone networks: Strategies to reduce the environmental impact. *Environmental Impact Assessment Review* 25 (5): 540–566.

Schinner, F. and Burgstaller, W. (1989). Extraction of zinc from industrial waste by a Penicillium sp. *Applied and Environmental Microbiology* 55 (5): 1153–1156.

Schmidt, I., Sliekers, O., Schmid, M. et al. (2003). New concepts of microbial treatment processes for the nitrogen removal in wastewater. *FEMS Microbiology Reviews* 27 (4): 481–492.

Selkov, E., Overbeek, R., Kogan, Y. et al. (2000). Functional analysis of gapped microbial genomes: amino acid metabolism of Thiobacillus ferrooxidans. *Proceedings of the National Academy of Sciences* 97 (7): 3509–3514.

Shah, M.B., Tipre, D.R., Purohit, M.S., and Dave, S.R. (2015). Development of two-step process for enhanced biorecovery of Cu–Zn–Ni from computer printed circuit boards. *Journal of Bioscience and Bioengineering* 120 (2): 167–173.

Sheng, P.P. and Etsell, T.H. (2007). Recovery of gold from computer circuit board scrap using aqua regia. *Waste Management & Research* 25 (4): 380–383.

Skippers, H.D. and Turco, R.F. (1995). Bioremediation: Science and applications. In: SSSA Spec. Publ. No. 43 ASA, CSSA and SSSA. Madison, Wisconsin.

Sodha, A.B., Shah, M.B., Qureshi, S.A. et al. (2019). Decouple and compare the role of abiotic factors and developed iron and sulphur oxidizers for enhanced extraction of metals from television printed circuit boards. *Separation Science and Technology* 54 (4): 591–601.

Suzuki, I. (2001). Microbial leaching of metals from sulfide minerals. *Biotechnology Advances* 19 (2): 119–132.

Tabak, H.H., Lens, P., Van Hullebusch, E.D., and Dejonghe, W. (2005). Developments in bioremediation of soils and sediments polluted with metals and radionuclides–1. Microbial processes and mechanisms affecting bioremediation of metal contamination and influencing metal toxicity and transport. *Reviews in Environmental Science and Bio/Technology* 4 (3): 115–156.

Thacker, S.C., Nayak, N.S., Tipre, D.R., and Dave, S.R. (2022). Multi-metal mining from waste cell phone printed circuit boards using lixiviant produced by a consortium of acidophilic iron oxidizers. *Environmental Engineering Science* 39 (3): 287–295.

Thiery, J.P., Acloque, H., Huang, R.Y., and Nieto, M.A. (2009). Epithelial-mesenchymal transitions in development and disease. *Cell* 139 (5): 871–890.

Tran, C.D., Lee, J.C., Pandey, B.D. et al. (2011). Bacterial cyanide generation in presence of metal ions (Na^{+}, Mg^{2+}, Fe^{2+}, Pb^{2+}) and gold bioleaching from waste PCBs. *Journal of Chemical Engineering of Japan* 1107120229.

Valix, M. (2017). Bioleaching of electronic waste: milestones and challenges. In: *Current Developments in Biotechnology and Bioengineering*, 407–442. Elsevier.

Vestola, E.A., Kuusenaho, M.K., Närhi, H.M. et al. (2010). Acid bioleaching of solid waste materials from copper, steel and recycling industries. *Hydrometallurgy* 103 (1–4): 74–79.

Wang, J., Bai, J., Xu, J., and Liang, B. (2009). Bioleaching of metals from printed wire boards by Acidithiobacillus ferrooxidans and Acidithiobacillus thiooxidans and their mixture. *Journal of Hazardous Materials* 172 (2–3): 1100–1105.

Wang, Y., Gao, H., Sun, J. et al. (2011). Selective reinforced competitive biosorption of Ag (I) and Cu (II) on *Magnetospirillum gryphiswaldense*. *Desalination* 270 (1–3): 258–263.

Watling, H.R. (2006). The bioleaching of sulphide minerals with emphasis on copper sulphides—a review. *Hydrometallurgy* 84 (1–2): 81–108.

Wu, C.B., Zeng, W.M., Zhou, H.B. et al. (2007). Bioleaching of chalcopyrite by mixed culture of moderately thermophilic microorganisms. *Journal of Central South University of Technology* 14 (4): 474–478.

Xiang, Y., Wu, P., Zhu, N. et al. (2010). Bioleaching of copper from waste printed circuit boards by bacterial consortium enriched from acid mine drainage. *Journal of Hazardous Materials* 184 (1–3): 812–818.

Yang, T., Xu, Z., Wen, J., and Yang, L. (2009). Factors influencing bioleaching copper from waste printed circuit boards by *Acidithiobacillus ferrooxidans*. *Hydrometallurgy* 97 (1–2): 29–32.

Yang, Z.Z., Zhao, X.R., Qin, Z.F. et al. (2009). Polybrominated diphenyl ethers in mudsnails (*Cipangopaludina cahayensis*) and sediments from an electronic waste recycling region in South China. *Bulletin of Environmental Contamination and Toxicology* 82 (2): 206–210.

Index

Management of Electronic Waste: Resource Recovery, Technology and Regulation, First Edition.
Edited by Anshu Priya.

Wiley The manufacturer's authorized representative according to the EU General Product Safety Regulation is Wiley-VCH GmbH, Boschstr. 12, 69469 Weinheim, Germany, e-mail: Product_Safety@wiley.com.

Printed and bound by CPI Group (UK) Ltd, Croydon, CR0 4YY
19/08/2026
02201089-0014